BUILDING CODE REQUIREMENTS FOR STRUCTURAL CONCRETE (ACI 318-05) AND COMMENTARY (ACI 318R-05)

REPORTED BY ACI COMMITTEE 318

ACI Committee 318
Structural Building Code

James K. Wight
Chair

Basile G. Rabbat
Secretary

Sergio M. Alcocer
Florian G. Barth
Roger J. Becker
Kenneth B. Bondy
John E. Breen
James R. Cagley
Michael P. Collins
W. Gene Corley
Charles W. Dolan
Anthony E. Fiorato
Catherine E. French
Luis E. Garcia
S. K. Ghosh
Lawrence G. Griffis
David P. Gustafson
D. Kirk Harman
James R. Harris
Neil M. Hawkins
Terence C. Holland
Kenneth C. Hover
Phillip J. Iverson
James O. Jirsa
Dominic J. Kelly
Gary J. Klein
Ronald Klemencic
Cary S. Kopczynski
H. S. Lew
Colin L. Lobo
Leslie D. Martin
Robert F. Mast
Steven L. McCabe
W. Calvin McCall
Jack P. Moehle
Myles A. Murray
Julio A. Ramirez
Thomas C. Schaeffer
Stephen J. Seguirant
Roberto Stark
Eric M. Tolles
Thomas D. Verti
Sharon L. Wood
Loring A. Wyllie
Fernando V. Yanez

Subcommittee Members

Neal S. Anderson
Mark A. Aschheim
John F. Bonacci
JoAnn P. Browning
Nicholas J. Carino
Ned M. Cleland
Ronald A. Cook
Juan P. Covarrubias
Robert J. Frosch
Harry A. Gleich
R. Doug Hooton
Javier F. Horvilleur†
L. S. Paul Johal
Michael E. Kreger
Daniel A. Kuchma
LeRoy A. Lutz
James G. MacGregor
Joe Maffei
Denis Mitchell
Vilas S. Mujumdar
Suzanne D. Nakaki
Theodore L. Neff
Andrzej S. Nowak
Randall W. Poston
Bruce W. Russell
Guillermo Santana
Andrew Scanlon
John F. Stanton
Fernando R. Stucchi
Raj Valluvan
John W. Wallace

Consulting Members

C. Raymond Hays
Richard C. Meininger
Charles G. Salmon

†Deceased

ACI 318-05 is deemed to satisfy ISO 19338, "Performance and Assessment Requirements for Design Standards on Structural Concrete," Reference Number ISO 19338.2003(E). Also Technical Corrigendum 1: 2004.

BUILDING CODE REQUIREMENTS FOR STRUCTURAL CONCRETE (ACI 318-05) AND COMMENTARY (ACI 318R-05)

REPORTED BY ACI COMMITTEE 318

PREFACE

The code portion of this document covers the design and construction of structural concrete used in buildings and where applicable in nonbuilding structures.

Among the subjects covered are: drawings and specifications; inspection; materials; durability requirements; concrete quality, mixing, and placing; formwork; embedded pipes; construction joints; reinforcement details; analysis and design; strength and serviceability; flexural and axial loads; shear and torsion; development and splices of reinforcement; slab systems; walls; footings; precast concrete; composite flexural members; prestressed concrete; shells and folded plate members; strength evaluation of existing structures; special provisions for seismic design; structural plain concrete; strut-and-tie modeling in Appendix A; alternative design provisions in Appendix B; alternative load and strength-reduction factors in Appendix C; and anchoring to concrete in Appendix D.

The quality and testing of materials used in construction are covered by reference to the appropriate ASTM standard specifications. Welding of reinforcement is covered by reference to the appropriate ANSI/AWS standard.

Uses of the code include adoption by reference in general building codes, and earlier editions have been widely used in this manner. The code is written in a format that allows such reference without change to its language. Therefore, background details or suggestions for carrying out the requirements or intent of the code portion cannot be included. The commentary is provided for this purpose. Some of the considerations of the committee in developing the code portion are discussed within the commentary, with emphasis given to the explanation of new or revised provisions. Much of the research data referenced in preparing the code is cited for the user desiring to study individual questions in greater detail. Other documents that provide suggestions for carrying out the requirements of the code are also cited.

Keywords: admixtures; aggregates; anchorage (structural); beam-column frame; beams (supports); **building codes;** cements; cold weather construction; columns (supports); combined stress; composite construction (concrete and steel); composite construction (concrete to concrete); compressive strength; **concrete construction; concretes;** concrete slabs; construction joints; continuity (structural); contraction joints; cover; curing; deep beams; deflections; drawings; earthquake resistant structures; embedded service ducts; flexural strength; floors; folded plates; footings; formwork (construction); frames; hot weather construction; inspection; isolation joints; joints (junctions); joists; lightweight concretes; loads (forces); load tests (structural); materials; mixing; mix proportioning; modulus of elasticity; moments; pipe columns; pipes (tubing); placing; plain concrete; precast concrete; prestressed concrete; prestressing steels; quality control; **reinforced concrete**; reinforcing steels; roofs; serviceability; shear strength; shearwalls; shells (structural forms); spans; specifications; splicing; strength; strength analysis; stresses; **structural analysis; structural concrete; structural design;** structural integrity; T-beams; torsion; walls; water; welded wire reinforcement.

ACI 318-05 was adopted as a standard of the American Concrete Institute October 27, 2004 to supersede ACI 318-02 in accordance with the Institute's standardization procedure.

A complete metric companion to ACI 318/318R has been developed, 318M/318RM; therefore no metric equivalents are included in this document.

CONTENTS

The ACI Building code and commentary are presented in a side-by-side column format, with code text placed in the left column and the corresponding commentary text aligned in the right column. To further distinguish the code from the commentary, the code has been printed in Helvetica, the same type face in which this paragraph is set.

This paragraph is set in Times Roman, and all portions of the text exclusive to the commentary are printed in this type face. Commentary section numbers are preceded by an "R" to further distinguish them from code section numbers.

Vertical lines in the margins indicate changes from the previous version. Changes to the notation and strictly editorial changes are not indicated with a vertical line.

INTRODUCTION

This commentary discusses some of the considerations of Committee 318 in developing the provisions contained in "Building Code Requirements for Structural Concrete (ACI 318-05)," hereinafter called the code or the 2005 code. Emphasis is given to the explanation of new or revised provisions that may be unfamiliar to code users. In addition, comments are included for some items contained in previous editions of the code to make the present commentary independent of the previous editions. Comments on specific provisions are made under the corresponding chapter and section numbers of the code.

The commentary is not intended to provide a complete historical background concerning the development of the ACI Building Code,* nor is it intended to provide a detailed résumé of the studies and research data reviewed by the committee in formulating the provisions of the code. However, references to some of the research data are provided for those who wish to study the background material in depth.

As the name implies, "Building Code Requirements for Structural Concrete" is meant to be used as part of a legally adopted building code and as such must differ in form and substance from documents that provide detailed specifications, recommended practice, complete design procedures, or design aids.

The code is intended to cover all buildings of the usual types, both large and small. Requirements more stringent than the code provisions may be desirable for unusual construction. The code and commentary cannot replace sound engineering knowledge, experience, and judgement.

A building code states only the minimum requirements necessary to provide for public health and safety. The code is based on this principle. For any structure, the owner or the structural designer may require the quality of materials and construction to be higher than the minimum requirements necessary to protect the public as stated in the code. However, lower standards are not permitted.

The commentary directs attention to other documents that provide suggestions for carrying out the requirements and intent of the code. However, those documents and the commentary are not a part of the code.

The code has no legal status unless it is adopted by the government bodies having the police power to regulate building design and construction. Where the code has not been adopted, it may serve as a reference to good practice even though it has no legal status.

The code provides a means of establishing minimum standards for acceptance of designs and construction by legally appointed building officials or their designated representatives. The code and commentary are not intended for use in settling disputes between the owner, engineer, architect, contractor, or their agents, subcontractors, material suppliers, or testing agencies. Therefore, the code cannot define the contract responsibility of each of the parties in usual construction. General references requiring compliance with the code in the project specifications should be avoided since the contractor is rarely in a position to accept responsibility for design details or construction requirements that depend on a detailed knowledge of the design. Design-build construction contractors, however, typically combine the design and construction responsibility. Generally, the drawings, specifications and contract documents should contain all of the necessary requirements to ensure compliance with the code. In part, this can be accomplished by reference to specific code sections in the project specifications. Other ACI publications, such as "Specifications for Structural Concrete (ACI 301)" are written specifically for use as contract documents for construction.

It is recommended to have testing and certification programs for the individual parties involved with the execution of work performed in accordance with this code. Available for this purpose are the plant certification programs of the Precast/Prestressed Concrete Institute, the Post-Tensioning Institute and the National Ready Mixed Concrete Association; the personnel certification programs of the American Concrete Institute and the Post-Tensioning Institute; and the Concrete Reinforcing Steel Institute's Voluntary Certification Program for Fusion-Bonded Epoxy Coating Applicator Plants. In addition, "Standard Specification for Agencies Engaged in the Testing and/or Inspection of Materials Used in Construction" (ASTM E 329-03) specifies performance requirements for inspection and testing agencies.

*For a history of the ACI Building Code see Kerekes, Frank, and Reid, Harold B., Jr., "Fifty Years of Development in Building Code Requirements for Reinforced Concrete," ACI JOURNAL, *Proceedings* V. 50, No. 6, Feb. 1954, p. 441. For a discussion of code philosophy, see Siess, Chester P., "Research, Building Codes, and Engineering Practice," ACI JOURNAL, *Proceedings* V. 56, No. 5, May 1960, p. 1105.

Design reference materials illustrating applications of the code requirements may be found in the following documents. The design aids listed may be obtained from the sponsoring organization.

Design aids:

"ACI Design Handbook," ACI Committee 340, Publication SP-17(97), American Concrete Institute, Farmington Hills, MI, 1997, 482 pp. (Provides tables and charts for design of eccentrically loaded columns by the Strength Design Method. Provides design aids for use in the engineering design and analysis of reinforced concrete slab systems carrying loads by two-way action. Design aids are also provided for the selection of slab thickness and for reinforcement required to control deformation and assure adequate shear and flexural strengths.)

"ACI Detailing Manual—2004," ACI Committee 315, Publication SP-66(04), American Concrete Institute, Farmington Hills, MI, 2004, 212 pp. (Includes the standard, ACI 315-99, and report, ACI 315R-04. Provides recommended methods and standards for preparing engineering drawings, typical details, and drawings placing reinforcing steel in reinforced concrete structures. Separate sections define responsibilities of both engineer and reinforcing bar detailer.)

"Guide to Durable Concrete (ACI 201.2R-92)," ACI Committee 201, American Concrete Institute, Farmington Hills, MI, 1992, 41 pp. (Describes specific types of concrete deterioration. It contains a discussion of the mechanisms involved in deterioration and the recommended requirements for individual components of the concrete, quality considerations for concrete mixtures, construction procedures, and influences of the exposure environment. Section R4.4.1 discusses the difference in chloride-ion limits between ACI 201.2R-92 and the code.)

"Guide for the Design of Durable Parking Structures (362.1R-97 (Reapproved 2002))," ACI Committee 362, American Concrete Institute, Farmington Hills, MI, 1997, 40 pp. (Summarizes practical information regarding design of parking structures for durability. It also includes information about design issues related to parking structure construction and maintenance.)

"CRSI Handbook," Concrete Reinforcing Steel Institute, Schaumburg, IL, 9th Edition, 2002, 648 pp. (Provides tabulated designs for structural elements and slab systems. Design examples are provided to show the basis of and use of the load tables. Tabulated designs are given for beams; square, round and rectangular columns; one-way slabs; and one-way joist construction. The design tables for two-way slab systems include flat plates, flat slabs and waffle slabs. The chapters on foundations provide design tables for square footings, pile caps, drilled piers (caissons) and cantilevered retaining walls. Other design aids are presented for crack control; and development of reinforcement and lap splices.)

"Reinforcement Anchorages and Splices," Concrete Reinforcing Steel Institute, Schaumberg, IL, 4th Edition, 1997, 100 pp. (Provides accepted practices in splicing reinforcement. The use of lap splices, mechanical splices, and welded splices are described. Design data are presented for development and lap splicing of reinforcement.)

"Structural Welded Wire Reinforcement Manual of Standard Practice," Wire Reinforcement Institute, Hartford, CT, 6th Edition, Apr. 2001, 38 pp. (Describes welded wire reinforcement material, gives nomenclature and wire size and weight tables. Lists specifications and properties and manufacturing limitations. Book has latest code requirements as code affects welded wire. Also gives development length and splice length tables. Manual contains customary units and soft metric units.)

"Structural Welded Wire Reinforcement Detailing Manual," Wire Reinforcement Institute, Hartford, CT, 1994, 252 pp. (Updated with current technical fact sheets inserted.) The manual, in addition to including ACI 318 provisions and design aids, also includes: detailing guidance on welded wire reinforcement in one-way and two-way slabs; precast/prestressed concrete components; columns and beams; cast-in-place walls; and slabs-on-ground. In addition, there are tables to compare areas and spacings of high-strength welded wire with conventional reinforcing.

"Strength Design of Reinforced Concrete Columns," Portland Cement Association, Skokie, IL, 1978, 48 pp. (Provides design tables of column strength in terms of load in kips versus moment in ft-kips for concrete strength of 5000 psi and Grade 60 reinforcement. Design examples are included. Note that the PCA design tables do not include the strength reduction factor ϕ in the tabulated values; M_u/ϕ and P_u/ϕ must be used when designing with this aid.

"PCI Design Handbook—Precast and Prestressed Concrete," Precast/Prestressed Concrete Institute, Chicago, IL, 5th Edition, 1999, 630 pp. (Provides load tables for common industry products, and procedures for design and analysis of precast and prestressed elements and structures composed of these elements. Provides design aids and examples.)

"Design and Typical Details of Connections for Precast and Prestressed Concrete," Precast/Prestressed Concrete Institute, Chicago, IL, 2nd Edition, 1988, 270 pp. (Updates available information on design of connections for both structural and architectural products, and presents a full spectrum of typical details. Provides design aids and examples.)

"Post-Tensioning Manual," Post-Tensioning Institute, Phoenix, AZ, 5th Edition, 1990, 406 pp. (Provides comprehensive coverage of post-tensioning systems, specifications, and design aid construction concepts.)

CHAPTER 1 — GENERAL REQUIREMENTS

CODE

1.1 — Scope

1.1.1 — This code provides minimum requirements for design and construction of structural concrete elements of any structure erected under requirements of the legally adopted general building code of which this code forms a part. In areas without a legally adopted building code, this code defines minimum acceptable standards of design and construction practice.

For structural concrete, f_c' shall not be less than 2500 psi. No maximum value of f_c' shall apply unless restricted by a specific code provision.

COMMENTARY

R1.1 — Scope

The American Concrete Institute **"Building Code Requirements for Structural Concrete (ACI 318-05),"** referred to as the code, provides minimum requirements for structural concrete design or construction.

The 2005 code revised the previous standard **"Building Code Requirements for Structural Concrete (ACI 318-02)."** This standard includes in one document the rules for all concrete used for structural purposes including both plain and reinforced concrete. The term "structural concrete" is used to refer to all plain or reinforced concrete used for structural purposes. This covers the spectrum of structural applications of concrete from nonreinforced concrete to concrete containing nonprestressed reinforcement, prestressing steel, or composite steel shapes, pipe, or tubing. Requirements for structural plain concrete are in Chapter 22.

Prestressed concrete is included under the definition of reinforced concrete. Provisions of the code apply to prestressed concrete except for those that are stated to apply specifically to nonprestressed concrete.

Chapter 21 of the code contains special provisions for design and detailing of earthquake resistant structures. See 1.1.8.

In the 1999 code and earlier editions, Appendix A contained provisions for an alternate method of design for nonprestressed reinforced concrete members using service loads (without load factors) and permissible service load stresses. The Alternate Design Method was intended to give results that were slightly more conservative than designs by the Strength Design Method of the code. The Alternate Design Method of the 1999 code may be used in place of applicable sections of this code.

Appendix A of the code contains provisions for the design of regions near geometrical discontinuities, or abrupt changes in loadings.

Appendix B of this code contains provisions for reinforcement limits based on $0.75\rho_b$, determination of the strength reduction factor ϕ, and moment redistribution that have been in the code for many years, including the 1999 code. The provisions are applicable to reinforced and prestressed concrete members. Designs made using the provisions of Appendix B are equally acceptable as those based on the body of the code, provided the provisions of Appendix B are used in their entirety.

Appendix C of the code allows the use of the factored load combinations given in Chapter 9 of the 1999 code.

CODE

COMMENTARY

Appendix D contains provisions for anchoring to concrete.

1.1.2 — This code supplements the general building code and shall govern in all matters pertaining to design and construction of structural concrete, except wherever this code is in conflict with requirements in the legally adopted general building code.

1.1.3 — This code shall govern in all matters pertaining to design, construction, and material properties wherever this code is in conflict with requirements contained in other standards referenced in this code.

1.1.4 — For special structures, such as arches, tanks, reservoirs, bins and silos, blast-resistant structures, and chimneys, provisions of this code shall govern where applicable. See also 22.1.2.

R1.1.2 — The American Concrete Institute recommends that the code be adopted in its entirety; however, it is recognized that when the code is made a part of a legally adopted general building code, the general building code may modify provisions of this code.

R1.1.4 — Some special structures involve unique design and construction problems that are not covered by the code. However, many code provisions, such as the concrete quality and design principles, are applicable for these structures. Detailed recommendations for design and construction of some special structures are given in the following ACI publications:

"Design and Construction of Reinforced Concrete Chimneys" reported by ACI Committee 307.[1.1] (Gives material, construction, and design requirements for circular cast-in-place reinforced chimneys. It sets forth minimum loadings for the design of reinforced concrete chimneys and contains methods for determining the stresses in the concrete and reinforcement required as a result of these loadings.)

"Standard Practice for Design and Construction of Concrete Silos and Stacking Tubes for Storing Granular Materials" reported by ACI Committee 313.[1.2] (Gives material, design, and construction requirements for reinforced concrete bins, silos, and bunkers and stave silos for storing granular materials. It includes recommended design and construction criteria based on experimental and analytical studies plus worldwide experience in silo design and construction.)

"Environmental Engineering Concrete Structures" reported by ACI Committee 350.[1.3] (Gives material, design and construction recommendations for concrete tanks, reservoirs, and other structures commonly used in water and waste treatment works where dense, impermeable concrete with high resistance to chemical attack is required. Special emphasis is placed on a structural design that minimizes the possibility of cracking and accommodates vibrating equipment and other special loads. Proportioning of concrete, placement, curing and protection against chemicals are also described. Design and spacing of joints receive special attention.)

"Code Requirements for Nuclear Safety Related Concrete Structures" reported by ACI Committee 349.[1.4] (Provides minimum requirements for design and construction of concrete structures that form part of a nuclear power plant and have nuclear safety related functions. The code does not cover concrete reactor vessels and concrete containment structures which are covered by ACI 359.)

CODE

COMMENTARY

"Code for Concrete Reactor Vessels and Containments" reported by ACI-ASME Committee 359.[1.5] (Provides requirements for the design, construction, and use of concrete reactor vessels and concrete containment structures for nuclear power plants.)

1.1.5 — This code does not govern design and installation of portions of concrete piles, drilled piers, and caissons embedded in ground except for structures in regions of high seismic risk or assigned to high seismic performance or design categories. See 21.10.4 for requirements for concrete piles, drilled piers, and caissons in structures in regions of high seismic risk or assigned to high seismic performance or design categories.

R1.1.5 — The design and installation of piling fully embedded in the ground is regulated by the general building code. For portions of piling in air or water, or in soil not capable of providing adequate lateral restraint throughout the piling length to prevent buckling, the design provisions of this code govern where applicable.

Recommendations for concrete piles are given in detail in **"Recommendations for Design, Manufacture, and Installation of Concrete Piles"** reported by ACI Committee 543.[1.6] (Provides recommendations for the design and use of most types of concrete piles for many kinds of construction.)

Recommendations for drilled piers are given in detail in **"Design and Construction of Drilled Piers"** reported by ACI Committee 336.[1.7] (Provides recommendations for design and construction of foundation piers 2-1/2 ft in diameter or larger made by excavating a hole in the soil and then filling it with concrete.)

Detailed recommendations for precast prestressed concrete piles are given in **"Recommended Practice for Design, Manufacture, and Installation of Prestressed Concrete Piling"** prepared by the PCI Committee on Prestressed Concrete Piling.[1.8]

1.1.6 — This code does not govern design and construction of soil-supported slabs, unless the slab transmits vertical loads or lateral forces from other portions of the structure to the soil.

R1.1.6 — Detailed recommendations for design and construction of soil-supported slabs and floors that do not transmit vertical loads or lateral forces from other portions of the structure to the soil, and residential post-tensioned slabs-on-ground, are given in the following publications:

"Design of Slabs on Grade" reported by ACI Committee 360.[1.9] (Presents information on the design of slabs on grade, primarily industrial floors and the slabs adjacent to them. The report addresses the planning, design, and detailing of the slabs. Background information on the design theories is followed by discussion of the soil support system, loadings, and types of slabs. Design methods are given for plain concrete, reinforced concrete, shrinkage-compensating concrete, and post-tensioned concrete slabs.)

"Design of Post-Tensioned Slabs-on-Ground," PTI[1.10] (Provides recommendations for post-tensioned slab-on-ground foundations. Presents guidelines for soil investigation, and design and construction of post-tensioned residential and light commercial slabs on expansive or compressible soils.)

CODE

1.1.7 — Concrete on steel form deck

1.1.7.1 — Design and construction of structural concrete slabs cast on stay-in-place, noncomposite steel form deck are governed by this code.

1.1.7.2 — This code does not govern the design of structural concrete slabs cast on stay-in-place, composite steel form deck. Concrete used in the construction of such slabs shall be governed by Chapters 1 through 7 of this code, where applicable.

1.1.8 — Special provisions for earthquake resistance

1.1.8.1 — In regions of low seismic risk, or for structures assigned to low seismic performance or design categories, provisions of Chapter 21 shall not apply.

1.1.8.2 — In regions of moderate or high seismic risk, or for structures assigned to intermediate or high seismic performance or design categories, provisions of Chapter 21 shall be satisfied. See 21.2.1.

COMMENTARY

R1.1.7 — Concrete on steel form deck

In steel framed structures, it is common practice to cast concrete floor slabs on stay-in-place steel form deck. In all cases, the deck serves as the form and may, in some cases, serve an additional structural function.

R1.1.7.1 — In its most basic application, the steel form deck serves as a form, and the concrete serves a structural function and, therefore, are to be designed to carry all superimposed loads.

R1.1.7.2 — Another type of steel form deck commonly used develops composite action between the concrete and steel deck. In this type of construction, the steel deck serves as the positive moment reinforcement. The design of composite slabs on steel deck is regulated by **"Standard for the Structural Design of Composite Slabs"** (ANSI/ASCE 3).[1.11] However, ANSI/ASCE 3 references the appropriate portions of ACI 318 for the design and construction of the concrete portion of the composite assembly. Guidelines for the construction of composite steel deck slabs are given in **"Standard Practice for the Construction and Inspection of Composite Slabs"** (ANSI/ASCE 9).[1.12]

R1.1.8 — Special provisions for earthquake resistance

Special provisions for seismic design were first introduced in Appendix A of the 1971 code and were continued without revision in the 1977 code. These provisions were originally intended to apply only to reinforced concrete structures located in regions of highest seismicity.

The special provisions were extensively revised in the 1983 code to include new requirements for certain earthquake-resisting systems located in regions of moderate seismicity. In the 1989 code, the special provisions were moved to Chapter 21.

R1.1.8.1 — For structures located in regions of low seismic risk, or for structures assigned to low seismic performance or design categories, no special design or detailing is required; the general requirements of the main body of the code apply for proportioning and detailing of reinforced concrete structures. It is the intent of Committee 318 that concrete structures proportioned by the main body of the code will provide a level of toughness adequate for low earthquake intensity.

R1.1.8.2 — For structures in regions of moderate seismic risk, or for structures assigned to intermediate seismic performance or design categories, reinforced concrete moment frames proportioned to resist seismic effects require special reinforcement details, as specified in 21.12. The special details apply only to beams, columns, and slabs to which the earthquake-induced forces have been assigned in design. The special reinforcement details will serve to provide a suitable level of inelastic behavior if the frame is subjected to an earthquake of such intensity as to require it to perform inelastically. There are no Chapter 21 require-

CODE | COMMENTARY

ments for cast-in-place structural walls provided to resist seismic effects, or for other structural components that are not part of the lateral-force-resisting system of structures in regions of moderate seismic risk, or assigned to intermediate seismic performance or design categories. For precast wall panels designed to resist forces induced by earthquake motions, special requirements are specified in 21.13 for connections between panels or between panels and the foundation. Cast-in-place structural walls proportioned to meet provisions of Chapters 1 through 18 and Chapter 22 are considered to have sufficient toughness at anticipated drift levels for these structures.

For structures located in regions of high seismic risk, or for structures assigned to high seismic performance or design categories, all building components that are part of the lateral-force-resisting system, including foundations (except plain concrete foundations as allowed by 22.10.1), should satisfy requirements of 21.2 through 21.10. In addition, frame members that are not assumed in the design to be part of the lateral-force-resisting system should comply with 21.11. The special proportioning and detailing requirements of Chapter 21 are intended to provide a monolithic reinforced concrete or precast concrete structure with adequate "toughness" to respond inelastically under severe earthquake motions. See also R21.2.1.

1.1.8.3 — The seismic risk level of a region, or seismic performance or design category of a structure, shall be regulated by the legally adopted general building code of which this code forms a part, or determined by local authority.

R1.1.8.3 — Seismic risk levels (Seismic Zone Maps) and seismic performance or design categories are under the jurisdiction of a general building code rather than ACI 318. Changes in terminology were made to the 1999 edition of the code to make it compatible with the latest editions of model building codes in use in the United States. For example, the phrase "seismic performance or design categories" was introduced. Over the past decade, the manner in which seismic risk levels have been expressed in United States building codes has changed. Previously they have been represented in terms of seismic zones. Recent editions of the "BOCA National Building Code" (NBC)[1.13] and "Standard Building Code" (SBC),[1.14] which are based on the 1991 NEHRP,[1.15] have expressed risk not only as a function of expected intensity of ground shaking on solid rock, but also on the nature of the occupancy and use of the structure. These two items are considered in assigning the structure to a Seismic Performance Category (SPC), which in turn is used to trigger different levels of detailing requirements for the structure. The 2000 and 2003 editions of the "International Building Code" (IBC)[1.16, 1.17] and the 2003 NFPA 5000 "Building Construction and Safety Code"[1.18] also consider the effects of soil amplification on the ground motion when assigning seismic risk. Under the IBC and NFPA codes, each structure is assigned a Seismic Design Category (SDC). Among its several uses, the SDC triggers different levels of detailing requirements. Table R1.1.8.3 correlates

CODE

COMMENTARY

TABLE R1.1.8.3—CORRELATION BETWEEN SEISMIC-RELATED TERMINOLOGY IN MODEL CODES

Code, standard, or resource document and edition	Level of seismic risk or assigned seismic performance or design categories as defined in the code section		
	Low (21.2.1.2)	Moderate/ intermediate (21.2.1.3)	High (21.2.1.4)
IBC 2000, 2003; NFPA 5000, 2003; ASCE 7-98, 7-02; NEHRP 1997, 2000	SDC* A, B	SDC C	SDC D, E, F
BOCA National Building Code 1993, 1996, 1999; Standard Building Code 1994, 1997, 1999; ASCE 7-93, 7-95; NEHRP 1991, 1994	SPC† A, B	SPC C	SPC D, E
Uniform Building Code 1991, 1994, 1997	Seismic Zone 0, 1	Seismic Zone 2	Seismic Zone 3, 4

*SDC = *Seismic Design Category* as defined in code, standard, or resource document.
†SPC = *Seismic Performance Category* as defined in code, standard, or resource document.

low, moderate/intermediate, and high seismic risk, which has been the terminology used in this code for several editions, to the various methods of assigning risk in use in the U.S. under the various model building codes, the ASCE 7 standard, and the NEHRP Recommended Provisions.

In the absence of a general building code that addresses earthquake loads and seismic zoning, it is the intent of Committee 318 that the local authorities (engineers, geologists, and building code officials) should decide on proper need and proper application of the special provisions for seismic design. Seismic ground-motion maps or zoning maps, such as recommended in References 1.17, 1.19, and 1.20, are suitable for correlating seismic risk.

1.2 — Drawings and specifications

1.2.1 — Copies of design drawings, typical details, and specifications for all structural concrete construction shall bear the seal of a registered engineer or architect. These drawings, details, and specifications shall show:

(a) Name and date of issue of code and supplement to which design conforms;

(b) Live load and other loads used in design;

(c) Specified compressive strength of concrete at stated ages or stages of construction for which each part of structure is designed;

(d) Specified strength or grade of reinforcement;

(e) Size and location of all structural elements, reinforcement, and anchors;

(f) Provision for dimensional changes resulting from creep, shrinkage, and temperature;

R1.2 — Drawings and specifications

R1.2.1 — The provisions for preparation of design drawings and specifications are, in general, consistent with those of most general building codes and are intended as supplements.

The code lists some of the more important items of information that should be included in the design drawings, details, or specifications. The code does not imply an all-inclusive list, and additional items may be required by the building official.

CODE

(g) Magnitude and location of prestressing forces;

(h) Anchorage length of reinforcement and location and length of lap splices;

(i) Type and location of mechanical and welded splices of reinforcement;

(j) Details and location of all contraction or isolation joints specified for plain concrete in Chapter 22;

(k) Minimum concrete compressive strength at time of post-tensioning;

(l) Stressing sequence for post-tensioning tendons;

(m) Statement if slab on grade is designed as a structural diaphragm, see 21.10.3.4.

1.2.2 — Calculations pertinent to design shall be filed with the drawings when required by the building official. Analyses and designs using computer programs shall be permitted provided design assumptions, user input, and computer-generated output are submitted. Model analysis shall be permitted to supplement calculations.

1.2.3 — Building official means the officer or other designated authority charged with the administration and enforcement of this code, or his duly authorized representative.

1.3 — Inspection

COMMENTARY

R1.2.2 — Documented computer output is acceptable in lieu of manual calculations. The extent of input and output information required will vary, according to the specific requirements of individual building officials. However, when a computer program has been used by the designer, only skeleton data should normally be required. This should consist of sufficient input and output data and other information to allow the building official to perform a detailed review and make comparisons using another program or manual calculations. Input data should be identified as to member designation, applied loads, and span lengths. The related output data should include member designation and the shears, moments, and reactions at key points in the span. For column design, it is desirable to include moment magnification factors in the output where applicable.

The code permits model analysis to be used to supplement structural analysis and design calculations. Documentation of the model analysis should be provided with the related calculations. Model analysis should be performed by an engineer or architect having experience in this technique.

R1.2.3 — Building official is the term used by many general building codes to identify the person charged with administration and enforcement of the provisions of the building code. However, such terms as building commissioner or building inspector are variations of the title, and the term building official as used in this code is intended to include those variations as well as others that are used in the same sense.

R1.3 — Inspection

The quality of concrete structures depends largely on workmanship in construction. The best of materials and design practices will not be effective unless the construction is performed well. Inspection is necessary to confirm that the construction is in accordance with the design drawings and project specifications. Proper performance of the structure

CODE | COMMENTARY

depends on construction that accurately represents the design and meets code requirements within the tolerances allowed. Qualification of the inspectors can be obtained from a certification program, such as the ACI Certification Program for Concrete Construction Special Inspector.

1.3.1 — Concrete construction shall be inspected as required by the legally adopted general building code. In the absence of such inspection requirements, concrete construction shall be inspected throughout the various work stages by or under the supervision of a registered design professional or by a qualified inspector.

R1.3.1 — Inspection of construction by or under the supervision of the registered design professional responsible for the design should be considered because the person in charge of the design is usually the best qualified to determine if construction is in conformance with construction documents. When such an arrangement is not feasible, inspection of construction through other registered design professionals or through separate inspection organizations with demonstrated capability for performing the inspection may be used.

Qualified inspectors should establish their qualification by becoming certified to inspect and record the results of concrete construction, including preplacement, placement, and postplacement operations through the ACI Inspector Certification Program: Concrete Construction Special Inspector.

When inspection is done independently of the registered design professional responsible for the design, it is recommended that the registered design professional responsible for the design be employed at least to oversee inspection and observe the work to see that the design requirements are properly executed.

In some jurisdictions, legislation has established special registration or licensing procedures for persons performing certain inspection functions. A check should be made in the general building code or with the building official to ascertain if any such requirements exist within a specific jurisdiction.

Inspection reports should be promptly distributed to the owner, registered design professional responsible for the design, contractor, appropriate subcontractors, appropriate suppliers, and the building official to allow timely identification of compliance or the need for corrective action.

Inspection responsibility and the degree of inspection required should be set forth in the contracts between the owner, architect, engineer, contractor, and inspector. Adequate fees should be provided consistent with the work and equipment necessary to properly perform the inspection.

1.3.2 — The inspector shall require compliance with design drawings and specifications. Unless specified otherwise in the legally adopted general building code, inspection records shall include:

(a) Quality and proportions of concrete materials and strength of concrete;

(b) Construction and removal of forms and reshoring;

R1.3.2 — By inspection, the code does not mean that the inspector should supervise the construction. Rather it means that the one employed for inspection should visit the project with the frequency necessary to observe the various stages of work and ascertain that it is being done in compliance with contract documents and code requirements. The frequency should be at least enough to provide general knowledge of each operation, whether this is several times a day or once in several days.

CODE

(c) Placing of reinforcement and anchors;

(d) Mixing, placing, and curing of concrete;

(e) Sequence of erection and connection of precast members;

(f) Tensioning of tendons;

(g) Any significant construction loadings on completed floors, members, or walls;

(h) General progress of work.

1.3.3 — When the ambient temperature falls below 40 F or rises above 95 F, a record shall be kept of concrete temperatures and of protection given to concrete during placement and curing.

1.3.4 — Records of inspection required in 1.3.2 and 1.3.3 shall be preserved by the inspecting engineer or architect for 2 years after completion of the project.

COMMENTARY

Inspection in no way relieves the contractor from his obligation to follow the plans and specifications and to provide the designated quality and quantity of materials and workmanship for all job stages. The inspector should be present as frequently as he or she deems necessary to judge whether the quality and quantity of the work complies with the contract documents; to counsel on possible ways of obtaining the desired results; to see that the general system proposed for formwork appears proper (though it remains the contractor's responsibility to design and build adequate forms and to leave them in place until it is safe to remove them); to see that reinforcement is properly installed; to see that concrete is of the correct quality, properly placed, and cured; and to see that tests for quality assurance are being made as specified.

The code prescribes minimum requirements for inspection of all structures within its scope. It is not a construction specification and any user of the code may require higher standards of inspection than cited in the legal code if additional requirements are necessary.

Recommended procedures for organization and conduct of concrete inspection are given in detail in **"Guide for Concrete Inspection,"** reported by ACI Committee 311.[1.21] (Sets forth procedures relating to concrete construction to serve as a guide to owners, architects, and engineers in planning an inspection program.)

Detailed methods of inspecting concrete construction are given in "**ACI Manual of Concrete Inspection"** (SP-2) reported by ACI Committee 311.[1.22] (Describes methods of inspecting concrete construction that are generally accepted as good practice. Intended as a supplement to specifications and as a guide in matters not covered by specifications.)

R1.3.3 — The term ambient temperature means the temperature of the environment to which the concrete is directly exposed. Concrete temperature as used in this section may be taken as the air temperature near the surface of the concrete; however, during mixing and placing it is practical to measure the temperature of the mixture.

R1.3.4 — A record of inspection in the form of a job diary is required in case questions subsequently arise concerning the performance or safety of the structure or members. Photographs documenting job progress may also be desirable.

Records of inspection should be preserved for at least 2 years after the completion of the project. The completion of the project is the date at which the owner accepts the project, or when a certificate of occupancy is issued, whichever date is later. The general building code or other legal requirements may require a longer preservation of such records.

CODE

1.3.5 — For special moment frames resisting seismic loads in regions of high seismic risk, or in structures assigned to high seismic performance or design categories, continuous inspection of the placement of the reinforcement and concrete shall be made by a qualified inspector. The inspector shall be under the supervision of the engineer responsible for the structural design or under the supervision of an engineer with demonstrated capability for supervising inspection of special moment frames resisting seismic loads in regions of high seismic risk, or in structures assigned to high seismic performance or design categories.

1.4 — Approval of special systems of design or construction

Sponsors of any system of design or construction within the scope of this code, the adequacy of which has been shown by successful use or by analysis or test, but which does not conform to or is not covered by this code, shall have the right to present the data on which their design is based to the building official or to a board of examiners appointed by the building official. This board shall be composed of competent engineers and shall have authority to investigate the data so submitted, to require tests, and to formulate rules governing design and construction of such systems to meet the intent of this code. These rules when approved by the building official and promulgated shall be of the same force and effect as the provisions of this code.

COMMENTARY

R1.3.5 — The purpose of this section is to ensure that the special detailing required in special moment frames is properly executed through inspection by personnel who are qualified to do this work. Qualifications of inspectors should be acceptable to the jurisdiction enforcing the general building code.

R1.4 — Approval of special systems of design or construction

New methods of design, new materials, and new uses of materials should undergo a period of development before being specifically covered in a code. Hence, good systems or components might be excluded from use by implication if means were not available to obtain acceptance.

For special systems considered under this section, specific tests, load factors, deflection limits, and other pertinent requirements should be set by the board of examiners, and should be consistent with the intent of the code.

The provisions of this section do not apply to model tests used to supplement calculations under 1.2.2 or to strength evaluation of existing structures under Chapter 20.

CHAPTER 2 — NOTATION AND DEFINITIONS

2.1 — Code notation

The terms in this list are used in the code and as needed in the commentary.

a = depth of equivalent rectangular stress block as defined in 10.2.7.1, in., Chapter 10

a_v = shear span, equal to distance from center of concentrated load to either (a) face of support for continuous or cantilevered members, or (b) center of support for simply supported members, in., Chapter 11, Appendix A

A_b = area of an individual bar or wire, in.^{2}, Chapters 10, 12

A_{brg} = bearing area of the head of stud or anchor bolt, in.^{2}, Appendix D

A_c = area of concrete section resisting shear transfer, in.^{2}, Chapter 11

A_{cf} = larger gross cross-sectional area of the slab-beam strips of the two orthogonal equivalent frames intersecting at a column of a two-way slab, in.^{2}, Chapter 18

A_{ch} = cross-sectional area of a structural member measured out-to-out of transverse reinforcement, in.^{2}, Chapters 10, 21

A_{cp} = area enclosed by outside perimeter of concrete cross section, in.^{2}, see 11.6.1, Chapter 11

A_{cs} = cross-sectional area at one end of a strut in a strut-and-tie model, taken perpendicular to the axis of the strut, in.^{2}, Appendix A

A_{ct} = area of that part of cross section between the flexural tension face and center of gravity of gross section, in.^{2}, Chapter 18

A_{cv} = gross area of concrete section bounded by web thickness and length of section in the direction of shear force considered, in.^{2}, Chapter 21

A_{cw} = area of concrete section of an individual pier, horizontal wall segment, or coupling beam resisting shear, in.^{2}, Chapter 21

A_f = area of reinforcement in bracket or corbel resisting factored moment, in.^{2}, see 11.9, Chapter 11

A_g = gross area of concrete section, in.^{2} For a hollow section, A_g is the area of the concrete only and does not include the area of the void(s), see 11.6.1, Chapters 9-11, 14-16, 21, 22, Appendixes B, C.

A_h = total area of shear reinforcement parallel to primary tension reinforcement in a corbel or bracket, in.^{2}, see 11.9, Chapter 11

A_j = effective cross-sectional area within a joint in a plane parallel to plane of reinforcement generating shear in the joint, in.^{2}, see 21.5.3.1, Chapter 21

A_ℓ = total area of longitudinal reinforcement to resist torsion, in.^{2}, Chapter 11

$A_{\ell,min}$ = minimum area of longitudinal reinforcement to resist torsion, in.^{2}, see 11.6.5.3, Chapter 11

A_n = area of reinforcement in bracket or corbel resisting tensile force N_{uc}, in.^{2}, see 11.9, Chapter 11

A_{nz} = area of a face of a nodal zone or a section through a nodal zone, in.^{2}, Appendix A

A_{Nc} = projected concrete failure area of a single anchor or group of anchors, for calculation of strength in tension, in.^{2}, see D.5.2.1, Appendix D

A_{Nco} = projected concrete failure area of a single anchor, for calculation of strength in tension if not limited by edge distance or spacing, in.^{2}, see D.5.2.1, Appendix D

A_o = gross area enclosed by shear flow path, in.^{2}, Chapter 11

A_{oh} = area enclosed by centerline of the outermost closed transverse torsional reinforcement, in.^{2}, Chapter 11

A_{ps} = area of prestressing steel in flexural tension zone, in.^{2}, Chapter 18, Appendix B

A_s = area of nonprestressed longitudinal tension reinforcement, in.^{2}, Chapters 10-12, 14, 15, 18, Appendix B

A_s' = area of longitudinal compression reinforcement, in.^{2}, Appendix A

A_{sc} = area of primary tension reinforcement in a corbel or bracket, in.^{2}, see 11.9.3.5, Chapter 11

A_{se} = effective cross-sectional area of anchor, in.^{2}, Appendix D

A_{sh} = total cross-sectional area of transverse reinforcement (including crossties) within spacing s and perpendicular to dimension b_c, in.^{2}, Chapter 21

A_{si} = total area of surface reinforcement at spacing s_i in the i-th layer crossing a strut, with reinforcement at an angle α_i to the axis of the strut, in.^{2}, Appendix A

$A_{s,min}$ = minimum area of flexural reinforcement, in.^{2}, see 10.5, Chapter 10

A_{st} = total area of nonprestressed longitudinal reinforcement, (bars or steel shapes), in.^{2}, Chapters 10, 21

A_{sx} = area of structural steel shape, pipe, or tubing

in a composite section, in.^{2}, Chapter 10

A_t = area of one leg of a closed stirrup resisting torsion within spacing s, in.^{2}, Chapter 11

A_{tp} = area of prestressing steel in a tie, in.^{2}, Appendix A

A_{tr} = total cross-sectional area of all transverse reinforcement within spacing s that crosses the potential plane of splitting through the reinforcement being developed, in.^{2}, Chapter 12

A_{ts} = area of nonprestressed reinforcement in a tie, in.^{2}, Appendix A

A_v = area of shear reinforcement spacing s, in.^{2}, Chapters 11, 17

A_{Vc} = projected concrete failure area of a single anchor or group of anchors, for calculation of strength in shear, in.^{2}, see D.6.2.1, Appendix D

A_{Vco} = projected concrete failure area of a single anchor, for calculation of strength in shear, if not limited by corner influences, spacing, or member thickness, in.^{2}, see D.6.2.1, Appendix D

A_{vd} = total area of reinforcement in each group of diagonal bars in a diagonally reinforced coupling beam, in.^{2}, Chapter 21

A_{vf} = area of shear-friction reinforcement, in.^{2}, Chapter 11

A_{vh} = area of shear reinforcement parallel to flexural tension reinforcement within spacing s_2, in.^{2}, Chapter 11

$A_{v,min}$ = minimum area of shear reinforcement within spacing s, in.^{2}, see 11.5.6.3 and 11.5.6.4, Chapter 11

A_1 = loaded area, in.^{2}, Chapters 10, 22

A_2 = area of the lower base of the largest frustum of a pyramid, cone, or tapered wedge contained wholly within the support and having for its upper base the loaded area, and having side slopes of 1 vertical to 2 horizontal, in.^{2}, Chapters 10, 22

b = width of compression face of member, in., Chapter 10, Appendix B

b_c = cross-sectional dimension of column core measured center-to-center of outer legs of the transverse reinforcement comprising area A_{sh}, in., Chapter 21

b_o = perimeter of critical section for shear in slabs and footings, in., see 11.12.1.2, Chapters 11, 22

b_s = width of strut, in., Appendix A

b_t = width of that part of cross section containing the closed stirrups resisting torsion, in., Chapter 11

b_v = width of cross section at contact surface being investigated for horizontal shear, in., Chapter 17

b_w = web width, or diameter of circular section, in., Chapters 10-12, 21, 22, Appendix B

b_1 = dimension of the critical section b_o measured in the direction of the span for which moments are determined, in., Chapter 13

b_2 = dimension of the critical section b_o measured in the direction perpendicular to b_1, in., Chapter 13

B_n = nominal bearing strength, lb, Chapter 22

B_u = factored bearing load, lb, Chapter 22

c = distance from extreme compression fiber to neutral axis, in., Chapters 9, 10, 14, 21

c_{ac} = critical edge distance required to develop the basic concrete breakout strength of a post-installed anchor in uncracked concrete without supplementary reinforcement to control splitting, in., see D.8.6, Appendix D

$c_{a,max}$ = maximum distance from center of an anchor shaft to the edge of concrete, in., Appendix D

$c_{a,min}$ = minimum distance from center of an anchor shaft to the edge of concrete, in., Appendix D

c_{a1} = distance from the center of an anchor shaft to the edge of concrete in one direction, in. If shear is applied to anchor, c_{a1} is taken in the direction of the applied shear. If the tension is applied to the anchor, c_{a1} is the minimum edge distance, Appendix D

c_{a2} = distance from center of an anchor shaft to the edge of concrete in the direction perpendicular to c_{a1}, in., Appendix D

c_b = smaller of (a) the distance from center of a bar or wire to nearest concrete surface, and (b) one-half the center-to-center spacing of bars or wires being developed, in., Chapter 12

c_c = clear cover of reinforcement, in., see 10.6.4, Chapter 10

c_t = distance from the interior face of the column to the slab edge measured parallel to c_1, but not exceeding c_1, in., Chapter 21

c_1 = dimension of rectangular or equivalent rectangular column, capital, or bracket measured in the direction of the span for which moments are being determined, in., Chapters 11, 13, 21

c_2 = dimension of rectangular or equivalent rectangular column, capital, or bracket measured in the direction perpendicular to c_1, in., Chapter 13

C = cross-sectional constant to define torsional properties of slab and beam, see 13.6.4.2, Chapter 13

C_m = factor relating actual moment diagram to an equivalent uniform moment diagram, Chapter 10

d = distance from extreme compression fiber to centroid of longitudinal tension reinforcement, in., Chapters 7, 9-12, 14, 17, 18, 21, Appendixes B, C

d' = distance from extreme compression fiber to centroid of longitudinal compression reinforcement, in., Chapters 9, 18, Appendix C

d_b = nominal diameter of bar, wire, or prestressing strand, in., Chapters 7, 12, 21

d_o = outside diameter of anchor or shaft diameter of headed stud, headed bolt, or hooked bolt, in., see D.8.4, Appendix D

d_o' = value substituted for d_o when an oversized anchor is used, in., see D.8.4, Appendix D

d_p = distance from extreme compression fiber to centroid of prestressing steel, in., Chapters 11,18, Appendix B

d_{pile} = diameter of pile at footing base, in., Chapter 15

d_t = distance from extreme compression fiber to centroid of extreme layer of longitudinal tension steel, in., Chapters 9, 10, Appendix C

D = dead loads, or related internal moments and forces, Chapters 8, 9, 20, 21, Appendix C

e = base of Napierian logarithms, Chapter 18

e_h = distance from the inner surface of the shaft of a J- or L-bolt to the outer tip of the J- or L-bolt, in., Appendix D

e'_N = distance between resultant tension load on a group of anchors loaded in tension and the centroid of the group of anchors loaded in tension, in.; e'_N is always positive, Appendix D

e'_V = distance between resultant shear load on a group of anchors loaded in shear in the same direction, and the centroid of the group of anchors loaded in shear in the same direction, in., e'_V is always positive, Appendix D

E = load effects of earthquake, or related internal moments and forces, Chapters 9, 21, Appendix C

E_c = modulus of elasticity of concrete, psi, see 8.5.1, Chapters 8-10, 14, 19

E_{cb} = modulus of elasticity of beam concrete, psi, Chapter 13

E_{cs} = modulus of elasticity of slab concrete, psi, Chapter 13

EI = flexural stiffness of compression member, in.^{2}-lb, see 10.12.3, Chapter 10

E_p = modulus of elasticity of prestressing steel, psi, see 8.5.3, Chapter 8

E_s = modulus of elasticity of reinforcement and structural steel, psi, see 8.5.2, Chapters 8, 10, 14

f_c' = specified compressive strength of concrete, psi, Chapters 4, 5, 8-12, 14, 18, 19, 21, 22, Appendixes A-D

$\sqrt{f_c'}$ = square root of specified compressive strength of concrete, psi, Chapters 8, 9, 11, 12, 18, 19, 21, 22, Appendix D

f_{ce} = effective compressive strength of the concrete in a strut or a nodal zone, psi, Chapter 15, Appendix A

f_{ci}' = specified compressive strength of concrete at time of initial prestress, psi, Chapters 7, 18

$\sqrt{f_{ci}'}$ = square root of specified compressive strength of concrete at time of initial prestress, psi, Chapter 18

f_{cr}' = required average compressive strength of concrete used as the basis for selection of concrete proportions, psi, Chapter 5

f_{ct} = average splitting tensile strength of lightweight concrete, psi, Chapters 5, 9, 11, 12, 22

f_d = stress due to unfactored dead load, at extreme fiber of section where tensile stress is caused by externally applied loads, psi, Chapter 11

f_{dc} = decompression stress; stress in the prestressing steel when stress is zero in the concrete at the same level as the centroid of the prestressing steel, psi, Chapter 18

f_{pc} = compressive stress in concrete (after allowance for all prestress losses) at centroid of cross section resisting externally applied loads or at junction of web and flange when the centroid lies within the flange, psi. (In a composite member, f_{pc} is the resultant compressive stress at centroid of composite section, or at junction of web and flange when the centroid lies within the flange, due to both prestress and moments resisted by precast member acting alone), Chapter 11

f_{pe} = compressive stress in concrete due to effective prestress forces only (after allowance for all prestress losses) at extreme fiber of section where tensile stress is caused by externally applied loads, psi, Chapter 11

f_{ps} = stress in prestressing steel at nominal flexural strength, psi, Chapters 12, 18

f_{pu} = specified tensile strength of prestressing steel, psi, Chapters 11, 18

f_{py} = specified yield strength of prestressing steel, psi, Chapter 18

f_r = modulus of rupture of concrete, psi, see 9.5.2.3, Chapters 9, 14, 18, Appendix B

f_s = calculated tensile stress in reinforcement at service loads, psi, Chapters 10, 18

f_s' = stress in compression reinforcement under factored loads, psi, Appendix A

f_{se} = effective stress in prestressing steel (after allowance for all prestress losses), psi, Chapters 12, 18, Appendix A

f_t = extreme fiber stress in tension in the precompressed tensile zone calculated at service loads using gross section properties, psi, see 18.3.3, Chapter 18

f_{uta} = specified tensile strength of anchor steel, psi, Appendix D

f_y = specified yield strength of reinforcement, psi, Chapters 3, 7, 9-12, 14, 17-19, 21, Appendixes A-C

f_{ya} = specified yield strength of anchor steel, psi,

Appendix D

f_{yt} = specified yield strength f_y of transverse reinforcement, psi, Chapters 10-12, 21

F = loads due to weight and pressures of fluids with well-defined densities and controllable maximum heights, or related internal moments and forces, Chapter 9, Appendix C

F_n = nominal strength of a strut, tie, or nodal zone, lb, Appendix A

F_{nn} = nominal strength at face of a nodal zone, lb, Appendix A

F_{ns} = nominal strength of a strut, lb, Appendix A

F_{nt} = nominal strength of a tie, lb, Appendix A

F_u = factored force acting in a strut, tie, bearing area, or nodal zone in a strut-and-tie model, lb, Appendix A

h = overall thickness or height of member, in., Chapters 9-12, 14, 17, 18, 20-22, Appendixes A, C

h_a = thickness of member in which an anchor is located, measured parallel to anchor axis, in., Appendix D

h_{ef} = effective embedment depth of anchor, in., see D.8.5, Appendix D

h_v = depth of shearhead cross section, in., Chapter 11

h_w = height of entire wall from base to top or height of the segment of wall considered, in., Chapters 11, 21

h_x = maximum center-to-center horizontal spacing of crossties or hoop legs on all faces of the column, in., Chapter 21

H = loads due to weight and pressure of soil, water in soil, or other materials, or related internal moments and forces, Chapter 9, Appendix C

I = moment of inertia of section about centroidal axis, in.^{4}, Chapters 10, 11

I_b = moment of inertia of gross section of beam about centroidal axis, in.^{4}, see 13.2.4, Chapter 13

I_{cr} = moment of inertia of cracked section transformed to concrete, in.^{4}, Chapters 9, 14

I_e = effective moment of inertia for computation of deflection, in.^{4}, see 9.5.2.3, Chapters 9, 14

I_g = moment of inertia of gross concrete section about centroidal axis, neglecting reinforcement, in.^{4},Chapters 9, 10

I_s = moment of inertia of gross section of slab about centroidal axis defined for calculating α_f and β_t, in.^{4}, Chapter 13

I_{se} = moment of inertia of reinforcement about centroidal axis of member cross section, in.^{4}, Chapter 10

I_{sx} = moment of inertia of structural steel shape, pipe, or tubing about centroidal axis of composite member cross section, in.^{4}, Chapter 10

k = effective length factor for compression members, Chapters 10, 14

k_c = coefficient for basic concrete breakout strength in tension, Appendix D

k_{cp} = coefficient for pryout strength, Appendix D

K = wobble friction coefficient per foot of tendon, Chapter 18

K_{tr} = transverse reinforcement index, see 12.2.3, Chapter 12

ℓ = span length of beam or one-way slab; clear projection of cantilever, in., see 8.7, Chapter 9

ℓ_a = additional embedment length beyond centerline of support or point of inflection, in., Chapter 12

ℓ_c = length of compression member in a frame, measured center-to-center of the joints in the frame, in., Chapters 10, 14, 22

ℓ_d = development length in tension of deformed bar, deformed wire, plain and deformed welded wire reinforcement, or pretensioned strand, in., Chapters 7, 12, 19, 21

ℓ_{dc} = development length in compression of deformed bars and deformed wire, in., Chapter 12

ℓ_{dh} = development length in tension of deformed bar or deformed wire with a standard hook, measured from critical section to outside end of hook (straight embedment length between critical section and start of hook [point of tangency] plus inside radius of bend and one bar diameter), in., see 12.5 and 21.5.4, Chapters 12, 21

ℓ_e = load bearing length of anchor for shear, in., see D.6.2.2, Appendix D

ℓ_n = length of clear span measured face-to-face of of supports, in., Chapters 8-11, 13, 16, 18, 21

ℓ_o = length, measured from joint face along axis of structural member, over which special transverse reinforcement must be provided, in., Chapter 21

ℓ_{px} = distance from jacking end of prestressing steel element to point under consideration, ft, see 18.6.2, Chapter 18

ℓ_t = span of member under load test, taken as the shorter span for two-way slab systems, in. Span is the smaller of (a) distance between centers of supports, and (b) clear distance between supports plus thickness h of member. Span for a cantilever shall be taken as twice the distance from face of support to cantilever end, Chapter 20

ℓ_u = unsupported length of compression member, in., see 10.11.3.1, Chapter 10

ℓ_v = length of shearhead arm from centroid of concentrated load or reaction, in., Chapter 11

ℓ_w = length of entire wall or length of segment of wall considered in direction of shear force, in., Chapters 11, 14, 21

ℓ_1 = length of span in direction that moments are being determined, measured center-to-center of supports, in., Chapter 13

ℓ_2 = length of span in direction perpendicular to ℓ_1, measured center-to-center of supports, in., see 13.6.2.3 and 13.6.2.4, Chapter 13

L = live loads, or related internal moments and forces, Chapters 8, 9, 20, 21, Appendix C

L_r = roof live load, or related internal moments and forces, Chapter 9

M = maximum unfactored moment due to service loads, including $P\Delta$ effects, in.-lb, Chapter 14

M_a = maximum unfactored moment in member at stage deflection is computed, in.-lb, Chapters 9, 14

M_c = factored moment amplified for the effects of member curvature used for design of compression member, in.-lb, see 10.12.3, Chapter 10

M_{cr} = cracking moment, in.-lb, see 9.5.2.3, Chapters 9, 14

M_{cre} = moment causing flexural cracking at section due to externally applied loads, in.-lb, Chapter 11

M_m = factored moment modified to account for effect of axial compression, in.-lb, see 11.3.2.2, Chapter 11

M_{max} = maximum factored moment at section due to externally applied loads, in.-lb, Chapter 11

M_n = nominal flexural strength at section, in.-lb, Chapters 11, 12, 14, 18, 21, 22

M_{nb} = nominal flexural strength of beam including slab where in tension, framing into joint, in.-lb, see 21.4.2.2, Chapter 21

M_{nc} = nominal flexural strength of column framing into joint, calculated for factored axial force, consistent with the direction of lateral forces considered, resulting in lowest flexural strength, in.-lb, see 21.4.2.2, Chapter 21

M_o = total factored static moment, in.-lb, Chapter 13

M_p = required plastic moment strength of shearhead cross section, in.-lb, Chapter 11

M_{pr} = probable flexural strength of members, with or without axial load, determined using the properties of the member at the joint faces assuming a tensile stress in the longitudinal bars of at least $1.25f_y$ and a strength reduction factor, ϕ, of 1.0, in.-lb, Chapter 21

M_s = factored moment due to loads causing appreciable sway, in.-lb, Chapter 10

M_{sa} = maximum unfactored applied moment due to service loads, not including $P\Delta$ effects, in.-lb, Chapter 14

M_{slab} = portion of slab factored moment balanced by support moment, in.-lb, Chapter 21

M_u = factored moment at section, in.-lb, Chapters 10, 11, 13, 14, 21, 22

M_{ua} = moment at the midheight section of the wall due to factored lateral and eccentric vertical loads, in.-lb, Chapter 14

M_v = moment resistance contributed by shearhead reinforcement, in.-lb, Chapter 11

M_1 = smaller factored end moment on a compression member, to be taken as positive if member is bent in single curvature, and negative if bent in double curvature, in.-lb, Chapter 10

M_{1ns} = factored end moment on a compression member at the end at which M_1 acts, due to loads that cause no appreciable sidesway, calculated using a first-order elastic frame analysis, in.-lb, Chapter 10

M_{1s} = factored end moment on compression member at the end at which M_1 acts, due to loads that cause appreciable sidesway, calculated using a first-order elastic frame analysis, in.-lb, Chapter 10

M_2 = larger factored end moment on compression member, always positive, in.-lb, Chapter 10

$M_{2,min}$ = minimum value of M_2, in.-lb, Chapter 10

M_{2ns} = factored end moment on compression member at the end at which M_2 acts, due to loads that cause no appreciable sidesway, calculated using a first-order elastic frame analysis, in.-lb, Chapter 10

M_{2s} = factored end moment on compression member at the end at which M_2 acts, due to loads that cause appreciable sidesway, calculated using a first-order elastic frame, in.-lb, Chapter 10

n = number of items, such as strength tests, bars, wires, monostrand anchorage devices, anchors, or shearhead arms, Chapters 5, 11, 12, 18, Appendix D

N_b = basic concrete breakout strength in tension of a single anchor in cracked concrete, lb, see D.5.2.2, Appendix D

N_c = tension force in concrete due to unfactored dead load plus live load, lb, Chapter 18

N_{cb} = nominal concrete breakout strength in tension of a single anchor, lb, see D.5.2.1, Appendix D

N_{cbg} = nominal concrete breakout strength in tension of a group of anchors, lb, see D.5.2.1, Appendix D

N_n = nominal strength in tension, lb, Appendix D

N_p = pullout strength in tension of a single anchor in cracked concrete, lb, see D.5.3.4 and D.5.3.5, Appendix D

N_{pn} = nominal pullout strength in tension of a single anchor, lb, see D.5.3.1, Appendix D

N_{sa} = nominal strength of a single anchor or group of anchors in tension as governed by the steel strength, lb, see D.5.1.1 and D.5.1.2, Appendix D

N_{sb} = side-face blowout strength of a single anchor, lb, Appendix D

N_{sbg} = side-face blowout strength of a group of anchors, lb, Appendix D

N_u = factored axial force normal to cross section occurring simultaneously with V_u or T_u; to be taken as positive for compression and negative for tension, lb, Chapter 11

N_{ua} = factored tensile force applied to anchor or group of anchors, lb, Appendix D

N_{uc} = factored horizontal tensile force applied at top of bracket or corbel acting simultaneously with V_u, to be taken as positive for tension, lb, Chapter 11

p_{cp} = outside perimeter of concrete cross section, in., see 11.6.1, Chapter 11

p_h = perimeter of centerline of outermost closed transverse torsional reinforcement, in., Chapter 11

P_b = nominal axial strength at balanced strain conditions, lb, see 10.3.2, Chapters 9, 10, Appendixes B, C

P_c = critical buckling load, lb, see 10.12.3, Chapter 10

P_n = nominal axial strength of cross section, lb, Chapters 9, 10, 14, 22, Appendixes B, C

$P_{n,max}$ = maximum allowable value of P_n, lb, see 10.3.6, Chapter 10

P_o = nominal axial strength at zero eccentricity, lb, Chapter 10

P_{pj} = prestressing force at jacking end, lb, Chapter 18

P_{pu} = factored prestressing force at anchorage device, lb, Chapter 18

P_{px} = prestressing force evaluated at distance ℓ_{px} from the jacking end, lb, Chapter 18

P_s = unfactored axial load at the design (midheight) section including effects of self-weight, lb, Chapter 14

P_u = factored axial force; to be taken as positive for compression and negative for tension, lb, Chapters 10, 14, 21, 22

q_{Du} = factored dead load per unit area, Chapter 13

q_{Lu} = factored live load per unit area, Chapter 13

q_u = factored load per unit area, Chapter 13

Q = stability index for a story, see 10.11.4, Chapter 10

r = radius of gyration of cross section of a compression member, in., Chapter 10

R = rain load, or related internal moments and forces, Chapter 9

s = center-to-center spacing of items, such as longitudinal reinforcement, transverse reinforcement, prestressing tendons, wires, or anchors, in., Chapters 10-12, 17-21, Appendix D

s_i = center-to-center spacing of reinforcement in the i-th layer adjacent to the surface of the member, in., Appendix A

s_o = center-to-center spacing of transverse reinforcement within the length ℓ_o, in., Chapter 21

s_s = sample standard deviation, psi, Chapter 5, Appendix D

s_2 = center-to-center spacing of longitudinal shear or torsion reinforcement, in., Chapter 11

S = snow load, or related internal moments and forces, Chapters 9, 21

S_e = moment, shear, or axial force at connection corresponding to development of probable strength at intended yield locations, based on the governing mechanism of inelastic lateral deformation, considering both gravity and earthquake load effects, Chapter 21

S_m = elastic section modulus, in.^{3}, Chapter 22

S_n nominal flexural, shear, or axial strength of connection, Chapter 21

S_y = yield strength of connection, based on f_y, for moment, shear, or axial force, Chapter 21

t = wall thickness of hollow section, in., Chapter 11

T = cumulative effect of temperature, creep, shrinkage, differential settlement, and shrinkage-compensating concrete, Chapter 9, Appendix C

T_n = nominal torsional moment strength, in.-lb, Chapter 11

T_u = factored torsional moment at section, in.-lb, Chapter 11

U = required strength to resist factored loads or related internal moments and forces, Chapter 9, Appendix C

v_n = nominal shear stress, psi, see 11.12.6.2, Chapters 11, 21

V_b = basic concrete breakout strength in shear of a single anchor in cracked concrete, lb, see D.6.2.2 and D.6.2.3, Appendix D

V_c = nominal shear strength provided by concrete, lb, Chapters 8, 11, 13, 21

V_{cb} = nominal concrete breakout strength in shear of a single anchor, lb, see D.6.2.1, Appendix D

V_{cbg} = nominal concrete breakout strength in shear of a group of anchors, lb, see D.6.2.1, Appendix D

V_{ci} = nominal shear strength provided by concrete when diagonal cracking results from combined shear and moment, lb, Chapter 11

V_{cp} = nominal concrete pryout strength of a single anchor, lb, see D.6.3, Appendix D

V_{cpg} = nominal concrete pryout strength of a group of anchors, lb, see D.6.3, Appendix D

V_{cw} = nominal shear strength provided by concrete when diagonal cracking results from high principal tensile stress in web, lb, Chapter 11

V_d = shear force at section due to unfactored dead load, lb, Chapter 11

V_e = design shear force corresponding to the development of the probable moment strength of the member, lb, see 21.3.4.1 and 21.4.5.1 Chapter 21

V_i = factored shear force at section due to externally applied loads occurring simultaneously with M_{max}, lb, Chapter 11

V_n = nominal shear strength, lb, Chapters 8, 10, 11, 21, 22, Appendix D

V_{nh} = nominal horizontal shear strength, lb, Chapter 17

V_p = vertical component of effective prestress force at section, lb, Chapter 11

V_s = nominal shear strength provided by shear reinforcement, lb, Chapter 11

V_{sa} = nominal strength in shear of a single anchor or group of anchors as governed by the steel strength, lb, see D.6.1.1 and D.6.1.2, Appendix D

V_u = factored shear force at section, lb, Chapters 11-13, 17, 21, 22

V_{ua} = factored shear force applied to a single anchor or group of anchors, lb, Appendix D

V_{us} = factored horizontal shear in a story, lb, Chapter 10

w_c = unit weight of concrete, lb/ft^3, Chapters 8, 9

w_u = factored load per unit length of beam or one-way slab, Chapter 8

W = wind load, or related internal moments and forces, Chapter 9, Appendix C

x = shorter overall dimension of rectangular part of cross section, in., Chapter 13

y = longer overall dimension of rectangular part of cross section, in., Chapter 13

y_t = distance from centroidal axis of gross section, neglecting reinforcement, to tension face, in., Chapters 9, 11

α = angle defining the orientation of reinforcement, Chapters 11, 21, Appendix A

α_c = coefficient defining the relative contribution of concrete strength to nominal wall shear strength, see 21.7.4.1, Chapter 21

α_f = ratio of flexural stiffness of beam section to flexural stiffness of a width of slab bounded laterally by centerlines of adjacent panels (if any) on each side of the beam, see 13.6.1.6, Chapters 9, 13

α_{fm} = average value of α_f for all beams on edges of a panel, Chapter 9

α_{f1} = α_f in direction of ℓ_1, Chapter 13

α_{f2} = α_f in direction of ℓ_2, Chapter 13

α_i = angle between the axis of a strut and the bars in the i-th layer of reinforcement crossing that strut, Appendix A

α_{px} = total angular change of tendon profile from tendon jacking end to point under consideration, radians, Chapter 18

α_s = constant used to compute V_c in slabs and footings, Chapter 11

α_v = ratio of flexural stiffness of shearhead arm to that of the surrounding composite slab section, see 11.12.4.5, Chapter 11

β = ratio of long to short dimensions: clear spans for two-way slabs, see 9.5.3.3 and 22.5.4; sides of column, concentrated load or reaction area, see 11.12.2.1; or sides of a footing, see 15.4.4.2, Chapters 9, 11, 15, 22

β_b = ratio of area of reinforcement cut off to total area of tension reinforcement at section, Chapter 12

β_d = ratio used to compute magnified moments in columns due to sustained loads, see 10.11.1 and 10.13.6, Chapter 10

β_n = factor to account for the effect of the anchorage of ties on the effective compressive strength of a nodal zone, Appendix A

β_p = factor used to compute V_c in prestressed slabs, Chapter 11

β_s = factor to account for the effect of cracking and confining reinforcement on the effective compressive strength of the concrete in a strut, Appendix A

β_t = ratio of torsional stiffness of edge beam section to flexural stiffness of a width of slab equal to span length of beam, center-to-center of supports, see 13.6.4.2, Chapter 13

β_1 = factor relating depth of equivalent rectangular compressive stress block to neutral axis depth, see 10.2.7.3, Chapters 10, 18, Appendix B

γ_f = factor used to determine the unbalanced moment transferred by flexure at slab-column connections, see 13.5.3.2, Chapters 11, 13, 21

γ_p = factor for type of prestressing steel, see 18.7.2, Chapter 18

γ_s = factor used to determine the portion of reinforcement located in center band of footing, see 15.4.4.2, Chapter 15

γ_v = factor used to determine the unbalanced moment transferred by eccentricity of shear at slab-column connections, see 11.12.6.1, Chapter 11

δ_{ns} = moment magnification factor for frames braced against sidesway, to reflect effects of member curvature between ends of compression member, Chapter 10

δ_s = moment magnification factor for frames not braced against sidesway, to reflect lateral drift resulting from lateral and gravity loads, Chapter 10

δ_u = design displacement, in., Chapter 21

Δf_p = increase in stress in prestressing steel due to factored loads, psi, Appendix A

Δf_{ps} = stress in prestressing steel at service loads less decompression stress, psi, Chapter 18

Δ_o = relative lateral deflection between the top and bottom of a story due to lateral forces computed using a first-order elastic frame analysis and stiffness values satisfying 10.11.1, in., Chapter 10

Δ_r = difference between initial and final (after load removal) deflections for load test or repeat load test, in., Chapter 20

Δ_s = maximum deflection at or near midheight due to service loads, in., Chapter 14

Δ_u = deflection at midheight of wall due to factored loads, in., Chapter 14

Δ_1 = measured maximum deflection during first load test, in., see 20.5.2, Chapter 20

Δ_2 = maximum deflection measured during second load test relative to the position of the structure at the beginning of second load test, in., see 20.5.2, Chapter 20

ε_t = net tensile strain in extreme layer of longitudinal tension steel at nominal strength, excluding strains due to effective prestress, creep, shrinkage, and temperature, Chapters 8-10, Appendix C

θ = angle between axis of strut, compression diagonal, or compression field and the tension chord of the member, Chapter 11,Appendix A

λ = modification factor related to unit weight of concrete, Chapters 11, 12, 17-19, Appendix A

λ_Δ = multiplier for additional deflection due to long-term effects, see 9.5.2.5, Chapter 9

μ = coefficient of friction, see 11.7.4.3, Chapter 11

μ_p = post-tensioning curvature friction coefficient, Chapter 18

ξ = time-dependent factor for sustained load, see 9.5.2.5, Chapter 9

ρ = ratio of $\boldsymbol{A_s}$ to $\boldsymbol{bd}$, Chapters 11, 13, 21, Appendix B

ρ' = ratio of $\boldsymbol{A_s}'$ to $\boldsymbol{bd}$, Chapter 9, Appendix B

ρ_b = ratio of $\boldsymbol{A_s}$ to $\boldsymbol{bd}$ producing balanced strain conditions, see 10.3.2, Chapters 10, 13, 14, Appendix B

ρ_ℓ = ratio of area of distributed longitudinal reinforcement to gross concrete area perpendicular to that reinforcement, Chapters 11,14, 21

ρ_p = ratio of $\boldsymbol{A_{ps}}$ to $\boldsymbol{bd_p}$, Chapter 18

ρ_s = ratio of volume of spiral reinforcement to total volume of core confined by the spiral (measured out-to-out of spirals), Chapters 10, 21

ρ_t = ratio of area distributed transverse reinforcement to gross concrete area perpendicular to that reinforcement, Chapters 11, 14, 21

ρ_v = ratio of tie reinforcement area to area of contact surface, see 17.5.3.3, Chapter 17

ρ_w = ratio of $\boldsymbol{A_s}$ to $\boldsymbol{b_w d}$, Chapter 11

ϕ = strength reduction factor, see 9.3, Chapters 8-11, 13, 14, 17-22, Appendixes A-D

$\psi_{c,N}$ = factor used to modify tensile strength of anchors based on presence or absence of cracks in concrete, see D.5.2.6, Appendix D

$\psi_{c,P}$ = factor used to modify pullout strength of anchors based on presence or absence of cracks in concrete, see D.5.3.6, Appendix D

$\psi_{c,V}$ = factor used to modify shear strength of anchors based on presence or absence of cracks in concrete and presence or absence of supplementary reinforcement, see D.6.2.7 for anchors in shear, Appendix D

$\psi_{cp,N}$ = factor used to modify tensile strength of post-installed anchors intended for use in uncracked concrete without supplementary reinforcement, see D.5.2.7, Appendix D

ψ_e = factor used to modify development length based on reinforcement coating, see 12.2.4, Chapter 12

$\psi_{ec,N}$ = factor used to modify tensile strength of anchors based on eccentricity of applied loads, see D.5.2.4, Appendix D

$\psi_{ec,V}$ = factor used to modify shear strength of anchors based on eccentricity of applied loads, see D.6.2.5, Appendix D

$\psi_{ed,N}$ = factor used to modify tensile strength of anchors based on proximity to edges of concrete member, see D.5.2.5, Appendix D

$\psi_{ed,V}$ = factor used to modify shear strength of anchors based on proximity to edges of concrete member, see D.6.2.6, Appendix D

ψ_s = factor used to modify development length based on reinforcement size, see 12.2.4, Chapter 12

ψ_t = factor used to modify development length based on reinforcement location, see 12.2.4, Chapter 12

ω = tension reinforcement index, see 18.7.2, Chapter 18, Appendix B

ω' = compression reinforcement index, see 18.7.2, Chapter 18, Appendix B

ω_p = prestressing steel index, see B.18.8.1, Appendix B

ω_{pw} = prestressing steel index for flanged sections, see B.18.8.1, Appendix B

ω_w = tensions reinforcement index for flanged sections, see B.18.8.1, Appendix B

ω_w' = compression reinforcement index for flanged sections, see B.18.8.1, Appendix B

R2.1 — Commentary notation

The terms used in this list are used in the commentary, but not in the code.

Units of measurement are given in the Notation to assist the user and are not intended to preclude the use of other correctly applied units for the same symbol, such as feet or kips.

c'_{a1} = limiting value of c_{a1} when anchors are located less than $1.5h_{ef}$ from three or more edges (see Fig. RD.6.2.4), Appendix D

C = compression force acting on a nodal zone, lb, Appendix A

f_{si} = stress in the i-th layer of surface reinforcement, psi, Appendix A

h_{anc} = dimension of anchorage device or single group of closely spaced devices in the direction of bursting being considered, in., Chapter 18

h'_{ef} = limiting value of h_{ef} when anchors are located less than $1.5h_{ef}$ from three or more edges (see Fig. RD.5.2.3), Appendix D

K_t = torsional stiffness of torsional member; moment per unit rotation, see R13.7.5, Chapter 13

K_{05} = coefficient associated with the 5 percent fractile, Appendix D

ℓ_{anc} = length along which anchorage of a tie must occur, in., Appendix A

ℓ_b = width of bearing, in., Appendix A

R = reaction, lb, Appendix A

T = tension force acting on a nodal zone, lb, Appendix A

w_s = width of a strut perpendicular to the axis of the strut, in., Appendix A

w_t = effective height of concrete concentric with a tie, used to dimension nodal zone, in., Appendix A

w_{tmax} = maximum effective height of concrete concentric with a tie, in., Appendix A

Δf_{pt} = f_{ps} at the section of maximum moment minus the stress in the prestressing steel due to prestressing and factored bending moments at the section under consideration, psi, see R11.6.3.10, Chapter 11

ϕ_K = stiffness reduction factor, see R10.12.3, Chapter 10

Ω_o = amplification factor to account for overstrength of the seismic-force-resisting system, specified in documents such as NEHRP,[21.1] SEI/ASCE,[21.48] IBC,[21.5] and UBC,[21.2] Chapter 21

SECTION 2.2, DEFINITIONS, BEGINS ON NEXT PAGE

CODE

2.2 — Definitions

The following terms are defined for general use in this code. Specialized definitions appear in individual chapters.

Admixture — Material other than water, aggregate, or hydraulic cement, used as an ingredient of concrete and added to concrete before or during its mixing to modify its properties.

Aggregate — Granular material, such as sand, gravel, crushed stone, and iron blast-furnace slag, used with a cementing medium to form a hydraulic cement concrete or mortar.

Aggregate, lightweight — Aggregate with a dry, loose weight of 70 lb/ft^3 or less.

Anchorage device — In post-tensioning, the hardware used for transferring a post-tensioning force from the prestressing steel to the concrete.

Anchorage zone — In post-tensioned members, the portion of the member through which the concentrated prestressing force is transferred to the concrete and distributed more uniformly across the section. Its extent is equal to the largest dimension of the cross section. For anchorage devices located away from the end of a member, the anchorage zone includes the disturbed regions ahead of and behind the anchorage devices.

Basic monostrand anchorage device — Anchorage device used with any single strand or a single 5/8 in. or smaller diameter bar that satisfies 18.21.1 and the anchorage device requirements of ACI 423.6, "Specification for Unbonded Single-Strand Tendons."

Basic multistrand anchorage device — Anchorage device used with multiple strands, bars, or wires, or with single bars larger than 5/8 in. diameter, that satisfies 18.21.1 and the bearing stress and minimum plate stiffness requirements of AASHTO Bridge Specifications, Division I, Articles 9.21.7.2.2 through 9.21.7.2.4.

Bonded tendon — Tendon in which prestressing steel is bonded to concrete either directly or through grouting.

Building official — See 1.2.3.

COMMENTARY

R2.2 — Definitions

For consistent application of the code, it is necessary that terms be defined where they have particular meanings in the code. The definitions given are for use in application of this code only and do not always correspond to ordinary usage. A glossary of most used terms relating to cement manufacturing, concrete design and construction, and research in concrete is contained in **"Cement and Concrete Terminology"** reported by ACI Committee 116.[2.1]

Anchorage device — Most anchorage devices for post-tensioning are standard manufactured devices available from commercial sources. In some cases, designers or constructors develop "special" details or assemblages that combine various wedges and wedge plates for anchoring prestressing steel with specialty end plates or diaphragms. These informal designations as standard anchorage devices or special anchorage devices have no direct relation to the ACI Building Code and AASHTO "Standard Specifications for Highway Bridges" classification of anchorage devices as Basic Anchorage Devices or Special Anchorage Devices.

Anchorage zone — The terminology "ahead of" and "behind" the anchorage device is illustrated in Fig. R18.13.1(b).

Basic anchorage devices — Devices that are so proportioned that they can be checked analytically for compliance with bearing stress and stiffness requirements without having to undergo the acceptance-testing program required of special anchorage devices.

CODE

Cementitious materials — Materials as specified in Chapter 3, which have cementing value when used in concrete either by themselves, such as portland cement, blended hydraulic cements, and expansive cement, or such materials in combination with fly ash, other raw or calcined natural pozzolans, silica fume, and/or ground granulated blast-furnace slag.

Column — Member with a ratio of height-to-least lateral dimension exceeding 3 used primarily to support axial compressive load.

Composite concrete flexural members — Concrete flexural members of precast or cast-in-place concrete elements, or both, constructed in separate placements but so interconnected that all elements respond to loads as a unit.

Compression-controlled section — A cross section in which the net tensile strain in the extreme tension steel at nominal strength is less than or equal to the compression-controlled strain limit.

Compression-controlled strain limit — The net tensile strain at balanced strain conditions. See 10.3.3.

Concrete — Mixture of portland cement or any other hydraulic cement, fine aggregate, coarse aggregate, and water, with or without admixtures.

Concrete, specified compressive strength of, (f_c') — Compressive strength of concrete used in design and evaluated in accordance with provisions of Chapter 5, expressed in pounds per square inch (psi). Whenever the quantity f_c' is under a radical sign, square root of numerical value only is intended, and result has units of pounds per square inch (psi).

COMMENTARY

Column — The term compression member is used in the code to define any member in which the primary stress is longitudinal compression. Such a member need not be vertical but may have any orientation in space. Bearing walls, columns, and pedestals qualify as compression members under this definition.

The differentiation between columns and walls in the code is based on the principal use rather than on arbitrary relationships of height and cross-sectional dimensions. The code, however, permits walls to be designed using the principles stated for column design (see 14.4), as well as by the empirical method (see 14.5).

While a wall always encloses or separates spaces, it may also be used to resist horizontal or vertical forces or bending. For example, a retaining wall or a basement wall also supports various combinations of loads.

A column is normally used as a main vertical member carrying axial loads combined with bending and shear. It may, however, form a small part of an enclosure or separation.

CODE

Concrete, structural lightweight — Concrete containing lightweight aggregate that conforms to 3.3 and has an equilibrium density as determined by "Test Method for Determining Density of Structural Lightweight Concrete" (ASTM C 567), not exceeding 115 lb/ft^3. In this code, a lightweight concrete without natural sand is termed "all-lightweight concrete" and lightweight concrete in which all of the fine aggregate consists of normal weight sand is termed "sand-lightweight concrete."

COMMENTARY

Concrete, structural lightweight — In 2000, ASTM C 567 adopted "equilibrium density" as the measure for determining compliance with specified in-service density requirements. According to ASTM C 657, equilibrium density may be determined by measurement or approximated by calculation using either the measured oven-dry density or the oven-dry density calculated from the mixture proportions. Unless specified otherwise, ASTM C 567 requires that equilibrium density be approximated by calculation.

By code definition, sand-lightweight concrete is structural lightweight concrete with all of the fine aggregate replaced by sand. This definition may not be in agreement with usage by some material suppliers or contractors where the majority, but not all, of the lightweight fines are replaced by sand. For proper application of the code provisions, the replacement limits should be stated, with interpolation when partial sand replacement is used.

Contraction joint — Formed, sawed, or tooled groove in a concrete structure to create a weakened plane and regulate the location of cracking resulting from the dimensional change of different parts of the structure.

Curvature friction — Friction resulting from bends or curves in the specified prestressing tendon profile.

Deformed reinforcement — Deformed reinforcing bars, bar mats, deformed wire, and welded wire reinforcement conforming to 3.5.3.

Deformed reinforcement — Deformed reinforcement is defined as that meeting the deformed reinforcement specifications of 3.5.3.1, or the specifications of 3.5.3.3, 3.5.3.4, 3.5.3.5, or 3.5.3.6. No other reinforcement qualifies. This definition permits accurate statement of anchorage lengths. Bars or wire not meeting the deformation requirements or welded wire reinforcement not meeting the spacing requirements are "plain reinforcement," for code purposes, and may be used only for spirals.

Development length — Length of embedded reinforcement, including pretensioned strand, required to develop the design strength of reinforcement at a critical section. See 9.3.3.

Drop panel — A projection below the slab at least one quarter of the slab thickness beyond the drop.

Duct — A conduit (plain or corrugated) to accommodate prestressing steel for post-tensioned installation. Requirements for post-tensioning ducts are given in 18.17.

CODE

Effective depth of section **(*d*)** — Distance measured from extreme compression fiber to centroid of longitudinal tension reinforcement.

Effective prestress — Stress remaining in prestressing steel after all losses have occurred.

Embedment length — Length of embedded reinforcement provided beyond a critical section.

Extreme tension steel — The reinforcement (prestressed or nonprestressed) that is the farthest from the extreme compression fiber.

Isolation joint — A separation between adjoining parts of a concrete structure, usually a vertical plane, at a designed location such as to interfere least with performance of the structure, yet such as to allow relative movement in three directions and avoid formation of cracks elsewhere in the concrete and through which all or part of the bonded reinforcement is interrupted.

Jacking force — In prestressed concrete, temporary force exerted by device that introduces tension into prestressing steel.

Load, dead — Dead weight supported by a member, as defined by general building code of which this code forms a part (without load factors).

Load, factored — Load, multiplied by appropriate load factors, used to proportion members by the strength design method of this code. See 8.1.1 and 9.2.

Load, live — Live load specified by general building code of which this code forms a part (without load factors).

Load, service — Load specified by general building code of which this code forms a part (without load factors).

Modulus of elasticity — Ratio of normal stress to corresponding strain for tensile or compressive stresses below proportional limit of material. See 8.5.

Moment frame — Frame in which members and joints resist forces through flexure, shear, and axial force. Moment frames shall be catergorized as follows:

> ***Intermediate moment frame*** — A cast-in-place frame complying with the requirements of

COMMENTARY

Loads — A number of definitions for loads are given as the code contains requirements that are to be met at various load levels. The terms dead load and live load refer to the unfactored loads (service loads) specified or defined by the general building code. Service loads (loads without load factors) are to be used where specified in the code to proportion or investigate members for adequate serviceability, as in 9.5, Control of Deflections. Loads used to proportion a member for adequate strength are defined as factored loads. Factored loads are service loads multiplied by the appropriate load factors specified in 9.2 for required strength. The term design loads, as used in the 1971 code edition to refer to loads multiplied by the appropriate load factors, was discontinued in the 1977 code to avoid confusion with the design load terminology used in general building codes to denote service loads, or posted loads in buildings. The factored load terminology, first adopted in the 1977 code, clarifies when the load factors are applied to a particular load, moment, or shear value as used in the code provisions.

CODE

21.2.2.3 and 21.12 in addition to the requirements for ordinary moment frames.

Ordinary moment frame — A cast-in-place or precast concrete frame complying with the requirements of Chapters 1 through 18.

Special moment frame — A cast-in-place frame complying with the requirements of 21.2 through 21.5, or a precast frame complying with the requirements of 21.2 through 21.6. In addition, the requirements for ordinary moment frames shall be satisfied.

Net tensile strain — The tensile strain at nominal strength exclusive of strains due to effective prestress, creep, shrinkage, and temperature.

Pedestal — Upright compression member with a ratio of unsupported height to average least lateral dimension not exceeding 3.

Plain concrete — Structural concrete with no reinforcement or with less reinforcement than the minimum amount specified for reinforced concrete.

Plain reinforcement — Reinforcement that does not conform to definition of deformed reinforcement. See 3.5.4.

Post-tensioning — Method of prestressing in which prestressing steel is tensioned after concrete has hardened.

Precast concrete — Structural concrete element cast elsewhere than its final position in the structure.

Precompressed tensile zone — Portion of a prestressed member where flexural tension, calculated using gross section properties, would occur under unfactored dead and live loads if the prestress force were not present.

Prestressed concrete — Structural concrete in which internal stresses have been introduced to reduce potential tensile stresses in concrete resulting from loads.

Prestressing steel — High-strength steel element such as wire, bar, or strand, or a bundle of such elements, used to impart prestress forces to concrete.

Pretensioning — Method of prestressing in which prestressing steel is tensioned before concrete is placed.

COMMENTARY

Prestressed concrete — Reinforced concrete is defined to include prestressed concrete. Although the behavior of a prestressed member with unbonded tendons may vary from that of members with continuously bonded tendons, bonded and unbonded prestressed concrete are combined with conventionally reinforced concrete under the generic term "reinforced concrete." Provisions common to both prestressed and conventionally reinforced concrete are integrated to avoid overlapping and conflicting provisions.

CODE

Registered design professional — An individual who is registered or licensed to practice the respective design profession as defined by the statutory requirements of the professional registration laws of the state or jurisdiction in which the project is to be constructed.

Reinforced concrete — Structural concrete reinforced with no less than the minimum amounts of prestressing steel or nonprestressed reinforcement specified in Chapters 1 through 21 and Appendices A through C.

Reinforcement — Material that conforms to 3.5, excluding prestressing steel unless specifically included.

Reshores — Shores placed snugly under a concrete slab or other structural member after the original forms and shores have been removed from a larger area, thus requiring the new slab or structural member to deflect and support its own weight and existing construction loads applied prior to the installation of the reshores.

Sheathing — A material encasing prestressing steel to prevent bonding of the prestressing steel with the surrounding concrete, to provide corrosion protection, and to contain the corrosion inhibiting coating.

Shores — Vertical or inclined support members designed to carry the weight of the formwork, concrete, and construction loads above.

Span length — See 8.7.

Special anchorage device — Anchorage device that satisfies 18.15.1 and the standardized acceptance tests of AASHTO "Standard Specifications for Highway Bridges," Division II, Article 10.3.2.3.

Spiral reinforcement — Continuously wound reinforcement in the form of a cylindrical helix.

Splitting tensile strength (f_{ct}) — Tensile strength of concrete determined in accordance with ASTM C 496 as described in "Standard Specification for Lightweight Aggregates for Structural Concrete" (ASTM C 330). See 5.1.4.

COMMENTARY

Sheathing — Typically, sheathing is a continuous, seamless, high-density polyethylene material extruded directly on the coated prestressing steel.

Special anchorage devices are any devices (monostrand or multistrand) that do not meet the relevant PTI or AASHTO bearing stress and, where applicable, stiffness requirements. Most commercially marketed multibearing surface anchorage devices are Special Anchorage Devices. As provided in 18.15.1, such devices can be used only when they have been shown experimentally to be in compliance with the AASHTO requirements. This demonstration of compliance will ordinarily be furnished by the device manufacturer.

CODE

Stirrup — Reinforcement used to resist shear and torsion stresses in a structural member; typically bars, wires, or welded wire reinforcement either single leg or bent into L, U, or rectangular shapes and located perpendicular to or at an angle to longitudinal reinforcement. (The term "stirrups" is usually applied to lateral reinforcement in flexural members and the term ties to those in compression members.) See also *Tie*.

Strength, design — Nominal strength multiplied by a strength reduction factor ϕ. See 9.3.

Strength, nominal — Strength of a member or cross section calculated in accordance with provisions and assumptions of the strength design method of this code before application of any strength reduction factors. See 9.3.1.

Strength, required — Strength of a member or cross section required to resist factored loads or related internal moments and forces in such combinations as are stipulated in this code. See 9.1.1.

Stress — Intensity of force per unit area.

COMMENTARY

Strength, nominal — Strength of a member or cross section calculated using standard assumptions and strength equations, and nominal (specified) values of material strengths and dimensions is referred to as "nominal strength." The subscript n is used to denote the nominal strengths; nominal axial load strength P_n, nominal moment strength M_n, and nominal shear strength V_n. "Design strength" or usable strength of a member or cross section is the nominal strength reduced by the strength reduction factor ϕ.

The required axial load, moment, and shear strengths used to proportion members are referred to either as factored axial loads, factored moments, and factored shears, or required axial loads, moments, and shears. The factored load effects are calculated from the applied factored loads and forces in such load combinations as are stipulated in the code (see 9.2).

The subscript u is used only to denote the required strengths; required axial load strength P_u, required moment strength M_u, and required shear strength V_u, calculated from the applied factored loads and forces.

The basic requirement for strength design may be expressed as follows:

Design strength ≥ Required strength

$$\phi P_n \geq P_u$$

$$\phi M_n \geq M_u$$

$$\phi V_n \geq V_u$$

For additional discussion on the concepts and nomenclature for strength design see commentary Chapter 9.

CODE

Structural concrete — All concrete used for structural purposes including plain and reinforced concrete.

Structural walls — Walls proportioned to resist combinations of shears, moments, and axial forces induced by earthquake motions. A shearwall is a structural wall. Structural walls shall be categorized as follows:

Intermediate precast structural wall — A wall complying with all applicable requirements of Chapters 1 through 18 in addition to 21.13.

Ordinary reinforced concrete structural wall — A wall complying with the requirements of Chapters 1 through 18.

Ordinary structural plain concrete wall — A wall complying with the requirements of Chapter 22.

Special precast structural wall — A precast wall complying with the requirements of 21.8. In addition, the requirements of ordinary reinforced concrete structural walls and the requirements of 21.2 shall be satisfied.

Special reinforced concrete structural wall — A cast-in-place wall complying with the requirements of 21.2 and 21.7 in addition to the requirements for ordinary reinforced concrete structural walls.

Tendon — In pretensioned applications, the tendon is the prestressing steel. In post-tensioned applications, the tendon is a complete assembly consisting of anchorages, prestressing steel, and sheathing with coating for unbonded applications or ducts with grout for bonded applications.

Tension-controlled section — A cross section in which the net tensile strain in the extreme tension steel at nominal strength is greater than or equal to 0.005.

Tie — Loop of reinforcing bar or wire enclosing longitudinal reinforcement. A continuously wound bar or wire in the form of a circle, rectangle, or other polygon shape without re-entrant corners is acceptable. See also *Stirrup*.

Transfer — Act of transferring stress in prestressing steel from jacks or pretensioning bed to concrete member.

Transfer length — Length of embedded pretensioned strand required to transfer the effective prestress to the concrete.

Unbonded tendon — Tendon in which the prestressing steel is prevented from bonding to the concrete and

COMMENTARY

CODE

COMMENTARY

is free to move relative to the concrete. The prestressing force is permanently transferred to the concrete at the tendon ends by the anchorages only.

Wall — Member, usually vertical, used to enclose or separate spaces.

Welded wire reinforcement — Reinforcing elements consisting of plain or deformed wires, conforming to ASTM A 82 or A 496, respectively, fabricated into sheets in accordance with ASTM A 185 or A 497, respectively.

Wobble friction — In prestressed concrete, friction caused by unintended deviation of prestressing sheath or duct from its specified profile.

Yield strength — Specified minimum yield strength or yield point of reinforcement. Yield strength or yield point shall be determined in tension according to applicable ASTM standards as modified by 3.5 of this code.

CHAPTER 3 — MATERIALS

CODE

3.1 — Tests of materials

3.1.1 — The building official shall have the right to order testing of any materials used in concrete construction to determine if materials are of quality specified.

3.1.2 — Tests of materials and of concrete shall be made in accordance with standards listed in 3.8.

3.1.3 — A complete record of tests of materials and of concrete shall be retained by the inspector for 2 years after completion of the project, and made available for inspection during the progress of the work.

3.2 — Cements

3.2.1 — Cement shall conform to one of the following specifications:

(a) "Standard Specification for Portland Cement" (ASTM C 150);

(b) "Standard Specification for Blended Hydraulic Cements" (ASTM C 595), excluding Types S and SA which are not intended as principal cementing constituents of structural concrete;

(c) "Standard Specification for Expansive Hydraulic Cement" (ASTM C 845);

(d) "Standard Performance Specification for Hydraulic Cement" (ASTM C 1157).

3.2.2 — Cement used in the work shall correspond to that on which selection of concrete proportions was based. See 5.2.

COMMENTARY

R3.1 — Tests of materials

R3.1.3 — The record of tests of materials and of concrete should be retained for at least 2 years after completion of the project. Completion of the project is the date at which the owner accepts the project or when the certificate of occupancy is issued, whichever date is later. Local legal requirements may require longer retention of such records.

R3.2 — Cements

R3.2.2 — Depending on the circumstances, the provision of 3.2.2 may require only the same type of cement or may require cement from the identical source. The latter would be the case if the sample standard deviation[3.1] of strength tests used in establishing the required strength margin was based on a cement from a particular source. If the sample standard deviation was based on tests involving a given type of cement obtained from several sources, the former interpretation would apply.

CODE

3.3 — Aggregates

3.3.1 — Concrete aggregates shall conform to one of the following specifications:

(a) "Standard Specification for Concrete Aggregates" (ASTM C 33);

(b) "Standard Specification for Lightweight Aggregates for Structural Concrete" (ASTM C 330).

Exception: Aggregates that have been shown by special test or actual service to produce concrete of adequate strength and durability and approved by the building official.

3.3.2 — Nominal maximum size of coarse aggregate shall be not larger than:

(a) 1/5 the narrowest dimension between sides of forms, nor

(b) 1/3 the depth of slabs, nor

(c) 3/4 the minimum clear spacing between individual reinforcing bars or wires, bundles of bars, individual tendons, bundled tendons, or ducts.

These limitations shall not apply if, in the judgment of the engineer, workability and methods of consolidation are such that concrete can be placed without honeycombs or voids.

3.4 — Water

3.4.1 — Water used in mixing concrete shall be clean and free from injurious amounts of oils, acids, alkalis, salts, organic materials, or other substances deleterious to concrete or reinforcement.

3.4.2 — Mixing water for prestressed concrete or for concrete that will contain aluminum embedments, including that portion of mixing water contributed in the form of free moisture on aggregates, shall not contain deleterious amounts of chloride ion. See 4.4.1.

3.4.3 — Nonpotable water shall not be used in concrete unless the following are satisfied:

3.4.3.1 — Selection of concrete proportions shall be based on concrete mixes using water from the same source.

COMMENTARY

R3.3 — Aggregates

R3.3.1 — Aggregates conforming to the ASTM specifications are not always economically available and, in some instances, noncomplying materials have a long history of satisfactory performance. Such nonconforming materials are permitted with special approval when acceptable evidence of satisfactory performance is provided. Satisfactory performance in the past, however, does not guarantee good performance under other conditions and in other localities. Whenever possible, aggregates conforming to the designated specifications should be used.

R3.3.2 — The size limitations on aggregates are provided to ensure proper encasement of reinforcement and to minimize honeycombing. Note that the limitations on maximum size of the aggregate may be waived if, in the judgment of the engineer, the workability and methods of consolidation of the concrete are such that the concrete can be placed without honeycombs or voids.

R3.4 — Water

R3.4.1 — Almost any natural water that is drinkable (potable) and has no pronounced taste or odor is satisfactory as mixing water for making concrete. Impurities in mixing water, when excessive, may affect not only setting time, concrete strength, and volume stability (length change), but may also cause efflorescence or corrosion of reinforcement. Where possible, water with high concentrations of dissolved solids should be avoided.

Salts or other deleterious substances contributed from the aggregate or admixtures are additive to the amount which might be contained in the mixing water. These additional amounts are to be considered in evaluating the acceptability of the total impurities that may be deleterious to concrete or steel.

CODE

3.4.3.2 — Mortar test cubes made with nonpotable mixing water shall have 7-day and 28-day strengths equal to at least 90 percent of strengths of similar specimens made with potable water. Strength test comparison shall be made on mortars, identical except for the mixing water, prepared and tested in accordance with "Standard Test Method for Compressive Strength of Hydraulic Cement Mortars (Using 2-in. or [50-mm] Cube Specimens)" (ASTM C 109).

3.5 — Steel reinforcement

3.5.1 — Reinforcement shall be deformed reinforcement, except that plain reinforcement shall be permitted for spirals or prestressing steel; and reinforcement consisting of structural steel, steel pipe, or steel tubing shall be permitted as specified in this code.

3.5.2 — Welding of reinforcing bars shall conform to "Structural Welding Code — Reinforcing Steel," ANSI/AWS D1.4 of the American Welding Society. Type and location of welded splices and other required welding of reinforcing bars shall be indicated on the design drawings or in the project specifications. ASTM reinforcing bar specifications, except for ASTM A 706, shall be supplemented to require a report of material properties necessary to conform to the requirements in ANSI/AWS D1.4.

COMMENTARY

R3.5 — Steel reinforcement

R3.5.1 — Fiber reinforced polymer (FRP) reinforcement is not addressed in this code. ACI Committee 440 has developed guidelines for the use of FRP reinforcement.[3.2, 3.3]

Materials permitted for use as reinforcement are specified. Other metal elements, such as inserts, anchor bolts, or plain bars for dowels at isolation or contraction joints, are not normally considered to be reinforcement under the provisions of this code.

R3.5.2 — When welding of reinforcing bars is required, the weldability of the steel and compatible welding procedures need to be considered. The provisions in ANSI/AWS D1.4 Welding Code cover aspects of welding reinforcing bars, including criteria to qualify welding procedures.

Weldability of the steel is based on its chemical composition or carbon equivalent (CE). The Welding Code establishes preheat and interpass temperatures for a range of carbon equivalents and reinforcing bar sizes. Carbon equivalent is calculated from the chemical composition of the reinforcing bars. The Welding Code has two expressions for calculating carbon equivalent. A relatively short expression, considering only the elements carbon and manganese, is to be used for bars other than ASTM A 706 material. A more comprehensive expression is given for ASTM A 706 bars. The CE formula in the Welding Code for A 706 bars is identical to the CE formula in the ASTM A 706 specification.

The engineer should realize that the chemical analysis, for bars other than A 706, required to calculate the carbon equivalent is not routinely provided by the producer of the reinforcing bars. For welding reinforcing bars other than A 706 bars, the design drawings or project specifications should specifically require results of the chemical analysis to be furnished.

The ASTM A 706 specification covers low-alloy steel reinforcing bars intended for applications requiring controlled tensile properties or welding. Weldability is accomplished in the A 706 specification by limits or controls on chemical composition and on carbon equivalent.[3.4] The producer is required by the A 706 specification to report the chemical composition and carbon equivalent.

CODE / COMMENTARY

The ANSI/AWS D1.4 Welding Code requires the contractor to prepare written welding procedure specifications conforming to the requirements of the Welding Code. Appendix A of the Welding Code contains a suggested form that shows the information required for such a specification for each joint welding procedure.

Often it is necessary to weld to existing reinforcing bars in a structure when no mill test report of the existing reinforcement is available. This condition is particularly common in alterations or building expansions. ANSI/AWS D1.4 states for such bars that a chemical analysis may be performed on representative bars. If the chemical composition is not known or obtained, the Welding Code requires a minimum preheat. For bars other than A 706 material, the minimum preheat required is 300 F for bars No. 6 or smaller, and 400 F for No. 7 bars or larger. The required preheat for all sizes of A 706 is to be the temperature given in the Welding Code's table for minimum preheat corresponding to the range of CE "over 45 percent to 55 percent." Welding of the particular bars should be performed in accordance with ANSI/AWS D 1.4. It should also be determined if additional precautions are in order, based on other considerations such as stress level in the bars, consequences of failure, and heat damage to existing concrete due to welding operations.

Welding of wire to wire, and of wire or welded wire reinforcement to reinforcing bars or structural steel elements is not covered by ANSI/AWS D1.4. If welding of this type is required on a project, the engineer should specify requirements or performance criteria for this welding. If cold drawn wires are to be welded, the welding procedures should address the potential loss of yield strength and ductility achieved by the cold working process (during manufacture) when such wires are heated by welding. Machine and resistance welding as used in the manufacture of welded plain and deformed wire reinforcement is covered by ASTM A 185 and A 497, respectively, and is not part of this concern.

3.5.3 — Deformed reinforcement

3.5.3.1 — Deformed reinforcing bars shall conform to the requirements for deformed bars in one of the following specifications:

(a) "Standard Specification for Deformed and Plain Carbon-Steel Bars for Concrete Reinforcement" (ASTM A 615);

(b) "Standard Specification for Low-Alloy Steel Deformed and Plain Bars for Concrete Reinforcement" (ASTM A 706);

(c) "Standard Specification for Rail-Steel and Axle-Steel Deformed Bars for Concrete Reinforcement" (ASTM A 996). Bars from rail-steel shall be Type R.

R3.5.3 — Deformed reinforcement

R3.5.3.1 — ASTM A 615 covers deformed carbon-steel reinforcing bars that are currently the most widely used type of steel bar in reinforced concrete construction in the United States. The specification requires that the bars be marked with the letter *S* for type of steel.

ASTM A 706 covers low-alloy steel deformed bars intended for applications where controlled tensile properties, restrictions on chemical composition to enhance weldability, or both, are required. The specification requires that the bars be marked with the letter *W* for type of steel.

Deformed bars produced to meet both ASTM A 615 and A 706 are required to be marked with the letters *S* and *W* for type of steel.

Rail-steel reinforcing bars used with this code are required to conform to ASTM A 996 including the provisions for Type R bars, and marked with the letter R for type of steel. Type R bars are required to meet more restrictive provisions for bend tests.

3.5.3.2 — Deformed reinforcing bars shall conform to one of the ASTM specifications listed in 3.5.3.1, except that for bars with f_y exceeding 60,000 psi, the yield strength shall be taken as the stress corresponding to a strain of 0.35 percent. See 9.4.

R3.5.3.2 — ASTM A 615 includes provisions for Grade 75 bars in sizes No. 6 through 18.

The 0.35 percent strain limit is necessary to ensure that the assumption of an elasto-plastic stress-strain curve in 10.2.4 will not lead to unconservative values of the member strength.

The 0.35 strain requirement is not applied to reinforcing bars having specified yield strengths of 60,000 psi or less. For steels having specified yield strengths of 40,000 psi, as were once used extensively, the assumption of an elasto-plastic stress-strain curve is well justified by extensive test data. For steels with specified yield strengths, up to 60,000 psi, the stress-strain curve may or may not be elasto-plastic as assumed in 10.2.4, depending on the properties of the steel and the manufacturing process. However, when the stress-strain curve is not elasto-plastic, there is limited experimental evidence to suggest that the actual steel stress at ultimate strength may not be enough less than the specified yield strength to warrant the additional effort of testing to the more restrictive criterion applicable to steels having specified yield strengths greater than 60,000 psi. In such cases, the ϕ-factor can be expected to account for the strength deficiency.

3.5.3.3 — Bar mats for concrete reinforcement shall conform to "Standard Specification for Welded Deformed Steel Bar Mats for Concrete Reinforcement" (ASTM A 184). Reinforcing bars used in bar mats shall conform to ASTM A 615 or A 706.

3.5.3.4 — Deformed wire for concrete reinforcement shall conform to "Standard Specification for Steel Wire, Deformed, for Concrete Reinforcement" (ASTM A 496), except that wire shall not be smaller than size D4 and for wire with f_y exceeding 60,000 psi, the yield strength shall be taken as the stress corresponding to a strain of 0.35 percent.

3.5.3.5 — Welded plain wire reinforcement shall conform to "Standard Specification for Steel Welded Wire Reinforcement, Plain, for Concrete" (ASTM A 185), except that for wire with f_y exceeding 60,000 psi, the yield strength shall be taken as the stress corresponding to a strain of 0.35 percent. Welded intersections shall not be spaced farther apart than 12 in. in direction of calculated stress, except for welded wire reinforcement used as stirrups in accordance with 12.13.2.

R3.5.3.5 — Welded plain wire reinforcement should be made of wire conforming to "Standard Specification for Steel Wire, Plain, for Concrete Reinforcement" (ASTM A 82). ASTM A 82 has a minimum yield strength of 70,000 psi. The code has assigned a yield strength value of 60,000 psi, but makes provision for the use of higher yield strengths provided the stress corresponds to a strain of 0.35 percent.

CODE

3.5.3.6 — Welded deformed wire reinforcement shall conform to "Standard Specification for Steel Welded Wire Reinforcement, Deformed, for Concrete" (ASTM A 497), except that for wire with f_y exceeding 60,000 psi, the yield strength shall be taken as the stress corresponding to a strain of 0.35 percent. Welded intersections shall not be spaced farther apart than 16 in. in direction of calculated stress, except for welded deformed wire reinforcement used as stirrups in accordance with 12.13.2.

3.5.3.7 — Galvanized reinforcing bars shall comply with "Standard Specification for Zinc-Coated (Galvanized) Steel Bars for Concrete Reinforcement" (ASTM A 767). Epoxy-coated reinforcing bars shall comply with "Standard Specification for Epoxy-Coated Steel Reinforcing Bars" (ASTM A 775) or with "Standard Specification for Epoxy-Coated Prefabricated Steel Reinforcing Bars" (ASTM A 934). Bars to be galvanized or epoxy-coated shall conform to one of the specifications listed in 3.5.3.1.

3.5.3.8 — Epoxy-coated wires and welded wire reinforcement shall comply with "Standard Specification for Epoxy-Coated Steel Wire and Welded Wire Reinforcement" (ASTM A 884). Wires to be epoxy-coated shall conform to 3.5.3.4 and welded wire reinforcement to be epoxy-coated shall conform to 3.5.3.5 or 3.5.3.6.

3.5.4 — Plain reinforcement

3.5.4.1 — Plain bars for spiral reinforcement shall conform to the specification listed in 3.5.3.1(a) or (b).

3.5.4.2 — Plain wire for spiral reinforcement shall conform to "Standard Specification for Steel Wire, Plain, for Concrete Reinforcement" (ASTM A 82), except that for wire with f_y exceeding 60,000 psi, the yield strength shall be taken as the stress corresponding to a strain of 0.35 percent.

3.5.5 — Prestressing steel

3.5.5.1 — Steel for prestressing shall conform to one of the following specifications:

(a) Wire conforming to "Standard Specification for Uncoated Stress-Relieved Steel Wire for Prestressed Concrete" (ASTM A 421);

(b) Low-relaxation wire conforming to "Standard Specification for Uncoated Stress-Relieved Steel Wire for Prestressed Concrete" including Supplement "Low-Relaxation Wire" (ASTM A 421);

COMMENTARY

R3.5.3.6 — Welded deformed wire reinforcement should be made of wire conforming to "Standard Specification for Steel Welded Wire Reinforcement, Deformed, for Concrete (ASTM A 497)." ASTM A 497 has a minimum yield strength of 70,000 psi. The code has assigned a yield strength value of 60,000 psi, but makes provision for the use of higher yield strengths provided the stress corresponds to a strain of 0.35 percent.

R3.5.3.7 — Galvanized reinforcing bars (A 767) and epoxy-coated reinforcing bars (A 775) were added to the 1983 code, and epoxy-coated prefabricated reinforcing bars (A 934) were added to the 1995 code recognizing their usage, especially for conditions where corrosion resistance of reinforcement is of particular concern. They have typically been used in parking decks, bridge decks, and other highly corrosive environments.

R3.5.4 — Plain reinforcement

Plain bars and plain wire are permitted only for spiral reinforcement (either as lateral reinforcement for compression members, for torsion members, or for confining reinforcement for splices).

R3.5.5 — Prestressing steel

R3.5.5.1 — Because low-relaxation prestressing steel is addressed in a supplement to ASTM A 421, which applies only when low-relaxation material is specified, the appropriate ASTM reference is listed as a separate entity.

CODE

(c) Strand conforming to "Standard Specification for Steel Strand, Uncoated Seven-Wire for Prestressed Concrete" (ASTM A 416);

(d) Bar conforming to "Standard Specification for Uncoated High-Strength Steel Bars for Prestressing Concrete" (ASTM A 722).

3.5.5.2 — Wire, strands, and bars not specifically listed in ASTM A 421, A 416, or A 722 are allowed provided they conform to minimum requirements of these specifications and do not have properties that make them less satisfactory than those listed in ASTM A 421, A 416, or A 722.

3.5.6 — Structural steel, steel pipe, or tubing

3.5.6.1 — Structural steel used with reinforcing bars in composite compression members meeting requirements of 10.16.7 or 10.16.8 shall conform to one of the following specifications:

(a) "Standard Specification for Carbon Structural Steel" (ASTM A 36);

(b) "Standard Specification for High-Strength Low-Alloy Structural Steel" (ASTM A 242);

(c) "Standard Specification for High-Strength Low-Alloy Columbium-Vanadium Structural Steel" (ASTM A 572);

(d) "Standard Specification for High-Strength Low-Alloy Structural Steel with 50 ksi (345 MPa) Minimum Yield Point to 4 in. (100 mm) Thick" (ASTM A 588);

(e) "Standard Specification for Structural Steel Shapes" (ASTM A 992).

3.5.6.2 — Steel pipe or tubing for composite compression members composed of a steel encased concrete core meeting requirements of 10.16.6 shall conform to one of the following specifications:

(a) Grade B of "Standard Specification for Pipe, Steel, Black and Hot-Dipped, Zinc-Coated Welded and Seamless" (ASTM A 53);

(b) "Standard Specification for Cold-Formed Welded and Seamless Carbon Steel Structural Tubing in Rounds and Shapes" (ASTM A 500);

(c) "Standard Specification for Hot-Formed Welded and Seamless Carbon Steel Structural Tubing" (ASTM A 501).

COMMENTARY

CODE

3.6 — Admixtures

3.6.1 — Admixtures to be used in concrete shall be subject to prior approval by the engineer.

3.6.2 — An admixture shall be shown capable of maintaining essentially the same composition and performance throughout the work as the product used in establishing concrete proportions in accordance with 5.2.

3.6.3 — Calcium chloride or admixtures containing chloride from other than impurities from admixture ingredients shall not be used in prestressed concrete, in concrete containing embedded aluminum, or in concrete cast against stay-in-place galvanized steel forms. See 4.3.2 and 4.4.1.

3.6.4 — Air-entraining admixtures shall conform to "Standard Specification for Air-Entraining Admixtures for Concrete" (ASTM C 260).

3.6.5 — Water-reducing admixtures, retarding admixtures, accelerating admixtures, water-reducing and retarding admixtures, and water-reducing and accelerating admixtures shall conform to "Standard Specification for Chemical Admixtures for Concrete" (ASTM C 494) or "Standard Specification for Chemical Admixtures for Use in Producing Flowing Concrete" (ASTM C 1017).

3.6.6 — Fly ash or other pozzolans used as admixtures shall conform to "Standard Specification for Coal Fly Ash and Raw or Calcined Natural Pozzolan for Use in Concrete" (ASTM C 618).

3.6.7 — Ground granulated blast-furnace slag used as an admixture shall conform to "Standard Specification for Ground Granulated Blast-Furnace Slag for Use in Concrete and Mortars" (ASTM C 989).

COMMENTARY

R3.6 — Admixtures

R3.6.3 — Admixtures containing any chloride, other than impurities from admixture ingredients, should not be used in prestressed concrete or in concrete with aluminum embedments. Concentrations of chloride ion may produce corrosion of embedded aluminum (e.g., conduit), especially if the aluminum is in contact with embedded steel and the concrete is in a humid environment. Serious corrosion of galvanized steel sheet and galvanized steel stay-in-place forms occurs, especially in humid environments or where drying is inhibited by the thickness of the concrete or coatings or impermeable coverings. See 4.4.1 for specific limits on chloride ion concentration in concrete.

R3.6.7 — Ground granulated blast-furnace slag conforming to ASTM C 989 is used as an admixture in concrete in much the same way as fly ash. Generally, it should be used with portland cements conforming to ASTM C 150, and only rarely would it be appropriate to use ASTM C 989 slag with an ASTM C 595 blended cement that already contains a pozzolan or slag. Such use with ASTM C 595 cements might be considered for massive concrete placements where slow strength gain can be tolerated and where low heat of hydration is of particular importance. ASTM C 989 includes appendices which discuss effects of ground granulated blast-furnace slag on concrete strength, sulfate resistance, and alkali-aggregate reaction.

CODE

3.6.8 — Admixtures used in concrete containing ASTM C 845 expansive cements shall be compatible with the cement and produce no deleterious effects.

3.6.9 — Silica fume used as an admixture shall conform to ASTM C 1240.

3.7 — Storage of materials

3.7.1 — Cementitious materials and aggregates shall be stored in such manner as to prevent deterioration or intrusion of foreign matter.

3.7.2 — Any material that has deteriorated or has been contaminated shall not be used for concrete.

3.8 — Referenced standards

3.8.1 — Standards of ASTM International referred to in this code are listed below with their serial designations, including year of adoption or revision, and are declared to be part of this code as if fully set forth herein:

A 36/ A 36M-04	Standard Specification for Carbon Structural Steel
A 53/ A 53M-02	Standard Specification for Pipe, Steel, Black and Hot-Dipped, Zinc-Coated, Welded and Seamless
A 82-02	Standard Specification for Steel Wire, Plain, for Concrete Reinforcement
A 184/ A 184M-01	Standard Specification for Welded Deformed Steel Bar Mats for Concrete Reinforcement
A 185-02	Standard Specification for Steel Welded Wire Reinforcement, Plain, for Concrete
A 242/ A 242M-04	Standard Specification for High-Strength Low-Alloy Structural Steel
A 307-04	Standard Specification for Carbon Steel Bolts and Studs, 60,000 psi Tensile Strength
A 416/ A 416M-02	Standard Specification for Steel Strand, Uncoated Seven-Wire for Prestressed Concrete
A 421/ A 421M-02	Standard Specification for Uncoated Stress-Relieved Steel Wire for Prestressed Concrete

COMMENTARY

R3.6.8 — The use of admixtures in concrete containing ASTM C 845 expansive cements has reduced levels of expansion or increased shrinkage values. See ACI 223.[3.5]

R3.8 — Referenced standards

The ASTM standard specifications listed are the latest editions at the time these code provisions were adopted. Since these specifications are revised frequently, generally in minor details only, the user of the code should check directly with the sponsoring organization if it is desired to reference the latest edition. However, such a procedure obligates the user of the specification to evaluate if any changes in the later edition are significant in the use of the specification.

Standard specifications or other material to be legally adopted by reference into a building code should refer to a specific document. This can be done by simply using the complete serial designation since the first part indicates the subject and the second part the year of adoption. All standard documents referenced in this code are listed in 3.8, with the title and complete serial designation. In other sections of the code, the designations do not include the date so that all may be kept up-to-date by simply revising 3.8.

ASTM standards are available from ASTM, 100 Barr Harbor Drive, West Conshohocken, Pa., 19428.

CODE	
A 496-02	Standard Specification for Steel Wire, Deformed, for Concrete Reinforcement
A 497/ A 497M-02	Standard Specification for Steel Welded Wire Reinforcement, Deformed, for Concrete
A 500-03a	Standard Specification for Cold-Formed Welded and Seamless Carbon Steel Structural Tubing in Rounds and Shapes
A 501-01	Standard Specification for Hot-Formed Welded and Seamless Carbon Steel Structural Tubing
A 572/ A 572M-04	Standard Specification for High-Strength Low-Alloy Columbium-Vanadium Structural Steel
A 588/ A 588M-04	Standard Specification for High-Strength Low-Alloy Structural Steel with 50 ksi [345 MPa] Minimum Yield Point to 4-in. [100-mm] Thick
A 615/ A 615M-04b	Standard Specification for Deformed and Plain Carbon Steel Bars for Concrete Reinforcement
A 706/ A 706M-04b	Standard Specification for Low-Alloy Steel Deformed and Plain Bars for Concrete Reinforcement
A 722/ A 722M-98(2003)	Standard Specification for Uncoated High-Strength Steel Bars for Prestressing Concrete
A 767/ A 767M-00b	Standard Specification for Zinc-Coated (Galvanized) Steel Bars for Concrete Reinforcement
A 775/ A 775M-04a	Standard Specification for Epoxy-Coated Steel Reinforcing Bars
A 884/ A 884M-04	Standard Specification for Epoxy-Coated Steel Wire and Welded Wire Reinforcement
A 934/ A 934M-04	Standard Specification for Epoxy-Coated Prefabricated Steel Reinforcing Bars
A 992/ A 992M-04	Standard Specification for Structural Steel Shapes
A 996/ A 996M-04	Standard Specification for Rail-Steel and Axle-Steel Deformed Bars for Concrete Reinforcement

COMMENTARY

Type R rail-steel bars are considered a mandatory requirement whenever ASTM A 996 is referenced in the code.

CODE

COMMENTARY

C 31/ C 31M-03a	Standard Practice for Making and Curing Concrete Test Specimens in the Field
C 33-03	Standard Specification for Concrete Aggregates
C 39/ C 39M-03	Standard Test Method for Compressive Strength of Cylindrical Concrete Specimens
C 42/ C 42M-04	Standard Test Method for Obtaining and Testing Drilled Cores and Sawed Beams of Concrete
C 94/ C 94M-04	Standard Specification for Ready-Mixed Concrete
C 109/ C 109M-02	Standard Test Method for Compressive Strength of Hydraulic Cement Mortars (Using 2-in. or [50-mm] Cube Specimens)
C 144-03	Standard Specification for Aggregate for Masonry Mortar
C 150-04a	Standard Specification for Portland Cement
C 172-04	Standard Practice for Sampling Freshly Mixed Concrete
C 192/ C 192M-02	Standard Practice for Making and Curing Concrete Test Specimens in the Laboratory
C 260-01	Standard Specification for Air-Entraining Admixtures for Concrete
C 330-04	Standard Specification for Lightweight Aggregates for Structural Concrete
C 494/ C 494M-04	Standard Specification for Chemical Admixtures for Concrete
C 496/ C 496M-04	Standard Test Method for Splitting Tensile Strength of Cylindrical Concrete Specimens
C 567-04	Standard Test Method for Determining Density of Structural Lightweight Concrete
C 595-03	Standard Specification for Blended Hydraulic Cements

CODE

C 618-03	Standard Specification for Coal Fly Ash and Raw or Calcined Natural Pozzolan for Use in Concrete
C 685/ C 685M-01	Standard Specification for Concrete Made by Volumetric Batching and Continuous Mixing
C 845-04	Standard Specification for Expansive Hydraulic Cement
C 989-04	Standard Specification for Ground Granulated Blast-Furnace Slag for Use in Concrete and Mortars
C 1017/ C 1017M-03	Standard Specification for Chemical Admixtures for Use in Producing Flowing Concrete
C 1157-03	Standard Performance Specification for Hydraulic Cement
C 1218/ C 1218M-99	Standard Test Method for Water-Soluble Chloride in Mortar and Concrete
C 1240-04	Standard Specification for Silica Fume Used in Cementitious Mixtures

3.8.2 — "Structural Welding Code—Reinforcing Steel" (ANSI/AWS D1.4-98) of the American Welding Society is declared to be part of this code as if fully set forth herein.

3.8.3 — Section 2.3.3 Load Combinations Including Flood Loads and 2.3.4 Load Combinations Including Atmospheric Ice Loads of "Minimum Design Loads for Buildings and Other Structures" (SEI/ASCE 7-02) is declared to be part of this code as if fully set forth herein, for the purpose cited in 9.2.4.

3.8.4 — "Specification for Unbonded Single Strand Tendons (ACI 423.6-01) and Commentary (423.6R-01)" is declared to be part of this code as if fully set forth herein.

3.8.5 — Articles 9.21.7.2 and 9.21.7.3 of Division I and Article 10.3.2.3 of Division II of AASHTO "Standard Specification for Highway Bridges" (AASHTO 17th Edition, 2002) are declared to be a part of this code as if fully set forth herein, for the purpose cited in 18.15.1.

3.8.6 — "Qualification of Post-Installed Mechanical Anchors in Concrete (ACI 355.2-04)" is declared to be part of this code as if fully set forth herein, for the purpose cited in Appendix D.

COMMENTARY

R3.8.3 — SEI/ASCE 7 is available from ASCE Book Orders, Box 79404, Baltimore, MD, 21279-0404.

R3.8.5 — The 2002 17th Edition of the AASHTO "Standard Specification for Highway Bridges" is available from AASHTO, 444 North Capitol Street, N.W., Suite 249, Washington, DC, 20001.

R3.8.6 — Parallel to development of the ACI 318-05 provisions for anchoring to concrete, ACI 355 developed a test method to define the level of performance required for post-installed anchors. This test method, ACI 355.2,

CODE

3.8.7 — “Structural Welding Code—Steel (AWS D 1.1/ D1.1M-2004)” of the American Welding Society is declared to be part of this code as if fully set forth herein.

3.8.8 — “Acceptance Criteria for Moment Frames Based on Structural Testing (ACI T1.1-01),” is declared to be part of this code as if fully set forth herein.

COMMENTARY

contains requirements for the testing and evaluation of post-installed anchors for both cracked and uncracked concrete applications.

Notes

CHAPTER 4 — DURABILITY REQUIREMENTS

CODE

4.1 — Water-cementitious material ratio

4.1.1 — The water-cementitious material ratios specified in Tables 4.2.2 and 4.3.1 shall be calculated using the weight of cement meeting ASTM C 150, C 595, C 845, or C 1157 plus the weight of fly ash and other pozzolans meeting ASTM C 618, slag meeting ASTM C 989, and silica fume meeting ASTM C 1240, if any, except that when concrete is exposed to deicing chemicals, 4.2.3 further limits the amount of fly ash, pozzolans, silica fume, slag, or the combination of these materials.

COMMENTARY

R4.1 — Water-cementitious material ratio

Chapters 4 and 5 of earlier editions of the code were reformatted in 1989 to emphasize the importance of considering durability requirements before the designer selects f_c' and cover over the reinforcing steel.

Maximum water-cementitious material ratios of 0.40 to 0.50 that may be required for concretes exposed to freezing and thawing, sulfate soils or waters, or for preventing corrosion of reinforcement will typically be equivalent to requiring an f_c' of 5000 to 4000 psi, respectively. Generally, the required average compressive strengths, f_{cr}', will be 500 to 700 psi higher than the specified compressive strength, f_c'. Since it is difficult to accurately determine the water-cementitious material ratio of concrete during production, the f_c' specified should be reasonably consistent with the water-cementitious material ratio required for durability. Selection of an f_c' that is consistent with the water-cementitious material ratio selected for durability will help ensure that the required water-cementitious material ratio is actually obtained in the field. Because the usual emphasis on inspection is for strength, test results substantially higher than the specified strength may lead to a lack of concern for quality and production of concrete that exceeds the maximum water-cementitious material ratio. Thus an f_c' of 3000 psi and a maximum water-cementitious material ratio of 0.45 should not be specified for a parking structure, if the structure will be exposed to deicing salts.

The code does not include provisions for especially severe exposures, such as acids or high temperatures, and is not concerned with aesthetic considerations such as surface finishes. These items are beyond the scope of the code and should be covered specifically in the project specifications. Concrete ingredients and proportions are to be selected to meet the minimum requirements stated in the code and the additional requirements of the contract documents.

R4.1.1 — For concrete exposed to deicing chemicals the quantity of fly ash, other pozzolans, silica fume, slag, or blended cements used in the concrete is subject to the percentage limits in 4.2.3. Further, in 4.3 for sulfate exposures,[4.1] the pozzolan should be Class F by ASTM C 618, or have been tested by ASTM C 1012[4.2] or determined by service record to improve sulfate resistance.

CODE

4.2 — Freezing and thawing exposures

4.2.1 — Normalweight and lightweight concrete exposed to freezing and thawing or deicing chemicals shall be air-entrained with air content indicated in Table 4.2.1. Tolerance on air content as delivered shall be ± 1.5 percent. For f_c' greater than 5000 psi, reduction of air content indicated in Table 4.2.1 by 1.0 percent shall be permitted.

TABLE 4.2.1—TOTAL AIR CONTENT FOR FROST-RESISTANT CONCRETE

Nominal maximum aggregate size, in.*	Air content, percent	
	Severe exposure	Moderate exposure
3/8	7.5	6
1/2	7	5.5
3/4	6	5
1	6	4.5
1-1/2	5.5	4.5
2†	5	4
3†	4.5	3.5

* See ASTM C 33 for tolerance on oversize for various nominal maximum size designations.

† These air contents apply to total mix, as for the preceding aggregate sizes. When testing these concretes, however, aggregate larger than 1-1/2 in. is removed by handpicking or sieving and air content is determined on the minus 1-1/2 in. fraction of mix (tolerance on air content as delivered applies to this value.). Air content of total mix is computed from value determined on the minus 1-1/2 in. fraction.

4.2.2 — Concrete that will be subject to the exposures given in Table 4.2.2 shall conform to the corresponding maximum water-cementitious material ratios and minimum f_c' requirements of that table. In addition, concrete that will be exposed to deicing chemicals shall conform to the limitations of 4.2.3.

TABLE 4.2.2—REQUIREMENTS FOR SPECIAL EXPOSURE CONDITIONS

Exposure condition	Maximum water-cementitious material ratio*, by weight, normalweight concrete	Minimum f_c', normalweight and lightweight concrete, psi*
Concrete intended to have low permeability when exposed to water	0.50	4000
Concrete exposed to freezing and thawing in a moist condition or to deicing chemicals	0.45	4500
For corrosion protection of reinforcement in concrete exposed to chlorides from deicing chemicals, salt, salt water, brackish water, seawater, or spray from these sources.	0.40	5000

* When both Table 4.3.1 and Table 4.2.2 are considered, the lowest applicable maximum water-cementitious material ratio and highest applicable minimum f_c' shall be used.

COMMENTARY

R4.2 — Freezing and thawing exposures

R4.2.1 — A table of required air contents for frost-resistant concrete is included in the code, based on **"Standard Practice for Selecting Proportions for Normal, Heavyweight, and Mass Concrete"** (ACI 211.1).[4.3] Values are provided for both severe and moderate exposures depending on the exposure to moisture or deicing salts. Entrained air will not protect concrete containing coarse aggregates that undergo disruptive volume changes when frozen in a saturated condition. In Table 4.2.1, a severe exposure is where the concrete in a cold climate may be in almost continuous contact with moisture prior to freezing, or where deicing salts are used. Examples are pavements, bridge decks, sidewalks, parking garages, and water tanks. A moderate exposure is where the concrete in a cold climate will be only occasionally exposed to moisture prior to freezing, and where no deicing salts are used. Examples are certain exterior walls, beams, girders, and slabs not in direct contact with soil. Section 4.2.1 permits 1 percent lower air content for concrete with f_c' greater than 5000 psi. Such high-strength concretes will have lower water-cementitious material ratios and porosity and, therefore, improved frost resistance.

R4.2.2 — Maximum water-cementitious material ratios are not specified for lightweight concrete because determination of the absorption of these aggregates is uncertain, making calculation of the water-cementitious material ratio uncertain. The use of a minimum specified compressive strength, f_c', will ensure the use of a high-quality cement paste. For normalweight concrete, use of both minimum strength and maximum water-cementitious material ratio provide additional assurance that this objective is met.

CODE

4.2.3 — For concrete exposed to deicing chemicals, the maximum weight of fly ash, other pozzolans, silica fume, or slag that is included in the concrete shall not exceed the percentages of the total weight of cementitious materials given in Table 4.2.3.

TABLE 4.2.3—REQUIREMENTS FOR CONCRETE EXPOSED TO DEICING CHEMICALS

Cementitious materials	Maximum percent of total cementitious materials by weight*
Fly ash or other pozzolans conforming to ASTM C 618	25
Slag conforming to ASTM C 989	50
Silica fume conforming to ASTM C 1240	10
Total of fly ash or other pozzolans, slag, and silica fume	50†
Total of fly ash or other pozzolans and silica fume	35†

* The total cementitious material also includes ASTM C 150, C 595, C 845, and C 1157 cement.
The maximum percentages above shall include:
(a) Fly ash or other pozzolans present in Type IP or I(PM) blended cement, ASTM C 595, or ASTM C 1157;
(b) Slag used in the manufacture of a IS or I(SM) blended cement, ASTM C 595, or ASTM C 1157;
(c) Silica fume, ASTM C 1240, present in a blended cement.
† Fly ash or other pozzolans and silica fume shall constitute no more than 25 and 10 percent, respectively, of the total weight of the cementitious materials.

4.3 — Sulfate exposures

4.3.1 — Concrete to be exposed to sulfate-containing solutions or soils shall conform to requirements of Table 4.3.1 or shall be concrete made with a cement that provides sulfate resistance and that has a maximum water-cementitious material ratio and minimum f_c' from Table 4.3.1.

COMMENTARY

R4.2.3 — Section 4.2.3 and Table 4.2.3 establish limitations on the amount of fly ash, other pozzolans, silica fume, and slag that can be included in concrete exposed to deicing chemicals.[4.4-4.6] Research has demonstrated that the use of fly ash, slag, and silica fume produce concrete with a finer pore structure and, therefore, lower permeability.[4.7-4.9]

R4.3 — Sulfate exposures

R4.3.1 — Concrete exposed to injurious concentrations of sulfates from soil and water should be made with a sulfate-resisting cement. Table 4.3.1 lists the appropriate types of cement and the maximum water-cementitious material ratios and minimum specified compressive strengths for various exposure conditions. In selecting a cement for sulfate resistance, the principal consideration is its tricalcium aluminate (C_3A) content. For moderate exposures, Type II cement is limited to a maximum C_3A content of 8.0 percent under ASTM C 150. The blended cements under ASTM C 595 with the MS designation are appropriate for use in moderate sulfate exposures. The appropriate types under ASTM C 595 are IP(MS), IS(MS), I(PM)(MS), and I(SM)(MS). For severe exposures, Type V cement with a maximum C_3A content of 5 percent is specified. In certain areas, the C_3A

TABLE 4.3.1—REQUIREMENTS FOR CONCRETE EXPOSED TO SULFATE-CONTAINING SOLUTIONS

Sulfate exposure	Water soluble sulfate (SO_4) in soil, percent by weight	Sulfate (SO_4) in water, ppm	Cement type	Maximum water-cementitious material ratio, by weight, normalweight concrete*	Minimum f_c', normalweight and lightweight concrete, psi*
Negligible	$0.00 \le SO_4 < 0.10$	$0 \le SO_4 < 150$	—	—	—
Moderate†	$0.10 \le SO_4 < 0.20$	$150 \le SO_4 < 1500$	II, IP(MS), IS(MS), P(MS), I(PM)(MS), I(SM)(MS)	0.50	4000
Severe	$0.20 \le SO_4 \le 2.00$	$1500 \le SO_4 \le 10,000$	V	0.45	4500
Very severe	$SO_4 > 2.00$	$SO_4 > 10,000$	V plus pozzolan‡	0.45	4500

* When both Table 4.3.1 and Table 4.2.2 are considered, the lowest applicable maximum water-cementitious material ratio and highest applicable minimum f_c' shall be used.
† Seawater.
‡ Pozzolan that has been determined by test or service record to improve sulfate resistance when used in concrete containing Type V cement.

CODE

COMMENTARY

content of other available types such as Type III or Type I may be less than 8 or 5 percent and are usable in moderate or severe sulfate exposures. Note that sulfate-resisting cement will not increase resistance to some chemically aggressive solutions, for example ammonium nitrate. The project specifications should cover all special cases.

Using fly ash (ASTM C 618, Class F) also has been shown to improve the sulfate resistance of concrete.[4.9] Certain Type IP cements made by blending Class F pozzolan with portland cement having a C_3A content greater than 8 percent can provide sulfate resistance for moderate exposures.

A note to Table 4.3.1 lists seawater as moderate exposure, even though it generally contains more than 1500 ppm SO_4. In seawater exposures, other types of cement with C_3A up to 10 percent may be used if the maximum water-cementitious material ratio is reduced to 0.40.

ASTM test method C 1012[4.2] can be used to evaluate the sulfate resistance of mixtures using combinations of cementitious materials.

In addition to the proper selection of cement, other requirements for durable concrete exposed to concentrations of sulfate are essential, such as, low water-cementitious material ratio, strength, adequate air entrainment, low slump, adequate consolidation, uniformity, adequate cover of reinforcement, and sufficient moist curing to develop the potential properties of the concrete.

4.3.2 — Calcium chloride as an admixture shall not be used in concrete to be exposed to severe or very severe sulfate-containing solutions, as defined in Table 4.3.1.

4.4 — Corrosion protection of reinforcement

4.4.1 — For corrosion protection of reinforcement in concrete, maximum water soluble chloride ion concentrations in hardened concrete at ages from 28 to 42 days contributed from the ingredients including water, aggregates, cementitious materials, and admixtures shall not exceed the limits of Table 4.4.1. When testing is performed to determine water soluble chloride ion content, test procedures shall conform to ASTM C 1218.

R4.4 — Corrosion protection of reinforcement

R4.4.1 — Additional information on the effects of chlorides on the corrosion of reinforcing steel is given in **"Guide to Durable Concrete"** reported by ACI Committee 201[4.10] and **"Corrosion of Metals in Concrete"** reported by ACI Committee 222.[4.11] Test procedures should conform to those given in ASTM C 1218. An initial evaluation may be obtained by testing individual concrete ingredients for total chloride ion content. If total chloride ion content, calculated on the basis of concrete proportions, exceeds those permitted in Table 4.4.1, it may be necessary to test samples of the hardened concrete for water-soluble chloride ion content described in the ACI 201 guide. Some of the total chloride ions present in the ingredients will either be insoluble or will react with the cement during hydration and become insoluble under the test procedures described in ASTM C 1218.

CODE

TABLE 4.4.1—MAXIMUM CHLORIDE ION CONTENT FOR CORROSION PROTECTION OF REINFORCEMENT

Type of member	Maximum water soluble chloride ion (Cl^-) in concrete, percent by weight of cement
Prestressed concrete	0.06
Reinforced concrete exposed to chloride in service	0.15
Reinforced concrete that will be dry or protected from moisture in service	1.00
Other reinforced concrete construction	0.30

4.4.2 — If concrete with reinforcement will be exposed to chlorides from deicing chemicals, salt, salt water, brackish water, seawater, or spray from these sources, requirements of Table 4.2.2 for maximum water-cementitious material ratio and minimum f_c', and the minimum concrete cover requirements of 7.7 shall be satisfied. See 18.16 for unbonded tendons.

COMMENTARY

When concretes are tested for soluble chloride ion content the tests should be made at an age of 28 to 42 days. The limits in Table 4.4.1 are to be applied to chlorides contributed from the concrete ingredients, not those from the environment surrounding the concrete.

The chloride ion limits in Table 4.4.1 differ from those recommended in ACI 201.2R[4.10] and ACI 222R.[4.11] For reinforced concrete that will be dry in service, a limit of 1 percent has been included to control total soluble chlorides. Table 4.4.1 includes limits of 0.15 and 0.30 percent for reinforced concrete that will be exposed to chlorides or will be damp in service, respectively. These limits compare to 0.10 and 0.15 recommended in ACI 201.2R.[4.10] ACI 222R[4.11] recommends limits of 0.08 and 0.20 percent by weight of cement for chlorides in prestressed and reinforced concrete, respectively, based on tests for acid soluble chlorides, not the test for water soluble chlorides required here.

When epoxy or zinc-coated bars are used, the limits in Table 4.4.1 may be more restrictive than necessary.

R4.4.2 — When concretes are exposed to external sources of chlorides, the water-cementitious material ratio and specified compressive strength f_c' of 4.2.2 are the minimum requirements that are to be considered. The designer should evaluate conditions in structures where chlorides may be applied, in parking structures where chlorides may be tracked in by vehicles, or in structures near seawater. Epoxy- or zinc-coated bars or cover greater than the minimum required in 7.7 may be desirable. Use of slag meeting ASTM C 989 or fly ash meeting ASTM C 618 and increased levels of specified strength provide increased protection. Use of silica fume meeting ASTM C 1240 with an appropriate high-range water reducer, ASTM C 494, Types F and G, or ASTM C 1017 can also provide additional protection.[4.12] The use of ASTM C 1202[4.13] to test concrete mixtures proposed for use will provide additional information on the performance of the mixtures.

Notes

CHAPTER 5 — CONCRETE QUALITY, MIXING, AND PLACING

CODE

5.1 — General

5.1.1 — Concrete shall be proportioned to provide an average compressive strength, f'_{cr}, as prescribed in 5.3.2 and shall satisfy the durability criteria of Chapter 4. Concrete shall be produced to minimize the frequency of strength tests below f_c', as prescribed in 5.6.3.3. For concrete designed and constructed in accordance with the code, f_c' shall not be less than 2500 psi.

5.1.2 — Requirements for f_c' shall be based on tests of cylinders made and tested as prescribed in 5.6.3.

5.1.3 — Unless otherwise specified, f_c' shall be based on 28-day tests. If other than 28 days, test age for f_c' shall be as indicated in design drawings or specifications.

5.1.4 — Where design criteria in 9.5.2.3, 11.2, and 12.2.4 provide for use of a splitting tensile strength value of concrete, laboratory tests shall be made in accordance with "Standard Specification for Lightweight Aggregates for Structural Concrete" (ASTM C 330) to establish a value of f_{ct} corresponding to f_c'.

5.1.5 — Splitting tensile strength tests shall not be used as a basis for field acceptance of concrete.

COMMENTARY

R5.1 — General

The requirements for proportioning concrete mixtures are based on the philosophy that concrete should provide both adequate durability (Chapter 4) and strength. The criteria for acceptance of concrete are based on the philosophy that the code is intended primarily to protect the safety of the public. Chapter 5 describes procedures by which concrete of adequate strength can be obtained, and provides procedures for checking the quality of the concrete during and after its placement in the work.

Chapter 5 also prescribes minimum criteria for mixing and placing concrete.

The provisions of 5.2, 5.3, and 5.4, together with Chapter 4, establish required mixture proportions. The basis for determining the adequacy of concrete strength is in 5.6.

R5.1.1 — The basic premises governing the designation and evaluation of concrete strength are presented. It is emphasized that the average compressive strength of concrete produced should always exceed the specified value of f_c' used in the structural design calculations. This is based on probabilistic concepts, and is intended to ensure that adequate concrete strength will be developed in the structure. The durability requirements prescribed in Chapter 4 are to be satisfied in addition to attaining the average concrete strength in accordance with 5.3.2.

R5.1.4 — Sections 9.5.2.3 (modulus of rupture), 11.2 (concrete shear strength) and 12.2.4 (development of reinforcement) require modification in the design criteria for the use of lightweight concrete. Two alternative modification procedures are provided. One alternative is based on laboratory tests to determine the relationship between average splitting tensile strength f_{ct} and specified compressive strength f_c' for the lightweight concrete. For a lightweight aggregate from a given source, it is intended that appropriate values of f_{ct} be obtained in advance of design.

R5.1.5 — Tests for splitting tensile strength of concrete (as required by 5.1.4) are not intended for control of, or acceptance of, the strength of concrete in the field. Indirect control will

CODE | COMMENTARY

be maintained through the normal compressive strength test requirements provided by 5.6.

5.2 — Selection of concrete proportions

R5.2 — Selection of concrete proportions

Recommendations for selecting proportions for concrete are given in detail in **"Standard Practice for Selecting Proportions for Normal, Heavyweight, and Mass Concrete"** (ACI 211.1).[5.1] (Provides two methods for selecting and adjusting proportions for normalweight concrete: the estimated weight and absolute volume methods. Example calculations are shown for both methods. Proportioning of heavyweight concrete by the absolute volume method is presented in an appendix.)

Recommendations for lightweight concrete are given in **"Standard Practice for Selecting Proportions for Structural Lightweight Concrete"** (ACI 211.2).[5.2] (Provides a method of proportioning and adjusting structural grade concrete containing lightweight aggregates.)

5.2.1 — Proportions of materials for concrete shall be established to provide:

(a) Workability and consistency to permit concrete to be worked readily into forms and around reinforcement under conditions of placement to be employed, without segregation or excessive bleeding;

(b) Resistance to special exposures as required by Chapter 4;

(c) Conformance with strength test requirements of 5.6.

R5.2.1 — The selected water-cementitious material ratio should be low enough, or in the case of lightweight concrete the compressive strength high enough to satisfy both the strength criteria (see 5.3 or 5.4) and the special exposure requirements (Chapter 4). The code does not include provisions for especially severe exposures, such as acids or high temperatures, and is not concerned with aesthetic considerations such as surface finishes. These items are beyond the scope of the code and should be covered specifically in the project specifications. Concrete ingredients and proportions are to be selected to meet the minimum requirements stated in the code and the additional requirements of the contract documents.

5.2.2 — Where different materials are to be used for different portions of proposed work, each combination shall be evaluated.

5.2.3 — Concrete proportions shall be established in accordance with 5.3 or, alternatively, 5.4, and shall meet applicable requirements of Chapter 4.

R5.2.3 — The code emphasizes the use of field experience or laboratory trial mixtures (see 5.3) as the preferred method for selecting concrete mixture proportions.

5.3 — Proportioning on the basis of field experience or trial mixtures, or both

R5.3 — Proportioning on the basis of field experience or trial mixtures, or both

In selecting a suitable concrete mixture there are three basic steps. The first is the determination of the sample standard deviation. The second is the determination of the required average compressive strength. The third is the selection of mixture proportions required to produce that average strength, either by conventional trial mixture procedures or by a suitable experience record. Fig. R5.3 is a flow chart outlining the mixture selection and documentation procedure.

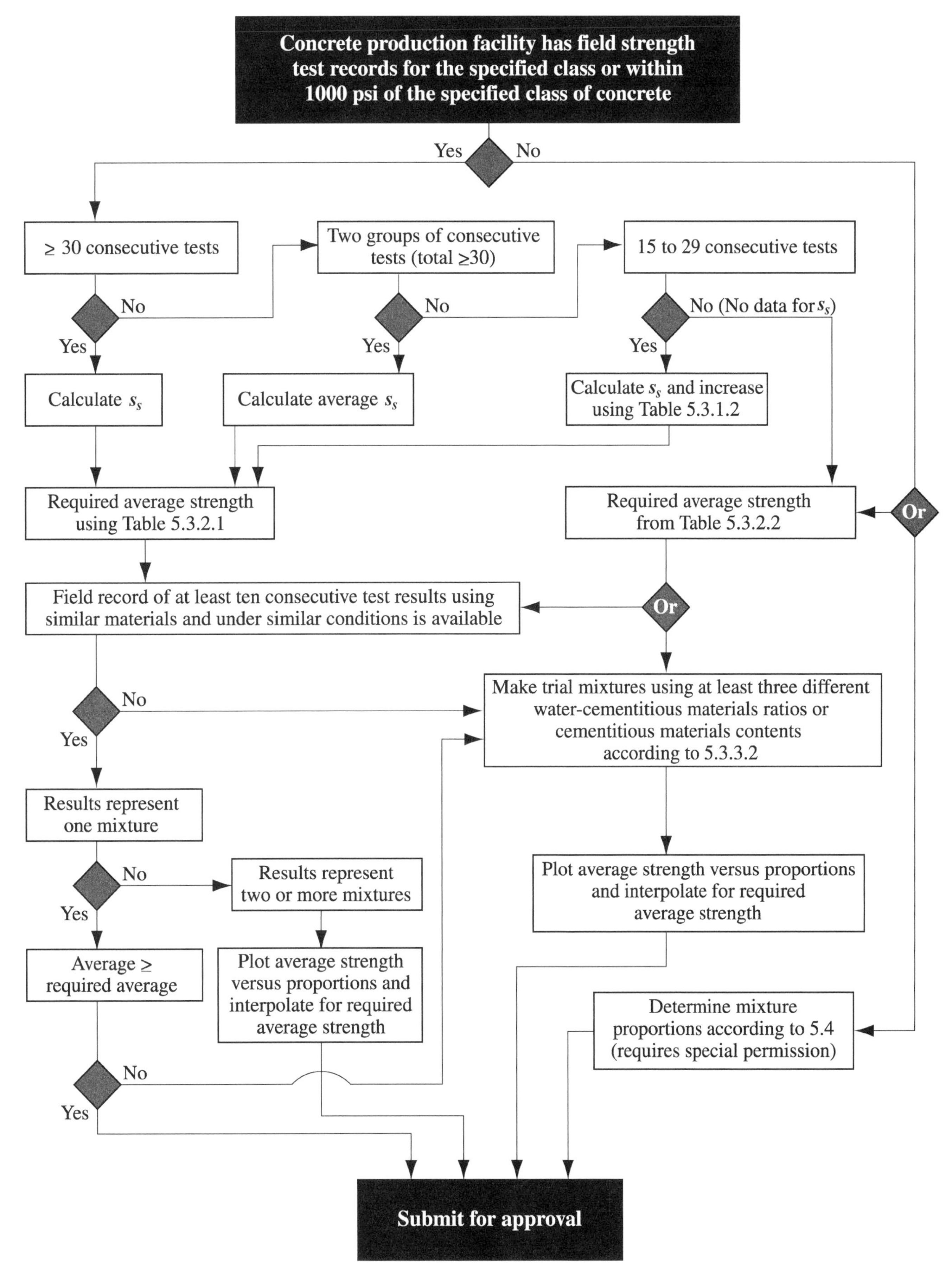

Fig. R5.3—Flow chart for selection and documentation of concrete proportions

CODE

COMMENTARY

The mixture selected should yield an average strength appreciably higher than the specified strength f_c'. The degree of mixture over design depends on the variability of the test results.

5.3.1 — Sample standard deviation

5.3.1.1 — Where a concrete production facility has test records, a sample standard deviation, s_s, shall be established. Test records from which s_s is calculated:

(a) Shall represent materials, quality control procedures, and conditions similar to those expected and changes in materials and proportions within the test records shall not have been more restricted than those for proposed work;

(b) Shall represent concrete produced to meet a specified compressive strength or strengths within 1000 psi of f_c';

(c) Shall consist of at least 30 consecutive tests or two groups of consecutive tests totaling at least 30 tests as defined in 5.6.2.4, except as provided in 5.3.1.2.

5.3.1.2 — Where a concrete production facility does not have test records meeting requirements of 5.3.1.1, but does have a record based on 15 to 29 consecutive tests, a sample standard deviation s_s shall be established as the product of the calculated sample standard deviation and modification factor of Table 5.3.1.2. To be acceptable, test records shall meet requirements (a) and (b) of 5.3.1.1, and represent only a single record of consecutive tests that span a period of not less than 45 calendar days.

TABLE 5.3.1.2—MODIFICATION FACTOR FOR SAMPLE STANDARD DEVIATION WHEN LESS THAN 30 TESTS ARE AVAILABLE

No. of tests*	Modification factor for sample standard deviation†
Less than 15	Use table 5.3.2.2
15	1.16
20	1.08
25	1.03
30 or more	1.00

*Interpolate for intermediate numbers of tests.

†Modified sample standard deviation, s_s, to be used to determine required average strength, f_{cr}', from 5.3.2.1.

R5.3.1 — Sample standard deviation

When a concrete production facility has a suitable record of 30 consecutive tests of similar materials and conditions expected, the sample standard deviation, s_s, is calculated from those results in accordance with the following formula:

$$s_s = \left[\frac{\Sigma(x_i - \bar{x})^2}{(n-1)}\right]^{1/2}$$

where:

s_s = sample standard deviation, psi
x_i = individual strength tests as defined in 5.6.2.4
$\bar{x}$ = average of n strength test results
n = number of consecutive strength tests

The sample standard deviation is used to determine the average strength required in 5.3.2.1.

If two test records are used to obtain at least 30 tests, the sample standard deviation used shall be the statistical average of the values calculated from each test record in accordance with the following formula:

$$\bar{s}_s = \left[\frac{(n_1 - 1)(s_{s1})^2 + (n_2 - 1)(s_{s2})^2}{(n_1 + n_2 - 2)}\right]^{1/2}$$

where:

$\bar{s}_s$ = statistical average standard deviation where two test records are used to estimate the sample standard deviation
s_{s1}, s_{s2} = sample standard deviations calculated from two test records, 1 and 2, respectively
n_1, n_2 = number of tests in each test record, respectively

If less than 30, but at least 15 tests are available, the calculated sample standard deviation is increased by the factor given in Table 5.3.1.2. This procedure results in a more conservative (increased) required average strength. The factors in Table 5.3.1.2 are based on the sampling distribution of the sample standard deviation and provide protection (equivalent to that from a record of 30 tests) against the possibility that the smaller sample underestimates the true or universe population standard deviation.

The sample standard deviation used in the calculation of required average strength should be developed under conditions "similar to those expected" [see 5.3.1.1(a)]. This requirement is important to ensure acceptable concrete.

CODE

COMMENTARY

Concrete for background tests to determine sample standard deviation is considered to be "similar" to that required if made with the same general types of ingredients under no more restrictive conditions of control over material quality and production methods than on the proposed work, and if its specified strength does not deviate more than 1000 psi from the f_c' required [see 5.3.1.1(b)]. A change in the type of concrete or a major increase in the strength level may increase the sample standard deviation. Such a situation might occur with a change in type of aggregate (i.e., from natural aggregate to lightweight aggregate or vice versa) or a change from non-air-entrained concrete to air-entrained concrete. Also, there may be an increase in sample standard deviation when the average strength level is raised by a significant amount, although the increment of increase in sample standard deviation should be somewhat less than directly proportional to the strength increase. When there is reasonable doubt, any estimated sample standard deviation used to calculate the required average strength should always be on the conservative (high) side.

Note that the code uses the sample standard deviation in pounds per square inch instead of the coefficient of variation in percent. The latter is equal to the former expressed as a percent of the average strength.

Even when the average strength and sample standard deviation are of the levels assumed, there will be occasional tests that fail to meet the acceptance criteria prescribed in 5.6.3.3 (perhaps 1 test in 100).

5.3.2 — Required average strength

5.3.2.1 — Required average compressive strength f_{cr}' used as the basis for selection of concrete proportions shall be determined from Table 5.3.2.1 using the sample standard deviation, s_s, calculated in accordance with 5.3.1.1 or 5.3.1.2.

TABLE 5.3.2.1—REQUIRED AVERAGE COMPRESSIVE STRENGTH WHEN DATA ARE AVAILABLE TO ESTABLISH A SAMPLE STANDARD DEVIATION

Specified compressive strength, psi	Required average compressive strength, psi
$f_c' \leq 5000$	Use the larger value computed from Eq. (5-1) and (5-2) $f_{cr}' = f_c' + 1.34s_s$ (5-1) $f_{cr}' = f_c' + 2.33s_s - 500$ (5-2)
$f_c' > 5000$	Use the larger value computed from Eq. (5-1) and (5-3) $f_{cr}' = f_c' + 1.34s_s$ (5-1) $f_{cr}' = 0.90f_c' + 2.33s_s$ (5-3)

R5.3.2 — Required average strength

R5.3.2.1 — Once the sample standard deviation has been determined, the required average compressive strength, f_{cr}', is obtained from the larger value computed from Eq. (5-1) and (5-2) for f_c' of 5000 psi or less, or the larger value computed from Eq. (5-1) and (5-3) for f_c' over 5000 psi. Equation (5-1) is based on a probability of 1-in-100 that the average of three consecutive tests may be below the specified compressive strength f_c'. Equation (5-2) is based on a similar probability that an individual test may be more than 500 psi below the specified compressive strength f_c'. Equation (5-3) is based on the same 1-in-100 probability that an individual test may be less than $0.90f_c'$. These equations assume that the sample standard deviation used is equal to the population value appropriate for an infinite or very large number of tests. For this reason, use of sample standard deviations estimated from records of 100 or more tests is desirable. When 30 tests are available, the probability of failure will likely be somewhat greater than 1-in-100. The additional refinements required to achieve the 1-in-100 probability are not considered necessary, because of the uncertainty inherent in assuming that conditions operating when the test record was accumulated will be similar to conditions when the concrete will be produced.

CODE

5.3.2.2 — When a concrete production facility does not have field strength test records for calculation of s_s meeting requirements of 5.3.1.1 or 5.3.1.2, f'_{cr} shall be determined from Table 5.3.2.2 and documentation of average strength shall be in accordance with requirements of 5.3.3.

TABLE 5.3.2.2—REQUIRED AVERAGE COMPRESSIVE STRENGTH WHEN DATA ARE NOT AVAILABLE TO ESTABLISH A SAMPLE STANDARD DEVIATION

Specified compressive strength, psi	Required average compressive strength, psi
$f'_c < 3000$	$f'_{cr} = f'_c + 1000$
$3000 \leq f'_c \leq 5000$	$f'_{cr} = f'_c + 1200$
$f'_c > 5000$	$f'_{cr} = 1.10f'_c + 700$

5.3.3 — Documentation of average compressive strength

Documentation that proposed concrete proportions will produce an average compressive strength equal to or greater than required average compressive strength f'_{cr} (see 5.3.2) shall consist of a field strength test record, several strength test records, or trial mixtures.

5.3.3.1 — When test records are used to demonstrate that proposed concrete proportions will produce f'_{cr} (see 5.3.2), such records shall represent materials and conditions similar to those expected. Changes in materials, conditions, and proportions within the test records shall not have been more restricted than those for proposed work. For the purpose of documenting average strength potential, test records consisting of

COMMENTARY

R.5.3.3 — Documentation of average compressive strength

Once the required average compressive strength f'_{cr} is known, the next step is to select mixture proportions that will produce an average strength at least as great as the required average strength, and also meet special exposure requirements of Chapter 4. The documentation may consist of a strength test record, several strength test records, or suitable laboratory or field trial mixtures. Generally, if a test record is used, it will be the same one that was used for computation of the standard deviation. However, if this test record shows either lower or higher average compressive strength than the required average compressive strength, different proportions may be necessary or desirable. In such instances, the average from a record of as few as 10 tests may be used, or the proportions may be established by interpolation between the strengths and proportions of two such records of consecutive tests. All test records for establishing proportions necessary to produce the average compressive strength are to meet the requirements of 5.3.3.1 for "similar materials and conditions."

For strengths over 5000 psi where the average compressive strength documentation is based on laboratory trial mixtures, it may be appropriate to increase f'_{cr} calculated in Table 5.3.2.2 to allow for a reduction in strength from laboratory trials to actual concrete production.

CODE

less than 30 but not less than 10 consecutive tests are acceptable provided test records encompass a period of time not less than 45 days. Required concrete proportions shall be permitted to be established by interpolation between the strengths and proportions of two or more test records, each of which meets other requirements of this section.

5.3.3.2 — When an acceptable record of field test results is not available, concrete proportions established from trial mixtures meeting the following restrictions shall be permitted:

(a) Materials shall be those for proposed work;

(b) Trial mixtures having proportions and consistencies required for proposed work shall be made using at least three different water-cementitious materials ratios or cementitious materials contents that will produce a range of strengths encompassing f'_{cr};

(c) Trial mixtures shall be designed to produce a slump within ± 0.75 in. of maximum permitted, and for air-entrained concrete, within ± 0.5 percent of maximum allowable air content;

(d) For each water-cementitious materials ratio or cementitious materials content, at least three test cylinders for each test age shall be made and cured in accordance with "Standard Practice for Making and Curing Concrete Test Specimens in the Laboratory" (ASTM C 192). Cylinders shall be tested at 28 days or at test age designated for determination of f_c';

(e) From results of cylinder tests a curve shall be plotted showing the relationship between water-cementitious materials ratio or cementitious materials content and compressive strength at designated test age;

(f) Maximum water-cementitious material ratio or minimum cementitious materials content for concrete to be used in proposed work shall be that shown by the curve to produce f'_{cr} required by 5.3.2, unless a lower water-cementitious materials ratio or higher strength is required by Chapter 4.

5.4 — Proportioning without field experience or trial mixtures

5.4.1 — If data required by 5.3 are not available, concrete proportions shall be based upon other experience or information, if approved by the registered design professional. The required average compressive strength f'_{cr} of concrete produced with materials

COMMENTARY

R5.4 — Proportioning without field experience or trial mixtures

R5.4.1 — When no prior experience (5.3.3.1) or trial mixture data (5.3.3.2) meeting the requirements of these sections is available, other experience may be used only when special permission is given. Because combinations of different ingredients may vary considerably in strength level, this

CODE

similar to those proposed for use shall be at least 1200 psi greater than f_c'. This alternative shall not be used if f_c' is greater than 5000 psi.

5.4.2 — Concrete proportioned by this section shall conform to the durability requirements of Chapter 4 and to compressive strength test criteria of 5.6.

COMMENTARY

procedure is not permitted for f_c' greater than 5000 psi and the required average compressive strength should exceed f_c' by 1200 psi. The purpose of this provision is to allow work to continue when there is an unexpected interruption in concrete supply and there is not sufficient time for tests and evaluation or in small structures where the cost of trial mixture data is not justified.

5.5 — Average compressive strength reduction

As data become available during construction, it shall be permitted to reduce the amount by which the required average concrete strength, f_{cr}', must exceed f_c', provided:

(a) Thirty or more test results are available and average of test results exceeds that required by 5.3.2.1, using a sample standard deviation calculated in accordance with 5.3.1.1; or

(b) Fifteen to 29 test results are available and average of test results exceeds that required by 5.3.2.1 using a sample standard deviation calculated in accordance with 5.3.1.2; and

(c) Special exposure requirements of Chapter 4 are met.

5.6 — Evaluation and acceptance of concrete

R5.6 — Evaluation and acceptance of concrete

Once the mixture proportions have been selected and the job started, the criteria for evaluation and acceptance of the concrete can be obtained from 5.6.

An effort has been made in the code to provide a clear-cut basis for judging the acceptability of the concrete, as well as to indicate a course of action to be followed when the results of strength tests are not satisfactory.

5.6.1 — Concrete shall be tested in accordance with the requirements of 5.6.2 through 5.6.5. Qualified field testing technicians shall perform tests on fresh concrete at the job site, prepare specimens required for curing under field conditions, prepare specimens required for testing in the laboratory, and record the temperature of the fresh concrete when preparing specimens for strength tests. Qualified laboratory technicians shall perform all required laboratory tests.

R5.6.1 — Laboratory and field technicians can establish qualifications by becoming certified through certification programs. Field technicians in charge of sampling concrete; testing for slump, unit weight, yield, air content, and temperature; and making and curing test specimens should be certified in accordance with the requirements of ACI Concrete Field Testing Technician—Grade 1 Certification Program, or the requirements of ASTM C 1077,[5.3] or an equivalent program. Concrete testing laboratory personnel should be certified in accordance with the requirements of ACI Concrete Laboratory Testing Technician, Concrete Strength Testing Technician, or the requirements of ASTM C 1077.

Testing reports should be promptly distributed to the owner, registered design professional responsible for the design, contractor, appropriate subcontractors, appropriate suppliers, and building official to allow timely identification of either compliance or the need for corrective action.

CODE

COMMENTARY

5.6.2 — Frequency of testing

5.6.2.1 — Samples for strength tests of each class of concrete placed each day shall be taken not less than once a day, nor less than once for each 150 yd^3 of concrete, nor less than once for each 5000 ft^2 of surface area for slabs or walls.

R5.6.2 — Frequency of testing

R5.6.2.1 — The following three criteria establish the required minimum sampling frequency for each class of concrete:

(a) Once each day a given class is placed, nor less than

(b) Once for each 150 yd^3 of each class placed each day, nor less than

(c) Once for each 5000 ft^2 of slab or wall surface area placed each day.

In calculating surface area, only one side of the slab or wall should be considered. Criteria (c) will require more frequent sampling than once for each 150 yd^3 placed if the average wall or slab thickness is less than 9-3/4 in.

5.6.2.2 — On a given project, if total volume of concrete is such that frequency of testing required by 5.6.2.1 would provide less than five strength tests for a given class of concrete, tests shall be made from at least five randomly selected batches or from each batch if fewer than five batches are used.

R5.6.2.2 — Samples for strength tests are to be taken on a strictly random basis if they are to measure properly the acceptability of the concrete. To be representative, the choice of times of sampling, or the batches of concrete to be sampled, are to be made on the basis of chance alone, within the period of placement. Batches should not be sampled on the basis of appearance, convenience, or other possibly biased criteria, because the statistical analyses will lose their validity. Not more than one test (average of two cylinders made from a sample, 5.6.2.4) should be taken from a single batch, and water may not be added to the concrete after the sample is taken.

ASTM D 3665[5.4] describes procedures for random selection of the batches to be tested.

5.6.2.3 — When total quantity of a given class of concrete is less than 50 yd^3, strength tests are not required when evidence of satisfactory strength is submitted to and approved by the building official.

5.6.2.4 — A strength test shall be the average of the strengths of two cylinders made from the same sample of concrete and tested at 28 days or at test age designated for determination of f_c'.

5.6.3 — Laboratory-cured specimens

R5.6.3 — Laboratory-cured specimens

5.6.3.1 — Samples for strength tests shall be taken in accordance with "Standard Practice for Sampling Freshly Mixed Concrete" (ASTM C 172).

5.6.3.2 — Cylinders for strength tests shall be molded and laboratory-cured in accordance with "Standard Practice for Making and Curing Concrete Test Specimens in the Field" (ASTM C 31) and tested in accordance with "Standard Test Method for Compressive Strength of Cylindrical Concrete Specimens" (ASTM C 39).

CODE

5.6.3.3 — Strength level of an individual class of concrete shall be considered satisfactory if both of the following requirements are met:

(a) Every arithmetic average of any three consecutive strength tests equals or exceeds f_c';

(b) No individual strength test (average of two cylinders) falls below f_c' by more than 500 psi when f_c' is 5000 psi or less; or by more than $0.10f_c'$ when f_c' is more than 5000 psi.

5.6.3.4 — If either of the requirements of 5.6.3.3 is not met, steps shall be taken to increase the average of subsequent strength test results. Requirements of 5.6.5 shall be observed if requirement of 5.6.3.3(b) is not met.

COMMENTARY

R5.6.3.3 — A single set of criteria is given for acceptability of strength and is applicable to all concrete used in structures designed in accordance with the code, regardless of design method used. The concrete strength is considered to be satisfactory as long as averages of any three consecutive strength tests remain above the specified f_c' and no individual strength test falls below the specified f_c' by more than 500 psi if f_c' is 5000 psi or less, or falls below f_c' by more than 10 percent if f_c' is over 5000 psi. Evaluation and acceptance of the concrete can be judged immediately as test results are received during the course of the work. Strength tests failing to meet these criteria will occur occasionally (probably about once in 100 tests) even though concrete strength and uniformity are satisfactory. Allowance should be made for such statistically expected variations in deciding whether the strength level being produced is adequate.

R5.6.3.4 — When concrete fails to meet either of the strength requirements of 5.6.3.3, steps should be taken to increase the average of the concrete test results. If sufficient concrete has been produced to accumulate at least 15 tests, these should be used to establish a new target average strength as described in 5.3.

If fewer than 15 tests have been made on the class of concrete in question, the new target strength level should be at least as great as the average level used in the initial selection of proportions. If the average of the available tests made on the project equals or exceeds the level used in the initial selection of proportions, a further increase in average level is required.

The steps taken to increase the average level of test results will depend on the particular circumstances, but could include one or more of the following:

(a) An increase in cementitious materials content;

(b) Changes in mixture proportions;

(c) Reductions in or better control of levels of slump supplied;

(d) A reduction in delivery time;

(e) Closer control of air content;

(f) An improvement in the quality of the testing, including strict compliance with standard test procedures.

Such changes in operating and testing procedures, or changes in cementitious materials content, or slump should not require a formal resubmission under the procedures of 5.3; however, important changes in sources of cement, aggregates, or admixtures should be accompanied by evidence that the average strength level will be improved.

Laboratories testing cylinders or cores to determine compliance with these requirements should be accredited or inspected for conformance to the requirement of ASTM C 1077[5.3] by a recognized agency such as the American Associ-

CODE

COMMENTARY

ation for Laboratory Accreditation (A2LA), AASHTO Materials Reference Laboratory (AMRL), National Voluntary Laboratory Accreditation Program (NVLAP), Cement and Concrete Reference Laboratory (CCRL), or their equivalent.

5.6.4 — Field-cured specimens

5.6.4.1 — If required by the building official, results of strength tests of cylinders cured under field conditions shall be provided.

5.6.4.2 — Field-cured cylinders shall be cured under field conditions in accordance with "Practice for Making and Curing Concrete Test Specimens in the Field" (ASTM C 31).

5.6.4.3 — Field-cured test cylinders shall be molded at the same time and from the same samples as laboratory-cured test cylinders.

5.6.4.4 — Procedures for protecting and curing concrete shall be improved when strength of field-cured cylinders at test age designated for determination of f_c' is less than 85 percent of that of companion laboratory-cured cylinders. The 85 percent limitation shall not apply if field-cured strength exceeds f_c' by more than 500 psi.

R5.6.4 — Field-cured specimens

R5.6.4.1 — Strength tests of cylinders cured under field conditions may be required to check the adequacy of curing and protection of concrete in the structure.

R5.6.4.4 — Positive guidance is provided in the code concerning the interpretation of tests of field-cured cylinders. Research has shown that cylinders protected and cured to simulate good field practice should test not less than about 85 percent of standard laboratory moist-cured cylinders. This percentage has been set as a rational basis for judging the adequacy of field curing. The comparison is made between the actual measured strengths of companion job-cured and laboratory-cured cylinders, not between job-cured cylinders and the specified value of f_c'. However, results for the job-cured cylinders are considered satisfactory if the job-cured cylinders exceed the specified f_c' by more than 500 psi, even though they fail to reach 85 percent of the strength of companion laboratory-cured cylinders.

5.6.5 — Investigation of low-strength test results

5.6.5.1 — If any strength test (see 5.6.2.4) of laboratory-cured cylinders falls below f_c' by more than the values given in 5.6.3.3(b) or if tests of field-cured cylinders indicate deficiencies in protection and curing (see 5.6.4.4), steps shall be taken to assure that load-carrying capacity of the structure is not jeopardized.

5.6.5.2 — If the likelihood of low-strength concrete is confirmed and calculations indicate that load-carrying capacity is significantly reduced, tests of cores drilled from the area in question in accordance with "Standard Test Method for Obtaining and Testing Drilled Cores and Sawed Beams of Concrete" (ASTM C 42) shall be permitted. In such cases, three cores shall be taken for each strength test that falls below the values given in 5.6.3.3(b).

R5.6.5 — Investigation of low-strength test results

Instructions are provided concerning the procedure to be followed when strength tests have failed to meet the specified acceptance criteria. For obvious reasons, these instructions cannot be dogmatic. The building official should apply judgment as to the significance of low test results and whether they indicate need for concern. If further investigation is deemed necessary, such investigation may include nondestructive tests, or in extreme cases, strength tests of cores taken from the structure.

Nondestructive tests of the concrete in place, such as by probe penetration, impact hammer, ultrasonic pulse velocity or pull out may be useful in determining whether or not a portion of the structure actually contains low-strength concrete. Such tests are of value primarily for comparisons within the same job rather than as quantitative measures of strength. For cores, if required, conservatively safe acceptance criteria are provided that should ensure structural ade-

CODE

5.6.5.3 — Cores shall be prepared for transport and storage by wiping drilling water from their surfaces and placing the cores in watertight bags or containers immediately after drilling. Cores shall be tested no earlier than 48 hours and not later than 7 days after coring unless approved by the registered design professional.

5.6.5.4 — Concrete in an area represented by core tests shall be considered structurally adequate if the average of three cores is equal to at least 85 percent of f_c' and if no single core is less than 75 percent of f_c'. Additional testing of cores extracted from locations represented by erratic core strength results shall be permitted.

5.6.5.5 — If criteria of 5.6.5.4 are not met and if the structural adequacy remains in doubt, the responsible authority shall be permitted to order a strength evaluation in accordance with Chapter 20 for the questionable portion of the structure, or take other appropriate action.

COMMENTARY

quacy for virtually any type of construction.[5.5-5.8] Lower strength may, of course, be tolerated under many circumstances, but this again becomes a matter of judgment on the part of the building official and design engineer. When the core tests fail to provide assurance of structural adequacy, it may be practical, particularly in the case of floor or roof systems, for the building official to require a load test (Chapter 20). Short of load tests, if time and conditions permit, an effort may be made to improve the strength of the concrete in place by supplemental wet curing. Effectiveness of such a treatment should be verified by further strength evaluation using procedures previously discussed.

A core obtained through the use of a water-cooled bit results in a moisture gradient between the exterior and interior of the core being created during drilling. This adversely affects the core's compressive strength.[5.9] The restriction on the commencement of core testing provides a minimum time for the moisture gradient to dissipate.

Core tests having an average of 85 percent of the specified strength are realistic. To expect core tests to be equal to f_c' is not realistic, since differences in the size of specimens, conditions of obtaining samples, and procedures for curing, do not permit equal values to be obtained.

The code, as stated, concerns itself with assuring structural safety, and the instructions in 5.6 are aimed at that objective. It is not the function of the code to assign responsibility for strength deficiencies, whether or not they are such as to require corrective measures.

Under the requirements of this section, cores taken to confirm structural adequacy will usually be taken at ages later than those specified for determination of f_c'.

5.7 — Preparation of equipment and place of deposit

5.7.1 — Preparation before concrete placement shall include the following:

(a) All equipment for mixing and transporting concrete shall be clean;

(b) All debris and ice shall be removed from spaces to be occupied by concrete;

(c) Forms shall be properly coated;

(d) Masonry filler units that will be in contact with concrete shall be well drenched;

(e) Reinforcement shall be thoroughly clean of ice or other deleterious coatings;

R5.7 — Preparation of equipment and place of deposit

Recommendations for mixing, handling and transporting, and placing concrete are given in detail in **"Guide for Measuring, Mixing, Transporting, and Placing Concrete"** reported by ACI Committee 304.[5.10] (Presents methods and procedures for control, handling and storage of materials, measurement, batching tolerances, mixing, methods of placing, transporting, and forms.)

Attention is directed to the need for using clean equipment and for cleaning forms and reinforcement thoroughly before beginning to deposit concrete. In particular, sawdust, nails, wood pieces, and other debris that may collect inside the forms should be removed. Reinforcement should be thoroughly cleaned of ice, dirt, loose rust, mill scale, or other coatings. Water should be removed from the forms.

CODE

(f) Water shall be removed from place of deposit before concrete is placed unless a tremie is to be used or unless otherwise permitted by the building official;

(g) All laitance and other unsound material shall be removed before additional concrete is placed against hardened concrete.

5.8 — Mixing

5.8.1 — All concrete shall be mixed until there is a uniform distribution of materials and shall be discharged completely before mixer is recharged.

5.8.2 — Ready-mixed concrete shall be mixed and delivered in accordance with requirements of "Standard Specification for Ready-Mixed Concrete" (ASTM C 94) or "Standard Specification for Concrete Made by Volumetric Batching and Continuous Mixing" (ASTM C 685).

5.8.3 — Job-mixed concrete shall be mixed in accordance with the following:

(a) Mixing shall be done in a batch mixer of approved type;

(b) Mixer shall be rotated at a speed recommended by the manufacturer;

(c) Mixing shall be continued for at least 1-1/2 minutes after all materials are in the drum, unless a shorter time is shown to be satisfactory by the mixing uniformity tests of "Standard Specification for Ready-Mixed Concrete" (ASTM C 94);

(d) Materials handling, batching, and mixing shall conform to applicable provisions of "Standard Specification for Ready-Mixed Concrete" (ASTM C 94);

(e) A detailed record shall be kept to identify:

(1) number of batches produced;

(2) proportions of materials used;

(3) approximate location of final deposit in structure;

(4) time and date of mixing and placing.

5.9 — Conveying

5.9.1 — Concrete shall be conveyed from mixer to place of final deposit by methods that will prevent separation or loss of materials.

COMMENTARY

R5.8 — Mixing

Concrete of uniform and satisfactory quality requires the materials to be thoroughly mixed until uniform in appearance and all ingredients are distributed. Samples taken from different portions of a batch should have essentially the same unit weight, air content, slump, and coarse aggregate content. Test methods for uniformity of mixing are given in ASTM C 94. The necessary time of mixing will depend on many factors including batch size, stiffness of the batch, size and grading of the aggregate, and the efficiency of the mixer. Excessively long mixing times should be avoided to guard against grinding of the aggregates.

R5.9 — Conveying

Each step in the handling and transporting of concrete needs to be controlled to maintain uniformity within a batch and from batch to batch. It is essential to avoid segregation of the coarse aggregate from the mortar or of water from the other ingredients.

CODE

5.9.2 — Conveying equipment shall be capable of providing a supply of concrete at site of placement without separation of ingredients and without interruptions sufficient to permit loss of plasticity between successive increments.

COMMENTARY

The code requires the equipment for handling and transporting concrete to be capable of supplying concrete to the place of deposit continuously and reliably under all conditions and for all methods of placement. The provisions of 5.9 apply to all placement methods, including pumps, belt conveyors, pneumatic systems, wheelbarrows, buggies, crane buckets, and tremies.

Serious loss in strength can result when concrete is pumped through pipe made of aluminum or aluminum alloy.[5.11] Hydrogen gas generated by the reaction between the cement alkalies and the aluminum eroded from the interior of the pipe surface has been shown to cause strength reduction as much as 50 percent. Hence, equipment made of aluminum or aluminum alloys should not be used for pump lines, tremies, or chutes other than short chutes such as those used to convey concrete from a truck mixer.

5.10 — Depositing

5.10.1 — Concrete shall be deposited as nearly as practical in its final position to avoid segregation due to rehandling or flowing.

5.10.2 — Concreting shall be carried on at such a rate that concrete is at all times plastic and flows readily into spaces between reinforcement.

5.10.3 — Concrete that has partially hardened or been contaminated by foreign materials shall not be deposited in the structure.

5.10.4 — Retempered concrete or concrete that has been remixed after initial set shall not be used unless approved by the engineer.

5.10.5 — After concreting is started, it shall be carried on as a continuous operation until placing of a panel or section, as defined by its boundaries or predetermined joints, is completed except as permitted or prohibited by 6.4.

5.10.6 — Top surfaces of vertically formed lifts shall be generally level.

5.10.7 — When construction joints are required, joints shall be made in accordance with 6.4.

5.10.8 — All concrete shall be thoroughly consolidated by suitable means during placement and shall be thoroughly worked around reinforcement and embedded fixtures and into corners of forms.

R5.10 — Depositing

Rehandling concrete can cause segregation of the materials. Hence the code cautions against this practice. Retempering of partially set concrete with the addition of water should not be permitted, unless authorized. This does not preclude the practice (recognized in ASTM C 94) of adding water to mixed concrete to bring it up to the specified slump range so long as prescribed limits on the maximum mixing time and water-cementitious materials ratio are not violated.

Section 5.10.4 of the 1971 code contained a requirement that "where conditions make consolidation difficult or where reinforcement is congested, batches of mortar containing the same proportions of cement, sand, and water as used in the concrete, shall first be deposited in the forms to a depth of at least 1 in." That requirement was deleted from the 1977 code since the conditions for which it was applicable could not be defined precisely enough to justify its inclusion as a code requirement. The practice, however, has merit and should be incorporated in job specifications where appropriate, with the specific enforcement the responsibility of the job inspector. The use of mortar batches aids in preventing honeycomb and poor bonding of the concrete with the reinforcement. The mortar should be placed immediately before depositing the concrete and should be plastic (neither stiff nor fluid) when the concrete is placed.

Recommendations for consolidation of concrete are given in detail in **"Guide for Consolidation of Concrete"** reported by ACI Committee 309.[5.12] (Presents current information on the mechanism of consolidation and gives recommendations on equipment characteristics and procedures for various classes of concrete.)

CODE

5.11 — Curing

5.11.1 — Concrete (other than high-early-strength) shall be maintained above 50 F and in a moist condition for at least the first 7 days after placement, except when cured in accordance with 5.11.3.

5.11.2 — High-early-strength concrete shall be maintained above 50 F and in a moist condition for at least the first 3 days, except when cured in accordance with 5.11.3.

5.11.3 — Accelerated curing

5.11.3.1 — Curing by high-pressure steam, steam at atmospheric pressure, heat and moisture, or other accepted processes, shall be permitted to accelerate strength gain and reduce time of curing.

5.11.3.2 — Accelerated curing shall provide a compressive strength of the concrete at the load stage considered at least equal to required design strength at that load stage.

5.11.3.3 — Curing process shall be such as to produce concrete with a durability at least equivalent to the curing method of 5.11.1 or 5.11.2.

5.11.4 — When required by the engineer or architect, supplementary strength tests in accordance with 5.6.4 shall be performed to assure that curing is satisfactory.

COMMENTARY

R5.11 — Curing

Recommendations for curing concrete are given in detail in **"Guide to Curing Concrete"** reported by ACI Committee 308.[5.13] (Presents basic principles of proper curing and describes the various methods, procedures, and materials for curing of concrete.)

R5.11.3 — Accelerated curing

The provisions of this section apply whenever an accelerated curing method is used, whether for precast or cast-in-place elements. The compressive strength of steam-cured concrete is not as high as that of similar concrete continuously cured under moist conditions at moderate temperatures. Also the modulus of elasticity E_c of steam-cured specimens may vary from that of specimens moist-cured at normal temperatures. When steam curing is used, it is advisable to base the concrete mixture proportions on steam-cured test cylinders.

Accelerated curing procedures require careful attention to obtain uniform and satisfactory results. Preventing moisture loss during the curing is essential.

R5.11.4 — In addition to requiring a minimum curing temperature and time for normal- and high-early-strength concrete, the code provides a specific criterion in 5.6.4 for judging the adequacy of field curing. At the test age for which the compressive strength is specified (usually 28 days), field-cured cylinders should produce strength not less than 85 percent of that of the standard, laboratory-cured cylinders. For a reasonably valid comparison to be made, field-cured cylinders and companion laboratory-cured cylinders should come from the same sample. Field-cured cylinders should be cured under conditions identical to those of the structure. If the structure is protected from the elements, the cylinder should be protected.

Cylinders related to members not directly exposed to weather should be cured adjacent to those members and provided with the same degree of protection and method of curing. The field cylinders should not be treated more favorably than the elements they represent. (See 5.6.4 for additional information.) If the field-cured cylinders do not provide satisfactory strength by this comparison, measures should be taken to improve the curing. If the tests indicate a possible serious deficiency in strength of concrete in the structure, core tests may be required, with or without supplemental wet curing, to check the structural adequacy, as provided in 5.6.5.

CODE

5.12 — Cold weather requirements

5.12.1 — Adequate equipment shall be provided for heating concrete materials and protecting concrete during freezing or near-freezing weather.

5.12.2 — All concrete materials and all reinforcement, forms, fillers, and ground with which concrete is to come in contact shall be free from frost.

5.12.3 — Frozen materials or materials containing ice shall not be used.

5.13 — Hot weather requirements

During hot weather, proper attention shall be given to ingredients, production methods, handling, placing, protection, and curing to prevent excessive concrete temperatures or water evaporation that could impair required strength or serviceability of the member or structure.

COMMENTARY

R5.12 — Cold weather requirements

Recommendations for cold weather concreting are given in detail in **"Cold Weather Concreting"** reported by ACI Committee 306.[5.14] (Presents requirements and methods for producing satisfactory concrete during cold weather.)

R5.13 — Hot weather requirements

Recommendations for hot weather concreting are given in detail in **"Hot Weather Concreting"** reported by ACI Committee 305.[5.15] (Defines the hot weather factors that effect concrete properties and construction practices and recommends measures to eliminate or minimize the undesirable effects.)

CHAPTER 6 — FORMWORK, EMBEDDED PIPES, AND CONSTRUCTION JOINTS

CODE

6.1 — Design of formwork

6.1.1 — Forms shall result in a final structure that conforms to shapes, lines, and dimensions of the members as required by the design drawings and specifications.

6.1.2 — Forms shall be substantial and sufficiently tight to prevent leakage of mortar.

6.1.3 — Forms shall be properly braced or tied together to maintain position and shape.

6.1.4 — Forms and their supports shall be designed so as not to damage previously placed structure.

6.1.5 — Design of formwork shall include consideration of the following factors:

(a) Rate and method of placing concrete;

(b) Construction loads, including vertical, horizontal, and impact loads;

(c) Special form requirements for construction of shells, folded plates, domes, architectural concrete, or similar types of elements.

6.1.6 — Forms for prestressed concrete members shall be designed and constructed to permit movement of the member without damage during application of prestressing force.

6.2 — Removal of forms, shores, and reshoring

6.2.1 — Removal of forms

Forms shall be removed in such a manner as not to impair safety and serviceability of the structure. Concrete exposed by form removal shall have sufficient strength not to be damaged by removal operation.

6.2.2 — Removal of shores and reshoring

The provisions of 6.2.2.1 through 6.2.2.3 shall apply to slabs and beams except where cast on the ground.

6.2.2.1 — Before starting construction, the contractor shall develop a procedure and schedule for removal of shores and installation of reshores and for calculating the loads transferred to the structure during the process.

COMMENTARY

R6.1 — Design of formwork

Only minimum performance requirements for formwork, necessary to provide for public health and safety, are prescribed in Chapter 6. Formwork for concrete, including proper design, construction, and removal, demands sound judgment and planning to achieve adequate forms that are both economical and safe. Detailed information on formwork for concrete is given in: **"Guide to Formwork for Concrete,"** reported by Committee 347.[6.1] (Provides recommendations for design, construction, and materials for formwork, forms for special structures, and formwork for special methods of construction. Directed primarily to contractors, the suggested criteria will aid engineers and architects in preparing job specifications for the contractors.)

Formwork for Concrete[6.2] prepared under the direction of ACI Committee 347. (A how-to-do-it handbook for contractors, engineers, and architects following the guidelines established in ACI 347R. Planning, building, and using formwork are discussed, including tables, diagrams, and formulas for form design loads.)

R6.2 — Removal of forms, shores, and reshoring

In determining the time for removal of forms, consideration should be given to the construction loads and to the possibilities of deflections.[6.3] The construction loads are frequently at least as great as the specified live loads. At early ages, a structure may be adequate to support the applied loads but may deflect sufficiently to cause permanent damage.

Evaluation of concrete strength during construction may be demonstrated by field-cured test cylinders or other procedures approved by the building official such as:

(a) Tests of cast-in-place cylinders in accordance with "Standard Test Method for Compressive Strength of Concrete Cylinders Cast-in-Place in Cylindrical Molds" (ASTM C 873[6.4]). (This method is limited to use in slabs where the depth of concrete is from 5 to 12 in.);

CODE

(a) The structural analysis and concrete strength data used in planning and implementing form removal and shoring shall be furnished by the contractor to the building official when so requested;

(b) No construction loads shall be supported on, nor any shoring removed from, any part of the structure under construction except when that portion of the structure in combination with remaining forming and shoring system has sufficient strength to support safely its weight and loads placed thereon;

(c) Sufficient strength shall be demonstrated by structural analysis considering proposed loads, strength of forming and shoring system, and concrete strength data. Concrete strength data shall be based on tests of field-cured cylinders or, when approved by the building official, on other procedures to evaluate concrete strength.

6.2.2.2 — No construction loads exceeding the combination of superimposed dead load plus specified live load shall be supported on any unshored portion of the structure under construction, unless analysis indicates adequate strength to support such additional loads.

6.2.2.3 — Form supports for prestressed concrete members shall not be removed until sufficient prestressing has been applied to enable prestressed members to carry their dead load and anticipated construction loads.

COMMENTARY

(b) Penetration resistance in accordance with "Standard Test Method for Penetration Resistance of Hardened Concrete" (ASTM C 803[6.5]);

(c) Pullout strength in accordance with "Standard Test Method for Pullout Strength of Hardened Concrete" (ASTM C 900[6.6]);

(d) Maturity factor measurements and correlation in accordance with ASTM C 1074.[6.7]

Procedures (b), (c), and (d) require sufficient data, using job materials, to demonstrate correlation of measurements on the structure with compressive strength of molded cylinders or drilled cores.

Where the structure is adequately supported on shores, the side forms of beams, girders, columns, walls, and similar vertical forms may generally be removed after 12 h of cumulative curing time, provided the side forms support no loads other than the lateral pressure of the plastic concrete. Cumulative curing time represents the sum of time intervals, not necessarily consecutive, during which the temperature of the air surrounding the concrete is above 50 F. The 12-h cumulative curing time is based on regular cements and ordinary conditions; the use of special cements or unusual conditions may require adjustment of the given limits. For example, concrete made with Type II or V (ASTM C 150) or ASTM C 595 cements, concrete containing retarding admixtures, and concrete to which ice was added during mixing (to lower the temperature of fresh concrete) may not have sufficient strength in 12 h and should be investigated before removal of formwork.

The removal of formwork for multistory construction should be a part of a planned procedure considering the temporary support of the whole structure as well as that of each individual member. Such a procedure should be worked out prior to construction and should be based on a structural analysis taking into account the following items, as a minimum:

(a) The structural system that exists at the various stages of construction and the construction loads corresponding to those stages;

(b) The strength of the concrete at the various ages during construction;

(c) The influence of deformations of the structure and shoring system on the distribution of dead loads and construction loads during the various stages of construction;

(d) The strength and spacing of shores or shoring systems used, as well as the method of shoring, bracing, shore removal, and reshoring including the minimum time

CODE / COMMENTARY

intervals between the various operations;

(e) Any other loading or condition that affects the safety or serviceability of the structure during construction.

For multistory construction, the strength of the concrete during the various stages of construction should be substantiated by field-cured test specimens or other approved methods.

6.3 — Conduits and pipes embedded in concrete

6.3.1 — Conduits, pipes, and sleeves of any material not harmful to concrete and within limitations of 6.3 shall be permitted to be embedded in concrete with approval of the engineer, provided they are not considered to replace structurally the displaced concrete, except as provided in 6.3.6.

R6.3 — Conduits and pipes embedded in concrete

R6.3.1 — Conduits, pipes, and sleeves not harmful to concrete can be embedded within the concrete, but the work should be done in such a manner that the structure will not be endangered. Empirical rules are given in 6.3 for safe installations under common conditions; for other than common conditions, special designs should be made. Many general building codes have adopted ANSI/ASME piping codes B 31.1 for power piping[6.8] and B 31.3 for chemical and petroleum piping.[6.9] The specifier should be sure that the appropriate piping codes are used in the design and testing of the system. The contractor should not be permitted to install conduits, pipes, ducts, or sleeves that are not shown on the plans or not approved by the engineer or architect.

For the integrity of the structure, it is important that all conduit and pipe fittings within the concrete be carefully assembled as shown on the plans or called for in the job specifications.

6.3.2 — Conduits and pipes of aluminum shall not be embedded in structural concrete unless effectively coated or covered to prevent aluminum-concrete reaction or electrolytic action between aluminum and steel.

R6.3.2 — The code prohibits the use of aluminum in structural concrete unless it is effectively coated or covered. Aluminum reacts with concrete and, in the presence of chloride ions, may also react electrolytically with steel, causing cracking and/or spalling of the concrete. Aluminum electrical conduits present a special problem since stray electric current accelerates the adverse reaction.

6.3.3 — Conduits, pipes, and sleeves passing through a slab, wall, or beam shall not impair significantly the strength of the construction.

6.3.4 — Conduits and pipes, with their fittings, embedded within a column shall not displace more than 4 percent of the area of cross section on which strength is calculated or which is required for fire protection.

6.3.5 — Except when drawings for conduits and pipes are approved by the structural engineer, conduits and pipes embedded within a slab, wall, or beam (other than those merely passing through) shall satisfy 6.3.5.1 through 6.3.5.3.

6.3.5.1 — They shall not be larger in outside dimension than 1/3 the overall thickness of slab, wall, or beam in which they are embedded.

CODE

COMMENTARY

6.3.5.2 — They shall not be spaced closer than 3 diameters or widths on center.

6.3.5.3 — They shall not impair significantly the strength of the construction.

6.3.6 — Conduits, pipes, and sleeves shall be permitted to be considered as replacing structurally in compression the displaced concrete provided in 6.3.6.1 through 6.3.6.3.

6.3.6.1 — They are not exposed to rusting or other deterioration.

6.3.6.2 — They are of uncoated or galvanized iron or steel not thinner than standard Schedule 40 steel pipe.

6.3.6.3 — They have a nominal inside diameter not over 2 in. and are spaced not less than 3 diameters on centers.

6.3.7 — Pipes and fittings shall be designed to resist effects of the material, pressure, and temperature to which they will be subjected.

R6.3.7 — The 1983 code limited the maximum pressure in embedded pipe to 200 psi, which was considered too restrictive. Nevertheless, the effects of such pressures and the expansion of embedded pipe should be considered in the design of the concrete member.

6.3.8 — No liquid, gas, or vapor, except water not exceeding 90 F nor 50 psi pressure, shall be placed in the pipes until the concrete has attained its design strength.

6.3.9 — In solid slabs, piping, unless it is for radiant heating or snow melting, shall be placed between top and bottom reinforcement.

6.3.10 — Concrete cover for pipes, conduits, and fittings shall not be less than 1-1/2 in. for concrete exposed to earth or weather, nor less than 3/4 in. for concrete not exposed to weather or in contact with ground.

6.3.11 — Reinforcement with an area not less than 0.002 times area of concrete section shall be provided normal to piping.

6.3.12 — Piping and conduit shall be so fabricated and installed that cutting, bending, or displacement of reinforcement from its proper location will not be required.

6.4 — Construction joints

R6.4 — Construction joints

6.4.1 — Surface of concrete construction joints shall be cleaned and laitance removed.

For the integrity of the structure, it is important that all construction joints be defined in construction documents and constructed as required. Any deviations should be approved by the engineer or architect.

CODE

6.4.2 — Immediately before new concrete is placed, all construction joints shall be wetted and standing water removed.

COMMENTARY

R6.4.2 — The requirements of the 1977 code for the use of neat cement on vertical joints have been removed, since it is rarely practical and can be detrimental where deep forms and steel congestion prevent proper access. Often wet blasting and other procedures are more appropriate. Because the code sets only minimum standards, the engineer may have to specify special procedures if conditions warrant. The degree to which mortar batches are needed at the start of concrete placement depend on concrete proportions, congestion of steel, vibrator access, and other factors.

6.4.3 — Construction joints shall be so made and located as not to impair the strength of the structure. Provision shall be made for transfer of shear and other forces through construction joints. See 11.7.9.

R6.4.3 — Construction joints should be located where they will cause the least weakness in the structure. When shear due to gravity load is not significant, as is usually the case in the middle of the span of flexural members, a simple vertical joint may be adequate. Lateral force design may require special design treatment of construction joints. Shear keys, intermittent shear keys, diagonal dowels, or the shear transfer method of 11.7 may be used whenever a force transfer is required.

6.4.4 — Construction joints in floors shall be located within the middle third of spans of slabs, beams, and girders.

6.4.5 — Construction joints in girders shall be offset a minimum distance of two times the width of intersecting beams.

6.4.6 — Beams, girders, or slabs supported by columns or walls shall not be cast or erected until concrete in the vertical support members is no longer plastic.

R6.4.6 — Delay in placing concrete in members supported by columns and walls is necessary to prevent cracking at the interface of the slab and supporting member caused by bleeding and settlement of plastic concrete in the supporting member.

6.4.7 — Beams, girders, haunches, drop panels, and capitals shall be placed monolithically as part of a slab system, unless otherwise shown in design drawings or specifications.

R6.4.7 — Separate placement of slabs and beams, haunches, and similar elements is permitted when shown on the drawings and where provision has been made to transfer forces as required in 6.4.3.

Notes

CHAPTER 7 — DETAILS OF REINFORCEMENT

CODE

7.1 — Standard hooks

The term standard hook as used in this code shall mean one of the following:

7.1.1 — 180-deg bend plus $4d_b$ extension, but not less than 2-1/2 in. at free end of bar.

7.1.2 — 90-deg bend plus $12d_b$ extension at free end of bar.

7.1.3 — For stirrup and tie hooks

(a) No. 5 bar and smaller, 90-deg bend plus $6d_b$ extension at free end of bar; or

(b) No. 6, No. 7, and No. 8 bar, 90-deg bend plus $12d_b$ extension at free end of bar; or

(c) No. 8 bar and smaller, 135-deg bend plus $6d_b$ extension at free end of bar.

7.1.4 — Seismic hooks as defined in 21.1.

7.2 — Minimum bend diameters

7.2.1 — Diameter of bend measured on the inside of the bar, other than for stirrups and ties in sizes No. 3 through No. 5, shall not be less than the values in Table 7.2.

7.2.2 — Inside diameter of bend for stirrups and ties shall not be less than $4d_b$ for No. 5 bar and smaller. For bars larger than No. 5, diameter of bend shall be in accordance with Table 7.2.

7.2.3 — Inside diameter of bend in welded wire reinforcement for stirrups and ties shall not be less than $4d_b$ for deformed wire larger than D6 and $2d_b$ for all other wires. Bends with inside diameter of less than $8d_b$ shall not be less than $4d_b$ from nearest welded intersection.

COMMENTARY

R7.1 — Standard hooks

Recommended methods and standards for preparing design drawings, typical details, and drawings for the fabrication and placing of reinforcing steel in reinforced concrete structures are given in the ***ACI Detailing Manual***, reported by ACI Committee 315.[7.1]

All provisions in the code relating to bar, wire, or strand diameter (and area) are based on the nominal dimensions of the reinforcement as given in the appropriate ASTM specification. Nominal dimensions are equivalent to those of a circular area having the same weight per foot as the ASTM designated bar, wire, or strand sizes. Cross-sectional area of reinforcement is based on nominal dimensions.

R7.1.3 — Standard stirrup and tie hooks are limited to No. 8 bars and smaller, and the 90-deg hook with $6d_b$ extension is further limited to No. 5 bars and smaller, in both cases as the result of research showing that larger bar sizes with 90-deg hooks and $6d_b$ extensions tend to pop out under high load.

R7.2 — Minimum bend diameters

Standard bends in reinforcing bars are described in terms of the inside diameter of bend since this is easier to measure than the radius of bend. The primary factors affecting the minimum bend diameter are feasibility of bending without breakage and avoidance of crushing the concrete inside the bend.

R7.2.2 — The minimum $4d_b$ bend for the bar sizes commonly used for stirrups and ties is based on accepted industry practice in the United States. Use of a stirrup bar size not greater than No. 5 for either the 90-deg or 135-deg standard stirrup hook will permit multiple bending on standard stirrup bending equipment.

R7.2.3 — Welded wire reinforcement can be used for stirrups and ties. The wire at welded intersections does not have the same uniform ductility and bendability as in areas which were not heated. These effects of the welding temperature are usually dissipated in a distance of approximately four wire diameters. Minimum bend diameters permitted

CODE

TABLE 7.2—MINIMUM DIAMETERS OF BEND

Bar size	Minimum diameter
No. 3 through No. 8	$6d_b$
No. 9, No. 10, and No. 11	$8d_b$
No. 14 and No. 18	$10d_b$

7.3 — Bending

7.3.1 — All reinforcement shall be bent cold, unless otherwise permitted by the engineer.

7.3.2 — Reinforcement partially embedded in concrete shall not be field bent, except as shown on the design drawings or permitted by the engineer.

7.4 — Surface conditions of reinforcement

7.4.1—At the time concrete is placed, reinforcement shall be free from mud, oil, or other nonmetallic coatings that decrease bond. Epoxy-coating of steel reinforcement in accordance with standards referenced in 3.5.3.7 and 3.5.3.8 shall be permitted.

7.4.2 — Except for prestressing steel, steel reinforcement with rust, mill scale, or a combination of both shall be considered satisfactory, provided the minimum dimensions (including height of deformations) and weight of a hand-wire-brushed test specimen comply with applicable ASTM specifications referenced in 3.5.

COMMENTARY

are in most cases the same as those required in the ASTM bend tests for wire material (ASTM A 82 and A 496).

R7.3 — Bending

R7.3.1 — The engineer may be the design engineer or architect or the engineer or architect employed by the owner to perform inspection. For unusual bends with inside diameters less than ASTM bend test requirements, special fabrication may be required.

R7.3.2 — Construction conditions may make it necessary to bend bars that have been embedded in concrete. Such field bending should not be done without authorization of the engineer. The engineer should determine whether the bars should be bent cold or if heating should be used. Bends should be gradual and should be straightened as required.

Tests[7.2,7.3] have shown that A 615 Grade 40 and Grade 60 reinforcing bars can be cold bent and straightened up to 90 deg at or near the minimum diameter specified in 7.2. If cracking or breakage is encountered, heating to a maximum temperature of 1500 F may avoid this condition for the remainder of the bars. Bars that fracture during bending or straightening can be spliced outside the bend region.

Heating should be performed in a manner that will avoid damage to the concrete. If the bend area is within approximately 6 in. of the concrete, some protective insulation may need to be applied. Heating of the bar should be controlled by temperature-indicating crayons or other suitable means. The heated bars should not be artificially cooled (with water or forced air) until after cooling to at least 600 F.

R7.4 — Surface conditions of reinforcement

Specific limits on rust are based on tests,[7.4] plus a review of earlier tests and recommendations. Reference 7.4 provides guidance with regard to the effects of rust and mill scale on bond characteristics of deformed reinforcing bars. Research has shown that a normal amount of rust increases bond. Normal rough handling generally removes rust that is loose enough to injure the bond between the concrete and reinforcement.

CODE

7.4.3 — Prestressing steel shall be clean and free of oil, dirt, scale, pitting and excessive rust. A light coating of rust shall be permitted.

7.5 — Placing reinforcement

7.5.1 — Reinforcement, including tendons, and post-tensioning ducts shall be accurately placed and adequately supported before concrete is placed, and shall be secured against displacement within tolerances permitted in 7.5.2.

7.5.2 — Unless otherwise specified by the registered design professional, reinforcement, including tendons, and post-tensioning ducts shall be placed within the tolerances in 7.5.2.1 and 7.5.2.2.

7.5.2.1 — Tolerance for ***d*** and minimum concrete cover in flexural members, walls, and compression members shall be as follows:

	Tolerance on ***d***	Tolerance on minimum concrete cover
$d \leq 8$ in.	±3/8 in.	–3/8 in.
$d > 8$ in.	±1/2 in.	–1/2 in.

except that tolerance for the clear distance to formed soffits shall be minus 1/4 in. and tolerance for cover shall not exceed minus 1/3 the minimum concrete cover required in the design drawings and specifications.

7.5.2.2 — Tolerance for longitudinal location of bends and ends of reinforcement shall be ±2 in., except the tolerance shall be ±1/2 in. at the discontinuous ends of brackets and corbels, and ±1 in. at the

COMMENTARY

R7.4.3 — Guidance for evaluating the degree of rusting on strand is given in Reference 7.5.

R7.5 — Placing reinforcement

R7.5.1 — Reinforcement, including tendons, and post-tensioning ducts should be adequately supported in the forms to prevent displacement by concrete placement or workers. Beam stirrups should be supported on the bottom form of the beam by positive supports such as continuous longitudinal beam bolsters. If only the longitudinal beam bottom reinforcement is supported, construction traffic can dislodge the stirrups as well as any prestressing tendons tied to the stirrups.

R7.5.2 — Generally accepted practice, as reflected in **"Standard Specifications for Tolerances for Concrete Construction and Materials,"** reported by ACI Committee 117.[7.6] has established tolerances on total depth (formwork or finish) and fabrication of truss bent reinforcing bars and closed ties, stirrups, and spirals. The engineer should specify more restrictive tolerances than those permitted by the code when necessary to minimize the accumulation of tolerances resulting in excessive reduction in effective depth or cover.

More restrictive tolerances have been placed on minimum clear distance to formed soffits because of its importance for durability and fire protection, and because bars are usually supported in such a manner that the specified tolerance is practical.

More restrictive tolerances than those required by the code may be desirable for prestressed concrete to achieve camber control within limits acceptable to the designer or owner. In such cases, the engineer should specify the necessary tolerances. Recommendations are given in Reference 7.7.

R7.5.2.1 — The code specifies a tolerance on depth ***d***, an essential component of strength of the member. Because reinforcing steel is placed with respect to edges of members and formwork surfaces, the depth ***d*** is not always conveniently measured in the field. Engineers should specify tolerances for bar placement, cover, and member size. See ACI 117.[7.6]

CODE | COMMENTARY

discontinuous ends of other members. The tolerance for minimum concrete cover of 7.5.2.1 shall also apply at discontinuous ends of members.

7.5.3 — Welded wire reinforcement (with wire size not greater than W5 or D5) used in slabs not exceeding 10 ft in span shall be permitted to be curved from a point near the top of slab over the support to a point near the bottom of slab at midspan, provided such reinforcement is either continuous over, or securely anchored at support.

7.5.4 — Welding of crossing bars shall not be permitted for assembly of reinforcement unless authorized by the engineer.

R7.5.4 — "Tack" welding (welding crossing bars) can seriously weaken a bar at the point welded by creating a metallurgical notch effect. This operation can be performed safely only when the material welded and welding operations are under continuous competent control, as in the manufacture of welded wire reinforcement.

7.6 — Spacing limits for reinforcement

7.6.1 — The minimum clear spacing between parallel bars in a layer shall be d_b, but not less than 1 in. See also 3.3.2.

7.6.2 — Where parallel reinforcement is placed in two or more layers, bars in the upper layers shall be placed directly above bars in the bottom layer with clear distance between layers not less than 1 in.

7.6.3 — In spirally reinforced or tied reinforced compression members, clear distance between longitudinal bars shall be not less than $1.5d_b$ nor less than 1-1/2 in. See also 3.3.2.

7.6.4 — Clear distance limitation between bars shall apply also to the clear distance between a contact lap splice and adjacent splices or bars.

7.6.5 — In walls and slabs other than concrete joist construction, primary flexural reinforcement shall not be spaced farther apart than three times the wall or slab thickness, nor farther apart than 18 in.

R7.6 — Spacing limits for reinforcement

Although the minimum bar spacings are unchanged in this code, the development lengths given in Chapter 12 became a function of the bar spacings since the 1989 code. As a result, it may be desirable to use larger than minimum bar spacings in some cases. The minimum limits were originally established to permit concrete to flow readily into spaces between bars and between bars and forms without honeycomb, and to ensure against concentration of bars on a line that may cause shear or shrinkage cracking. Use of nominal bar diameter to define minimum spacing permits a uniform criterion for all bar sizes.

7.6.6 — Bundled bars

7.6.6.1 — Groups of parallel reinforcing bars bundled in contact to act as a unit shall be limited to four in any one bundle.

7.6.6.2 — Bundled bars shall be enclosed within stirrups or ties.

7.6.6.3 — Bars larger than No. 11 shall not be bundled in beams.

7.6.6.4 — Individual bars within a bundle terminated within the span of flexural members shall terminate at different points with at least $40d_b$ stagger.

R7.6.6 — Bundled bars

Bond research[7.8] showed that bar cutoffs within bundles should be staggered. Bundled bars should be tied, wired, or otherwise fastened together to ensure remaining in position whether vertical or horizontal.

A limitation that bars larger than No. 11 not be bundled in beams or girders is a practical limit for application to building size members. (The "**Standard Specifications for Highway Bridges**"[7.9] permits two-bar bundles for No. 14 and No. 18 bars in bridge girders.) Conformance to the crack control requirements of 10.6 will effectively preclude bundling of bars larger than No. 11 as tensile reinforcement. The code phrasing "bundled in contact to act as a unit," is intended to

CODE

7.6.6.5 — Where spacing limitations and minimum concrete cover are based on bar diameter, d_b, a unit of bundled bars shall be treated as a single bar of a diameter derived from the equivalent total area.

7.6.7 — Tendons and ducts

7.6.7.1 — Center-to-center spacing of pretensioning tendons at each end of a member shall be not less than $4d_b$ for strands, or $5d_b$ for wire, except that if specified compressive strength of concrete at time of initial prestress, f'_{ci}, is 4000 psi or more, minimum center-to-center spacing of strands shall be 1-3/4 in. for strands of 1/2 in. nominal diameter or smaller and 2 in. for strands of 0.6 in. nominal diameter. See also 3.3.2. Closer vertical spacing and bundling of tendons shall be permitted in the middle portion of a span.

7.6.7.2 — Bundling of post-tensioning ducts shall be permitted if shown that concrete can be satisfactorily placed and if provision is made to prevent the prestressing steel, when tensioned, from breaking through the duct.

7.7 — Concrete protection for reinforcement

7.7.1 — Cast-in-place concrete (nonprestressed)

The following minimum concrete cover shall be provided for reinforcement, but shall not be less than required by 7.7.5 and 7.7.7:

	Minimum cover, in.
(a) Concrete cast against and permanently exposed to earth	3
(b) Concrete exposed to earth or weather:	
No. 6 through No. 18 bars	2
No. 5 bar, W31 or D31 wire, and smaller	1-1/2
(c) Concrete not exposed to weather or in contact with ground:	
Slabs, walls, joists:	
No. 14 and No. 18 bars	1-1/2
No. 11 bar and smaller	3/4

COMMENTARY

preclude bundling more than two bars in the same plane. Typical bundle shapes are triangular, square, or L-shaped patterns for three- or four-bar bundles. As a practical caution, bundles more than one bar deep in the plane of bending should not be hooked or bent as a unit. Where end hooks are required, it is preferable to stagger the individual bar hooks within a bundle.

R7.6.7 — Tendons and ducts

R7.6.7.1 — The allowed decreased spacing in this section for transfer strengths of 4000 psi or greater is based on Reference 7.10, 7.11.

R7.6.7.2 — When ducts for prestressing steel in a beam are arranged closely together vertically, provision should be made to prevent the prestressing steel from breaking through the duct when tensioned. Horizontal disposition of ducts should allow proper placement of concrete. A clear spacing of one and one-third times the size of the coarse aggregate, but not less than 1 in., has proven satisfactory. Where concentration of tendons or ducts tends to create a weakened plane in the concrete cover, reinforcement should be provided to control cracking.

R7.7 — Concrete protection for reinforcement

Concrete cover as protection of reinforcement against weather and other effects is measured from the concrete surface to the outermost surface of the steel to which the cover requirement applies. Where minimum cover is prescribed for a class of structural member, it is measured to the outer edge of stirrups, ties, or spirals if transverse reinforcement encloses main bars; to the outermost layer of bars if more than one layer is used without stirrups or ties; or to the metal end fitting or duct on post-tensioned prestressing steel.

The condition "concrete surfaces exposed to earth or weather" refers to direct exposure to moisture changes and not just to temperature changes. Slab or thin shell soffits are not usually considered directly exposed unless subject to alternate wetting and drying, including that due to condensation conditions or direct leakage from exposed top surface, run off, or similar effects.

Alternative methods of protecting the reinforcement from weather may be provided if they are equivalent to the additional concrete cover required by the code. When approved by the building official under the provisions of 1.4, reinforcement with alternative protection from the weather may

CODE

Beams, columns:
Primary reinforcement, ties, stirrups, spirals 1-1/2

Shells, folded plate members:
No. 6 bar and larger.............................. 3/4
No. 5 bar, W31 or D31 wire, and smaller 1/2

7.7.2 — Cast-in-place concrete (prestressed)

The following minimum concrete cover shall be provided for prestressed and nonprestressed reinforcement, ducts, and end fittings, but shall not be less than required by 7.7.5, 7.7.5.1, and 7.7.7:

Minimum cover, in.

(a) Concrete cast against and permanently exposed to earth 3

(b) Concrete exposed to earth or weather:

Wall panels, slabs, joists................................ 1
Other members.. 1-1/2

(c) Concrete not exposed to weather or in contact with ground:

Slabs, walls, joists 3/4
Beams, columns:
Primary reinforcement.......................... 1-1/2
Ties, stirrups, spirals 1

Shells, folded plate members:
No. 5 bar, W31 or D31 wire, and smaller 3/8
Other reinforcement .... d_b but not less than 3/4

7.7.3 — Precast concrete (manufactured under plant control conditions)

The following minimum concrete cover shall be provided for prestressed and nonprestressed reinforcement, ducts, and end fittings, but shall not be less than required by 7.7.5, 7.7.5.1, and 7.7.7:

Minimum cover, in.

(a) Concrete exposed to earth or weather:

Wall panels:
No. 14 and No. 18 bars, prestressing tendons larger than 1-1/2 in. diameter... 1-1/2
No. 11 bar and smaller, prestressing tendons 1-1/2 in. diameter and smaller, W31 and D31 wire and smaller............ 3/4

COMMENTARY

have concrete cover not less than the cover required for reinforcement not exposed to weather.

The development length given in Chapter 12 is now a function of the bar cover. As a result, it may be desirable to use larger than minimum cover in some cases.

R7.7.3 — Precast concrete (manufactured under plant control conditions)

The lesser cover thicknesses for precast construction reflect the greater convenience of control for proportioning, placing, and curing inherent in precasting. The term "manufactured under plant control conditions" does not specifically imply that precast members should be manufactured in a plant. Structural elements precast at the job site will also qualify under this section if the control of form dimensions, placing of reinforcement, quality control of concrete, and curing procedure are equal to that normally expected in a plant.

Concrete cover to pretensioned strand as described in this section is intended to provide minimum protection against weather and other effects. Such cover may not be sufficient to transfer or develop the stress in the strand, and it may be necessary to increase the cover accordingly.

CODE

Other members:
No. 14 and No. 18 bars, prestressing tendons larger than 1-1/2 in. diameter 2
No. 6 through No. 11 bars, prestressing tendons larger than 5/8 in. diameter through 1-1/2 in. diameter 1-1/2
No. 5 bar and smaller, prestressing tendons 5/8 in. diameter and smaller, W31 and D31 wire, and smaller 1-1/4

(b) Concrete not exposed to weather or in contact with ground:

Slabs, walls, joists:
No. 14 and No. 18 bars, prestressing tendons larger than 1-1/2 in. diameter 1-1/4
Prestressing tendons 1-1/2 in. diameter and smaller............................. 3/4
No. 11 bar and smaller, W31 or D31 wire, and smaller.............. 5/8

Beams, columns:
Primary reinforcement ***........*** d_b but not less than 5/8 and need not exceed 1-1/2
Ties, stirrups, spirals 3/8

Shells, folded plate members:
Prestressing tendons 3/4
No. 6 bar and larger 5/8
No. 5 bar and smaller, W31 or D31 wire, and smaller............... 3/8

7.7.4 — Bundled bars

For bundled bars, minimum concrete cover shall be equal to the equivalent diameter of the bundle, but need not be greater than 2 in.; except for concrete cast against and permanently exposed to earth, where minimum cover shall be 3 in.

7.7.5— Corrosive environments

In corrosive environments or other severe exposure conditions, amount of concrete protection shall be suitably increased, and denseness and nonporosity of protecting concrete shall be considered, or other protection shall be provided.

COMMENTARY

R7.7.5 — Corrosive environments

Where concrete will be exposed to external sources of chlorides in service, such as deicing salts, brackish water, seawater, or spray from these sources, concrete should be proportioned to satisfy the special exposure requirements of Chapter 4. These include minimum air content, maximum water-cementitious materials ratio, minimum strength for normal weight and lightweight concrete, maximum chloride ion in concrete, and cement type. Additionally, for corrosion protection, a minimum concrete cover for reinforcement of 2 in. for walls and slabs and 2-1/2 in. for other members is recommended. For precast concrete manufactured under plant control conditions, a minimum cover of 1-1/2 and 2 in., respectively, is recommended.

CODE

7.7.5.1 — For prestressed concrete members exposed to corrosive environments or other severe exposure conditions, and which are classified as Class T or C in 18.3.3, minimum cover to the prestressed reinforcement shall be increased 50 percent. This requirement shall be permitted to be waived if the precompressed tensile zone is not in tension under sustained loads.

7.7.6 — Future extensions

Exposed reinforcement, inserts, and plates intended for bonding with future extensions shall be protected from corrosion.

7.7.7 — Fire protection

When the general building code (of which this code forms a part) requires a thickness of cover for fire protection greater than the minimum concrete cover specified in 7.7, such greater thicknesses shall be used.

7.8 — Special reinforcement details for columns

7.8.1 — Offset bars

Offset bent longitudinal bars shall conform to the following:

7.8.1.1 — Slope of inclined portion of an offset bar with axis of column shall not exceed 1 in 6.

7.8.1.2 — Portions of bar above and below an offset shall be parallel to axis of column.

7.8.1.3 — Horizontal support at offset bends shall be provided by lateral ties, spirals, or parts of the floor construction. Horizontal support provided shall be designed to resist 1-1/2 times the horizontal component of the computed force in the inclined portion of an offset bar. Lateral ties or spirals, if used, shall be placed not more than 6 in. from points of bend.

7.8.1.4 — Offset bars shall be bent before placement in the forms. See 7.3.

7.8.1.5 — Where a column face is offset 3 in. or greater, longitudinal bars shall not be offset bent. Separate dowels, lap spliced with the longitudinal bars adjacent to the offset column faces, shall be provided. Lap splices shall conform to 12.17.

COMMENTARY

R7.7.5.1 — Corrosive environments are defined in sections 4.4.2 and R4.4.2. Additional information on corrosion in parking structures is given in ACI 362.1R-97,[7.12] "Design of Parking Structures," pp. 21-26.

R7.8 — Special reinforcement details for columns

CODE

7.8.2 — Steel cores

Load transfer in structural steel cores of composite compression members shall be provided by the following:

7.8.2.1 — Ends of structural steel cores shall be accurately finished to bear at end bearing splices, with positive provision for alignment of one core above the other in concentric contact.

7.8.2.2 — At end bearing splices, bearing shall be considered effective to transfer not more than 50 percent of the total compressive stress in the steel core.

7.8.2.3 — Transfer of stress between column base and footing shall be designed in accordance with 15.8.

7.8.2.4 — Base of structural steel section shall be designed to transfer the total load from the entire composite member to the footing; or, the base shall be designed to transfer the load from the steel core only, provided ample concrete section is available for transfer of the portion of the total load carried by the reinforced concrete section to the footing by compression in the concrete and by reinforcement.

COMMENTARY

R7.8.2 — Steel cores

The 50 percent limit on transfer of compressive load by end bearing on ends of structural steel cores is intended to provide some tensile capacity at such splices (up to 50 percent), since the remainder of the total compressive stress in the steel core are to be transmitted by dowels, splice plates, welds, etc. This provision should ensure that splices in composite compression members meet essentially the same tensile capacity as required for conventionally reinforced concrete compression members.

7.9 — Connections

7.9.1 — At connections of principal framing elements (such as beams and columns), enclosure shall be provided for splices of continuing reinforcement and for anchorage of reinforcement terminating in such connections.

7.9.2 — Enclosure at connections shall consist of external concrete or internal closed ties, spirals, or stirrups.

R7.9 — Connections

Confinement is essential at connections to ensure that the flexural capacity of the members can be developed without deterioration of the joint under repeated loadings.[7.13,7.14]

7.10 — Lateral reinforcement for compression members

7.10.1 — Lateral reinforcement for compression members shall conform to the provisions of 7.10.4 and 7.10.5 and, where shear or torsion reinforcement is required, shall also conform to provisions of Chapter 11.

7.10.2 — Lateral reinforcement requirements for composite compression members shall conform to 10.16. Lateral reinforcement requirements for tendons shall conform to 18.11.

7.10.3 — It shall be permitted to waive the lateral reinforcement requirements of 7.10, 10.16, and 18.11 where tests and structural analysis show adequate strength and feasibility of construction.

R7.10 — Lateral reinforcement for compression members

R7.10.3 — Precast columns with cover less than 1-1/2 in., prestressed columns without longitudinal bars, columns smaller than minimum dimensions prescribed in earlier code editions, columns of concrete with small size coarse aggregate, wall-like columns, and other special cases may require special designs for lateral reinforcement. Wire, W4, D4, or larger, may be used for ties or spirals. If such special columns are considered as spiral columns for load strength in design, the volumetric reinforcement ratio for the spiral, ρ_s, is to conform to 10.9.3.

7.10.4 — Spirals

Spiral reinforcement for compression members shall conform to 10.9.3 and to the following:

7.10.4.1 — Spirals shall consist of evenly spaced continuous bar or wire of such size and so assembled to permit handling and placing without distortion from designed dimensions.

7.10.4.2 — For cast-in-place construction, size of spirals shall not be less than 3/8 in. diameter.

7.10.4.3 — Clear spacing between spirals shall not exceed 3 in., nor be less than 1 in. See also 3.3.2.

7.10.4.4 — Anchorage of spiral reinforcement shall be provided by 1-1/2 extra turns of spiral bar or wire at each end of a spiral unit.

7.10.4.5 — Spiral reinforcement shall be spliced, if needed, by any one of the following methods:

(a) Lap splices not less than the larger of 12 in. and the length indicated in one of (1) through (5) below:

(1) deformed uncoated bar or wire.......... $48d_b$
(2) plain uncoated bar or wire .... $72d_b$
(3) epoxy-coated deformed bar or wire... $72d_b$
(4) plain uncoated bar or wire with a standard stirrup or tie hook in accordance with 7.1.3 at ends of lapped spiral reinforcement. The hooks shall be embedded within the core confined by the spiral reinforcement .... $48d_b$
(5) epoxy-coated deformed bar or wire with a standard stirrup or tie hook in accordance with 7.1.3 at ends of lapped spiral reinforcement. The hooks shall be embedded within the core confined by the spiral reinforcement.................................... $48d_b$

(b) Full mechanical or welded splices in accordance with 12.14.3.

7.10.4.6 — Spirals shall extend from top of footing or slab in any story to level of lowest horizontal reinforcement in members supported above.

R7.10.4 — Spirals

For practical considerations in cast-in-place construction, the minimum diameter of spiral reinforcement is 3/8 in. (3/8 in. round, No. 3 bar, or equivalent deformed or plain wire). This is the smallest size that can be used in a column with 1-1/2 in. or more cover and having concrete compressive strengths of 3000 psi or more if the minimum clear spacing for placing concrete is to be maintained.

Standard spiral sizes are 3/8, 1/2, and 5/8 in. diameter for hot rolled or cold drawn material, plain or deformed.

The code allows spirals to be terminated at the level of lowest horizontal reinforcement framing into the column. However, if one or more sides of the column are not enclosed by beams or brackets, ties are required from the termination of the spiral to the bottom of the slab or drop panel. If beams or brackets enclose all sides of the column but are of different depths, the ties should extend from the spiral to the level of the horizontal reinforcement of the shallowest beam or bracket framing into the column. These additional ties are to enclose the longitudinal column reinforcement and the portion of bars from beams bent into the column for anchorage. See also 7.9.

Spirals should be held firmly in place, at proper pitch and alignment, to prevent displacement during concrete placement. The code has traditionally required spacers to hold the fabricated spiral cage in place but was changed in 1989 to allow alternate methods of installation. When spacers are used, the following may be used for guidance: For spiral bar or wire smaller than 5/8 in. diameter, a minimum of two spacers should be used for spirals less than 20 in. in diameter, three spacers for spirals 20 to 30 in. in diameter, and four spacers for spirals greater than 30 in. in diameter. For spiral bar or wire 5/8 in. diameter or larger, a minimum of three spacers should be used for spirals 24 in. or less in diameter, and four spacers for spirals greater than 24 in. in diameter. The project specifications or subcontract agreements should be clearly written to cover the supply of spacers or field tying of the spiral reinforcement. In the 1999 code, splice requirements were modified for epoxy-coated and plain spirals and to allow mechanical splices.

CODE

7.10.4.7 — Where beams or brackets do not frame into all sides of a column, ties shall extend above termination of spiral to bottom of slab or drop panel.

7.10.4.8 — In columns with capitals, spirals shall extend to a level at which the diameter or width of capital is two times that of the column.

7.10.4.9 — Spirals shall be held firmly in place and true to line.

7.10.5 — Ties

Tie reinforcement for compression members shall conform to the following:

7.10.5.1 — All nonprestressed bars shall be enclosed by lateral ties, at least No. 3 in size for longitudinal bars No. 10 or smaller, and at least No. 4 in size for No. 11, No. 14, No. 18, and bundled longitudinal bars. Deformed wire or welded wire reinforcement of equivalent area shall be permitted.

7.10.5.2 — Vertical spacing of ties shall not exceed 16 longitudinal bar diameters, 48 tie bar or wire diameters, or least dimension of the compression member.

7.10.5.3 — Ties shall be arranged such that every corner and alternate longitudinal bar shall have lateral support provided by the corner of a tie with an included angle of not more than 135 deg and no bar shall be farther than 6 in. clear on each side along the tie from such a laterally supported bar. Where longitudinal bars are located around the perimeter of a circle, a complete circular tie shall be permitted.

7.10.5.4 — Ties shall be located vertically not more than one-half a tie spacing above the top of footing or slab in any story, and shall be spaced as provided herein to not more than one-half a tie spacing below the lowest horizontal reinforcement in slab or drop panel above.

COMMENTARY

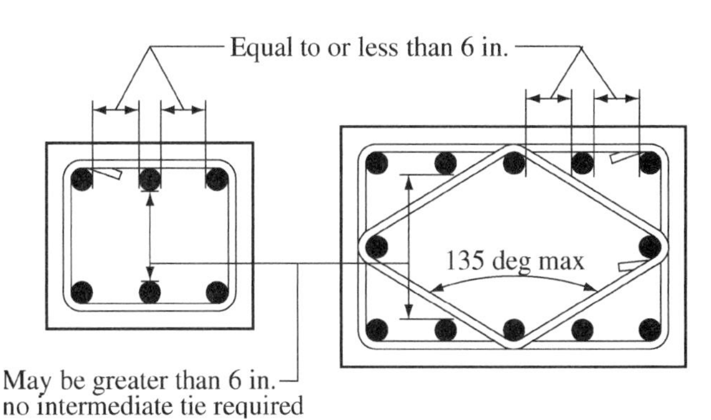

Fig. R7.10.5—Sketch to clarify measurements between laterally supported column bars

R7.10.5 — Ties

All longitudinal bars in compression should be enclosed within lateral ties. Where longitudinal bars are arranged in a circular pattern, only one circular tie per specified spacing is required. This requirement can be satisfied by a continuous circular tie (helix) at larger pitch than required for spirals under 10.9.3, the maximum pitch being equal to the required tie spacing (see also 7.10.4.3).

The 1956 code required "lateral support equivalent to that provided by a 90-deg corner of a tie," for every vertical bar. Tie requirements were liberalized in 1963 by increasing the permissible included angle from 90 to 135 deg and exempting bars that are located within 6 in. clear on each side along the tie from adequately tied bars (see Fig. R7.10.5). Limited tests[7.15] on full-size, axially-loaded, tied columns containing full-length bars (without splices) showed no appreciable difference between ultimate strengths of columns with full tie requirements and no ties at all.

Since spliced bars and bundled bars were not included in the tests of Reference 7.15, is prudent to provide a set of ties at each end of lap spliced bars, above and below end-bearing splices, and at minimum spacings immediately below sloping regions of offset bent bars.

Standard tie hooks are intended for use with deformed bars only, and should be staggered where possible. See also 7.9.

Continuously wound bars or wires can be used as ties provided their pitch and area are at least equivalent to the area and spacing of separate ties. Anchorage at the end of a continuously wound bar or wire should be by a standard hook as for separate bars or by one additional turn of the tie pattern. A circular continuously wound bar or wire is considered a spiral if it conforms to 7.10.4, otherwise it is considered a tie.

CODE

7.10.5.5 — Where beams or brackets frame from four directions into a column, termination of ties not more than 3 in. below lowest reinforcement in shallowest of such beams or brackets shall be permitted.

7.10.5.6 — Where anchor bolts are placed in the top of columns or pedestals, the bolts shall be enclosed by lateral reinforcement that also surrounds at least four vertical bars of the column or pedestal. The lateral reinforcement shall be distributed within 5 in. of the top of the column or pedestal, and shall consist of at least two No. 4 or three No. 3 bars.

7.11 — Lateral reinforcement for flexural members

7.11.1 — Compression reinforcement in beams shall be enclosed by ties or stirrups satisfying the size and spacing limitations in 7.10.5 or by welded wire reinforcement of equivalent area. Such ties or stirrups shall be provided throughout the distance where compression reinforcement is required.

7.11.2 — Lateral reinforcement for flexural framing members subject to stress reversals or to torsion at supports shall consist of closed ties, closed stirrups, or spirals extending around the flexural reinforcement.

7.11.3 — Closed ties or stirrups shall be formed in one piece by overlapping standard stirrup or tie end hooks around a longitudinal bar, or formed in one or two pieces lap spliced with a Class B splice (lap of $\mathbf{1.3}\ell_d$) or anchored in accordance with 12.13.

7.12 — Shrinkage and temperature reinforcement

7.12.1 — Reinforcement for shrinkage and temperature stresses normal to flexural reinforcement shall be provided in structural slabs where the flexural reinforcement extends in one direction only.

7.12.1.1 — Shrinkage and temperature reinforcement shall be provided in accordance with either 7.12.2 or 7.12.3.

7.12.1.2 — Where shrinkage and temperature movements are significantly restrained, the requirements of 8.2.4 and 9.2.3 shall be considered.

COMMENTARY

R7.10.5.5 — With the 1983 code, the wording of this section was modified to clarify that ties may be terminated only when elements frame into all four sides of square and rectangular columns; for round or polygonal columns, such elements frame into the column from four directions.

R7.10.5.6 — Provisions for confinement of anchor bolts that are placed in the top of columns or pedestals were added in the 2002 code. Confinement improves load transfer from the anchor bolts to the column or pier for situations where the concrete cracks in the vicinity of the bolts. Such cracking can occur due to unanticipated forces caused by temperature, restrained shrinkage, and similar effects.

R7.11 — Lateral reinforcement for flexural members

R7.11.1 — Compression reinforcement in beams and girders should be enclosed to prevent buckling; similar requirements for such enclosure have remained essentially unchanged through several editions of the code, except for minor clarification.

R7.12 — Shrinkage and temperature reinforcement

R7.12.1 — Shrinkage and temperature reinforcement is required at right angles to the principal reinforcement to minimize cracking and to tie the structure together to ensure its acting as assumed in the design. The provisions of this section are intended for structural slabs only; they are not intended for soil-supported slabs on grade.

R7.12.1.2 — The area of shrinkage and temperature reinforcement required by 7.12 has been satisfactory where shrinkage and temperature movements are permitted to occur. For cases where structural walls or large columns provide significant restraint to shrinkage and temperature movements, it may be necessary to increase the amount of reinforcement normal to the flexural reinforcement in 7.12.1.2 (see Reference 7.16). Top and bottom reinforcement are both effective in controlling cracks. Control strips during

CODE

COMMENTARY

the construction period, which permit initial shrinkage to occur without causing an increase in stresses, are also effective in reducing cracks caused by restraint.

7.12.2 — Deformed reinforcement conforming to 3.5.3 used for shrinkage and temperature reinforcement shall be provided in accordance with the following:

7.12.2.1 — Area of shrinkage and temperature reinforcement shall provide at least the following ratios of reinforcement area to gross concrete area, but not less than 0.0014:

(a) Slabs where Grade 40 or 50 deformed bars are used 0.0020

(b) Slabs where Grade 60 deformed bars or welded wire reinforcement are used................................. 0.0018

(c) Slabs where reinforcement with yield stress exceeding 60,000 psi measured at a yield strain of 0.35 percent is used .... $\frac{0.0018 \times 60{,}000}{f_y}$

7.12.2.2 — Shrinkage and temperature reinforcement shall be spaced not farther apart than five times the slab thickness, nor farther apart than 18 in.

7.12.2.3 — At all sections where required, reinforcement to resist shrinkage and temperature stresses shall develop f_y in tension in accordance with Chapter 12.

R7.12.2 — The amounts specified for deformed bars and welded wire reinforcement are empirical but have been used satisfactorily for many years. Splices and end anchorages of shrinkage and temperature reinforcement are to be designed for the full specified yield strength in accordance with 12.1, 12.15, 12.18, and 12.19.

7.12.3 — Prestressing steel conforming to 3.5.5 used for shrinkage and temperature reinforcement shall be provided in accordance with the following:

7.12.3.1 — Tendons shall be proportioned to provide a minimum average compressive stress of 100 psi on gross concrete area using effective prestress, after losses, in accordance with 18.6.

7.12.3.2 — Spacing of tendons shall not exceed 6 ft.

7.12.3.3 — When spacing of tendons exceeds 54 in., additional bonded shrinkage and temperature reinforcement conforming to 7.12.2 shall be provided between the tendons at slab edges extending from the slab edge for a distance equal to the tendon spacing.

R7.12.3 — Prestressed reinforcement requirements have been selected to provide an effective force on the slab approximately equal to the yield strength force for nonprestressed shrinkage and temperature reinforcement. This amount of prestressing, 100 psi on the gross concrete area, has been successfully used on a large number of projects. When the spacing of tendons used for shrinkage and temperature reinforcement exceeds 54 in., additional bonded reinforcement is required at slab edges where the prestressing forces are applied in order to adequately reinforce the area between the slab edge and the point where compressive stresses behind individual anchorages have spread sufficiently such that the slab is uniformly in compression. Application of the provisions of 7.12.3 to monolithic cast-in-place post-tensioned beam and slab construction is illustrated in Fig. R7.12.3.

Tendons used for shrinkage and temperature reinforcement should be positioned vertically in the slab as close as practicable to the center of the slab. In cases where the shrinkage and temperature tendons are used for supporting the principal tendons, variations from the slab centroid are permissible; however, the resultant of the shrinkage and temperature tendons should not fall outside the kern area of the slab.

CODE | COMMENTARY

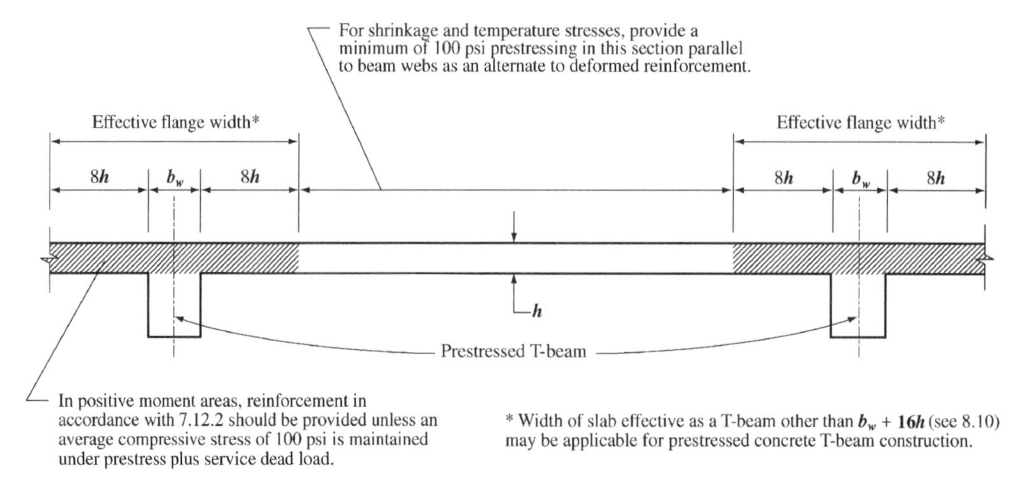

Fig. R7.12.3—Prestressing used for shrinkage and temperature.

The designer should evaluate the effects of slab shortening to ensure proper action. In most cases, the low level of prestressing recommended should not cause difficulties in a properly detailed structure. Special attention may be required where thermal effects become significant.

7.13 — Requirements for structural integrity

7.13.1 — In the detailing of reinforcement and connections, members of a structure shall be effectively tied together to improve integrity of the overall structure.

7.13.2 — For cast-in-place construction, the following shall constitute minimum requirements:

7.13.2.1 — In joist construction, at least one bottom bar shall be continuous or shall be spliced with a Class A tension splice or a mechanical or welded splice satisfying 12.14.3 and at noncontinuous supports shall be terminated with a standard hook.

7.13.2.2 — Beams along the perimeter of the structure shall have continuous reinforcement consisting of:

(a) at least one-sixth of the tension reinforcement required for negative moment at the support, but not less than two bars; and

(b) at least one-quarter of the tension reinforcement required for positive moment at midspan, but not less than two bars.

7.13.2.3 — Where splices are needed to provide the required continuity, the top reinforcement shall be spliced at or near midspan and bottom reinforcement

R7.13 — Requirements for structural integrity

Experience has shown that the overall integrity of a structure can be substantially enhanced by minor changes in detailing of reinforcement. It is the intent of this section of the code to improve the redundancy and ductility in structures so that in the event of damage to a major supporting element or an abnormal loading event, the resulting damage may be confined to a relatively small area and the structure will have a better chance to maintain overall stability.

R7.13.2 — With damage to a support, top reinforcement that is continuous over the support, but not confined by stirrups, will tend to tear out of the concrete and will not provide the catenary action needed to bridge the damaged support. By making a portion of the bottom reinforcement continuous, catenary action can be provided.

Requiring continuous top and bottom reinforcement in perimeter or spandrel beams provides a continuous tie around the structure. It is not the intent to require a tensile tie of continuous reinforcement of constant size around the entire perimeter of a structure, but simply to require that one half of the top flexural reinforcement required to extend past the point of inflection by 12.12.3 be further extended and spliced at or near midspan. Similarly, the bottom reinforcement required to extend into the support by 12.11.1 should be made continuous or spliced with bottom reinforcement from the adjacent span. If the depth of a continuous beam changes at a support, the bottom reinforcement in the deeper member should be terminated with a standard hook and bottom reinforcement in the shallower member should be extended into and fully developed in the deeper member.

CODE

shall be spliced at or near the support. Splices shall be Class A tension splices or mechanical or welded splices satisfying 12.14.3. The continuous reinforcement required in 7.13.2.2(a) and 7.13.2.2(b) shall be enclosed by the corners of U-stirrups having not less than 135-deg hooks around the continuous top bars, or by one-piece closed stirrups with not less than 135-deg hooks around one of the continuous top bars. Stirrups need not be extended through any joints.

7.13.2.4 — In other than perimeter beams, when stirrups as defined in 7.13.2.3 are not provided, at least one-quarter of the positive moment reinforcement required at midspan, but not less than two bars, shall be continuous or shall be spliced over or near the support with a Class A tension splice or a mechanical or welded splice satisfying 12.14.3, and at noncontinuous supports shall be terminated with a standard hook.

7.13.2.5 — For two-way slab construction, see 13.3.8.5.

7.13.3 — For precast concrete construction, tension ties shall be provided in the transverse, longitudinal, and vertical directions and around the perimeter of the structure to effectively tie elements together. The provisions of 16.5 shall apply.

7.13.4 — For lift-slab construction, see 13.3.8.6 and 18.12.6.

COMMENTARY

In the 2002 code, provisions were added to permit the use of mechanical or welded splices for splicing reinforcement, and the detailing requirements for the longitudinal reinforcement and stirrups in beams were revised. Section 7.13.2 was revised in 2002 to require U-stirrups with not less than 135-deg hooks around the continuous bars, or one-piece closed stirrups, because a crosstie forming the top of a two-piece closed stirrup is ineffective in preventing the top continuous bars from tearing out of the top of the beam.

R7.13.3 — The code requires tension ties for precast concrete buildings of all heights. Details should provide connections to resist applied loads. Connection details that rely solely on friction caused by gravity forces are not permitted.

Connection details should be arranged so as to minimize the potential for cracking due to restrained creep, shrinkage and temperature movements. For information on connections and detailing requirements, see Reference 7.17.

Reference 7.18 recommends minimum tie requirements for precast concrete bearing wall buildings.

Notes

CHAPTER 8 — ANALYSIS AND DESIGN — GENERAL CONSIDERATIONS

CODE

8.1 — Design methods

8.1.1 — In design of structural concrete, members shall be proportioned for adequate strength in accordance with provisions of this code, using load factors and strength reduction factors ϕ specified in Chapter 9.

8.1.2 — Design of reinforced concrete using the provisions of Appendix B, Alternative Provisions for Reinforced and Prestressed Concrete Flexural and Compression Members, shall be permitted.

8.1.3 — Anchors within the scope of Appendix D, Anchoring to Concrete, installed in concrete to transfer loads between connected elements shall be designed using Appendix D.

8.2 — Loading

8.2.1 — Design provisions of this code are based on the assumption that structures shall be designed to resist all applicable loads.

8.2.2 — Service loads shall be in accordance with the general building code of which this code forms a part, with such live load reductions as are permitted in the general building code.

COMMENTARY

R8.1 — Design methods

R8.1.1 — The strength design method requires service loads or related internal moments and forces to be increased by specified load factors (required strength) and computed nominal strengths to be reduced by specified strength reduction factors ϕ (design strength).

R8.1.2 — Designs in accordance with Appendix B are equally acceptable, provided the provisions of Appendix B are used in their entirety.

An appendix may be judged not to be an official part of a legal document unless specifically adopted. Therefore, specific reference is made to Appendix B in the main body of the code, to make it a legal part of the code.

R8.1.3 — The code has included specific provisions for anchoring to concrete for the first time in the 2002 edition. As has been done in the past with a number of new sections and chapters, the new material has been presented as an appendix.

An appendix may be judged not to be an official part of a legal document unless specifically adopted. Therefore, specific reference is made to Appendix D in the main part of the code to make it a legal part of the code.

R8.2 — Loading

The provisions in the code are for live, wind, and earthquake loads such as those recommended in **"Minimum Design Loads for Buildings and Other Structures,"** (SEI/ASCE 7),[1.19] formerly known as ANSI A58.1. If the service loads specified by the general building code (of which ACI 318 forms a part) differ from those of ASCE 7, the general building code governs. However, if the nature of the loads contained in a general building code differs considerably from ASCE 7 loads, some provisions of this code may need modification to reflect the difference.

Roofs should be designed with sufficient slope or camber to ensure adequate drainage accounting for any long-term deflection of the roof due to the dead loads, or the loads should be increased to account for all likely accumulations of water. If deflection of roof members may result in ponding of water accompanied by increased deflection and additional ponding, the design should ensure that this process is self-limiting.

CODE

8.2.3 — In design for wind and earthquake loads, integral structural parts shall be designed to resist the total lateral loads.

8.2.4 — Consideration shall be given to effects of forces due to prestressing, crane loads, vibration, impact, shrinkage, temperature changes, creep, expansion of shrinkage-compensating concrete, and unequal settlement of supports.

8.3 — Methods of analysis

8.3.1 — All members of frames or continuous construction shall be designed for the maximum effects of factored loads as determined by the theory of elastic analysis, except as modified according to 8.4. It shall be permitted to simplify design by using the assumptions specified in 8.6 through 8.9.

8.3.2 — Except for prestressed concrete, approximate methods of frame analysis shall be permitted for buildings of usual types of construction, spans, and story heights.

8.3.3 — As an alternate to frame analysis, the following approximate moments and shears shall be permitted for design of continuous beams and one-way slabs (slabs reinforced to resist flexural stresses in only one direction), provided:

(a) There are two or more spans;

(b) Spans are approximately equal, with the larger of two adjacent spans not greater than the shorter by more than 20 percent;

(c) Loads are uniformly distributed;

(d) Unfactored live load, L, does not exceed three times unfactored dead load, D; and

(e) Members are prismatic.

For calculating negative moments, ℓ_n is taken as the average of the adjacent clear span lengths.

COMMENTARY

R8.2.3 — Any reinforced concrete wall that is monolithic with other structural elements is considered to be an "integral part." Partition walls may or may not be integral structural parts. If partition walls may be removed, the primary lateral load resisting system should provide all of the required resistance without contribution of the removable partition. However, the effects of all partition walls attached to the structure should be considered in the analysis of the structure because they may lead to increased design forces in some or all elements. Special provisions for seismic design are given in Chapter 21.

R8.2.4 — Information is accumulating on the magnitudes of these various effects, especially the effects of column creep and shrinkage in tall structures,[8.1] and on procedures for including the forces resulting from these effects in design.

R8.3 — Methods of analysis

R8.3.1 — Factored loads are service loads multiplied by appropriate load factors. For the strength design method, elastic analysis is used to obtain moments, shears, and reactions.

R8.3.3 — The approximate moments and shears give reasonably conservative values for the stated conditions if the flexural members are part of a frame or continuous construction. Because the load patterns that produce critical values for moments in columns of frames differ from those for maximum negative moments in beams, column moments should be evaluated separately.

CODE

Positive moment	
End spans	
Discontinuous end unrestrained	$w_u \ell_n^2/11$
Discontinuous end integral with support	$w_u \ell_n^2/14$
Interior spans	$w_u \ell_n^2/16$
Negative moments at exterior face of first interior support	
Two spans	$w_u \ell_n^2/9$
More than two spans	$w_u \ell_n^2/10$
Negative moment at other faces of interior supports	$w_u \ell_n^2/11$
Negative moment at face of all supports for	
Slabs with spans not exceeding 10 ft; and beams where ratio of sum of column stiffnesses to beam stiffness exceeds eight at each end of the span	$w_u \ell_n^2/12$
Negative moment at interior face of exterior support for members built integrally with supports	
Where support is spandrel beam	$w_u \ell_n^2/24$
Where support is a column	$w_u \ell_n^2/16$
Shear in end members at face of first interior support	$1.15\ w_u \ell_n/2$
Shear at face of all other supports	$w_u \ell_n/2$

8.3.4 — Strut-and-tie models shall be permitted to be used in the design of structural concrete. See Appendix A.

COMMENTARY

R8.3.4 — The strut-and-tie model in Appendix A is based on the assumption that portions of concrete structures can be analyzed and designed using hypothetical pin-jointed trusses consisting of struts and ties connected at nodes. This design method can be used in the design of regions where the basic assumptions of flexure theory are not applicable, such as regions near force discontinuities arising from concentrated forces or reactions, and regions near geometric discontinuities, such as abrupt changes in cross section.

CODE

8.4 — Redistribution of negative moments in continuous flexural members

8.4.1 — Except where approximate values for moments are used, it shall be permitted to increase or decrease negative moments calculated by elastic theory at supports of continuous flexural members for any assumed loading arrangement by not more than $1000\varepsilon_t$ percent, with a maximum of 20 percent.

8.4.2 — The modified negative moments shall be used for calculating moments at sections within the spans.

8.4.3 — Redistribution of negative moments shall be made only when ε_t is equal to or greater than 0.0075 at the section at which moment is reduced.

COMMENTARY

R8.4 — Redistribution of negative moments in continuous flexural members

Moment redistribution is dependent on adequate ductility in plastic hinge regions. These plastic hinge regions develop at points of maximum moment and cause a shift in the elastic moment diagram. The usual result is a reduction in the values of negative moments in the plastic hinge region and an increase in the values of positive moments from those computed by elastic analysis. Because negative moments are determined for one loading arrangement and positive moments for another, each section has a reserve capacity that is not fully utilized for any one loading condition. The plastic hinges permit the utilization of the full capacity of more cross sections of a flexural member at ultimate loads.

Using conservative values of limiting concrete strains and lengths of plastic hinges derived from extensive tests, flexural members with small rotation capacity were analyzed for moment redistribution up to 20 percent, depending on the reinforcement ratio. The results were found to be conservative (see Fig. R8.4). Studies by Cohn[8.2] and Mattock[8.3] support this conclusion and indicate that cracking and deflection of beams designed for moment redistribution are not significantly greater at service loads than for beams designed by the elastic theory distribution of moments. Also, these studies indicated that adequate rotation capacity for the moment redistribution allowed by the code is available if the members satisfy the code requirements.

Moment redistribution may not be used for slab systems designed by the Direct Design Method (see 13.6.1.7).

In previous codes, Section 8.4 specified the permissible redistribution percentage in terms of reinforcement indices. The 2002 code specifies the permissible redistribution percentage in terms of the net tensile strain in extreme tension steel at nominal strength, ε_t. See Reference 8.4 for a comparison of these moment redistribution provisions.

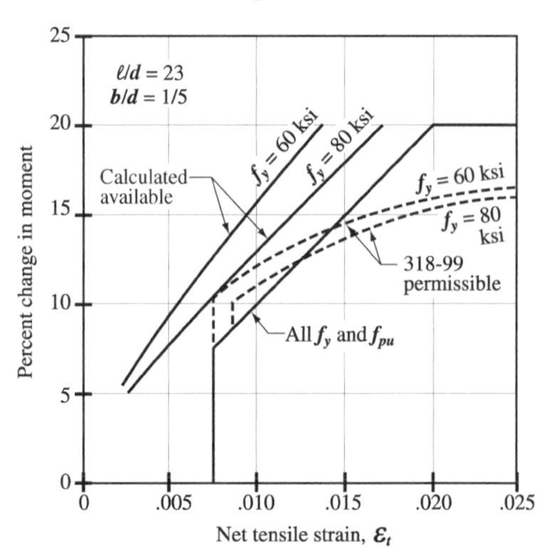

Fig. R8.4—Permissible moment redistribution for minimum rotation capacity

CODE

COMMENTARY

8.5 — Modulus of elasticity

8.5.1 — Modulus of elasticity, E_c, for concrete shall be permitted to be taken as $w_c^{1.5}\ 33\sqrt{f_c'}$ (in psi) for values of w_c between 90 and 155 lb/ft.[3] For normal weight concrete, E_c shall be permitted to be taken as $57{,}000\sqrt{f_c'}$.

8.5.2 — Modulus of elasticity, E_s, for nonprestressed reinforcement shall be permitted to be taken as 29,000,000 psi.

8.5.3 — Modulus of elasticity, E_p, for prestressing steel shall be determined by tests or reported by the manufacturer.

R8.5 — Modulus of elasticity

R8.5.1 — Studies leading to the expression for modulus of elasticity of concrete in 8.5.1 are summarized in Reference 8.5 where E_c was defined as the slope of the line drawn from a stress of zero to a compressive stress of $0.45f_c'$. The modulus of elasticity for concrete is sensitive to the modulus of elasticity of the aggregate and may differ from the specified value. Measured values range typically from 120 to 80 percent of the specified value. Methods for determining the modulus of elasticity for concrete are described in Reference 8.6.

8.6 — Stiffness

8.6.1 — Use of any set of reasonable assumptions shall be permitted for computing relative flexural and torsional stiffnesses of columns, walls, floors, and roof systems. The assumptions adopted shall be consistent throughout analysis.

8.6.2 — Effect of haunches shall be considered both in determining moments and in design of members.

R8.6 — Stiffness

R8.6.1 — Ideally, the member stiffnesses E_cI and GJ should reflect the degree of cracking and inelastic action that has occurred along each member before yielding. However, the complexities involved in selecting different stiffnesses for all members of a frame would make frame analyses inefficient in design offices. Simpler assumptions are required to define flexural and torsional stiffnesses.

For braced frames, relative values of stiffness are important. Two usual assumptions are to use gross E_cI values for all members or, to use half the gross E_cI of the beam stem for beams and the gross E_cI for the columns.

For frames that are free to sway, a realistic estimate of E_cI is desirable and should be used if second-order analyses are carried out. Guidance for the choice of E_cI for this case is given in R10.11.1.

Two conditions determine whether it is necessary to consider torsional stiffness in the analysis of a given structure: (1) the relative magnitude of the torsional and flexural stiffnesses, and (2) whether torsion is required for equilibrium of the structure (equilibrium torsion) or is due to members twisting to maintain deformation compatibility (compatibility torsion). In the case of compatibility torsion, the torsional stiffness may be neglected. For cases involving equilibrium torsion, torsional stiffness should be considered.

R8.6.2 — Stiffness and fixed-end moment coefficients for haunched members may be obtained from Reference 8.7.

CODE

8.7 — Span length

8.7.1 — Span length of members not built integrally with supports shall be considered as the clear span plus the depth of the member, but need not exceed distance between centers of supports.

8.7.2 — In analysis of frames or continuous construction for determination of moments, span length shall be taken as the distance center-to-center of supports.

8.7.3 — For beams built integrally with supports, design on the basis of moments at faces of support shall be permitted.

8.7.4 — It shall be permitted to analyze solid or ribbed slabs built integrally with supports, with clear spans not more than 10 ft, as continuous slabs on knife edge supports with spans equal to the clear spans of the slab and width of beams otherwise neglected.

8.8 — Columns

8.8.1 — Columns shall be designed to resist the axial forces from factored loads on all floors or roof and the maximum moment from factored loads on a single adjacent span of the floor or roof under consideration. Loading condition giving the maximum ratio of moment to axial load shall also be considered.

8.8.2 — In frames or continuous construction, consideration shall be given to the effect of unbalanced floor or roof loads on both exterior and interior columns and of eccentric loading due to other causes.

8.8.3 — In computing gravity load moments in columns, it shall be permitted to assume far ends of columns built integrally with the structure to be fixed.

8.8.4 — Resistance to moments at any floor or roof level shall be provided by distributing the moment between columns immediately above and below the given floor in proportion to the relative column stiffnesses and conditions of restraint.

8.9 — Arrangement of live load

8.9.1 — It shall be permitted to assume that:

(a) The live load is applied only to the floor or roof under consideration; and

(b) The far ends of columns built integrally with the structure are considered to be fixed.

COMMENTARY

R8.7 — Span length

Beam moments calculated at support centers may be reduced to the moments at support faces for design of beams. Reference 8.8 provides an acceptable method of reducing moments at support centers to those at support faces.

R8.8 — Columns

Section 8.8 has been developed with the intent of making certain that the most demanding combinations of axial load and moments be identified for design.

Section 8.8.4 has been included to make certain that moments in columns are recognized in the design if the girders have been proportioned using 8.3.3. The moment in 8.8.4 refers to the difference between the moments in a given vertical plane, exerted at column centerline by members framing into that column.

R8.9 — Arrangement of live load

For determining column, wall, and beam moments and shears caused by gravity loads, the code permits the use of a model limited to the beams in the level considered and the columns above and below that level. Far ends of columns are to be considered as fixed for the purpose of analysis under gravity loads. This assumption does not apply to lateral load analysis. However in analysis for lateral loads,

CODE

8.9.2 — It shall be permitted to assume that the arrangement of live load is limited to combinations of:

(a) Factored dead load on all spans with full factored live load on two adjacent spans; and

(b) Factored dead load on all spans with full factored live load on alternate spans.

8.10 — T-beam construction

8.10.1 — In T-beam construction, the flange and web shall be built integrally or otherwise effectively bonded together.

8.10.2 — Width of slab effective as a T-beam flange shall not exceed one-quarter of the span length of the beam, and the effective overhanging flange width on each side of the web shall not exceed:

(a) eight times the slab thickness; and

(b) one-half the clear distance to the next web.

8.10.3 — For beams with a slab on one side only, the effective overhanging flange width shall not exceed:

(a) one-twelfth the span length of the beam;

(b) six times the slab thickness; and

(c) one-half the clear distance to the next web.

8.10.4 — Isolated beams, in which the T-shape is used to provide a flange for additional compression area, shall have a flange thickness not less than one-half the width of web and an effective flange width not more than four times the width of web.

8.10.5 — Where primary flexural reinforcement in a slab that is considered as a T-beam flange (excluding joist construction) is parallel to the beam, reinforcement perpendicular to the beam shall be provided in the top of the slab in accordance with the following:

8.10.5.1 — Transverse reinforcement shall be designed to carry the factored load on the overhanging slab width assumed to act as a cantilever. For isolated beams, the full width of overhanging flange shall be

COMMENTARY

simplified methods (such as the portal method) may be used to obtain the moments, shears, and reactions for structures that are symmetrical and satisfy the assumptions used for such simplified methods. For unsymmetrical and high-rise structures, rigorous methods recognizing all structural displacements should be used.

The engineer is expected to establish the most demanding sets of design forces by investigating the effects of live load placed in various critical patterns.

Most approximate methods of analysis neglect effects of deflections on geometry and axial flexibility. Therefore, beam and column moments may have to be amplified for column slenderness in accordance with 10.11, 10.12, and 10.13.

R8.10 — T-beam construction

This section contains provisions identical to those of previous codes for limiting dimensions related to stiffness and flexural calculations. Special provisions related to T-beams and other flanged members are stated in 11.6.1 with regard to torsion.

CODE | COMMENTARY

considered. For other T-beams, only the effective overhanging slab width need be considered.

8.10.5.2 — Transverse reinforcement shall be spaced not farther apart than five times the slab thickness, nor farther apart than 18 in.

8.11 — Joist construction

8.11.1 — Joist construction consists of a monolithic combination of regularly spaced ribs and a top slab arranged to span in one direction or two orthogonal directions.

8.11.2 — Ribs shall be not less than 4 in. in width, and shall have a depth of not more than 3-1/2 times the minimum width of rib.

8.11.3 — Clear spacing between ribs shall not exceed 30 in.

8.11.4 — Joist construction not meeting the limitations of 8.11.1 through 8.11.3 shall be designed as slabs and beams.

8.11.5 — When permanent burned clay or concrete tile fillers of material having a unit compressive strength at least equal to f_c' in the joists are used:

8.11.5.1 — For shear and negative moment strength computations, it shall be permitted to include the vertical shells of fillers in contact with the ribs. Other portions of fillers shall not be included in strength computations.

8.11.5.2 — Slab thickness over permanent fillers shall be not less than one-twelfth the clear distance between ribs, nor less than 1-1/2 in.

8.11.5.3 — In one-way joists, reinforcement normal to the ribs shall be provided in the slab as required by 7.12.

8.11.6 — When removable forms or fillers not complying with 8.11.5 are used:

8.11.6.1 — Slab thickness shall be not less than one-twelfth the clear distance between ribs, nor less than 2 in.

8.11.6.2 — Reinforcement normal to the ribs shall be provided in the slab as required for flexure, considering load concentrations, if any, but not less than required by 7.12.

R8.11 — Joist construction

The size and spacing limitations for concrete joist construction meeting the limitations of 8.11.1 through 8.11.3 are based on successful performance in the past.

R8.11.3 — A limit on the maximum spacing of ribs is required because of the special provisions permitting higher shear strengths and less concrete protection for the reinforcement for these relatively small, repetitive members.

CODE

8.11.7 — Where conduits or pipes as permitted by 6.3 are embedded within the slab, slab thickness shall be at least 1 in. greater than the total overall depth of the conduits or pipes at any point. Conduits or pipes shall not impair significantly the strength of the construction.

8.11.8 — For joist construction, V_c shall be permitted to be 10 percent more than that specified in Chapter 11. It shall be permitted to increase V_n using shear reinforcement or by widening the ends of ribs.

8.12 — Separate floor finish

8.12.1 — A floor finish shall not be included as part of a structural member unless placed monolithically with the floor slab or designed in accordance with requirements of Chapter 17.

8.12.2 — It shall be permitted to consider all concrete floor finishes as part of required cover or total thickness for nonstructural considerations.

COMMENTARY

R8.11.8 — The increase in shear strength permitted by 8.11.8 is justified on the basis of: (1) satisfactory performance of joist construction with higher shear strengths, designed under previous codes, which allowed comparable shear stresses, and (2) redistribution of local overloads to adjacent joists.

R8.12 — Separate floor finish

The code does not specify an additional thickness for wearing surfaces subjected to unusual conditions of wear. The need for added thickness for unusual wear is left to the discretion of the designer.

As in previous editions of the code, a floor finish may be considered for strength purposes only if it is cast monolithically with the slab. Permission is given to include a separate finish in the structural thickness if composite action is provided for in accordance with Chapter 17.

All floor finishes may be considered for nonstructural purposes such as cover for reinforcement, fire protection, etc. Provisions should be made, however, to ensure that the finish will not spall off, thus causing decreased cover. Furthermore, development of reinforcement considerations requires minimum monolithic concrete cover according to 7.7.

Notes

CHAPTER 9 — STRENGTH AND SERVICEABILITY REQUIREMENTS

CODE

9.1 — General

9.1.1 — Structures and structural members shall be designed to have design strengths at all sections at least equal to the required strengths calculated for the factored loads and forces in such combinations as are stipulated in this code.

9.1.2 — Members also shall meet all other requirements of this code to ensure adequate performance at service load levels.

9.1.3 — Design of structures and structural members using the load factor combinations and strength reduction factors of Appendix C shall be permitted. Use of load factor combinations from this chapter in conjunction with strength reduction factors of Appendix C shall not be permitted.

9.2 — Required strength

9.2.1 — Required strength U shall be at least equal to the effects of factored loads in Eq. (9-1) through (9-7). The effect of one or more loads not acting simultaneously shall be investigated.

$$U = 1.4(D + F) \quad (9\text{-}1)$$

$$U = 1.2(D + F + T) + 1.6(L + H) + 0.5(L_r \text{ or } S \text{ or } R) \quad (9\text{-}2)$$

$$U = 1.2D + 1.6(L_r \text{ or } S \text{ or } R) + (1.0L \text{ or } 0.8W) \quad (9\text{-}3)$$

$$U = 1.2D + 1.6W + 1.0L + 0.5(L_r \text{ or } S \text{ or } R) \quad (9\text{-}4)$$

COMMENTARY

R9.1 — General

In the 2002 code, the load factor combinations and strength reduction factors of the 1999 code were revised and moved to Appendix C. The 1999 combinations have been replaced with those of SEI/ASCE 7-02.[9.1] The strength reduction factors were replaced with those of the 1999 Appendix C, except that the factor for flexure was increased.

The changes were made to further unify the design profession on one set of load factors and combinations, and to facilitate the proportioning of concrete building structures that include members of materials other than concrete. When used with the strength reduction factors in 9.3, the designs for gravity loads will be comparable to those obtained using the strength reduction and load factors of the 1999 and earlier codes. For combinations with lateral loads, some designs will be different, but the results of either set of load factors are considered acceptable.

Chapter 9 defines the basic strength and serviceability conditions for proportioning structural concrete members.

The basic requirement for strength design may be expressed as follows:

$$\text{Design Strength} \geq \text{Required Strength}$$

$$\phi\,(\text{Nominal Strength}) \geq U$$

In the strength design procedure, the margin of safety is provided by multiplying the service load by a load factor and the nominal strength by a strength reduction factor.

R9.2 — Required strength

The required strength U is expressed in terms of factored loads, or related internal moments and forces. Factored loads are the loads specified in the general building code multiplied by appropriate load factors.

The factor assigned to each load is influenced by the degree of accuracy to which the load effect usually can be calculated and the variation that might be expected in the load during the lifetime of the structure. Dead loads, because they are more accurately determined and less variable, are assigned a lower load factor than live loads. Load factors also account for variability in the structural analysis used to compute moments and shears.

The code gives load factors for specific combinations of loads. In assigning factors to combinations of loading, some

CODE

$$U = 1.2D + 1.0E + 1.0L + 0.2S \quad (9\text{-}5)$$

$$U = 0.9D + 1.6W + 1.6H \quad (9\text{-}6)$$

$$U = 0.9D + 1.0E + 1.6H \quad (9\text{-}7)$$

except as follows:

(a) The load factor on the live load L in Eq. (9-3) to (9-5) shall be permitted to be reduced to 0.5 except for garages, areas occupied as places of public assembly, and all areas where L is greater than 100 lb/ft^2.

(b) Where wind load W has not been reduced by a directionality factor, it shall be permitted to use $1.3W$ in place of $1.6W$ in Eq. (9-4) and (9-6).

(c) Where E, the load effects of earthquake, is based on service-level seismic forces, $1.4E$ shall be used in place of $1.0E$ in Eq. (9-5) and (9-7).

(d) The load factor on H, loads due to weight and pressure of soil, water in soil, or other materials, shall be set equal to zero in Eq. (9-6) and (9-7) if the structural action due to H counteracts that due to W or E. Where lateral earth pressure provides resistance to structural actions from other forces, it shall not be included in H but shall be included in the design resistance.

9.2.2 — If resistance to impact effects is taken into account in design, such effects shall be included with L.

9.2.3 — Estimations of differential settlement, creep, shrinkage, expansion of shrinkage-compensating concrete, or temperature change shall be based on a realistic assessment of such effects occurring in service.

COMMENTARY

consideration is given to the probability of simultaneous occurrence. While most of the usual combinations of loadings are included, the designer should not assume that all cases are covered.

Due regard is to be given to sign in determining U for combinations of loadings, as one type of loading may produce effects of opposite sense to that produced by another type. The load combinations with $0.9D$ are specifically included for the case where a higher dead load reduces the effects of other loads. The loading case may also be critical for tension-controlled column sections. In such a case, a reduction in axial load and an increase in moment may result in a critical load combination.

Consideration should be given to various combinations of loading to determine the most critical design condition. This is particularly true when strength is dependent on more than one load effect, such as strength for combined flexure and axial load or shear strength in members with axial load.

If special circumstances require greater reliance on the strength of particular members than encountered in usual practice, some reduction in the stipulated strength reduction factors ϕ or increase in the stipulated load factors may be appropriate for such members.

The wind load equation in SEI/ASCE 7-02[9.1] and IBC 2003[9.2] includes a factor for wind directionality that is equal to 0.85 for buildings. The corresponding load factor for wind in the load combination equations was increased accordingly (1.3/0.85 = 1.53 rounded up to 1.6). The code allows use of the previous wind load factor of 1.3 when the design wind load is obtained from other sources that do not include the wind directionality factor.

Model building codes and design load references have converted earthquake forces to strength level, and reduced the earthquake load factor to 1.0 (ASCE 7-93[9.3]; BOCA/NBC 93[9.4]; SBC 94[9.5]; UBC 97[9.6]; and IBC 2000[9.2]). The code requires use of the previous load factor for earthquake loads, approximately 1.4, when service-level earthquake forces from earlier editions of these references are used.

R9.2.2 — If the live load is applied rapidly, as may be the case for parking structures, loading docks, warehouse floors, elevator shafts, etc., impact effects should be considered. In all equations, substitute (L + impact) for L when impact should be considered.

R9.2.3 — The designer should consider the effects of differential settlement, creep, shrinkage, temperature, and shrinkage-compensating concrete. The term realistic assessment is used to indicate that the most probable values rather than the upper bound values of the variables should be used.

CODE

9.2.4 — If a structure is in a flood zone, or is subjected to forces from atmospheric ice loads, the flood or ice loads and the appropriate load combinations of SEI/ASCE 7 shall be used.

9.2.5 — For post-tensioned anchorage zone design, a load factor of 1.2 shall be applied to the maximum prestressing steel jacking force.

COMMENTARY

R9.2.4 — Areas subject to flooding are defined by flood hazard maps, usually maintained by local governmental jurisdictions.

R9.2.5 — The load factor of 1.2 applied to the maximum tendon jacking force results in a design load of about 113 percent of the specified prestressing steel yield strength but not more than 96 percent of the nominal ultimate strength of the prestressing steel. This compares well with the maximum attainable jacking force, which is limited by the anchor efficiency factor.

9.3 — Design strength

9.3.1 — Design strength provided by a member, its connections to other members, and its cross sections, in terms of flexure, axial load, shear, and torsion, shall be taken as the nominal strength calculated in accordance with requirements and assumptions of this code, multiplied by the strength reduction factors ϕ in 9.3.2, 9.3.4, and 9.3.5.

9.3.2 — Strength reduction factor ϕ shall be as given in 9.3.2.1 through 9.3.2.7:

9.3.2.1 — Tension-controlled sections as defined in 10.3.4 .... 0.90 (See also 9.3.2.7)

9.3.2.2 — Compression-controlled sections, as defined in 10.3.3:

(a) Members with spiral reinforcement conforming to 10.9.3.. 0.70

(b) Other reinforced members 0.65

R9.3 — Design strength

R9.3.1 — The design strength of a member refers to the nominal strength calculated in accordance with the requirements stipulated in this code multiplied by a strength reduction factor ϕ, which is always less than one.

The purposes of the strength reduction factor ϕ are (1) to allow for the probability of under-strength members due to variations in material strengths and dimensions, (2) to allow for inaccuracies in the design equations, (3) to reflect the degree of ductility and required reliability of the member under the load effects being considered, and (4) to reflect the importance of the member in the structure.[9.7,9.8]

In the 2002 code, the strength reduction factors were adjusted to be compatible with the SEI/ASCE 7[9.1] load combinations, which were the basis for the required factored load combinations in model building codes at that time. These factors are essentially the same as those published in Appendix C of the 1995 edition, except the factor for flexure/tension controlled limits is increased from 0.80 to 0.90. This change is based on past[9.7] and current reliability analyses,[9.9] statistical study of material properties, as well as the opinion of the committee that the historical performance of concrete structures supports $\phi = 0.90$.

R9.3.2.1 — In applying 9.3.2.1 and 9.3.2.2, the axial tensions and compressions to be considered are those caused by external forces. Effects of prestressing forces are not included.

R9.3.2.2 — Before the 2002 edition, the code specified the magnitude of the ϕ-factor for cases of axial load or flexure, or both, in terms of the type of loading. For these cases, the ϕ-factor is now determined by the strain conditions at a cross section, at nominal strength.

A lower ϕ-factor is used for compression-controlled sections than is used for tension-controlled sections because

CODE

For sections in which the net tensile strain in the extreme tension steel at nominal strength, ε_t, is between the limits for compression-controlled and tension-controlled sections, ϕ shall be permitted to be linearly increased from that for compression-controlled sections to 0.90 as ε_t increases from the compression-controlled strain limit to 0.005.

Alternatively, when Appendix B is used, for members in which f_y does not exceed 60,000 psi, with symmetric reinforcement, and with $(d - d')/h$ not less than 0.70, ϕ shall be permitted to be increased linearly to 0.90 as ϕP_n decreases from $0.10f_c' A_g$ to zero. For other reinforced members, ϕ shall be permitted to be increased linearly to 0.90 as ϕP_n decreases from $0.10f_c' A_g$ or ϕP_b, whichever is smaller, to zero.

9.3.2.3 — Shear and torsion 0.75

9.3.2.4 — Bearing on concrete (except for post-tensioned anchorage zones and strut-and-tie models) 0.65

COMMENTARY

compression-controlled sections have less ductility, are more sensitive to variations in concrete strength, and generally occur in members that support larger loaded areas than members with tension-controlled sections. Members with spiral reinforcement are assigned a higher ϕ than tied columns since they have greater ductility or toughness.

For sections subjected to axial load with flexure, design strengths are determined by multiplying both P_n and M_n by the appropriate single value of ϕ. Compression-controlled and tension-controlled sections are defined in 10.3.3 and 10.3.4 as those that have net tensile strain in the extreme tension steel at nominal strength less than or equal to the compression-controlled strain limit, and equal to or greater than 0.005, respectively. For sections with net tensile strain ε_t in the extreme tension steel at nominal strength between the above limits, the value of ϕ may be determined by linear interpolation, as shown in Fig. R9.3.2. The concept of net tensile strain ε_t is discussed in R10.3.3.

Since the compressive strain in the concrete at nominal strength is assumed in 10.2.3 to be 0.003, the net tensile strain limits for compression-controlled members may also be stated in terms of the ratio c/d_t, where c is the depth of the neutral axis at nominal strength, and d_t is the distance from the extreme compression fiber to the extreme tension steel. The c/d_t limits for compression-controlled and tension-controlled sections are 0.6 and 0.375, respectively. The 0.6 limit applies to sections reinforced with Grade 60 steel and to prestressed sections. Fig. R9.3.2 also gives equations for ϕ as a function of c/d_t.

The net tensile strain limit for tension-controlled sections may also be stated in terms of the ρ/ρ_b as defined in the 1999 and earlier editions of the code. The net tensile strain limit of 0.005 corresponds to a ρ/ρ_b ratio of 0.63 for rectangular sections with Grade 60 reinforcement. For a comparison of these provisions with the 1999 code Section 9.3, see Reference 9.10.

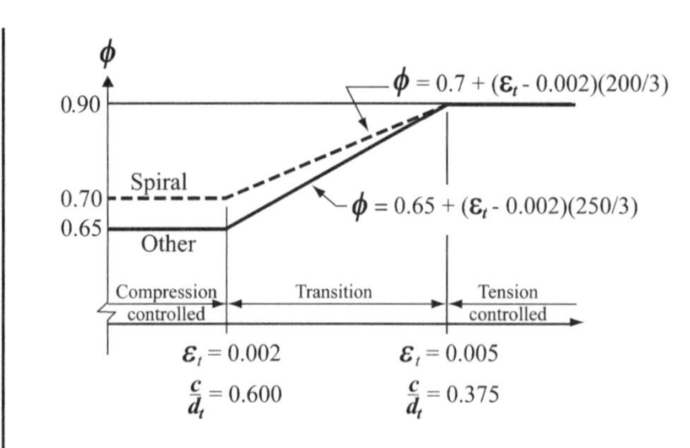

Fig. R9.3.2—Variation of ϕ with net tensile strain in extreme tension steel, ε_t, and c/d_t for Grade 60 reinforcement and for prestressing steel.

CODE

9.3.2.5 — Post-tensioned anchorage zones..... 0.85

9.3.2.6 — Strut-and-tie models (Appendix A), and struts, ties, nodal zones, and bearing areas in such models.. 0.75

9.3.2.7 — Flexural sections in pretensioned members where strand embedment is less than the development length as provided in 12.9.1.1:

(a) From the end of the member to the end of the transfer length.. 0.75

(b) From the end of the transfer length to the end of the development length ϕ shall be permitted to be linearly increased from 0.75 to 0.9.

Where bonding of a strand does not extend to the end of the member, strand embedment shall be assumed to begin at the end of the debonded length. See also 12.9.3.

COMMENTARY

R9.3.2.5 — The ϕ-factor of 0.85 reflects the wide scatter of results of experimental anchorage zone studies. Since 18.13.4.2 limits the nominal compressive strength of unconfined concrete in the general zone to $\mathbf{0.7\lambda f'_{ci}}$, the effective design strength for unconfined concrete is $\mathbf{0.85 \times 0.7\lambda f'_{ci} \approx 0.6\lambda f'_{ci}}$.

R9.3.2.6 — The ϕ factor used in strut-and-tie models is taken equal to the ϕ factor for shear. The value of ϕ for strut-and-tie models is applied to struts, ties, and bearing areas in such models.

R9.3.2.7 — If a critical section occurs in a region where strand is not fully developed, failure may be by bond slip. Such a failure resembles a brittle shear failure; hence, the requirements for a reduced ϕ. For sections between the end of the transfer length and the end of the development length, the value of ϕ may be determined by linear interpolation, as shown in Fig. R9.3.2.7(a) and (b).

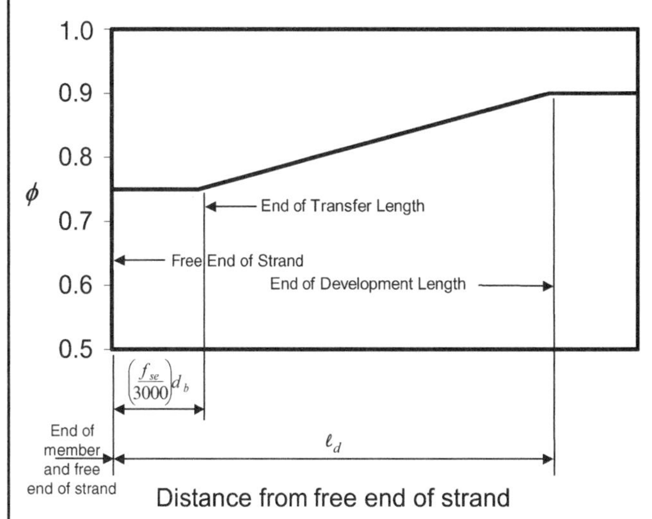

Fig. R9.3.2.7(a)—Variation of ϕ with distance from the free end of strand in pretensioned members with fully bonded strands.

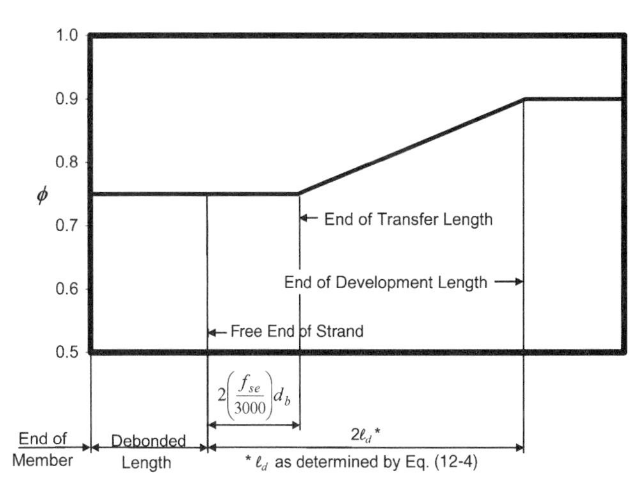

Fig. R9.3.2.7(b)—Variation of ϕ with distance from the free end of strand in pretensioned members with debonded strands where 12.9.3 applies.

CODE

COMMENTARY

Where bonding of one or more strands does not extend to the end of the member, instead of a more rigorous analysis, ϕ may be conservatively taken as 0.75 from the end of the member to the end of the transfer length of the strand with the longest debonded length. Beyond this point, ϕ may be varied linearly to 0.9 at the location where all strands are developed, as shown in Fig. R9.3.2.7(b). Alternatively, the contribution of the debonded strands may be ignored until they are fully developed. Embedment of debonded strand is considered to begin at the termination of the debonding sleeves. Beyond this point, the provisions of 12.9.3 are applicable.

9.3.3 — Development lengths specified in Chapter 12 do not require a ϕ-factor.

9.3.4 — For structures that rely on special moment resisting frames or special reinforced concrete structural walls to resist earthquake effects, ***E***, ϕ shall be modified as given in (a) through (c):

(a) For any structural member that is designed to resist ***E***, ϕ for shear shall be 0.60 if the nominal shear strength of the member is less than the shear corresponding to the development of the nominal flexural strength of the member. The nominal flexural strength shall be determined considering the most critical factored axial loads and including ***E***;

(b) For diaphragms, ϕ for shear shall not exceed the minimum ϕ for shear used for the vertical components of the primary lateral-force-resisting system;

(c) For joints and diagonally reinforced coupling beams, ϕ for shear shall be 0.85.

R9.3.4 — Strength reduction factors in 9.3.4 are intended to compensate for uncertainties in estimation of strength of structural members in buildings. They are based primarily on experience with constant or steadily increasing applied load. For construction in regions of high seismic risk, some of the strength reduction factors have been modified in 9.3.4 to account for the effects of displacement reversals into the nonlinear range of response on strength.

Section 9.3.4(a) refers to brittle members such as low-rise walls, portions of walls between openings, or diaphragms that are impractical to reinforce to raise their nominal shear strength above nominal flexural strength for the pertinent loading conditions.

Short structural walls were the primary vertical elements of the lateral-force-resisting system in many of the parking structures that sustained damage during the 1994 Northridge earthquake. Section 9.3.4(b) requires the shear strength reduction factor for diaphragms to be 0.60 if the shear strength reduction factor for the walls is 0.60.

9.3.5 — In Chapter 22, ϕ shall be 0.55 for flexure, compression, shear, and bearing of structural plain concrete.

R9.3.5 — The strength reduction factor ϕ for structural plain concrete design is the same for all strength conditions. Since both flexural tension strength and shear strength for plain concrete depend on the tensile strength characteristics of the concrete, with no reserve strength or ductility possible due to the absence of reinforcement, equal strength reduction factors for both bending and shear are considered appropriate.

9.4 — Design strength for reinforcement

The values of f_y and f_{yt} used in design calculations shall not exceed 80,000 psi, except for prestressing steel and for spiral transverse reinforcement in 10.9.3.

R9.4 — Design strength for reinforcement

In addition to the upper limit of 80,000 psi for yield strength of nonprestressed reinforcement, there are limitations on yield strength in other sections of the code.

In 11.5.2, 11.6.3.4, and 11.7.6, the maximum value of f_y that may be used in design for shear and torsion reinforcement is 60,000 psi, except that f_y up to 80,000 psi may be used for shear reinforcement meeting the requirements of ASTM A 497.

CODE

COMMENTARY

In 19.3.2 and 21.2.5, the maximum specified yield strength f_y is 60,000 psi in shells, folded plates, and structures governed by the special seismic provisions of Chapter 21.

The deflection provisions of 9.5 and the limitations on distribution of flexural reinforcement of 10.6 become increasingly critical as f_y increases.

9.5 — Control of deflections

9.5.1 — Reinforced concrete members subjected to flexure shall be designed to have adequate stiffness to limit deflections or any deformations that adversely affect strength or serviceability of a structure.

9.5.2 — One-way construction (nonprestressed)

9.5.2.1 — Minimum thickness stipulated in Table 9.5(a) shall apply for one-way construction not supporting or attached to partitions or other construction likely to be damaged by large deflections, unless computation of deflection indicates a lesser thickness can be used without adverse effects.

R9.5 — Control of deflections

R9.5.1 — The provisions of 9.5 are concerned only with deflections or deformations that may occur at service load levels. When long-term deflections are computed, only the dead load and that portion of the live load that is sustained need be considered.

Two methods are given for controlling deflections.[9.11] For nonprestressed beams and one-way slabs, and for composite members, provision of a minimum overall thickness as required by Table 9.5(a) will satisfy the requirements of the code for members not supporting or attached to partitions or other construction likely to be damaged by large deflections. For nonprestressed two-way construction, minimum thickness as required by 9.5.3.1, 9.5.3.2, and 9.5.3.3 will satisfy the requirements of the code.

For nonprestressed members that do not meet these minimum thickness requirements, or that support or are attached to partitions or other construction likely to be damaged by large deflections, and for all prestressed concrete flexural members, deflections should be calculated by the procedures described or referred to in the appropriate sections of the code, and are limited to the values in Table 9.5(b).

R9.5.2 — One-way construction (nonprestressed)

R9.5.2.1 — The minimum thicknesses of Table 9.5(a) apply for nonprestressed beams and one-way slabs (see 9.5.2), and for composite members (see 9.5.5). These minimum thicknesses apply only to members not supporting or attached to partitions and other construction likely to be damaged by deflection.

Values of minimum thickness should be modified if other than normalweight concrete and Grade 60 reinforcement are used. The notes beneath the table are essential to its use for reinforced concrete members constructed with structural lightweight concrete or with reinforcement having a specified yield strength, f_y, other than 60,000 psi. If both of these conditions exist, the corrections in footnotes (a) and (b) should both be applied.

The modification for lightweight concrete in footnote (a) is based on studies of the results and discussions in Reference 9.12. No correction is given for concretes weighing between 120 and 145 lb/ft^3 because the correction term would be close to unity in this range.

CODE

TABLE 9.5(a)—MINIMUM THICKNESS OF NONPRESTRESSED BEAMS OR ONE-WAY SLABS UNLESS DEFLECTIONS ARE CALCULATED

	Minimum thickness, h			
	Simply supported	One end continuous	Both ends continuous	Cantilever
Member	Members not supporting or attached to partitions or other construction likely to be damaged by large deflections.			
Solid one-way slabs	$\ell/20$	$\ell/24$	$\ell/28$	$\ell/10$
Beams or ribbed one-way slabs	$\ell/16$	$\ell/18.5$	$\ell/21$	$\ell/8$

Notes:
Values given shall be used directly for members with normalweight concrete (w_c = 145 lb/ft^3) and Grade 60 reinforcement. For other conditions, the values shall be modified as follows:
a) For structural lightweight concrete having unit weight, w_c, in the range 90-120 lb/ft^3, the values shall be multiplied by ($1.65 - 0.005w_c$) but not less than 1.09.
b) For f_y other than 60,000 psi, the values shall be multiplied by ($0.4 + f_y/100{,}000$).

9.5.2.2 — Where deflections are to be computed, deflections that occur immediately on application of load shall be computed by usual methods or formulas for elastic deflections, considering effects of cracking and reinforcement on member stiffness.

9.5.2.3 — Unless stiffness values are obtained by a more comprehensive analysis, immediate deflection shall be computed with the modulus of elasticity for concrete, E_c, as specified in 8.5.1 (normalweight or lightweight concrete) and with the effective moment of inertia, I_e, as follows, but not greater than I_g

$$I_e = \left(\frac{M_{cr}}{M_a}\right)^3 I_g + \left[1 - \left(\frac{M_{cr}}{M_a}\right)^3\right] I_{cr} \quad (9\text{-}8)$$

where:

$$M_{cr} = \frac{f_r I_g}{y_t} \quad (9\text{-}9)$$

and for normalweight concrete,

$$f_r = 7.5\sqrt{f_c'} \quad (9\text{-}10)$$

When lightweight aggregate concrete is used, one of the following modifications shall apply:

(a) When f_{ct} is specified and concrete is proportioned in accordance with 5.2, f_r shall be modified by substituting $f_{ct}/6.7$ for $\sqrt{f_c'}$, but the value of $f_{ct}/6.7$ shall not exceed $\sqrt{f_c'}$;

(b) When f_{ct} is not specified, f_r shall be multiplied by 0.75 for all-lightweight concrete, and 0.85 for sand-

COMMENTARY

The modification for f_y in footnote (b) is approximate but should yield conservative results for the type of members considered in the table, for typical reinforcement ratios, and for values of f_y between 40,000 and 80,000 psi.

R9.5.2.2 — For calculation of immediate deflections of uncracked prismatic members, the usual methods or formulas for elastic deflections may be used with a constant value of $E_c I_g$ along the length of the member. However, if the member is cracked at one or more sections, or if its depth varies along the span, a more exact calculation becomes necessary.

R9.5.2.3 — The effective moment of inertia procedure described in the code and developed in Reference 9.13 was selected as being sufficiently accurate for use to control deflections.[9.14-9.16] The effective moment of inertia I_e was developed to provide a transition between the upper and lower bounds of I_g and I_{cr} as a function of the ratio M_{cr}/M_a. For most cases I_e will be less than I_g.

CODE

lightweight concrete. Linear interpolation shall be permitted if partial sand replacement is used.

9.5.2.4 — For continuous members, I_e shall be permitted to be taken as the average of values obtained from Eq. (9-8) for the critical positive and negative moment sections. For prismatic members, I_e shall be permitted to be taken as the value obtained from Eq. (9-8) at midspan for simple and continuous spans, and at support for cantilevers.

9.5.2.5 — Unless values are obtained by a more comprehensive analysis, additional long-term deflection resulting from creep and shrinkage of flexural members (normalweight or lightweight concrete) shall be determined by multiplying the immediate deflection caused by the sustained load considered, by the factor λ_Δ

$$\lambda_\Delta = \frac{\xi}{1 + 50\rho'} \quad (9\text{-}11)$$

where ρ' shall be the value at midspan for simple and continuous spans, and at support for cantilevers. It shall be permitted to assume ξ, the time-dependent factor for sustained loads, to be equal to:

5 years or more.. 2.0
12 months... 1.4
6 months... 1.2
3 months... 1.0

COMMENTARY

R9.5.2.4 — For continuous members, the code procedure suggests a simple averaging of I_e values for the positive and negative moment sections. The use of the midspan section properties for continuous prismatic members is considered satisfactory in approximate calculations primarily because the midspan rigidity (including the effect of cracking) has the dominant effect on deflections, as shown by ACI Committee 435[9.17,9.18] and SP-43.[9.11]

R9.5.2.5 — Shrinkage and creep due to sustained loads cause additional long-term deflections over and above those which occur when loads are first placed on the structure. Such deflections are influenced by temperature, humidity, curing conditions, age at time of loading, quantity of compression reinforcement, and magnitude of the sustained load. The expression given in this section is considered satisfactory for use with the code procedures for the calculation of immediate deflections, and with the limits given in Table 9.5(b). The deflection computed in accordance with this section is the additional long-term deflection due to the dead load and that portion of the live load that will be sustained for a sufficient period to cause significant time-dependent deflections.

Eq. (9-11) was developed in Reference 9.19. In Eq. (9-11) the multiplier on ξ accounts for the effect of compression reinforcement in reducing long-term deflections. $\xi = 2.0$ represents a nominal time-dependent factor for 5 years duration of loading. The curve in Fig. R9.5.2.5 may be used to estimate values of ξ for loading periods less than five years.

If it is desired to consider creep and shrinkage separately, approximate equations provided in References 9.13, 9.14, 9.19, and 9.20 may be used.

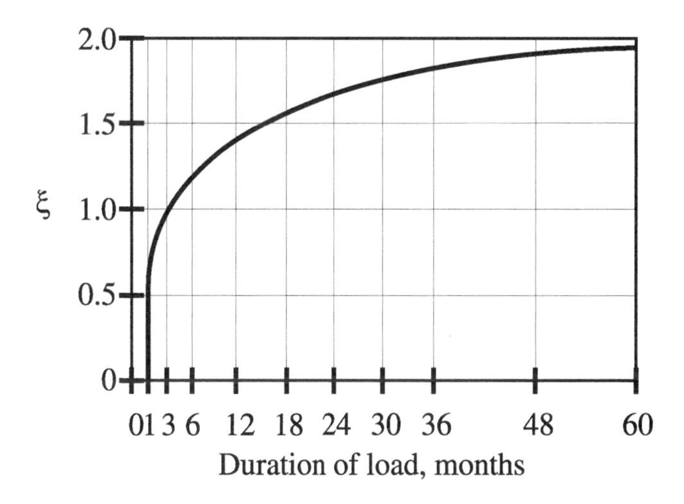

Fig. R9.5.2.5—Multipliers for long-term deflections

CODE

9.5.2.6 — Deflection computed in accordance with 9.5.2.2 through 9.5.2.5 shall not exceed limits stipulated in Table 9.5(b).

COMMENTARY

R9.5.2.6 — It should be noted that the limitations given in this table relate only to supported or attached nonstructural elements. For those structures in which structural members are likely to be affected by deflection or deformation of members to which they are attached in such a manner as to affect adversely the strength of the structure, these deflections and the resulting forces should be considered explicitly in the analysis and design of the structures as required by 9.5.1. (See Reference 9.16.)

Where long-term deflections are computed, the portion of the deflection before attachment of the nonstructural elements may be deducted. In making this correction use may be made of the curve in Fig. R9.5.2.5 for members of usual sizes and shapes.

9.5.3 — Two-way construction (nonprestressed)

9.5.3.1 — Section 9.5.3 shall govern the minimum thickness of slabs or other two-way construction designed in accordance with the provisions of Chapter 13 and conforming with the requirements of 13.6.1.2. The thickness of slabs without interior beams spanning between the supports on all sides shall satisfy the requirements of 9.5.3.2 or 9.5.3.4. The thickness of slabs with beams spanning between the supports on all sides shall satisfy requirements of 9.5.3.3 or 9.5.3.4.

9.5.3.2 — For slabs without interior beams spanning between the supports and having a ratio of long to short span not greater than 2, the minimum thickness shall be in accordance with the provisions of Table 9.5(c) and shall not be less than the following values:

(a) Slabs without drop panels as defined in 13.2.5 .... 5 in.;

(b) Slabs with drop panels as defined in 13.2.5 .... 4 in.

9.5.3.3 — For slabs with beams spanning between the supports on all sides, the minimum thickness, h, shall be as follows:

(a) For α_{fm} equal to or less than 0.2, the provisions of 9.5.3.2 shall apply;

(b) For α_{fm} greater than 0.2 but not greater than 2.0, h shall not be less than

$$h = \frac{\ell_n\left(0.8 + \frac{f_y}{200{,}000}\right)}{36 + 5\beta(\alpha_{fm} - 0.2)} \qquad (9\text{-}12)$$

and not less than 5 in.;

R9.5.3 — Two-way construction (nonprestressed)

R9.5.3.2 — The minimum thicknesses in Table 9.5(c) are those that have been developed through the years. Slabs conforming to those limits have not resulted in systematic problems related to stiffness for short- and long-term loads. These limits apply to only the domain of previous experience in loads, environment, materials, boundary conditions, and spans.

R9.5.3.3 — For panels having a ratio of long to short span greater than 2, the use of Eq. (9-12) and (9-13), which express the minimum thickness as a fraction of the long span, may give unreasonable results. For such panels, the rules applying to one-way construction in 9.5.2 should be used.

The requirement in 9.5.3.3(a) for α_{fm} equal to 0.2 made it possible to eliminate Eq. (9-13) of the 1989 code. That equation gave values essentially the same as those in Table 9.5(c), as does Eq. (9-12) at a value of α_{fm} equal to 0.2.

CODE COMMENTARY

TABLE 9.5(b) — MAXIMUM PERMISSIBLE COMPUTED DEFLECTIONS

Type of member	Deflection to be considered	Deflection limitation
Flat roofs not supporting or attached to nonstructural elements likely to be damaged by large deflections	Immediate deflection due to live load ***L***	$\ell/180$*
Floors not supporting or attached to nonstructural elements likely to be damaged by large deflections	Immediate deflection due to live load ***L***	$\ell/360$
Roof or floor construction supporting or attached to nonstructural elements likely to be damaged by large deflections	That part of the total deflection occurring after attachment of nonstructural elements (sum of the long-term deflection due to all sustained loads and the immediate deflection due to any additional live load)†	$\ell/480$‡
Roof or floor construction supporting or attached to nonstructural elements not likely to be damaged by large deflections		$\ell/240$§

*Limit not intended to safeguard against ponding. Ponding should be checked by suitable calculations of deflection, including added deflections due to ponded water, and considering long-term effects of all sustained loads, camber, construction tolerances, and reliability of provisions for drainage.

†Long-term deflection shall be determined in accordance with 9.5.2.5 or 9.5.4.3, but may be reduced by amount of deflection calculated to occur before attachment of nonstructural elements. This amount shall be determined on basis of accepted engineering data relating to time-deflection characteristics of members similar to those being considered.

‡Limit may be exceeded if adequate measures are taken to prevent damage to supported or attached elements.

§Limit shall not be greater than tolerance provided for nonstructural elements. Limit may be exceeded if camber is provided so that total deflection minus camber does not exceed limit.

TABLE 9.5(c)—MINIMUM THICKNESS OF SLABS WITHOUT INTERIOR BEAMS*

	Without drop panels‡			With drop panels‡		
	Exterior panels		Interior panels	Exterior panels		Interior panels
f_y, psi†	Without edge beams	With edge beams§		Without edge beams	With edge beams§	
40,000	$\frac{\ell_n}{33}$	$\frac{\ell_n}{36}$	$\frac{\ell_n}{36}$	$\frac{\ell_n}{36}$	$\frac{\ell_n}{40}$	$\frac{\ell_n}{40}$
60,000	$\frac{\ell_n}{30}$	$\frac{\ell_n}{33}$	$\frac{\ell_n}{33}$	$\frac{\ell_n}{33}$	$\frac{\ell_n}{36}$	$\frac{\ell_n}{36}$
75,000	$\frac{\ell_n}{28}$	$\frac{\ell_n}{31}$	$\frac{\ell_n}{31}$	$\frac{\ell_n}{31}$	$\frac{\ell_n}{34}$	$\frac{\ell_n}{34}$

*For two-way construction, ℓ_n is the length of clear span in the long direction, measured face-to-face of supports in slabs without beams and face-to-face of beams or other supports in other cases.

†For f_y between the values given in the table, minimum thickness shall be determined by linear interpolation.

‡Drop panels as defined in 13.2.5.

§Slabs with beams between columns along exterior edges. The value of α_f for the edge beam shall not be less than 0.8.

(c) For α_{fm} greater than 2.0, h shall not be less than

$$h = \frac{\ell_n\left(0.8 + \frac{f_y}{200{,}000}\right)}{36 + 9\beta} \quad (9\text{-}13)$$

and not less than 3.5 in.;

(d) At discontinuous edges, an edge beam shall be provided with a stiffness ratio α_f not less than 0.80 or the minimum thickness required by Eq. (9-12) or (9-13) shall be increased by at least 10 percent in the panel with a discontinuous edge.

Term ℓ_n in (b) and (c) is length of clear span in long direction measured face-to-face of beams. Term β in (b) and (c) is ratio of clear spans in long to short direction of slab.

CODE

9.5.3.4 — Slab thickness less than the minimum required by 9.5.3.1, 9.5.3.2, and 9.5.3.3 shall be permitted where computed deflections do not exceed the limits of Table 9.5(b). Deflections shall be computed taking into account size and shape of the panel, conditions of support, and nature of restraints at the panel edges. The modulus of elasticity of concrete, E_c, shall be as specified in 8.5.1. The effective moment of inertia, I_e, shall be that given by Eq. (9-8); other values shall be permitted to be used if they result in computed deflections in reasonable agreement with results of comprehensive tests. Additional long-term deflection shall be computed in accordance with 9.5.2.5.

9.5.4 — Prestressed concrete construction

9.5.4.1 — For flexural members designed in accordance with provisions of Chapter 18, immediate deflection shall be computed by usual methods or formulas for elastic deflections, and the moment of inertia of the gross concrete section, I_g, shall be permitted to be used for Class U flexural members, as defined in 18.3.3.

9.5.4.2 — For Class C and Class T flexural members, as defined in 18.3.3, deflection calculations shall be based on a cracked transformed section analysis. It shall be permitted to base computations on a bilinear moment-deflection relationship, or an effective moment of inertia, I_e, as defined by Eq. (9-8).

9.5.4.3 — Additional long-term deflection of prestressed concrete members shall be computed taking into account stresses in concrete and steel under sustained load and including effects of creep and shrinkage of concrete and relaxation of steel.

COMMENTARY

R9.5.3.4 — The calculation of deflections for slabs is complicated even if linear elastic behavior can be assumed. For immediate deflections, the values of E_c and I_e specified in 9.5.2.3 may be used.[9.16] However, other procedures and other values of the stiffness E_cI may be used if they result in predictions of deflection in reasonable agreement with the results of comprehensive tests.

Since available data on long-term deflections of slabs are too limited to justify more elaborate procedures, the additional long-term deflection for two-way construction is required to be computed using the multipliers given in 9.5.2.5.

R9.5.4 — Prestressed concrete construction

The code requires deflections for all prestressed concrete flexural members to be computed and compared with the allowable values in Table 9.5(b).

R9.5.4.1 — Immediate deflections of Class U prestressed concrete members may be calculated by the usual methods or formulas for elastic deflections using the moment of inertia of the gross (uncracked) concrete section and the modulus of elasticity for concrete specified in 8.5.1.

R9.5.4.2 — Class C and Class T prestressed flexural members are defined in 18.3.3. Reference 9.21 gives information on deflection calculations using a bilinear moment-deflection relationship and using an effective moment of inertia. Reference 9.22 gives additional information on deflection of cracked prestressed concrete members.

Reference 9.23 shows that the I_e method can be used to compute deflections of Class T prestressed members loaded above the cracking load. For this case, the cracking moment should take into account the effect of prestress. A method for predicting the effect of nonprestressed tension steel in reducing creep camber is also given in Reference 9.23, with approximate forms given in References 9.16 and 9.24.

R9.5.4.3 — Calculation of long-term deflections of prestressed concrete flexural members is complicated. The calculations should consider not only the increased deflections due to flexural stresses, but also the additional long-term deflections resulting from time-dependent shortening of the flexural member.

Prestressed concrete members shorten more with time than similar nonprestressed members due to the precompression in the slab or beam which causes axial creep. This creep together with concrete shrinkage results in significant shortening of the flexural members that continues for several years after construction and should be considered in design. The shortening tends to reduce the tension in the prestressing

CODE

COMMENTARY

steel, reducing the precompression in the member and thereby causing increased long-term deflections.

Another factor that can influence long-term deflections of prestressed flexural members is adjacent concrete or masonry that is nonprestressed in the direction of the prestressed member. This can be a slab nonprestressed in the beam direction adjacent to a prestressed beam or a nonprestressed slab system. As the prestressed member tends to shrink and creep more than the adjacent nonprestressed concrete, the structure will tend to reach a compatibility of the shortening effects. This results in a reduction of the precompression in the prestressed member as the adjacent concrete absorbs the compression. This reduction in precompression of the prestressed member can occur over a period of years and will result in additional long-term deflections and in increase tensile stresses in the prestressed member.

Any suitable method for calculating long-term deflections of prestressed members may be used, provided all effects are considered. Guidance may be found in References 9.16, 9.25, 9.26, and 9.27.

9.5.4.4 — Deflection computed in accordance with 9.5.4.1 or 9.5.4.2, and 9.5.4.3 shall not exceed limits stipulated in Table 9.5(b).

9.5.5 — Composite construction

9.5.5.1 — Shored construction

If composite flexural members are supported during construction so that, after removal of temporary supports, dead load is resisted by the full composite section, it shall be permitted to consider the composite member equivalent to a monolithically cast member for computation of deflection. For nonprestressed members, the portion of the member in compression shall determine whether values in Table 9.5(a) for normalweight or lightweight concrete shall apply. If deflection is computed, account shall be taken of curvatures resulting from differential shrinkage of precast and cast-in-place components, and of axial creep effects in a prestressed concrete member.

9.5.5.2 — Unshored construction

If the thickness of a nonprestressed precast flexural member meets the requirements of Table 9.5(a), deflection need not be computed. If the thickness of a nonprestressed composite member meets the requirements of Table 9.5(a), it is not required to compute deflection occurring after the member becomes composite, but the long-term deflection of the precast member shall be investigated for magnitude and duration of load prior to beginning of effective composite action.

R9.5.5 — Composite construction

Since few tests have been made to study the immediate and long-term deflections of composite members, the rules given in 9.5.5.1 and 9.5.5.2 are based on the judgment of ACI Committee 318 and on experience.

If any portion of a composite member is prestressed or if the member is prestressed after the components have been cast, the provisions of 9.5.4 apply and deflections are to be calculated. For nonprestressed composite members, deflections need to be calculated and compared with the limiting values in Table 9.5(b) only when the thickness of the member or the precast part of the member is less than the minimum thickness given in Table 9.5(a). In unshored construction the thickness of concern depends on whether the deflection before or after the attainment of effective composite action is being considered. (In Chapter 17, it is stated that distinction need not be made between shored and unshored members. This refers to strength calculations, not to deflections.)

CODE

9.5.5.3 — Deflection computed in accordance with 9.5.5.1 or 9.5.5.2 shall not exceed limits stipulated in Table 9.5(b).

COMMENTARY

CHAPTER 10 — FLEXURE AND AXIAL LOADS

10.1 — Scope

Provisions of Chapter 10 shall apply for design of members subject to flexure or axial loads or to combined flexure and axial loads.

10.2 — Design assumptions

10.2.1 — Strength design of members for flexure and axial loads shall be based on assumptions given in 10.2.2 through 10.2.7, and on satisfaction of applicable conditions of equilibrium and compatibility of strains.

10.2.2 — Strain in reinforcement and concrete shall be assumed directly proportional to the distance from the neutral axis, except that, for deep beams as defined in 10.7.1, an analysis that considers a nonlinear distribution of strain shall be used. Alternatively, it shall be permitted to use a strut-and-tie model. See 10.7, 11.8, and Appendix A.

10.2.3 — Maximum usable strain at extreme concrete compression fiber shall be assumed equal to 0.003.

10.2.4 — Stress in reinforcement below f_y shall be taken as E_s times steel strain. For strains greater than that corresponding to f_y, stress in reinforcement shall be considered independent of strain and equal to f_y.

COMMENTARY

R10.2 — Design assumptions

R10.2.1 — The strength of a member computed by the strength design method of the code requires that two basic conditions be satisfied: (1) static equilibrium, and (2) compatibility of strains. Equilibrium between the compressive and tensile forces acting on the cross section at nominal strength should be satisfied. Compatibility between the stress and strain for the concrete and the reinforcement at nominal strength conditions should also be established within the design assumptions allowed by 10.2.

R10.2.2—Many tests have confirmed that the distribution of strain is essentially linear across a reinforced concrete cross section, even near ultimate strength.

The strain in both reinforcement and in concrete is assumed to be directly proportional to the distance from the neutral axis. This assumption is of primary importance in design for determining the strain and corresponding stress in the reinforcement.

R10.2.3 — The maximum concrete compressive strain at crushing of the concrete has been observed in tests of various kinds to vary from 0.003 to higher than 0.008 under special conditions. However, the strain at which ultimate moments are developed is usually about 0.003 to 0.004 for members of normal proportions and materials.

R10.2.4 — For deformed reinforcement, it is reasonably accurate to assume that the stress in reinforcement is proportional to strain below the specified yield strength f_y. The increase in strength due to the effect of strain hardening of the reinforcement is neglected for strength computations. In strength computations, the force developed in tensile or compressive reinforcement is computed as,

when $\varepsilon_s < \varepsilon_y$ (yield strain)

$$A_s f_s = A_s E_s \varepsilon_s$$

when $\varepsilon_s \geq \varepsilon_y$

$$A_s f_s = A_s f_y$$

CODE

COMMENTARY

where ε_s is the value from the strain diagram at the location of the reinforcement. For design, the modulus of elasticity of steel reinforcement E_s may be taken as 29,000,000 psi (see 8.5.2).

10.2.5 — Tensile strength of concrete shall be neglected in axial and flexural calculations of reinforced concrete, except when meeting requirements of 18.4.

R10.2.5 — The tensile strength of concrete in flexure (modulus of rupture) is a more variable property than the compressive strength and is about 10 to 15 percent of the compressive strength. Tensile strength of concrete in flexure is neglected in strength design. For members with normal percentages of reinforcement, this assumption is in good agreement with tests. For very small percentages of reinforcement, neglect of the tensile strength at ultimate is usually correct.

The strength of concrete in tension, however, is important in cracking and deflection considerations at service loads.

10.2.6 — The relationship between concrete compressive stress distribution and concrete strain shall be assumed to be rectangular, trapezoidal, parabolic, or any other shape that results in prediction of strength in substantial agreement with results of comprehensive tests.

R10.2.6 — This assumption recognizes the inelastic stress distribution of concrete at high stress. As maximum stress is approached, the stress-strain relationship for concrete is not a straight line but some form of a curve (stress is not proportional to strain). The general shape of a stress-strain curve is primarily a function of concrete strength and consists of a rising curve from zero to a maximum at a compressive strain between 0.0015 and 0.002 followed by a descending curve to an ultimate strain (crushing of the concrete) from 0.003 to higher than 0.008. As discussed under R10.2.3, the code sets the maximum usable strain at 0.003 for design.

The actual distribution of concrete compressive stress is complex and usually not known explicitly. Research has shown that the important properties of the concrete stress distribution can be approximated closely using any one of several different assumptions as to the form of stress distribution. The code permits any particular stress distribution to be assumed in design if shown to result in predictions of ultimate strength in reasonable agreement with the results of comprehensive tests. Many stress distributions have been proposed. The three most common are the parabola, trapezoid, and rectangle.

10.2.7 — Requirements of 10.2.6 are satisfied by an equivalent rectangular concrete stress distribution defined by the following:

10.2.7.1 — Concrete stress of $0.85f_c'$ shall be assumed uniformly distributed over an equivalent compression zone bounded by edges of the cross section and a straight line located parallel to the neutral axis at a distance $a = \beta_1 c$ from the fiber of maximum compressive strain.

10.2.7.2 — Distance from the fiber of maximum strain to the neutral axis, c, shall be measured in a direction perpendicular to the neutral axis.

R10.2.7 — For design, the code allows the use of an equivalent rectangular compressive stress distribution (stress block) to replace the more exact concrete stress distribution. In the equivalent rectangular stress block, an average stress of $0.85f_c'$ is used with a rectangle of depth $a = \beta_1 c$. The β_1 of 0.85 for concrete with $f_c' \leq 4000$ psi and 0.05 less for each 1000 psi of f_c' in excess of 4000 was determined experimentally.

In the 1976 supplement to the 1971 code, a lower limit of β_1 equal to 0.65 was adopted for concrete strengths greater than 8000 psi. Research data from tests with high strength concretes[10.1,10.2] supported the equivalent rectangular stress block for concrete strengths exceeding 8000 psi, with a β_1 equal to 0.65. Use of the equivalent rectangular stress distribution specified in the 1971 code, with no lower limit on β_1,

CODE

10.2.7.3 — For f_c' between 2500 and 4000 psi, β_1 shall be taken as 0.85. For f_c' above 4000 psi, β_1 shall be reduced linearly at a rate of 0.05 for each 1000 psi of strength in excess of 4000 psi, but β_1 shall not be taken less than 0.65.

COMMENTARY

resulted in inconsistent designs for high strength concrete for members subject to combined flexure and axial load.

The equivalent rectangular stress distribution does not represent the actual stress distribution in the compression zone at ultimate, but does provide essentially the same results as those obtained in tests.[10.3]

10.3 — General principles and requirements

10.3.1 — Design of cross sections subject to flexure or axial loads, or to combined flexure and axial loads, shall be based on stress and strain compatibility using assumptions in 10.2.

10.3.2 — Balanced strain conditions exist at a cross section when tension reinforcement reaches the strain corresponding to f_y just as concrete in compression reaches its assumed ultimate strain of 0.003.

10.3.3 — Sections are compression-controlled if the net tensile strain in the extreme tension steel, ε_t, is equal to or less than the compression-controlled strain limit when the concrete in compression reaches its assumed strain limit of 0.003. The compression-controlled strain limit is the net tensile strain in the reinforcement at balanced strain conditions. For Grade 60 reinforcement, and for all prestressed reinforcement, it shall be permitted to set the compression-controlled strain limit equal to 0.002.

R10.3 — General principles and requirements

R10.3.1 — Design strength equations for members subject to flexure or combined flexure and axial load are derived in the paper, "Rectangular Concrete Stress Distribution in Ultimate Strength Design."[10.3] Reference 10.3 and previous editions of this commentary also give the derivations of strength equations for cross sections other than rectangular.

R10.3.2 — A balanced strain condition exists at a cross section when the maximum strain at the extreme compression fiber just reaches 0.003 simultaneously with the first yield strain f_y/E_s in the tension reinforcement. The reinforcement ratio ρ_b, which produces balanced strain conditions under flexure, depends on the shape of the cross section and the location of the reinforcement.

R10.3.3 — The nominal flexural strength of a member is reached when the strain in the extreme compression fiber reaches the assumed strain limit 0.003. The net tensile strain ε_t is the tensile strain in the extreme tension steel at nominal strength, exclusive of strains due to prestress, creep, shrinkage, and temperature. The net tensile strain in the extreme tension steel is determined from a linear strain distribution at nominal strength, shown in Fig. R10.3.3, using similar triangles.

When the net tensile strain in the extreme tension steel is sufficiently large (equal to or greater than 0.005), the section is defined as tension-controlled where ample warning of failure with excessive deflection and cracking may be expected. When the net tensile strain in the extreme tension

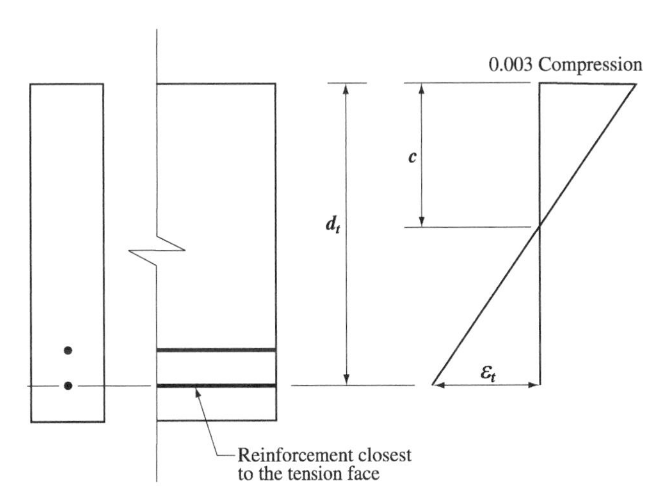

Fig. R10.3.3— Strain distribution and net tensile strain.

CODE / **COMMENTARY**

steel is small (less than or equal to the compression-controlled strain limit), a brittle failure condition may be expected, with little warning of impending failure. Flexural members are usually tension-controlled, whereas compression members are usually compression-controlled. Some sections, such as those with small axial load and large bending moment, will have net tensile strain in the extreme tension steel between the above limits. These sections are in a transition region between compression- and tension-controlled sections. Section 9.3.2 specifies the appropriate strength reduction factors for tension-controlled and compression-controlled sections, and for intermediate cases in the transition region.

Before the development of these provisions, the limiting tensile strain for flexural members was not stated, but was implicit in the maximum tension reinforcement ratio that was given as a fraction of ρ_b, which was dependent on the yield strength of the reinforcement. The net tensile strain limit of 0.005 for tension-controlled sections was chosen to be a single value that applies to all types of steel (prestressed and nonprestressed) permitted by this code.

Unless unusual amounts of ductility are required, the 0.005 limit will provide ductile behavior for most designs. One condition where greater ductile behavior is required is in design for redistribution of moments in continuous members and frames. Section 8.4 permits redistribution of negative moments. Since moment redistribution is dependent on adequate ductility in hinge regions, moment redistribution is limited to sections that have a net tensile strain of at least 0.0075.

For beams with compression reinforcement, or T-beams, the effects of compression reinforcement and flanges are automatically accounted for in the computation of net tensile strain ε_t.

10.3.4 — Sections are tension-controlled if the net tensile strain in the extreme tension steel, ε_t, is equal to or greater than 0.005 when the concrete in compression reaches its assumed strain limit of 0.003. Sections with ε_t between the compression-controlled strain limit and 0.005 constitute a transition region between compression-controlled and tension-controlled sections.

10.3.5 — For nonprestressed flexural members and nonprestressed members with factored axial compressive load less than $\mathbf{0.10 f_c' A_g}$, ε_t at nominal strength shall not be less than 0.004.

10.3.5.1 — Use of compression reinforcement shall be permitted in conjunction with additional tension reinforcement to increase the strength of flexural members.

R10.3.5 — The effect of this limitation is to restrict the reinforcement ratio in nonprestressed beams to about the same ratio as in editions of the code prior to 2002. The reinforcement limit of $\mathbf{0.75\rho_b}$ results in a net tensile strain in extreme tension steel at nominal strength of 0.00376. The proposed limit of 0.004 is slightly more conservative. This limitation does not apply to prestressed members.

CODE

10.3.6 — Design axial strength ϕP_n of compression members shall not be taken greater than $\phi P_{n,max}$, computed by Eq. (10-1) or (10-2).

10.3.6.1 — For nonprestressed members with spiral reinforcement conforming to 7.10.4 or composite members conforming to 10.16:

$$\phi P_{n,max} = 0.85\phi[0.85f_c'(A_g - A_{st}) + f_y A_{st}] \quad (10\text{-}1)$$

10.3.6.2—For nonprestressed members with tie reinforcement conforming to 7.10.5:

$$\phi P_{n,max} = 0.80\phi[0.85f_c'(A_g - A_{st}) + f_y A_{st}] \quad (10\text{-}2)$$

10.3.6.3 — For prestressed members, design axial strength, ϕP_n, shall not be taken greater than 0.85 (for members with spiral reinforcement) or 0.80 (for members with tie reinforcement) of the design axial strength at zero eccentricity, ϕP_o.

10.3.7 — Members subject to compressive axial load shall be designed for the maximum moment that can accompany the axial load. The factored axial force P_u at given eccentricity shall not exceed that given in 10.3.6. The maximum factored moment M_u shall be magnified for slenderness effects in accordance with 10.10.

COMMENTARY

R10.3.6 and R10.3.7 — The minimum design eccentricities included in the 1963 and 1971 codes were deleted from the 1977 code except for consideration of slenderness effects in compression members with small or zero computed end moments (see 10.12.3.2). The specified minimum eccentricities were originally intended to serve as a means of reducing the axial load design strength of a section in pure compression to account for accidental eccentricities not considered in the analysis that may exist in a compression member, and to recognize that concrete strength may be less than f_c' under sustained high loads. The primary purpose of the minimum eccentricity requirement was to limit the maximum design axial load strength of a compression member. This is now accomplished directly in 10.3.6 by limiting the design axial strength of a section in pure compression to 85 or 80 percent of the nominal strength. These percentage values approximate the axial strengths at eccentricity to depth ratios of 0.05 and 0.10, specified in the earlier codes for the spirally reinforced and tied members, respectively. The same axial load limitation applies to both cast-in-place and precast compression members. Design aids and computer programs based on the minimum eccentricity requirement of the 1963 and 1971 codes are equally applicable.

For prestressed members, the design axial load strength in pure compression is computed by the strength design methods of Chapter 10, including the effect of the prestressing force.

Compression member end moments should be considered in the design of adjacent flexural members. In nonsway frames, the effects of magnifying the end moments need not be considered in the design of the adjacent beams. In sway frames, the magnified end moments should be considered in designing the flexural members, as required in 10.13.7.

Corner and other columns exposed to known moments about each axis simultaneously should be designed for biaxial bending and axial load. Satisfactory methods are available in the *ACI Design Handbook*[10.4] and the *CRSI Handbook*.[10.5] The reciprocal load method[10.6] and the load contour method[10.7] are the methods used in those two handbooks. Research[10.8,10.9] indicates that using the equivalent rectangular stress block provisions of 10.2.7 produces satisfactory strength estimates for doubly symmetric sections. A simple and somewhat conservative estimate of nominal strength P_{ni} can be obtained from the reciprocal load relationship[10.6]

$$\frac{1}{P_{ni}} = \frac{1}{P_{nx}} + \frac{1}{P_{ny}} - \frac{1}{P_o}$$

where:

P_{ni} = nominal axial load strength at given eccentricity along both axes

CODE

COMMENTARY

P_o = nominal axial load strength at zero eccentricity

P_{nx} = nominal axial load strength at given eccentricity along x-axis

P_{ny} = nominal axial load strength at given eccentricity along y-axis

This relationship is most suitable when values P_{nx} and P_{ny} are greater than the balanced axial force P_b for the particular axis.

10.4 — Distance between lateral supports of flexural members

10.4.1 — Spacing of lateral supports for a beam shall not exceed 50 times b, the least width of compression flange or face.

10.4.2 — Effects of lateral eccentricity of load shall be taken into account in determining spacing of lateral supports.

R10.4 — Distance between lateral supports of flexural members

Tests[10.10,10.11] have shown that laterally unbraced reinforced concrete beams of any reasonable dimensions, even when very deep and narrow, will not fail prematurely by lateral buckling provided the beams are loaded without lateral eccentricity that causes torsion.

Laterally unbraced beams are frequently loaded off center (lateral eccentricity) or with slight inclination. Stresses and deformations set up by such loading become detrimental for narrow, deep beams, the more so as the unsupported length increases. Lateral supports spaced closer than $50b$ may be required by loading conditions.

10.5 — Minimum reinforcement of flexural members

10.5.1 — At every section of a flexural member where tensile reinforcement is required by analysis, except as provided in 10.5.2, 10.5.3, and 10.5.4, A_s provided shall not be less than that given by

$$A_{s,min} = \frac{3\sqrt{f_c'}}{f_y}b_w d \qquad (10\text{-}3)$$

and not less than $200b_w d/f_y$.

10.5.2 — For statically determinate members with a flange in tension, $A_{s,min}$ shall not be less than the value given by Eq. (10-3), except that b_w is replaced by either $2b_w$ or the width of the flange, whichever is smaller.

R10.5 — Minimum reinforcement of flexural members

The provision for a minimum amount of reinforcement applies to flexural members, which for architectural or other reasons, are larger in cross section than required for strength. With a very small amount of tensile reinforcement, the computed moment strength as a reinforced concrete section using cracked section analysis becomes less than that of the corresponding unreinforced concrete section computed from its modulus of rupture. Failure in such a case can be sudden.

To prevent such a failure, a minimum amount of tensile reinforcement is required by 10.5.1 in both positive and negative moment regions. When concrete strength higher than about 5000 psi is used, the $200/f_y$ value previously prescribed may not be sufficient. Equation (10-3) gives the same amount of reinforcement as $200b_w d/f_y$ when f_c' equals 4440 psi. When the flange of a section is in tension, the amount of tensile reinforcement needed to make the strength of the reinforced section equal that of the unreinforced section is about twice that for a rectangular section or that of a flanged section with the flange in compression. A higher amount of minimum tensile reinforcement is particularly necessary in cantilevers and other statically determinate members where there is no possibility for redistribution of moments.

CODE

10.5.3 — The requirements of 10.5.1 and 10.5.2 need not be applied if at every section A_s provided is at least one-third greater than that required by analysis.

10.5.4 — For structural slabs and footings of uniform thickness, $A_{s,min}$ in the direction of the span shall be the same as that required by 7.12. Maximum spacing of this reinforcement shall not exceed three times the thickness, nor 18 in.

10.6 — Distribution of flexural reinforcement in beams and one-way slabs

10.6.1 — This section prescribes rules for distribution of flexural reinforcement to control flexural cracking in beams and in one-way slabs (slabs reinforced to resist flexural stresses in only one direction).

COMMENTARY

R10.5.3 — The minimum reinforcement required by Eq. (10-3) is to be provided wherever reinforcement is needed, except where such reinforcement is at least one-third greater than that required by analysis. This exception provides sufficient additional reinforcement in large members where the amount required by 10.5.1 or 10.5.2 would be excessive.

R10.5.4 — The minimum reinforcement required for slabs should be equal to the same amount as that required by 7.12 for shrinkage and temperature reinforcement.

Soil-supported slabs such as slabs on grade are not considered to be structural slabs in the context of this section, unless they transmit vertical loads from other parts of the structure to the soil. Reinforcement, if any, in soil-supported slabs should be proportioned with due consideration of all design forces. Mat foundations and other slabs that help support the structure vertically should meet the requirements of this section.

In reevaluating the overall treatment of 10.5, the maximum spacing for reinforcement in structural slabs (including footings) was reduced from the $5h$ for temperature and shrinkage reinforcement to the compromise value of $3h$, which is somewhat larger than the $2h$ limit of 13.3.2 for two-way slab systems.

R10.6 — Distribution of flexural reinforcement in beams and one-way slabs

R10.6.1 — Many structures designed by working stress methods and with low steel stress served their intended functions with very limited flexural cracking. When high strength reinforcing steels are used at high service load stresses, however, visible cracks should be expected, and steps should be taken in detailing of the reinforcement to control cracking. For reasons of durability and appearance, many fine cracks are preferable to a few wide cracks.

Control of cracking is particularly important when reinforcement with a yield strength in excess of 40,000 psi is used. Current good detailing practices will usually lead to adequate crack control even when reinforcement of 60,000 psi yield strength is used.

Extensive laboratory work[10.12-10.14] involving deformed bars has confirmed that crack width at service loads is proportional to steel stress. The significant variables reflecting steel detailing were found to be thickness of concrete cover and the spacing of reinforcement.

Crack width is inherently subject to wide scatter even in careful laboratory work and is influenced by shrinkage and other time-dependent effects. Improved crack control is obtained when the steel reinforcement is well distributed over the zone of maximum concrete tension.

CODE

10.6.2 — Distribution of flexural reinforcement in two-way slabs shall be as required by 13.3.

10.6.3 — Flexural tension reinforcement shall be well distributed within maximum flexural tension zones of a member cross section as required by 10.6.4.

10.6.4 — The spacing of reinforcement closest to the tension face, s, shall not exceed that given by

$$s = 15\left(\frac{40{,}000}{f_s}\right) - 2.5c_c \qquad (10\text{-}4)$$

but not greater than $12(40{,}000/f_s)$, where c_c is the least distance from surface of reinforcement or prestressing steel to the tension face. If there is only one bar or wire nearest to the extreme tension face, s used in Eq. (10-4) is the width of the extreme tension face.

Calculated stress f_s in reinforcement closest to the tension face at service load shall be computed based on the unfactored moment. It shall be permitted to take f_s as $2/3f_y$.

10.6.5 — Provisions of 10.6.4 are not sufficient for structures subject to very aggressive exposure or designed to be watertight. For such structures, special investigations and precautions are required.

10.6.6 — Where flanges of T-beam construction are in tension, part of the flexural tension reinforcement shall be distributed over an effective flange width as defined in 8.10, or a width equal to one-tenth the span, whichever is smaller. If the effective flange width exceeds one-tenth the span, some longitudinal reinforcement shall be provided in the outer portions of the flange.

10.6.7 — Where h of a beam or joist exceeds 36 in., longitudinal skin reinforcement shall be uniformly distributed along both side faces of the member. Skin reinforcement shall extend for a distance $h/2$ from the tension face. The spacing s shall be as provided in 10.6.4, where c_c is the least distance from

COMMENTARY

R10.6.3 — Several bars at moderate spacing are much more effective in controlling cracking than one or two larger bars of equivalent area.

R10.6.4 — This section was updated in the 2005 edition to reflect the higher service stresses that occur in flexural reinforcement with the use of the load combinations introduced in the 2002 code. The maximum bar spacing is specified directly to control cracking.[10.15,10.16,10.17] For the usual case of beams with Grade 60 reinforcement and 2 in. clear cover to the main reinforcement, with f_s = 40,000 psi, the maximum bar spacing is 10 in.

Crack widths in structures are highly variable. In codes before the 1999 edition, provisions were given for distribution of reinforcement that were based on empirical equations using a calculated maximum crack width of 0.016 in. The current provisions for spacing are intended to limit surface cracks to a width that is generally acceptable in practice but may vary widely in a given structure.

The role of cracks in the corrosion of reinforcement is controversial. Research[10.18,10.19] shows that corrosion is not clearly correlated with surface crack widths in the range normally found with reinforcement stresses at service load levels. For this reason, the former distinction between interior and exterior exposure has been eliminated.

R10.6.5 — Although a number of studies have been conducted, clear experimental evidence is not available regarding the crack width beyond which a corrosion danger exists. Exposure tests indicate that concrete quality, adequate compaction, and ample concrete cover may be of greater importance for corrosion protection than crack width at the concrete surface.

R10.6.6 — In major T-beams, distribution of the negative reinforcement for control of cracking should take into account two considerations: (1) wide spacing of the reinforcement across the full effective width of flange may cause some wide cracks to form in the slab near the web and, (2) close spacing near the web leaves the outer regions of the flange unprotected. The one-tenth limitation is to guard against too wide a spacing, with some additional reinforcement required to protect the outer portions of the flange.

R10.6.7 — For relatively deep flexural members, some reinforcement should be placed near the vertical faces of the tension zone to control cracking in the web.[10.20, 10.21] (See Fig. R10.6.7.) Without such auxiliary steel, the width of the cracks in the web may exceed the crack widths at the level of the flexural tension reinforcement. This section was

CODE

the surface of the skin reinforcement or prestressing steel to the side face. It shall be permitted to include such reinforcement in strength computations if a strain compatibility analysis is made to determine stress in the individual bars or wires.

10.7 — Deep beams

10.7.1 — Deep beams are members loaded on one face and supported on the opposite face so that compression struts can develop between the loads and the supports, and have either:

(a) clear spans, ℓ_n, equal to or less than four times the overall member depth; or

(b) regions with concentrated loads within twice the member depth from the face of the support.

Deep beams shall be designed either taking into account nonlinear distribution of strain, or by Appendix A. (See also 11.8.1 and 12.10.6.) Lateral buckling shall be considered.

10.7.2 — V_n of deep beams shall be in accordance with 11.8.

10.7.3 — Minimum area of flexural tension reinforcement, $A_{s,min}$, shall conform to 10.5.

10.7.4 — Minimum horizontal and vertical reinforcement in the side faces of deep beams shall satisfy either A.3.3 or 11.8.4 and 11.8.5.

COMMENTARY

modified in the 2005 edition to make the skin reinforcement spacing consistent with that of the flexural reinforcement. The size of the skin reinforcement is not specified; research has indicated that the spacing rather than bar size is of primary importance.[10.21] Bar sizes No. 3 to No. 5 (or welded wire reinforcement with a minimum area of 0.1 in.^{2} per foot of depth) are typically provided.

Where the provisions for deep beams, walls, or precast panels require more reinforcement, those provisions (along with their spacing requirements) will govern.

R10.7 — Deep beams

The span-to-depth ratios used to define deep beams in the 1999 and earlier codes were based on papers published in 1946 and 1953. The definitions of deep beams given in 10.7.1 and 11.7.1 of these earlier codes were different from each other and different from the current code definition that is based on D-region behavior (see Appendix A). Since 2002, the definitions of deep beams in Sections 10.7.1 and 11.7.1 are consistent with each other.

This code does not contain detailed requirements for designing deep beams for flexure except that nonlinearity of strain distribution and lateral buckling is to be considered. Suggestions for the design of deep beams for flexure are given in References 10.22, 10.23, and 10.24.

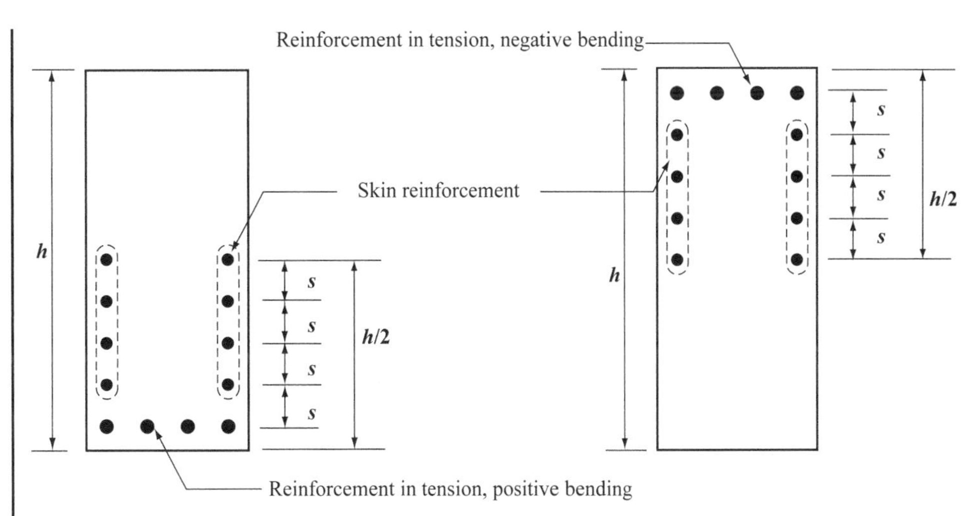

Fig. R10.6.7—Skin reinforcement for beams and joists with **h** *> 36 in.*

CODE

10.8 — Design dimensions for compression members

10.8.1 — Isolated compression member with multiple spirals

Outer limits of the effective cross section of a compression member with two or more interlocking spirals shall be taken at a distance outside the extreme limits of the spirals equal to the minimum concrete cover required by 7.7.

10.8.2 — Compression member built monolithically with wall

Outer limits of the effective cross section of a spirally reinforced or tied reinforced compression member built monolithically with a concrete wall or pier shall be taken not greater than 1-1/2 in. outside the spiral or tie reinforcement.

10.8.3 — Equivalent circular compression member

As an alternative to using the full gross area for design of a compression member with a square, octagonal, or other shaped cross section, it shall be permitted to use a circular section with a diameter equal to the least lateral dimension of the actual shape. Gross area considered, required percentage of reinforcement, and design strength shall be based on that circular section.

10.8.4 — Limits of section

For a compression member with a cross section larger than required by considerations of loading, it shall be permitted to base the minimum reinforcement and strength on a reduced effective area A_g not less than one-half the total area. This provision shall not apply in regions of high seismic risk.

10.9 — Limits for reinforcement of compression members

10.9.1 — Area of longitudinal reinforcement, A_{st}, for noncomposite compression members shall be not less than $0.01A_g$ or more than $0.08A_g$.

COMMENTARY

R10.8 — Design dimensions for compression members

With the 1971 code, minimum sizes for compression members were eliminated to allow wider utilization of reinforced concrete compression members in smaller size and lightly loaded structures, such as low-rise residential and light office buildings. The engineer should recognize the need for careful workmanship, as well as the increased significance of shrinkage stresses with small sections.

R10.8.2, R10.8.3, and R10.8.4 — For column design,[10.25] the code provisions for quantity of reinforcement, both vertical and spiral, are based on the gross column area and core area, and the design strength of the column is based on the gross area of the column section. In some cases, however, the gross area is larger than necessary to carry the factored load. The basis of 10.8.2, 10.8.3, and 10.8.4 is that it is satisfactory to design a column of sufficient size to carry the factored load and then simply add concrete around the designed section without increasing the reinforcement to meet the minimum percentages required by 10.9.1. The additional concrete should not be considered as carrying load; however, the effects of the additional concrete on member stiffness should be included in the structural analysis. The effects of the additional concrete also should be considered in design of the other parts of the structure that interact with the oversize member.

R10.9 — Limits for reinforcement of compression members

R10.9.1 — This section prescribes the limits on the amount of longitudinal reinforcement for noncomposite compression members. If the use of high reinforcement ratios would involve practical difficulties in the placing of concrete, a lower percentage and hence a larger column, or higher strength concrete or reinforcement (see R9.4) should be considered. The percentage of reinforcement in columns should usually not exceed 4 percent if the column bars are required to be lap spliced.

Minimum reinforcement — Since the design methods for columns incorporate separate terms for the load carried by concrete and by reinforcement, it is necessary to specify

CODE

COMMENTARY

some minimum amount of reinforcement to ensure that only reinforced concrete columns are designed by these procedures. Reinforcement is necessary to provide resistance to bending, which may exist whether or not computations show that bending exists, and to reduce the effects of creep and shrinkage of the concrete under sustained compressive stresses. Tests have shown that creep and shrinkage tend to transfer load from the concrete to the reinforcement, with a consequent increase in stress in the reinforcement, and that this increase is greater as the ratio of reinforcement decreases. Unless a lower limit is placed on this ratio, the stress in the reinforcement may increase to the yield level under sustained service loads. This phenomenon was emphasized in the report of ACI Committee 105[10.26] and minimum reinforcement ratios of 0.01 and 0.005 were recommended for spiral and tied columns, respectively. However, in all editions of the code since 1936, the minimum ratio has been 0.01 for both types of laterally reinforced columns.

Maximum reinforcement — Extensive tests of the ACI column investigation[10.26] included reinforcement ratios no greater than 0.06. Although other tests with as much as 17 percent reinforcement in the form of bars produced results similar to those obtained previously, it is necessary to note that the loads in these tests were applied through bearing plates on the ends of the columns and the problem of transferring a proportional amount of the load to the bars was thus minimized or avoided. Maximum ratios of 0.08 and 0.03 were recommended by ACI Committee 105[10.26] for spiral and tied columns, respectively. In the 1936 code, these limits were made 0.08 and 0.04, respectively. In the 1956 code, the limit for tied columns with bending was raised to 0.08. Since the 1963 code, it has been required that bending be considered in the design of all columns, and the maximum ratio of 0.08 has been applied to both types of columns. This limit can be considered a practical maximum for reinforcement in terms of economy and requirements for placing.

10.9.2 — Minimum number of longitudinal bars in compression members shall be 4 for bars within rectangular or circular ties, 3 for bars within triangular ties, and 6 for bars enclosed by spirals conforming to 10.9.3.

R10.9.2 — For compression members, a minimum of four longitudinal bars are required when bars are enclosed by rectangular or circular ties. For other shapes, one bar should be provided at each apex or corner and proper lateral reinforcement provided. For example, tied triangular columns require three longitudinal bars, one at each apex of the triangular ties. For bars enclosed by spirals, six bars are required.

When the number of bars in a circular arrangement is less than eight, the orientation of the bars will affect the moment strength of eccentrically loaded columns and should be considered in design.

10.9.3 — Volumetric spiral reinforcement ratio, ρ_s, shall be not less than the value given by

$$\rho_s = 0.45\left(\frac{A_g}{A_{ch}} - 1\right)\frac{f_c'}{f_{yt}} \qquad (10\text{-}5)$$

R10.9.3 — The effect of spiral reinforcement in increasing the load-carrying strength of the concrete within the core is not realized until the column has been subjected to a load and deformation sufficient to cause the concrete shell outside the core to spall off. The amount of spiral reinforcement

CODE

where the value of f_{yt} used in Eq. (10-5) shall not exceed 100,000 psi. For f_{yt} greater than 60,000 psi, lap splices according to 7.10.4.5(a) shall not be used.

COMMENTARY

required by Eq. (10-5) is intended to provide additional load-carrying strength for concentrically loaded columns equal to or slightly greater than the strength lost when the shell spalls off. This principle was recommended by ACI Committee 105[10.26] and has been a part of the code since 1963. The derivation of Eq. (10-5) is given in the ACI Committee 105 report. Tests and experience show that columns containing the amount of spiral reinforcement required by this section exhibit considerable toughness and ductility. Research[10.27-10.29] has indicated that 100,000 psi yield strength reinforcement can be used for confinement. For the 2005 code, the limit in yield strength for spiral reinforcement was increased from 60,000 to 100,000 psi.

10.10 — Slenderness effects in compression members

10.10.1 — Except as allowed in 10.10.2, the design of compression members, restraining beams, and other supporting members shall be based on the factored forces and moments from a second-order analysis considering material nonlinearity and cracking, as well as the effects of member curvature and lateral drift, duration of the loads, shrinkage and creep, and interaction with the supporting foundation. The dimensions of each member cross section used in the analysis shall be within 10 percent of the dimensions of the members shown on the design drawings or the analysis shall be repeated. The analysis procedure shall have been shown to result in prediction of strength in substantial agreement with the results of comprehensive tests of columns in statically indeterminate reinforced concrete structures.

10.10.2 — As an alternate to the procedure prescribed in 10.10.1, it shall be permitted to base the design of compression members, restraining beams, and other supporting members on axial forces and moments from the analyses described in 10.11.

R10.10 — Slenderness effects in compression members

Provisions for slenderness effects in compression members and frames were revised in the 1995 code to better recognize the use of second-order analyses and to improve the arrangement of the provisions dealing with sway (unbraced) and nonsway (braced) frames.[10.30] The use of a refined nonlinear second-order analysis is permitted in 10.10.1. Sections 10.11, 10.12, and 10.13 present an approximate design method based on the traditional moment magnifier method. For sway frames, the magnified sway moment $\delta_s M_s$ may be calculated using a second-order elastic analysis, by an approximation to such an analysis, or by the traditional sway moment magnifier.

R10.10.1 — Two limits are placed on the use of the refined second-order analysis. First, the structure that is analyzed should have members similar to those in the final structure. If the members in the final structure have cross-sectional dimensions more than 10 percent different from those assumed in the analysis, new member properties should be computed and the analysis repeated. Second, the refined second-order analysis procedure should have been shown to predict ultimate loads within 15 percent of those reported in tests of indeterminate reinforced concrete structures. At the very least, the comparison should include tests of columns in planar nonsway frames, sway frames, and frames with varying column stiffnesses. To allow for variability in the actual member properties and in the analysis, the member properties used in analysis should be multiplied by a stiffness reduction factor ϕ_K less than one. For consistency with the second-order analysis in 10.13.4.1, the stiffness reduction factor ϕ_K can be taken as 0.80. The concept of a stiffness reduction factor ϕ_K is discussed in R10.12.3.

R10.10.2 — As an alternate to the refined second-order analysis of 10.10.1, design may be based on elastic analyses and the moment magnifier approach.[10.31,10.32] For sway frames the magnified sway moments may be calculated using a second-order elastic analysis based on realistic stiffness values. See R10.13.4.1.

CODE

10.11 — Magnified moments — General

10.11.1 — Factored axial forces P_u, factored moments M_1 and M_2 at the ends of the column, and, where required, relative lateral story deflections, Δ_o, shall be computed using an elastic first-order frame analysis with the section properties determined taking into account the influence of axial loads, the presence of cracked regions along the length of the member, and effects of duration of the loads. Alternatively, it shall be permitted to use the following properties for the members in the structure:

(a) Modulus of elasticity .... E_c from 8.5.1

(b) Moments of inertia, I

- Beams .... $0.35I_g$
- Columns .... $0.70I_g$
- Walls—Uncracked .... $0.70I_g$
- —Cracked .... $0.35I_g$
- Flat plates and flat slabs .... $0.25I_g$

(c) Area .... $1.0A_g$

In (b), I shall be divided by $(1 + \beta_d)$ when sustained lateral loads act or for stability checks made in accordance with 10.13.6. For nonsway frames, β_d is ratio of maximum factored axial sustained load to maximum factored axial load associated with the same load combination. For sway frames except as specified in 10.13.6, β_d is ratio of maximum factored sustained shear within a story to the maximum factored shear in that story.

COMMENTARY

R10.11 — Magnified moments — General

This section describes an approximate design procedure that uses the moment magnifier concept to account for slenderness effects. Moments computed using an ordinary first-order frame analysis are multiplied by a moment magnifier that is a function of the factored axial force P_u and the critical buckling load P_c for the column. Nonsway and sway frames are treated separately in 10.12 and 10.13. Provisions applicable to both nonsway and sway columns are given in 10.11. A first-order frame analysis is an elastic analysis that does not include the internal force effects resulting from deflections.

R10.11.1 — The stiffnesses EI used in an elastic analysis used for strength design should represent the stiffnesses of the members immediately prior to failure. This is particularly true for a second-order analysis that should predict the lateral deflections at loads approaching ultimate. The EI values should not be based totally on the moment-curvature relationship for the most highly loaded section along the length of each member. Instead, they should correspond to the moment-end rotation relationship for a complete member.

The alternative values of E_c, I_g, and A_g given in 10.11.1 have been chosen from the results of frame tests and analyses and include an allowance for the variability of the computed deflections. The modulus of elasticity of the concrete E_c is based on the specified concrete compressive strength while the sway deflections are a function of the average concrete strength, which is higher. The moments of inertia were taken as 0.875 times those in Reference 10.33. These two effects result in an overestimation of the second-order deflections in the order of 20 to 25 percent, corresponding to an implicit stiffness reduction factor ϕ_K of 0.80 to 0.85 on the stability calculation. The concept of a stiffness reduction factor ϕ_K is discussed in R10.12.3

The moment of inertia of T-beams should be based on the effective flange width defined in 8.10. It is generally sufficiently accurate to take I_g of a T-beam as two times the I_g for the web, $2(b_w h^3/12)$.

If the factored moments and shears from an analysis based on the moment of inertia of a wall taken equal to $0.70I_g$ indicate that the wall will crack in flexure, based on the modulus of rupture, the analysis should be repeated with $I = 0.35I_g$ in those stories where cracking is predicted at factored loads.

The alternative values of the moments of inertia given in 10.11.1 were derived for nonprestressed members. For prestressed members, the moments of inertia may differ from the values in 10.11.1 depending on the amount, location, and type of the reinforcement and the degree of cracking prior to ultimate. The stiffness values for

CODE

COMMENTARY

prestressed concrete members should include an allowance for the variability of the stiffnesses.

Sections 10.11 through 10.13 provide requirements for strength and assume frame analyses will be carried out using factored loads. Analyses of deflections, vibrations, and building periods are needed at various service (unfactored) load levels[10.34,10.35] to determine the serviceability of the structure and to estimate the wind forces in wind tunnel laboratories. The seismic base shear is also based on the service load periods of vibration. The magnified service loads and deflections by a second-order analysis should also be computed using service loads. The moments of inertia of the structural members in the service load analyses should, therefore, be representative of the degree of cracking at the various service load levels investigated. Unless a more accurate estimate of the degree of cracking at design service load level is available, it is satisfactory to use **1/0.70 = 1.43** times the moments of inertia given in 10.11.1 for service load analyses.

The unusual case of sustained lateral loads might exist, for example, if there were permanent lateral loads resulting from unequal earth pressures on two sides of a building.

10.11.2 — It shall be permitted to take the radius of gyration, r, equal to 0.30 times the overall dimension in the direction stability is being considered for rectangular compression members and 0.25 times the diameter for circular compression members. For other shapes, it shall be permitted to compute r for the gross concrete section.

10.11.3 — Unsupported length of compression members

10.11.3.1 — The unsupported length of a compression member, ℓ_u, shall be taken as the clear distance between floor slabs, beams, or other members capable of providing lateral support in the direction being considered.

10.11.3.2 — Where column capitals or haunches are present, ℓ_u shall be measured to the lower extremity of the capital or haunch in the plane considered.

10.11.4 — Columns and stories in structures shall be designated as nonsway or sway columns or stories. The design of columns in nonsway frames or stories shall be based on 10.12. The design of columns in sway frames or stories shall be based on 10.13.

10.11.4.1 — It shall be permitted to assume a column in a structure is nonsway if the increase in column end moments due to second-order effects does not exceed 5 percent of the first-order end moments.

R10.11.4 — The moment magnifier design method requires the designer to distinguish between nonsway frames, which are designed according to 10.12, and sway frames, which are designed according to 10.13. Frequently this can be done by inspection by comparing the total lateral stiffness of the columns in a story to that of the bracing elements. A compression member may be assumed nonsway by inspection if it is located in a story in which the bracing elements (shearwalls, shear trusses, or other types of lateral bracing) have such substantial lateral stiffness to resist the lateral

CODE

10.11.4.2 — It also shall be permitted to assume a story within a structure is nonsway if:

$$Q = \frac{\Sigma P_u \Delta_o}{V_{us} \ell_c} \quad (10\text{-}6)$$

is less than or equal to 0.05, where ΣP_u and V_{us} are the total factored vertical load and the horizontal story shear, respectively, in the story being evaluated, and Δ_o is the first-order relative lateral deflection between the top and bottom of that story due to V_{us}.

10.11.5 — Where an individual compression member in the frame has a slenderness $k\ell_u/r$ of more than 100, 10.10.1 shall be used to compute the forces and moments in the frame.

10.11.6 — For compression members subject to bending about both principal axes, the moment about each axis shall be magnified separately based on the conditions of restraint corresponding to that axis.

10.12 — Magnified moments — Nonsway frames

10.12.1 — For compression members in nonsway frames, the effective length factor, k, shall be taken as 1.0, unless analysis shows that a lower value is justified. The calculation of k shall be based on the values of E_c and I given in 10.11.1.

COMMENTARY

deflections of the story that any resulting lateral deflection is not large enough to affect the column strength substantially. If not readily apparent by inspection, 10.11.4.1 and 10.11.4.2 give two possible ways of doing this. In 10.11.4.1, a story in a frame is said to be nonsway if the increase in the lateral load moments resulting from $P\Delta$ effects does not exceed 5 percent of the first-order moments.[10.33] Section 10.11.4.2 gives an alternative method of determining this based on the stability index for a story Q. In computing Q, ΣP_u should correspond to the lateral loading case for which ΣP_u is greatest. A frame may contain both nonsway and sway stories. This test would not be suitable if V_{us} is zero.

If the lateral load deflections of the frame have been computed using service loads and the service load moments of inertia given in 10.11.1, it is permissible to compute Q in Eq. (10-6) using 1.2 times the sum of the service gravity loads, the service load story shear, and 1.43 times the first-order service load story deflections.

R10.11.5 — An upper limit is imposed on the slenderness ratio of columns designed by the moment magnifier method of 10.11 to 10.13. No similar limit is imposed if design is carried out according to 10.10.1. The limit of $k\ell_u/r = 100$ represents the upper range of actual tests of slender compression members in frames.

R10.11.6 — When biaxial bending occurs in a compression member, the computed moments about each principal axes should be magnified. The magnification factors δ are computed considering the critical buckling load P_c about each axis separately based on the appropriate effective length $k\ell_u$ and the stiffness EI. If the buckling capacities are different about the two axes, different magnification factors will result.

R10.12 — Magnified moments — Nonsway frames

R10.12.1 — The moment magnifier equations were derived for hinged end columns and should be modified to account for the effect of end restraints. This is done by using an effective length $k\ell_u$ in the computation of P_c.

The primary design aid to estimate the effective length factor k is the Jackson and Moreland Alignment Charts (Fig. R10.12.1), which allow a graphical determination of k for a column of constant cross section in a multibay frame.[10.36,10.37]

The effective length is a function of the relative stiffness at each end of the compression member. Studies have indicated that the effects of varying beam and column reinforcement percentages and beam cracking should be considered in determining the relative end stiffnesses. In determining ψ for use in evaluating the effective length factor k, the rigidity of the flexural members may be calculated on the basis of

COMMENTARY

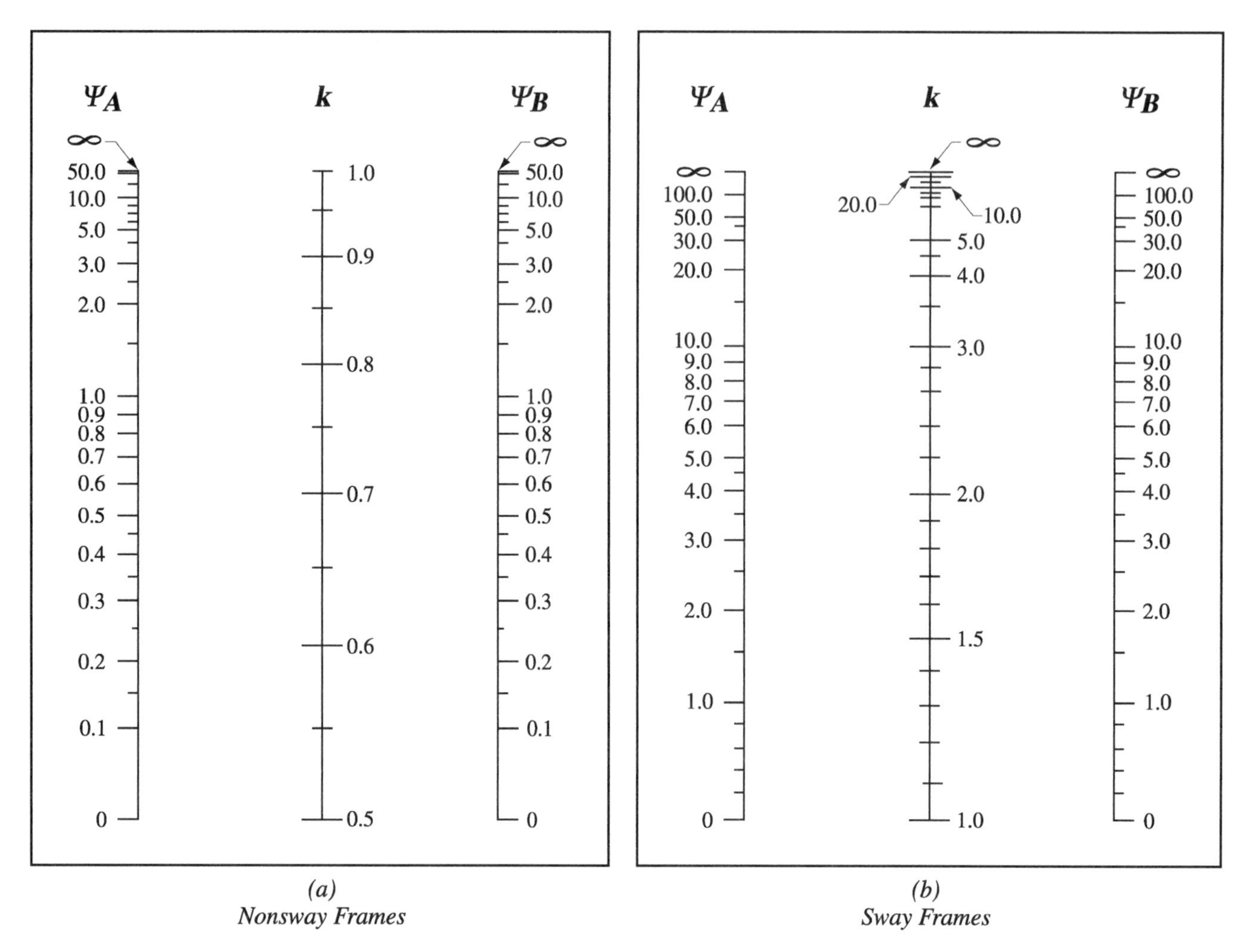

Ψ = ratio of Σ(EI/ℓc) of compression members to Σ(EI/ℓ) of flexural members in a plane at one end of a compression member
ℓ = span length of flexural member measured center to center of joints

Fig. R10.12.1—Effective length factors, k

$\mathbf{0.35}\boldsymbol{I_g}$ for flexural members to account for the effect of cracking and reinforcement on relative stiffness, and $\mathbf{0.70}\boldsymbol{I_g}$ for compression members.

The simplified equations (A-E), listed below for computing the effective length factors for nonsway and sway members, may be used. Eq. (A), (B), and (E) are taken from the 1972 British Standard Code of Practice.[10.38,10.39] Eq. (C) and (D) for sway members were developed in Reference 10.25.

For compression members in a nonsway frame, an upper bound to the effective length factor may be taken as the smaller of the following two expressions:

$$k = 0.7 + 0.05\,(\psi_A + \psi_B) \leq 1.0 \qquad \text{(A)}$$

$$k = 0.85 + 0.05\,\psi_{min} \leq 1.0 \qquad \text{(B)}$$

where ψ_A and ψ_B are the values of ψ at the two ends of the column and ψ_{min} is the smaller of the two values.

CODE

COMMENTARY

For compression members in a sway frame, restrained at both ends, the effective length factor may be taken as:

For $\psi_m < 2$

$$k = \frac{20 - \psi_m}{20}\sqrt{1 + \psi_m} \qquad \text{(C)}$$

For $\psi_m \geq 2$

$$k = 0.9\sqrt{1 + \psi_m} \qquad \text{(D)}$$

where ψ_m is the average of the ψ-values at the two ends of the compression member.

For compression members in a sway frame, hinged at one end, the effective length factor may be taken as:

$$k = 2.0 + 0.3\psi \qquad \text{(E)}$$

where ψ is the value at the restrained end.

The use of the charts in Fig. R10.12.1, or the equations in this section, may be considered as satisfying the requirements of the code to justify k less than 1.0.

10.12.2 — In nonsway frames it shall be permitted to ignore slenderness effects for compression members that satisfy:

$$\frac{k\ell_u}{r} \leq 34 - 12(M_1/M_2) \qquad \text{(10-7)}$$

where the term [$34 - 12M_1/M_2$] shall not be taken greater than 40. The term M_1/M_2 is positive if the member is bent in single curvature, and negative if the member is bent in double curvature.

R10.12.2 — Eq. (10-7) is derived from Eq. (10-9) assuming that a 5 percent increase in moments due to slenderness is acceptable.[10.31] The derivation did not include ϕ in the calculation of the moment magnifier. As a first approximation, k may be taken equal to 1.0 in Eq. (10-7).

10.12.3 — Compression members shall be designed for factored axial force P_u and the moment amplified for the effects of member curvature M_c as follows:

$$M_c = \delta_{ns} M_2 \qquad \text{(10-8)}$$

where

$$\delta_{ns} = \frac{C_m}{1 - \dfrac{P_u}{0.75P_c}} \geq 1.0 \qquad \text{(10-9)}$$

$$P_c = \frac{\pi^2 EI}{(k\ell_u)^2} \qquad \text{(10-10)}$$

R10.12.3 — The ϕ-factors used in the design of slender columns represent two different sources of variability. First, the stiffness reduction ϕ-factors in the magnifier equations in the 1989 and earlier codes were intended to account for the variability in the stiffness EI and the moment magnification analysis. Second, the variability of the strength of the cross section is accounted for by strength reduction ϕ-factors of 0.70 for tied columns and 0.75 for spiral columns. Studies reported in Reference 10.40 indicate that the stiffness reduction factor ϕ_K, and the cross-sectional strength reduction ϕ-factors do not have the same values, contrary to the assumption in the 1989 and earlier codes. These studies suggest the stiffness reduction factor ϕ_K for an isolated column should be 0.75 for both tied and spiral columns. The 0.75 factors in Eq. (10-9) and (10-18) are stiffness reduction factors ϕ_K and replace the ϕ-factors in these equations in the 1989 and earlier codes. This has been done to avoid confusion between a

CODE

EI shall be taken as

$$EI = \frac{(0.2E_cI_g + E_sI_{se})}{1 + \beta_d} \quad (10\text{-}11)$$

or

$$EI = \frac{0.4E_cI_g}{1 + \beta_d} \quad (10\text{-}12)$$

10.12.3.1 — For members without transverse loads between supports, C_m shall be taken as

$$C_m = 0.6 + 0.4\frac{M_1}{M_2} \geq 0.4 \quad (10\text{-}13)$$

where M_1/M_2 is positive if the column is bent in single curvature. For members with transverse loads between supports, C_m shall be taken as 1.0.

10.12.3.2 — Factored moment, M_2, in Eq. (10-8) shall not be taken less than

$$M_{2,min} = P_u(0.6 + 0.03h) \quad (10\text{-}14)$$

about each axis separately, where 0.6 and h are in inches. For members for which $M_{2,min}$ exceeds M_2,

COMMENTARY

stiffness reduction factor ϕ_K in Eq. (10-9) and (10-18), and the cross-sectional strength reduction ϕ-factors.

In defining the critical load, the main problem is the choice of a stiffness EI that reasonably approximates the variations in stiffness due to cracking, creep, and the nonlinearity of the concrete stress-strain curve. Eq. (10-11) was derived for small eccentricity ratios and high levels of axial load where the slenderness effects are most pronounced.

Creep due to sustained load will increase the lateral deflections of a column and hence the moment magnification. This is approximated for design by reducing the stiffness EI used to compute P_c and hence δ_{ns} by dividing EI by $(1 + \beta_d)$. Both the concrete and steel terms in Eq. (10-11) are divided by $(1 + \beta_d)$. This reflects the premature yielding of steel in columns subjected to sustained load.

Either Eq. (10-11) or (10-12) may be used to compute EI. Eq. (10-12) is a simplified approximation to Eq. (10-11). It is less accurate than Eq. (10-11).[10.41] Eq. (10-12) may be simplified further by assuming $\beta_d = 0.6$. When this is done Eq. (10-12) becomes

$$EI = 0.25E_cI_g \quad (F)$$

The term β_d is defined differently for nonsway and sway frames. See 2.1. For nonsway frames, β_d is the ratio of the maximum factored axial sustained load to the maximum factored axial load.

R10.12.3.1 — The factor C_m is a correction factor relating the actual moment diagram to an equivalent uniform moment diagram. The derivation of the moment magnifier assumes that the maximum moment is at or near midheight of the column. If the maximum moment occurs at one end of the column, design should be based on an equivalent uniform moment C_mM_2 that would lead to the same maximum moment when magnified.[10.31]

In the case of compression members that are subjected to transverse loading between supports, it is possible that the maximum moment will occur at a section away from the end of the member. If this occurs, the value of the largest calculated moment occurring anywhere along the member should be used for the value of M_2 in Eq. (10-8). In accordance with the last sentence of 10.12.3.1, C_m is to be taken as 1.0 for this case.

R10.12.3.2 — In the code, slenderness is accounted for by magnifying the column end moments. If the factored column moments are very small or zero, the design of slender columns should be based on the minimum eccentricity given in this section. It is not intended that the minimum eccentricity be applied about both axes simultaneously.

CODE

the value of C_m in Eq. (10-13) shall either be taken equal to 1.0, or shall be based on the ratio of the computed end moments M_1 to M_2.

COMMENTARY

The factored column end moments from the structural analysis are used in Eq. (10-13) in determining the ratio M_1/M_2 for the column when the design should be based on minimum eccentricity. This eliminates what would otherwise be a discontinuity between columns with computed eccentricities less than the minimum eccentricity and columns with computed eccentricities equal to or greater than the minimum eccentricity.

10.13 — Magnified moments — Sway frames

R10.13 — Magnified moments — Sway frames

The design of sway frames for slenderness was revised in the 1995 code. The revised procedure consists of three steps:

(1) The magnified sway moments $\delta_s M_s$ are computed. This should be done in one of three ways. First, a second-order elastic frame analysis may be used (10.13.4.1). Second, an approximation to such analysis (10.13.4.2) may be used. The third option is to use the sway magnifier δ_s from previous editions of the code (10.13.4.3);

(2) The magnified sway moments $\delta_s M_s$ are added to the unmagnified nonsway moment M_{ns} at each end of each column (10.13.3). The nonsway moments may be computed using a first-order elastic analysis;

(3) If the column is slender and heavily loaded, it is checked to see whether the moments at points between the ends of the column exceed those at the ends of the column. As specified in 10.13.5 this is done using the nonsway frame magnifier δ_{ns} with P_c computed assuming k = **1.0** or less.

10.13.1 — For compression members not braced against sidesway, the effective length factor k shall be determined using the values of E_c and I given in 10.11.1 and shall not be less than 1.0.

R10.13.1 — See R10.12.1.

10.13.2 — For compression members not braced against sidesway, it shall be permitted to neglect the effects of slenderness when $k\ell_u/r$ is less than 22.

10.13.3 — Moments M_1 and M_2 at the ends of an individual compression member shall be taken as

$$M_1 = M_{1ns} + \delta_s M_{1s} \quad (10\text{-}15)$$

$$M_2 = M_{2ns} + \delta_s M_{2s} \quad (10\text{-}16)$$

where $\delta_s M_{1s}$ and $\delta_s M_{2s}$ shall be computed according to 10.13.4.

R10.13.3 — The analysis described in this section deals only with plane frames subjected to loads causing deflections in that plane. If torsional displacements are significant, a three-dimensional second-order analysis should be used.

CODE

10.13.4 — Calculation of $\delta_s M_s$

10.13.4.1 — Magnified sway moments $\delta_s M_s$ shall be taken as the column end moments calculated using a second-order elastic analysis based on the member stiffnesses given in 10.11.1.

10.13.4.2 — Alternatively, it shall be permitted to calculate $\delta_s M_s$ as

$$\delta_s M_s = \frac{M_s}{1 - Q} \geq M_s \tag{10-17}$$

If δ_s calculated in this way exceeds 1.5, $\delta_s M_s$ shall be calculated using 10.13.4.1 or 10.13.4.3.

COMMENTARY

R10.13.4 — Calculation of $\delta_s M_s$

R10.13.4.1 — A second-order analysis is a frame analysis that includes the internal force effects resulting from deflections. When a second-order elastic analysis is used to compute $\delta_s M_s$, the deflections should be representative of the stage immediately prior to the ultimate load. For this reason the reduced $E_c I_g$ values given in 10.11.1 should be used in the second-order analysis.

The term β_d is defined differently for nonsway and sway frames. See 2.1. Sway deflections due to short-term loads such as wind or earthquake are a function of the short-term stiffness of the columns following a period of sustained gravity load. For this case the definition of β_d in 10.0 gives $\beta_d = 0$. In the unusual case of a sway frame where the lateral loads are sustained, β_d will not be zero. This might occur if a building on a sloping site is subjected to earth pressure on one side but not on the other.

In a second-order analysis the axial loads in all columns that are not part of the lateral load resisting elements and depend on these elements for stability should be included.

In the 1989 and earlier codes, the moment magnifier equations for δ_b and δ_s included a stiffness reduction factor ϕ_K to cover the variability in the stability calculation. The second-order analysis method is based on the values of E_c and I from 10.11.1. These lead to a 20 to 25 percent overestimation of the lateral deflections that corresponds to a stiffness reduction factor ϕ_K between 0.80 and 0.85 on the $P\Delta$ moments. No additional ϕ-factor is needed in the stability calculation. Once the moments are established, selection of the cross sections of the columns involves the strength reduction factors ϕ from 9.3.2.2.

R10.13.4.2 — The iterative $P\Delta$ analysis for second-order moments can be represented by an infinite series. The solution of this series is given by Eq. (10-17).[10.33] Reference 10.42 shows that Eq. (10-17) closely predicts the second-order moments in a sway frame until δ_s exceeds 1.5.

The $P\Delta$ moment diagrams for deflected columns are curved, with Δ related to the deflected shape of the columns. Eq. (10-17) and most commercially available second-order frame analyses have been derived assuming that the $P\Delta$ moments result from equal and opposite forces of $P\Delta/\ell_c$ applied at the bottom and top of the story. These forces give a straight line $P\Delta$ moment diagram. The curved $P\Delta$ moment diagrams lead to lateral displacements on the order of 15 percent larger than those from the straight line $P\Delta$ moment diagrams. This effect can be included in Eq. (10-17) by writing the denominator as $(1 - 1.15Q)$ rather than $(1 - Q)$. The 1.15 factor has been left out of Eq. (10-17) to maintain consistency with available computer programs.

CODE

COMMENTARY

If deflections have been calculated using service loads, Q in Eq. (10-17) should be calculated in the manner explained in R10.11.4.

In the 1989 and earlier codes, the moment magnifier equations for δ_b and δ_s included a stiffness reduction factor ϕ_K to cover the variability in the stability calculation. The Q factor analysis is based on deflections calculated using the values of E_c and I_g from 10.11.1, which include the equivalent of a stiffness reduction factor ϕ_K as explained in R10.13.4.1. As a result, no additional ϕ-factor is needed in the stability calculation. Once the moments are established using Eq. (10-17), selection of the cross sections of the columns involves the strength reduction factors ϕ from 9.3.2.2.

10.13.4.3 — Alternatively, it shall be permitted to calculate $\delta_s M_s$ as

$$\delta_s M_s = \frac{M_s}{1 - \frac{\Sigma P_u}{0.75 \Sigma P_c}} \geq M_s \quad \text{(10-18)}$$

where ΣP_u is the summation for all the factored vertical loads in a story and ΣP_c is the summation for all sway resisting columns in a story. P_c is calculated using Eq. (10-10) with k from 10.13.1 and EI from Eq. (10-11) or Eq. (10-12).

R10.13.4.3 — To check the effects of story stability, δ_s is computed as an averaged value for the entire story based on use of $\Sigma P_u / \Sigma P_c$. This reflects the interaction of all sway resisting columns in the story in the $P\Delta$ effects since the lateral deflection of all columns in the story should be equal in the absence of torsional displacements about a vertical axis. In addition, it is possible that a particularly slender individual column in a sway frame could have substantial midheight deflections even if adequately braced against lateral end deflections by other columns in the story. Such a column will have ℓ_u / r greater than the value given in Eq. (10-19) and should be checked using 10.13.5.

If the lateral load deflections involve a significant torsional displacement, the moment magnification in the columns farthest from the center of twist may be underestimated by the moment magnifier procedure. In such cases, a three-dimensional second-order analysis should be considered.

The 0.75 in the denominator of Eq. (10-18) is a stiffness reduction factor ϕ_K as explained in R10.12.3.

In the calculation of EI, β_d will normally be zero for a sway frame because the lateral loads are generally of short duration. (See R10.13.4.1).

10.13.5 — If an individual compression member has

$$\frac{\ell_u}{r} > \frac{35}{\sqrt{\frac{P_u}{f_c' A_g}}} \quad \text{(10-19)}$$

it shall be designed for factored axial force P_u, and moment, M_c, calculated using 10.12.3 in which M_1 and M_2 are computed in accordance with 10.13.3, β_d as defined for the load combination under consideration, and k as defined in 10.12.1.

R10.13.5 — The unmagnified nonsway moments at the ends of the columns are added to the magnified sway moments at the same points. Generally, one of the resulting end moments is the maximum moment in the column. However, for slender columns with high axial loads the point of maximum moment may be between the ends of the column so that the end moments are no longer the maximum moments. If ℓ_u / r is less than the value given by Eq. (10-19) the maximum moment at any point along the height of such a column will be less than 1.05 times the maximum end moment. When ℓ_u / r exceeds the value given by Eq. (10-19), the maximum moment will occur at a point between the ends of the column and will exceed the maximum end moment by more than 5 percent.[10.30] In such a case the maximum moment is calculated by magnifying the end moments using Eq. (10-8).

CODE

10.13.6 — In addition to load combinations involving lateral loads, the strength and stability of the structure as a whole under factored gravity loads shall be considered.

(a) When $\delta_s M_s$ is computed from 10.13.4.1, the ratio of second-order lateral deflections to first-order lateral deflections for factored dead and live loads plus factored lateral load applied to the structure shall not exceed 2.5;

(b) When $\delta_s M_s$ is computed according to 10.13.4.2, the value of Q computed using ΣP_u for factored dead and live loads shall not exceed 0.60;

(c) When $\delta_s M_s$ is computed from 10.13.4.3, δ_s computed using ΣP_u and ΣP_c corresponding to the factored dead and live loads shall be positive and shall not exceed 2.5.

In (a), (b), and (c) above, β_d shall be taken as the ratio of the maximum factored sustained axial load to the maximum factored axial load.

COMMENTARY

R10.13.6 — The possibility of sidesway instability under gravity loads alone should be investigated. When using second-order analyses to compile $\delta_s M_s$ (10.13.4.1), the frame should be analyzed twice for the case of factored gravity loads plus a lateral load applied to the frame. This load may be the lateral load used in design or it may be a single lateral load applied to the top of the frame. The first analysis should be a first-order analysis; the second analysis should be a second-order analysis. The deflection from the second-order analysis should not exceed 2.5 times the deflection from the first-order analysis. If one story is much more flexible than the others, the deflection ratio should be computed in that story. The lateral load should be large enough to give deflections of a magnitude that can be compared accurately. In unsymmetrical frames that deflect laterally under gravity loads alone, the lateral load should act in the direction for which it will increase the lateral deflections.

When using 10.13.4.2 to compute $\delta_s M_s$, the value of Q evaluated using factored gravity loads should not exceed 0.60. This is equivalent to $\delta_s = 2.5$. The values of V_u and Δ_o used to compute Q can result from assuming any real or arbitrary set of lateral loads provided that V_u and Δ_o are both from the same loading. If Q as computed in 10.11.4.2 is 0.2 or less, the stability check in 10.13.6 is satisfied.

When $\delta_s M_s$ is computed using Eq. (10-18), an upper limit of 2.5 is placed on δ_s. For higher δ_s values, the frame will be very susceptible to variations in EI and foundation rotations. If δ_s exceeds 2.5, the frame should be stiffened to reduce δ_s. ΣP_u shall include the axial load in all columns and walls including columns that are not part of the lateral load resisting system. The value $\delta_s = 2.5$ is a very high magnifier. It has been chosen to offset the conservatism inherent in the moment magnifier procedure.

For nonsway frames, β_d is the ratio of the maximum factored axial sustained load to the maximum factored axial load associated with the same load combination.

10.13.7 — In sway frames, flexural members shall be designed for the total magnified end moments of the compression members at the joint.

R10.13.7 — The strength of a sway frame is governed by the stability of the columns and by the degree of end restraint provided by the beams in the frame. If plastic hinges form in the restraining beam, the structure approaches a failure mechanism and its axial load capacity is drastically reduced. Section 10.13.7 provides that the designer make certain that the restraining flexural members have the capacity to resist the magnified column moments.

10.14 — Axially loaded members supporting slab system

Axially loaded members supporting a slab system included within the scope of 13.1 shall be designed as provided in Chapter 10 and in accordance with the additional requirements of Chapter 13.

CODE

10.15 — Transmission of column loads through floor system

If f_c' of a column is greater than 1.4 times that of the floor system, transmission of load through the floor system shall be provided by 10.15.1, 10.15.2, or 10.15.3.

10.15.1 — Concrete of strength specified for the column shall be placed in the floor at the column location. Top surface of the column concrete shall extend 2 ft into the slab from face of column. Column concrete shall be well integrated with floor concrete, and shall be placed in accordance with 6.4.6 and 6.4.7.

10.15.2 — Strength of a column through a floor system shall be based on the lower value of concrete strength with vertical dowels and spirals as required.

10.15.3 — For columns laterally supported on four sides by beams of approximately equal depth or by slabs, it shall be permitted to base strength of the column on an assumed concrete strength in the column joint equal to 75 percent of column concrete strength plus 35 percent of floor concrete strength. In the application of 10.15.3, the ratio of column concrete strength to slab concrete strength shall not be taken greater than 2.5 for design.

COMMENTARY

R10.15 — Transmission of column loads through floor system

The requirements of this section are based on a paper on the effect of floor concrete strength on column strength.[10.43] The provisions mean that when the column concrete strength does not exceed the floor concrete strength by more than 40 percent, no special precautions need be taken. For higher column concrete strengths, methods in 10.15.1 or 10.15.2 should be used for corner or edge columns. Methods in 10.15.1, 10.15.2, or 10.15.3 should be used for interior columns with adequate restraint on all four sides.

R10.15.1 — Application of the concrete placement procedure described in 10.15.1 requires the placing of two different concrete mixtures in the floor system. The lower strength mixture should be placed while the higher strength concrete is still plastic and should be adequately vibrated to ensure the concretes are well integrated. This requires careful coordination of the concrete deliveries and the possible use of retarders. In some cases, additional inspection services will be required when this procedure is used. It is important that the higher strength concrete in the floor in the region of the column be placed before the lower strength concrete in the remainder of the floor to prevent accidental placing of the low strength concrete in the column area. It is the designer's responsibility to indicate on the drawings where the high and low strength concretes are to be placed.

With the 1983 code, the amount of column concrete to be placed within the floor is expressed as a simple 2-ft extension from face of the column. Since the concrete placement requirement should be carried out in the field, it is now expressed in a way that is directly evident to workers. The new requirement will also locate the interface between column and floor concrete farther out into the floor, away from regions of very high shear.

R10.15.3 — Research[10.44] has shown that heavily loaded slabs do not provide as much confinement as lightly loaded slabs when ratios of column concrete strength to slab concrete strength exceed about 2.5. Consequently, a limit is placed on the concrete strength ratio assumed in design.

CODE

10.16 — Composite compression members

10.16.1 — Composite compression members shall include all such members reinforced longitudinally with structural steel shapes, pipe, or tubing with or without longitudinal bars.

10.16.2 — Strength of a composite member shall be computed for the same limiting conditions applicable to ordinary reinforced concrete members.

10.16.3 — Any axial load strength assigned to concrete of a composite member shall be transferred to the concrete by members or brackets in direct bearing on the composite member concrete.

10.16.4 — All axial load strength not assigned to concrete of a composite member shall be developed by direct connection to the structural steel shape, pipe, or tube.

10.16.5 — For evaluation of slenderness effects, radius of gyration, r, of a composite section shall be not greater than the value given by

$$r = \sqrt{\frac{(E_c I_g/5) + E_s I_{sx}}{(E_c A_g/5) + E_s A_{sx}}} \qquad (10\text{-}20)$$

and, as an alternative to a more accurate calculation, EI in Eq. (10-10) shall be taken either as Eq. (10-11) or

$$EI = \frac{(E_c I_g/5)}{1 + \beta_d} + E_s I_{sx} \qquad (10\text{-}21)$$

10.16.6 — Structural steel encased concrete core

10.16.6.1 — For a composite member with a concrete core encased by structural steel, the thickness of the steel encasement shall be not less than

$$b\sqrt{\frac{f_y}{3E_s}} \text{ for each face of width } b$$

nor

$$h\sqrt{\frac{f_y}{8E_s}} \text{ for circular sections of diameter } h$$

COMMENTARY

R10.16 — Composite compression members

R10.16.1 — Composite columns are defined without reference to classifications of combination, composite, or concrete-filled pipe column. Reference to other metals used for reinforcement has been omitted because they are seldom used in concrete construction.

R10.16.2 — The same rules used for computing the load-moment interaction strength for reinforced concrete sections can be applied to composite sections. Interaction charts for concrete-filled tubing would have a form identical to those of ACI SP-7[10.45] and the *ACI Design Handbook*[10.37] but with γ slightly greater than 1.0.

R10.16.3 and R10.16.4 — Direct bearing or direct connection for transfer of forces between steel and concrete can be developed through lugs, plates, or reinforcing bars welded to the structural shape or tubing before the concrete is cast. Flexural compressive stress need not be considered a part of direct compression load to be developed by bearing. A concrete encasement around a structural steel shape may stiffen the shape, but it would not necessarily increase its strength.

R10.16.5 — Eq. (10-20) is given because the rules of 10.11.2 for estimating the radius of gyration are overly conservative for concrete filled tubing and are not applicable for members with enclosed structural shapes.

In reinforced concrete columns subject to sustained loads, creep transfers some of the load from the concrete to the steel, increasing the steel stresses. In the case of lightly reinforced columns, this load transfer may cause the compression steel to yield prematurely, resulting in a loss in the effective EI. Accordingly, both the concrete and steel terms in Eq. (10-11) are reduced to account for creep. For heavily reinforced columns or for composite columns in which the pipe or structural shape makes up a large percentage of the cross section, the load transfer due to creep is not significant. Accordingly, Eq. (10-21) was revised in the 1980 code supplement so that only the EI of the concrete is reduced for sustained load effects.

R10.16.6 — Structural steel encased concrete core

Steel encased concrete sections should have a metal wall thickness large enough to attain longitudinal yield stress before buckling outward.

CODE

10.16.6.2 — Longitudinal bars located within the encased concrete core shall be permitted to be used in computing A_{sx} and I_{sx}.

10.16.7 — Spiral reinforcement around structural steel core

A composite member with spirally reinforced concrete around a structural steel core shall conform to 10.16.7.1 through 10.16.7.5.

10.16.7.1 — Specified compressive strength, f_c', shall not be less than that given in 1.1.1.

10.16.7.2 — Design yield strength of structural steel core shall be the specified minimum yield strength for the grade of structural steel used but not to exceed 50,000 psi.

10.16.7.3 — Spiral reinforcement shall conform to 10.9.3.

10.16.7.4 — Longitudinal bars located within the spiral shall be not less than 0.01 nor more than 0.08 times net area of concrete section.

10.16.7.5 — Longitudinal bars located within the spiral shall be permitted to be used in computing A_{sx} and I_{sx}.

10.16.8 — Tie reinforcement around structural steel core

A composite member with laterally tied concrete around a structural steel core shall conform to 10.16.8.1 through 10.16.8.8.

10.16.8.1 — Specified compressive strength, f_c', shall not be less than that given in 1.1.1.

10.16.8.2 — Design yield strength of structural steel core shall be the specified minimum yield strength for the grade of structural steel used but not to exceed 50,000 psi.

10.16.8.3 — Lateral ties shall extend completely around the structural steel core.

10.16.8.4 — Lateral ties shall have a diameter not less than 0.02 times the greatest side dimension of composite member, except that ties shall not be smaller than No. 3 and are not required to be larger than No. 5. Welded wire reinforcement of equivalent area shall be permitted.

10.16.8.5 — Vertical spacing of lateral ties shall not exceed 16 longitudinal bar diameters, 48 tie bar diameters, or 0.5 times the least side dimension of the composite member.

COMMENTARY

R10.16.7 — Spiral reinforcement around structural steel core

Concrete that is laterally confined by a spiral has increased load-carrying strength, and the size of the spiral required can be regulated on the basis of the strength of the concrete outside the spiral the same reasoning that applies for columns reinforced only with longitudinal bars. The radial pressure provided by the spiral ensures interaction between concrete, reinforcing bars, and steel core such that longitudinal bars will both stiffen and strengthen the cross section.

R10.16.8 — Tie reinforcement around structural steel core

Concrete that is laterally confined by tie bars is likely to be rather thin along at least one face of a steel core section. Therefore, complete interaction between the core, the concrete, and any longitudinal reinforcement should not be assumed. Concrete will probably separate from smooth faces of the steel core. To maintain the concrete around the structural steel core, it is reasonable to require more lateral ties than needed for ordinary reinforced concrete columns. Because of probable separation at high strains between the steel core and the concrete, longitudinal bars will be ineffective in stiffening cross sections even though they would be useful in sustaining compression forces. The yield strength of the steel core should be limited to that which exists at strains below those that can be sustained without spalling of the concrete. It has been assumed that axially compressed concrete will not spall at strains less than 0.0018. The yield strength of $0.0018 \times 29{,}000{,}000$, or 52,000 psi, represents an upper limit of the useful maximum steel stress.

CODE

COMMENTARY

10.16.8.6 — Longitudinal bars located within the ties shall be not less than 0.01 nor more than 0.08 times net area of concrete section.

10.16.8.7 — A longitudinal bar shall be located at every corner of a rectangular cross section, with other longitudinal bars spaced not farther apart than one-half the least side dimension of the composite member.

10.16.8.8 — Longitudinal bars located within the ties shall be permitted to be used in computing A_{sx} for strength but not in computing I_{sx} for evaluation of slenderness effects.

10.17 — Bearing strength

10.17.1 — Design bearing strength of concrete shall not exceed $\phi(0.85f_c'A_1)$, except when the supporting surface is wider on all sides than the loaded area, then the design bearing strength of the loaded area shall be permitted to be multiplied by $\sqrt{A_2/A_1}$ but not more than 2.

10.17.2 — Section 10.17 does not apply to post-tensioning anchorages.

R10.17 — Bearing strength

R10.17.1 — This section deals with bearing strength of concrete supports. The permissible bearing stress of $0.85f_c'$ is based on tests reported in Reference 10.46. (See also 15.8).

When the supporting area is wider than the loaded area on all sides, the surrounding concrete confines the bearing area, resulting in an increase in bearing strength. No minimum depth is given for a supporting member. The minimum depth of support will be controlled by the shear requirements of 11.12.

When the top of the support is sloped or stepped, advantage may still be taken of the condition that the supporting member is larger than the loaded area, provided the supporting member does not slope at too great an angle. Fig. R10.17 illustrates the application of the frustum to find A_2. The frustum should not be confused with the path by which a load spreads out as it travels downward through the support. Such a load path would have steeper sides. However, the frustum described has somewhat flat side slopes to ensure that there is concrete immediately surrounding the zone of high stress at the bearing. A_1 is the loaded area but not greater than the bearing plate or bearing cross-sectional area.

R10.17.2 — Post-tensioning anchorages are usually laterally reinforced, in accordance with 18.13.

COMMENTARY

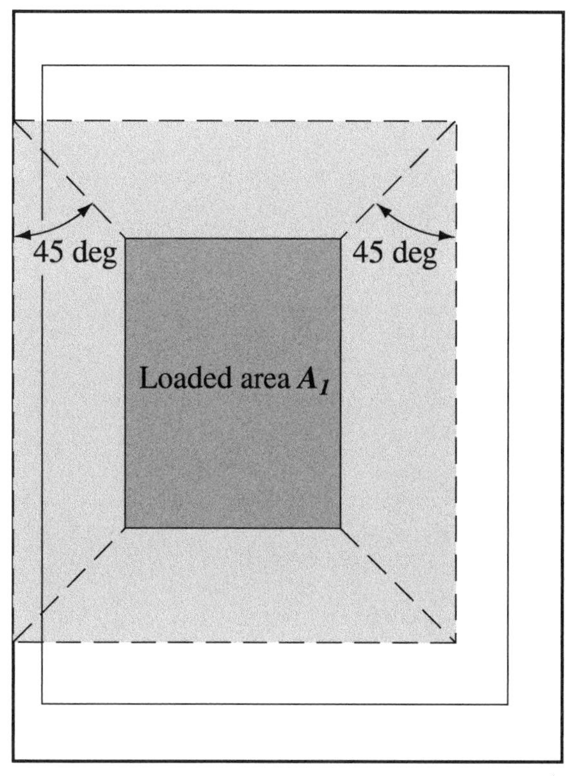

Plan

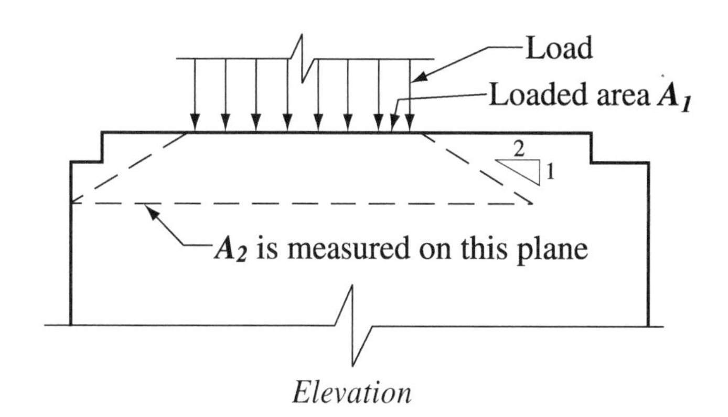

Elevation

Fig. R10.17—Application of frustum to find $\mathbf{A_2}$ *in stepped or sloped supports*

Notes

CHAPTER 11 — SHEAR AND TORSION

CODE

11.1 — Shear strength

11.1.1 — Except for members designed in accordance with Appendix A, design of cross sections subject to shear shall be based on:

$$\phi V_n \geq V_u \quad (11\text{-}1)$$

where V_u is the factored shear force at the section considered and V_n is nominal shear strength computed by:

$$V_n = V_c + V_s \quad (11\text{-}2)$$

where V_c is nominal shear strength provided by concrete calculated in accordance with 11.3, 11.4, or 11.12, and V_s is nominal shear strength provided by shear reinforcement calculated in accordance with 11.5, 11.10.9, or 11.12.

11.1.1.1 — In determining V_n, the effect of any openings in members shall be considered.

11.1.1.2 — In determining V_c, whenever applicable, effects of axial tension due to creep and shrinkage in restrained members shall be considered and effects of inclined flexural compression in variable depth members shall be permitted to be included.

11.1.2 — The values of $\sqrt{f_c'}$ used in this chapter shall not exceed 100 psi except as allowed in 11.1.2.1.

COMMENTARY

R11.1 — Shear strength

This chapter includes shear and torsion provisions for both nonprestressed and prestressed concrete members. The shear-friction concept (11.7) is particularly applicable to design of reinforcement details in precast structures. Special provisions are included for deep flexural members (11.8), brackets and corbels (11.9), and shearwalls (11.10). Shear provisions for slabs and footings are given in 11.12.

The shear strength is based on an average shear stress on the full effective cross section $b_w d$. In a member without shear reinforcement, shear is assumed to be carried by the concrete web. In a member with shear reinforcement, a portion of the shear strength is assumed to be provided by the concrete and the remainder by the shear reinforcement.

The shear strength provided by concrete V_c is assumed to be the same for beams with and without shear reinforcement and is taken as the shear causing significant inclined cracking. These assumptions are discussed in References 11.1, 11.2, and 11.3.

Appendix A allows the use of strut-and-tie models in the shear design of disturbed regions. The traditional shear design procedures, which ignore D-regions, are acceptable in shear spans that include B-regions.

R11.1.1.1 — Openings in the web of a member can reduce its shear strength. The effects of openings are discussed in Section 4.7 of Reference 11.1 and in References 11.4 and 11.5.

R11.1.1.2 — In a member of variable depth, the internal shear at any section is increased or decreased by the vertical component of the inclined flexural stresses. Computation methods are outlined in various textbooks and in the 1940 Joint Committee Report.[11.6]

R11.1.2 — Because of a lack of test data and practical experience with concretes having compressive strengths greater than 10,000 psi, the 1989 edition of the code imposed a maximum value of 100 psi on $\sqrt{f_c'}$ for use in the calculation of shear strength of concrete beams, joists, and slabs. Exceptions to this limit were permitted in beams and joists when the transverse reinforcement satisfied an increased value for the minimum amount of web reinforcement. There are limited test data on the two-way shear strength of high-strength concrete slabs. Until more experience is obtained

CODE

COMMENTARY

for two-way slabs built with concretes that have strengths greater than 10,000 psi, it is prudent to limit $\sqrt{f_c'}$ to 100 psi for the calculation of shear strength.

11.1.2.1 — Values of $\sqrt{f_c'}$ greater than 100 psi shall be permitted in computing V_c, V_{ci}, and V_{cw} for reinforced or prestressed concrete beams and concrete joist construction having minimum web reinforcement in accordance with 11.5.6.3, 11.5.6.4, or 11.6.5.2.

R11.1.2.1 — Based on the test results in References 11.7, 11.8, 11.9, 11.10, and 11.11, an increase in the minimum amount of transverse reinforcement is required for high-strength concrete. These tests indicated a reduction in the reserve shear strength as f_c' increased in beams reinforced with the specified minimum amount of transverse reinforcement, which is equivalent to an effective shear stress of 50 psi. A provision introduced in the 1989 edition of the code required an increase in the minimum amount of transverse reinforcement for concrete strengths between 10,000 and 15,000 psi. This provision, which led to a sudden increase in the minimum amount of transverse reinforcement at a compressive strength of 10,000 psi, has been replaced by a gradual increase in the minimum A_v as f_c' increases, as given by Eq. (11-13).

11.1.3 — Computation of maximum V_u at supports in accordance with 11.1.3.1 or 11.1.3.2 shall be permitted if all conditions (a), (b), and (c) are satisfied:

(a) Support reaction, in direction of applied shear, introduces compression into the end regions of member;

(b) Loads are applied at or near the top of the member;

(c) No concentrated load occurs between face of support and location of critical section defined in 11.1.3.1 or 11.1.3.2.

11.1.3.1 — For nonprestressed members, sections located less than a distance d from face of support shall be permitted to be designed for V_u computed at a distance d.

R11.1.3.1 — The closest inclined crack to the support of the beam in Fig. R11.1.3.1(a) will extend upwards from the face of the support reaching the compression zone about d from the face of the support. If loads are applied to the top of this beam, the stirrups across this crack are stressed by loads acting on the lower freebody in Fig. R11.1.3.1(a). The loads applied to the beam between the face of the column and the point d away from the face are transferred directly to the support by compression in the web above the crack. Accordingly, the code permits design for a maximum factored shear force V_u at a distance d from the support for nonprestressed members, and at a distance $h/2$ for prestressed members. Two things are emphasized: first, stirrups are required across the potential crack designed for the shear at d from the support, and second, a tension force exists in the longitudinal reinforcement at the face of the support.

In Fig. R11.1.3.1(b), loads are shown acting near the bottom of a beam. In this case, the critical section is taken at the face of the support. Loads acting near the support should be transferred across the inclined crack extending upward from the support

CODE

COMMENTARY

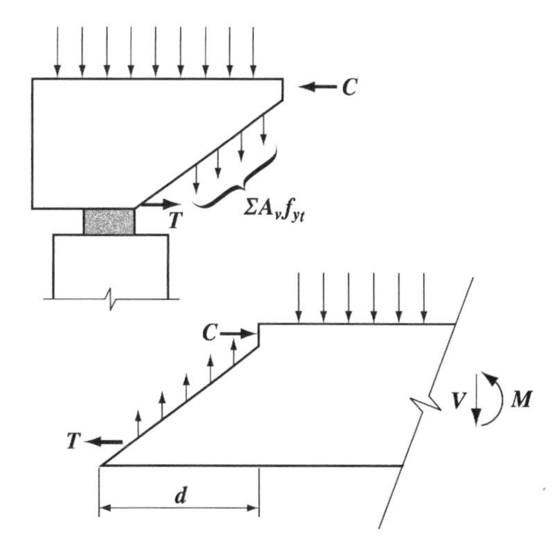

Fig. R11.1.3.1(a)—Free body diagrams of the end of a beam

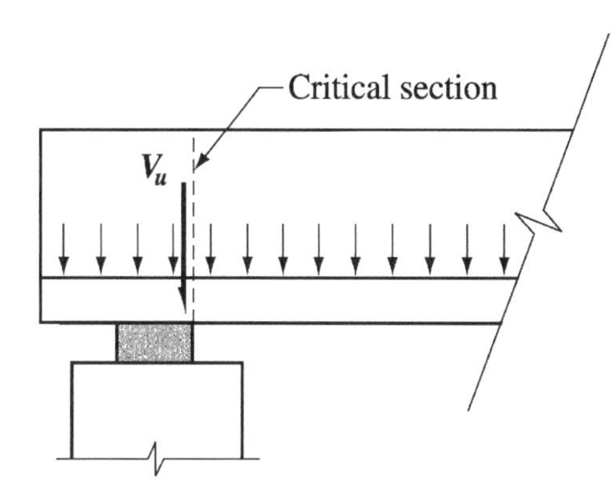

Fig. R11.1.3.1(b)—Location of critical section for shear in a member loaded near bottom.

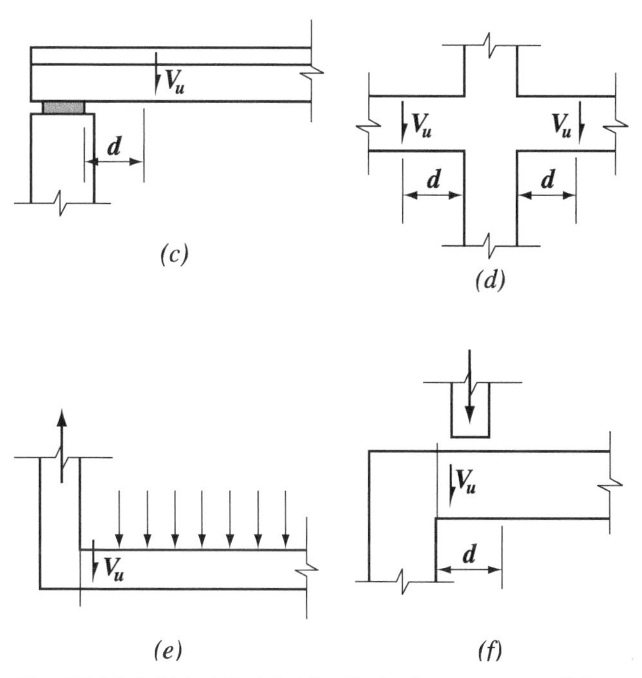

Fig. R11.1.3.1(c), (d), (e), (f)—Typical support conditions for locating factored shear force $\mathbf{V_u}$.

face. The shear force acting on the critical section should include all loads applied below the potential inclined crack.

Typical support conditions where the shear force at a distance d from the support may be used include: (1) members supported by bearing at the bottom of the member, such as shown in Fig. R11.1.3.1(c); and (2) members framing monolithically into another member as illustrated in Fig. R11.1.3.1(d).

Support conditions where this provision should not be applied include: (1) Members framing into a supporting member in tension, such as shown in Fig. R11.1.3.1(e). For this case, the critical section for shear should be taken at the face of the support. Shear within the connection should also be investigated and special corner reinforcement should be provided. (2) Members for which loads are not applied at or near the top of the member. This is the condition referred to in Fig. 11.1.3.1(b). For such cases the critical section is taken at the face of the support. Loads acting near the support should be transferred across the inclined crack extending upward from the support face. The shear force acting on the critical section should include all loads applied below the potential inclined crack. (3) Members loaded such that the shear at sections between the support and a distance d from the support differs radically from the shear at distance d. This commonly occurs in brackets and in beams where a concentrated load is located close to the support, as shown in Fig. R11.1.3.1(f) or in footings supported on piles. In this case the shear at the face of the support should be used.

11.1.3.2 — For prestressed members, sections located less than a distance $h/2$ from face of support shall be permitted to be designed for V_u computed at a distance $h/2$.

R11.1.3.2 — Because d frequently varies in prestressed members, the location of the critical section has arbitrarily been taken as $h/2$ from the face of the support.

11.1.4 — For deep beams, brackets and corbels, walls, and slabs and footings, the special provisions of 11.8 through 11.12 shall apply.

11.2 — Lightweight concrete

11.2.1 — Provisions for shear and torsion strength apply to normalweight concrete. When lightweight aggregate concrete is used, one of the following modifications shall apply to $\sqrt{f_c'}$ throughout Chapter 11, except 11.5.5.3, 11.5.7.9, 11.6.3.1, 11.12.3.2, and 11.12.4.8

11.2.1.1 — When f_{ct} is specified and concrete is proportioned in accordance with 5.2, $f_{ct}/6.7$ shall be substituted for $\sqrt{f_c'}$, but the value of $f_{ct}/6.7$ shall not exceed $\sqrt{f_c'}$.

R11.2 — Lightweight concrete

Two alternative procedures are provided to modify the provisions for shear and torsion when lightweight aggregate concrete is used. The lightweight concrete modification applies only to the terms containing $\sqrt{f_c'}$ in the equations of Chapter 11.

R11.2.1.1 — The first alternative bases the modification on laboratory tests to determine the relationship between average splitting tensile strength f_{ct} and the specified compressive strength f_c' for the lightweight concrete being used. For normalweight concrete, the average splitting tensile strength f_{ct} is approximately equal to $6.7\sqrt{f_c'}$.[11.10,11.11]

CODE

11.2.1.2 — When f_{ct} is not specified, all values of $\sqrt{f_c'}$ shall be multiplied by 0.75 for all-lightweight concrete and 0.85 for sand-lightweight concrete. Linear interpolation shall be permitted when partial sand replacement is used.

COMMENTARY

R11.2.1.2 — The second alternative bases the modification on the assumption that the tensile strength of lightweight concrete is a fixed fraction of the tensile strength of normalweight concrete.[11.12] The multipliers are based on data from tests[11.13] on many types of structural lightweight aggregate concrete.

11.3 — Shear strength provided by concrete for nonprestressed members

11.3.1 — V_c shall be computed by provisions of 11.3.1.1 through 11.3.1.3, unless a more detailed calculation is made in accordance with 11.3.2.

11.3.1.1 — For members subject to shear and flexure only,

$$V_c = 2\sqrt{f_c'}\,b_w d \qquad (11\text{-}3)$$

11.3.1.2 — For members subject to axial compression,

$$V_c = 2\left(1 + \frac{N_u}{2000A_g}\right)\sqrt{f_c'}\,b_w d \qquad (11\text{-}4)$$

Quantity N_u/A_g shall be expressed in psi.

11.3.1.3 — For members subject to significant axial tension, V_c shall be taken as zero unless a more detailed analysis is made using 11.3.2.3.

11.3.2 — V_c shall be permitted to be computed by the more detailed calculation of 11.3.2.1 through 11.3.2.3.

11.3.2.1 — For members subject to shear and flexure only,

$$V_c = \left(1.9\sqrt{f_c'} + 2500\rho_w \frac{V_u d}{M_u}\right) b_w d \qquad (11\text{-}5)$$

but not greater than $3.5\sqrt{f_c'}\,b_w d$. When computing V_c by Eq. (11-5), $V_u d/M_u$ shall not be taken greater than 1.0, where M_u occurs simultaneously with V_u at section considered.

11.3.2.2 — For members subject to axial compression, it shall be permitted to compute V_c using Eq. (11-5) with M_m substituted for M_u and $V_u d/M_u$ not then limited to 1.0, where

R11.3 — Shear strength provided by concrete for nonprestressed members

R11.3.1.1 — See R11.3.2.1.

R11.3.1.2 and R11.3.1.3 — See R11.3.2.2.

R11.3.2.1 — Eq. (11-5) is the basic expression for shear strength of members without shear reinforcement.[11.3] Designers should recognize that the three variables in Eq. (11-5), $\sqrt{f_c'}$ (as a measure of concrete tensile strength), ρ_w, and $V_u d/M_u$, are known to affect shear strength, although some research data[11.1,11.14] indicate that Eq. (11-5) overestimates the influence of f_c' and underestimates the influence of ρ_w and $V_u d/M_u$. Further information[11.15] has indicated that shear strength decreases as the overall depth of the member increases.

The minimum value of M_u equal to $V_u d$ in Eq. (11-5) is to limit V_c near points of inflection.

For most designs, it is convenient to assume that the second term of Eq. (11-5) equals $0.1\sqrt{f_c'}$ and use V_c equal to $2\sqrt{f_c'}\,b_w d$ as permitted in 11.3.1.1.

R11.3.2.2 — Eq. (11-6) and (11-7), for members subject to axial compression in addition to shear and flexure are derived in the ACI-ASCE Committee 326 report.[11.3] As N_u is increased, the value of V_c computed from Eq. (11-5) and

CODE

$$M_m = M_u - N_u\frac{(4h-d)}{8} \quad (11\text{-}6)$$

However, V_c shall not be taken greater than

$$V_c = 3.5\sqrt{f_c'}b_w d\sqrt{1+\frac{N_u}{500A_g}} \quad (11\text{-}7)$$

N_u/A_g shall be expressed in psi. When M_m as computed by Eq. (11-6) is negative, V_c shall be computed by Eq. (11-7).

11.3.2.3 — For members subject to significant axial tension,

$$V_c = 2\left(1+\frac{N_u}{500A_g}\right)\sqrt{f_c'}b_w d \quad (11\text{-}8)$$

but not less than zero, where N_u is negative for tension. N_u/A_g shall be expressed in psi.

11.3.3 — For circular members, the area used to compute V_c shall be taken as the product of the diameter and effective depth of the concrete section. It shall be permitted to take d as 0.80 times the diameter of the concrete section.

COMMENTARY

(11-6) will exceed the upper limit given by Eq. (11-7) before the value of M_m given by Eq. (11-6) becomes negative. The value of V_c obtained from Eq. (11-5) has no physical significance if a negative value of M_m is substituted. For this condition, Eq. (11-7) or Eq. (11-4) should be used to calculate V_c. Values of V_c for members subject to shear and axial load are illustrated in Fig. R11.3.2.2. The background for these equations is discussed and comparisons are made with test data in Reference 11.2.

Because of the complexity of Eq. (11-5) and (11-6), an alternative design provision, Eq. (11-4), is permitted.

R11.3.2.3 — Eq. (11-8) may be used to compute V_c for members subject to significant axial tension. Shear reinforcement may then be designed for $V_n - V_c$. The term significant is used to recognize that a designer must use judgment in deciding whether axial tension needs to be considered. Low levels of axial tension often occur due to volume changes, but are not important in structures with adequate expansion joints and minimum reinforcement. It may be desirable to design shear reinforcement to carry total shear if there is uncertainty about the magnitude of axial tension.

R11.3.3 — Shear tests of members with circular sections indicate that the effective area can be taken as the gross area of the section or as an equivalent rectangular area.[11.1, 11.16, 11.17]

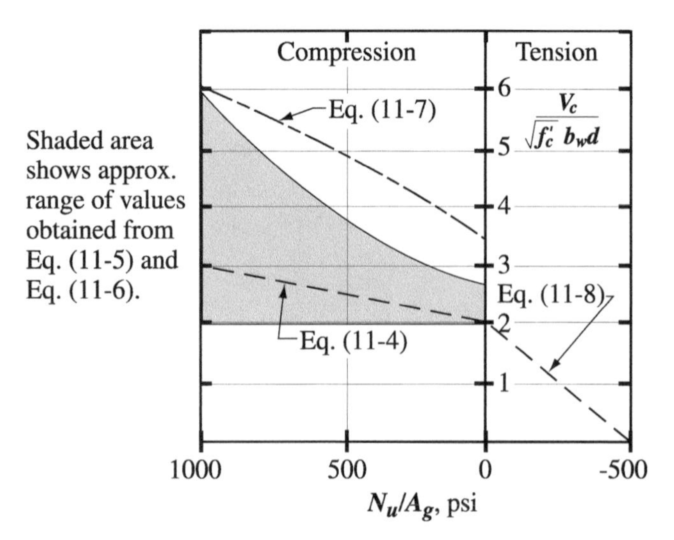

Fig. R11.3.2.2—Comparison of shear strength equations for members subject to axial load

CODE

11.4 — Shear strength provided by concrete for prestressed members

11.4.1 — For the provisions of 11.4, d shall be taken as the distance from extreme compression fiber to centroid of prestressed and nonprestressed longitudinal tension reinforcement, if any, but need not be taken less than $0.80h$.

11.4.2 — For members with effective prestress force not less than 40 percent of the tensile strength of flexural reinforcement, unless a more detailed calculation is made in accordance with 11.4.3,

$$V_c = \left(0.6\sqrt{f_c'} + 700\frac{V_u d_p}{M_u}\right) b_w d \qquad (11\text{-}9)$$

but V_c need not be taken less than $2\sqrt{f_c'}\, b_w d$. V_c shall not be taken greater than $5\sqrt{f_c'}\, b_w d$ or the value given in 11.4.4 or 11.4.5. $V_u d_p/M_u$ shall not be taken greater than 1.0, where M_u occurs simultaneously with V_u at the section considered.

11.4.3 — V_c shall be permitted to be computed in accordance with 11.4.3.1 and 11.4.3.2, where V_c shall be the lesser of V_{ci} and V_{cw}.

11.4.3.1—V_{ci} shall be computed by

$$V_{ci} = 0.6\sqrt{f_c'} b_w d_p + V_d + \frac{V_i M_{cre}}{M_{max}} \qquad (11\text{-}10)$$

where d_p need not be taken less than $0.80h$ and

$$M_{cre} = (I/y_t)(6\sqrt{f_c'} + f_{pe} - f_d) \qquad (11\text{-}11)$$

and values of M_{max} and V_i shall be computed from the load combination causing maximum factored moment to occur at the section. V_{ci} need not be taken less than $1.7\sqrt{f_c'}\, b_w d$.

11.4.3.2—V_{cw} shall be computed by

$$V_{cw} = (3.5\sqrt{f_c'} + 0.3 f_{pc}) b_w d_p + V_p \qquad (11\text{-}12)$$

where d_p need not be taken less than $0.80h$.

COMMENTARY

R11.4 — Shear strength provided by concrete for prestressed members

R11.4.2 — Eq. (11-9) offers a simple means of computing V_c for prestressed concrete beams.[11.2] It may be applied to beams having prestressed reinforcement only, or to members reinforced with a combination of prestressed reinforcement and nonprestressed deformed bars. Eq. (11-9) is most applicable to members subject to uniform loading and may give conservative results when applied to composite girders for bridges.

In applying Eq. (11-9) to simply supported members subject to uniform loads $V_u d_p/M_u$ can be expressed as:

$$\frac{V_u d_p}{M_u} = \frac{d_p(\ell - 2x)}{x(\ell - x)}$$

where ℓ is the span length and x is the distance from the section being investigated to the support. For concrete with f_c' equal to 5000 psi, V_c from 11.4.2 varies as shown in Fig. R11.4.2. Design aids based on this equation are given in Reference 11.18.

R11.4.3 — Two types of inclined cracking occur in concrete beams: web-shear cracking and flexure-shear cracking. These two types of inclined cracking are illustrated in Fig. R11.4.3.

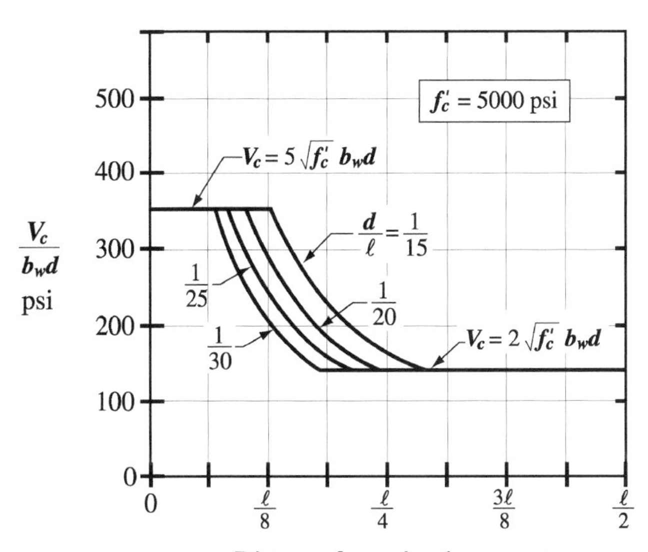

Fig. R11.4.2—Application of Eq. (11-9) to uniformly loaded prestressed members

CODE

Alternatively, V_{cw} shall be computed as the shear force corresponding to dead load plus live load that results in a principal tensile stress of $4\sqrt{f_c'}$ at the centroidal axis of member, or at the intersection of flange and web when the centroidal axis is in the flange. In composite members, the principal tensile stress shall be computed using the cross section that resists live load.

COMMENTARY

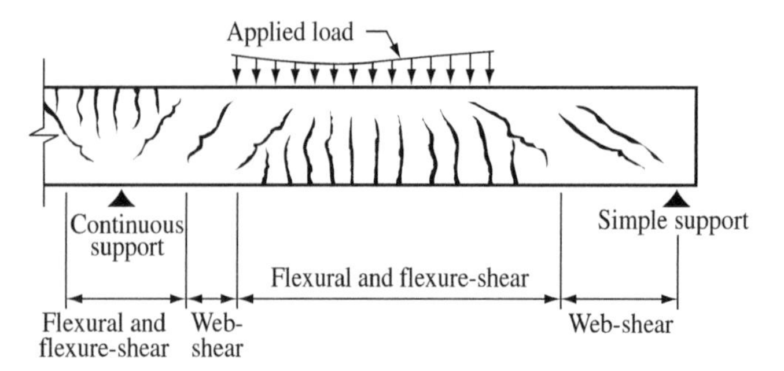

Fig. R11.4.3—Types of cracking in concrete beams

Web-shear cracking begins from an interior point in a member when the principal tensile stresses exceed the tensile strength of the concrete. Flexure-shear cracking is initiated by flexural cracking. When flexural cracking occurs, the shear stresses in the concrete above the crack are increased. The flexure-shear crack develops when the combined shear and tensile stress exceeds the tensile strength of the concrete.

Eq. (11-10) and (11-12) may be used to determine the shear forces causing flexure-shear and web-shear cracking, respectively. The nominal shear strength provided by the concrete V_c is assumed equal to the lesser of V_{ci} and V_{cw}. The derivations of Eq. (11-10) and (11-12) are summarized in Reference 11.19.

In deriving Eq. (11-10) it was assumed that V_{ci} is the sum of the shear required to cause a flexural crack at the point in question given by:

$$V = \frac{V_i M_{cre}}{M_{max}}$$

plus an additional increment of shear required to change the flexural crack to a flexure-shear crack. The externally applied factored loads, from which V_i and M_{max} are determined, include superimposed dead load, earth pressure, and live load. In computing M_{cre} for substitution into Eq. (11-10), I and y_t are the properties of the section resisting the externally applied loads.

For a composite member, where part of the dead load is resisted by only a part of the section, appropriate section properties should be used to compute f_d. The shear due to dead loads, V_d and that due to other loads V_i are separated in this case. V_d is then the total shear force due to unfactored dead load acting on that part of the section carrying the dead loads acting prior to composite action plus the unfactored superimposed dead load acting on the composite member. The terms V_i and M_{max} may be taken as:

$$V_i = V_u - V_d$$

$$M_{max} = M_u - M_d$$

CODE

COMMENTARY

where V_u and M_u are the factored shear and moment due to the total factored loads, and M_d is the moment due to unfactored dead load (the moment corresponding to f_d).

For noncomposite, uniformly loaded beams, the total cross section resists all the shear and the live and dead load shear force diagrams are similar. In this case Eq. (11-10) reduces to:

$$V_{ci} = 0.6\sqrt{f_c'}b_w d + \frac{V_u M_{ct}}{M_u}$$

where:

$$M_{ct} = (I/y_t)(6\sqrt{f_c'} + f_{pe})$$

The symbol M_{ct} in the two preceding equations represents the total moment, including dead load, required to cause cracking at the extreme fiber in tension. This is not the same as M_{cre} in code Eq. (11-10) where the cracking moment is that due to all loads except the dead load. In Eq. (11-10) the dead load shear is added as a separate term.

M_u is the factored moment on the beam at the section under consideration, and V_u is the factored shear force occurring simultaneously with M_u. Since the same section properties apply to both dead and live load stresses, there is no need to compute dead load stresses and shears separately. The cracking moment M_{ct} reflects the total stress change from effective prestress to a tension of $6\sqrt{f_c'}$, assumed to cause flexural cracking.

Eq. (11-12) is based on the assumption that web-shear cracking occurs due to the shear causing a principal tensile stress of approximately $4\sqrt{f_c'}$ at the centroidal axis of the cross section. V_p is calculated from the effective prestress force without load factors.

11.4.4 — In a pretensioned member in which the section at a distance ***h*/2** from face of support is closer to the end of member than the transfer length of the prestressing steel, the reduced prestress shall be considered when computing V_{cw}. This value of V_{cw} shall also be taken as the maximum limit for Eq. (11-9). The prestress force shall be assumed to vary linearly from zero at end of the prestressing steel, to a maximum at a distance from end of the prestressing steel equal to the transfer length, assumed to be 50 diameters for strand and 100 diameters for single wire.

R11.4.4 and R11.4.5 — The effect of the reduced prestress near the ends of pretensioned beams on the shear strength should be taken into account. Section 11.4.4 relates to the shear strength at sections within the transfer length of prestressing steel when bonding of prestressing steel extends to the end of the member.

Section 11.4.5 relates to the shear strength at sections within the length over which some of the prestressing steel is not bonded to the concrete, or within the transfer length of the prestressing steel for which bonding does not extend to the end of the beam.

11.4.5—In a pretensioned member where bonding of some tendons does not extend to the end of member, a reduced prestress shall be considered when computing V_c in accordance with 11.4.2 or 11.4.3. The value of V_{cw} calculated using the reduced prestress

CODE

shall also be taken as the maximum limit for Eq. (11-9). The prestress force due to tendons for which bonding does not extend to the end of member shall be assumed to vary linearly from zero at the point at which bonding commences to a maximum at a distance from this point equal to the transfer length, assumed to be 50 diameters for strand and 100 diameters for single wire.

11.5 — Shear strength provided by shear reinforcement

11.5.1 — Types of shear reinforcement

11.5.1.1 — Shear reinforcement consisting of the following shall be permitted:

(a) Stirrups perpendicular to axis of member;

(b) Welded wire reinforcement with wires located perpendicular to axis of member;

(c) Spirals, circular ties, or hoops.

11.5.1.2 — For nonprestressed members, shear reinforcement shall be permitted to also consist of:

(a) Stirrups making an angle of 45 deg or more with longitudinal tension reinforcement;

(b) Longitudinal reinforcement with bent portion making an angle of 30 deg or more with the longitudinal tension reinforcement;

(c) Combinations of stirrups and bent longitudinal reinforcement.

11.5.2 — The values of f_y and f_{yt} used in design of shear reinforcement shall not exceed 60,000 psi, except the value shall not exceed 80,000 psi for welded deformed wire reinforcement.

11.5.3 — Where the provisions of 11.5 are applied to prestressed members, d shall be taken as the distance from extreme compression fiber to centroid of the prestressed and nonprestressed longitudinal tension reinforcement, if any, but need not be taken less than $0.80h$.

COMMENTARY

R11.5 — Shear strength provided by shear reinforcement

R11.5.2 — Limiting the values of f_y and f_{yt} used in design of shear reinforcement to 60,000 psi provides a control on diagonal crack width. In the 1995 code, the limitation of 60,000 psi for shear reinforcement was raised to 80,000 psi for welded deformed wire reinforcement. Research[11.20-11.22] has indicated that the performance of higher strength steels as shear reinforcement has been satisfactory. In particular, full-scale beam tests described in Reference 11.21 indicated that the widths of inclined shear cracks at service load levels were less for beams reinforced with smaller diameter welded deformed wire reinforcement cages designed on the basis of a yield strength of 75 ksi than beams reinforced with deformed Grade 60 stirrups.

R11.5.3 — Although the value of d may vary along the span of a prestressed beam, studies[11.2] have shown that, for prestressed concrete members, d need not be taken less than $0.80h$. The beams considered had some straight tendons or reinforcing bars at the bottom of the section and had stirrups that enclosed the steel.

CODE

11.5.4 — Stirrups and other bars or wires used as shear reinforcement shall extend to a distance d from extreme compression fiber and shall be developed at both ends according to 12.13.

11.5.5 — Spacing limits for shear reinforcement

11.5.5.1 — Spacing of shear reinforcement placed perpendicular to axis of member shall not exceed $d/2$ in nonprestressed members or $0.75h$ in prestressed members, nor 24 in.

11.5.5.2 — Inclined stirrups and bent longitudinal reinforcement shall be so spaced that every 45 degree line, extending toward the reaction from mid-depth of member $d/2$ to longitudinal tension reinforcement, shall be crossed by at least one line of shear reinforcement.

11.5.5.3 — Where V_s exceeds $4\sqrt{f_c'}\,b_w d$, maximum spacings given in 11.5.5.1 and 11.5.5.2 shall be reduced by one-half.

11.5.6 — Minimum shear reinforcement

11.5.6.1 — A minimum area of shear reinforcement, $A_{v,min}$, shall be provided in all reinforced concrete flexural members (prestressed and nonprestressed) where V_u exceeds $0.5\phi V_c$, except:

(a) Slabs and footings;

(b) Concrete joist construction defined by 8.11;

(c) Beams with h not greater than the largest of 10 in., 2.5 times thickness of flange, or 0.5 the width of web.

COMMENTARY

R11.5.4 — It is essential that shear (and torsion) reinforcement be adequately anchored at both ends to be fully effective on either side of any potential inclined crack. This generally requires a hook or bend at the end of the reinforcement as provided by 12.13.

R11.5.6 — Minimum shear reinforcement

R11.5.6.1 — Shear reinforcement restrains the growth of inclined cracking. Ductility is increased and a warning of failure is provided. In an unreinforced web, the sudden formation of inclined cracking might lead directly to failure without warning. Such reinforcement is of great value if a member is subjected to an unexpected tensile force or an overload. Accordingly, a minimum area of shear reinforcement not less than that given by Eq. (11-13) or (11-14) is required wherever V_u is greater than $0.5\phi V_c$. Slabs, footings and joists are excluded from the minimum shear reinforcement requirement because there is a possibility of load sharing between weak and strong areas. However, research results[11.23] have shown that deep, lightly reinforced one-way slabs, particularly if constructed with high-strength concrete, may fail at shear loads less than V_c, calculated from Eq. (11-3).

Tests of hollow core units[11.24,11.25] with h values of 12.5 in. and less have shown shear strengths greater than those calculated by Eq. (11-12) and (11-10). Unpublished test results of precast prestressed concrete hollow core units with greater depths have shown that web-shear cracking strengths in end regions can be less than strengths computed by Eq. (11-12). By contrast, flexure-shear cracking strengths in those tests equaled or exceeded strengths computed by Eq. (11-10).

Even when V_u is less than $0.5\phi V_c$, the use of some web reinforcement is recommended in all thin-web post-tensioned prestressed concrete members (joists, waffle slabs, beams, and T-beams) to reinforce against tensile forces in webs resulting from local deviations from the design tendon pro-

CODE

COMMENTARY

file, and to provide a means of supporting the tendons in the design profile during construction. If sufficient support is not provided, lateral wobble and local deviations from the smooth parabolic tendon profile assumed in design may result during placement of the concrete. In such cases, the deviations in the tendons tend to straighten out when the tendons are stressed. This process may impose large tensile stresses in webs, and severe cracking may develop if no web reinforcement is provided. Unintended curvature of the tendons, and the resulting tensile stresses in webs, may be minimized by securely tying tendons to stirrups that are rigidly held in place by other elements of the reinforcing cage and held down in the forms. The maximum spacing of stirrups used for this purpose should not exceed the smaller of $1.5h$ or 4 ft. When applicable, the shear reinforcement provisions of 11.5.5 and 11.5.6 will require closer stirrup spacings.

For repeated loading of flexural members, the possibility of inclined diagonal tension cracks forming at stresses appreciably smaller than under static loading should be taken into account in the design. In these instances, it would be prudent to use at least the minimum shear reinforcement expressed by Eq. (11-13) or (11-14), even though tests or calculations based on static loads show that shear reinforcement is not required.

11.5.6.2 — Minimum shear reinforcement requirements of 11.5.6.1 shall be permitted to be waived if shown by test that required M_n and V_n can be developed when shear reinforcement is omitted. Such tests shall simulate effects of differential settlement, creep, shrinkage, and temperature change, based on a realistic assessment of such effects occurring in service.

R11.5.6.2 — When a member is tested to demonstrate that its shear and flexural strengths are adequate, the actual member dimensions and material strengths are known. The strength used as a basis for comparison should therefore be that corresponding to a strength reduction factor of unity ($\phi = 1.0$), i.e. the required nominal strength V_n and M_n. This ensures that if the actual material strengths in the field were less than specified, or the member dimensions were in error such as to result in a reduced member strength, a satisfactory margin of safety will be retained.

11.5.6.3 — Where shear reinforcement is required by 11.5.6.1 or for strength and where 11.6.1 allows torsion to be neglected, $A_{v,min}$ for prestressed (except as provided in 11.5.6.4) and nonprestressed members shall be computed by

$$A_{v,min} = 0.75\sqrt{f_c'}\frac{b_w s}{f_{yt}} \qquad (11\text{-}13)$$

but shall not be less than $(50 b_w s)/f_{yt}$.

R11.5.6.3 — Previous versions of the code have required a minimum area of transverse reinforcement that is independent of concrete strength. Tests[11.9] have indicated the need to increase the minimum area of shear reinforcement as concrete strength increases to prevent sudden shear failures when inclined cracking occurs. Equation (11-13) provides for a gradual increase in the minimum area of transverse reinforcement, while maintaining the previous minimum value.

11.5.6.4 — For prestressed members with an effective prestress force not less than 40 percent of the tensile strength of the flexural reinforcement, $A_{v,min}$ shall not be less than the smaller value from Eq. (11-13) and (11-14).

R11.5.6.4 — Tests[11.26] of prestressed beams with minimum web reinforcement based on Eq. (11-13) and (11-14) indicated that the smaller A_v from these two equations was sufficient to develop ductile behavior.

Eq. (11-14) may be used only for prestressed members meeting the minimum prestress force requirements given in 11.5.6.4. This equation is discussed in Reference 11.26.

CODE

$$A_{v,min} = \frac{A_{ps}f_{pu}s}{80f_{yt}d}\sqrt{\frac{d}{b_w}} \quad (11\text{-}14)$$

11.5.7 — Design of shear reinforcement

11.5.7.1 — Where V_u exceeds ϕV_c, shear reinforcement shall be provided to satisfy Eq. (11-1) and (11-2), where V_s shall be computed in accordance with 11.5.7.2 through 11.5.7.9.

11.5.7.2 — Where shear reinforcement perpendicular to axis of member is used,

$$V_s = \frac{A_v f_{yt} d}{s} \quad (11\text{-}15)$$

where A_v is the area of shear reinforcement within spacing s.

11.5.7.3 — Where circular ties, hoops, or spirals are used as shear reinforcement, V_s shall be computed using Eq. (11-15) where d is defined in 11.3.3 for circular members, A_v shall be taken as two times the area of the bar in a circular tie, hoop, or spiral at a spacing s, s is measured in a direction parallel to longitudinal reinforcement, and f_{yt} is the specified yield strength of circular tie, hoop, or spiral reinforcement.

11.5.7.4 — Where inclined stirrups are used as shear reinforcement,

$$V_s = \frac{A_v f_{yt}(\sin\alpha + \cos\alpha)d}{s} \quad (11\text{-}16)$$

where α is angle between inclined stirrups and longitudinal axis of the member, and s is measured in direction parallel to longitudinal reinforcement.

11.5.7.5 — Where shear reinforcement consists of a single bar or a single group of parallel bars, all bent up at the same distance from the support,

$$V_s = A_v f_y \sin\alpha \quad (11\text{-}17)$$

but not greater than $3\sqrt{f_c'}b_w d$, where α is angle

COMMENTARY

R11.5.7 — Design of shear reinforcement

Design of shear reinforcement is based on a modified truss analogy. The truss analogy assumes that the total shear is carried by shear reinforcement. However, considerable research on both nonprestressed and prestressed members has indicated that shear reinforcement needs to be designed to carry only the shear exceeding that which causes inclined cracking, provided the diagonal members in the truss are assumed to be inclined at 45 degrees.

Eq. (11-15), (11-16), and (11-17) are presented in terms of nominal shear strength provided by shear reinforcement V_s. When shear reinforcement perpendicular to axis of member is used, the required area of shear reinforcement A_v and its spacing s are computed by

$$\frac{A_v}{s} = \frac{(V_u - \phi V_c)}{\phi f_{yt} d}$$

Research[11.27,11.28] has shown that shear behavior of wide beams with substantial flexural reinforcement is improved if the transverse spacing of stirrup legs across the section is reduced.

R11.5.7.3 — Although the transverse reinforcement in a circular section may not consist of straight legs, tests indicate that Eq. (11-15) is conservative if d is taken as defined in 11.3.3.[11.16, 11.17]

CODE

between bent-up reinforcement and longitudinal axis of the member.

11.5.7.6 — Where shear reinforcement consists of a series of parallel bent-up bars or groups of parallel bent-up bars at different distances from the support, V_s shall be computed by Eq. (11-16).

11.5.7.7 — Only the center three-fourths of the inclined portion of any longitudinal bent bar shall be considered effective for shear reinforcement.

11.5.7.8 — Where more than one type of shear reinforcement is used to reinforce the same portion of a member, V_s shall be computed as the sum of the values computed for the various types of shear reinforcement.

11.5.7.9 — V_s shall not be taken greater than $8\sqrt{f_c'}\,b_w d$.

11.6 — Design for torsion

Design for torsion shall be in accordance with 11.6.1 through 11.6.6, or 11.6.7.

COMMENTARY

R11.6 — Design for torsion

The design for torsion in 11.6.1 through 11.6.6 is based on a thin-walled tube, space truss analogy. A beam subjected to torsion is idealized as a thin-walled tube with the core concrete cross section in a solid beam neglected as shown in Fig. R11.6(a). Once a reinforced concrete beam has cracked in torsion, its torsional resistance is provided primarily by closed stirrups and longitudinal bars located near the surface of the member. In the thin-walled tube analogy the resistance is assumed to be provided by the outer skin of the cross section roughly centered on the closed stirrups. Both hollow and solid sections are idealized as thin-walled tubes both before and after cracking.

In a closed thin-walled tube, the product of the shear stress τ and the wall thickness t at any point in the perimeter is

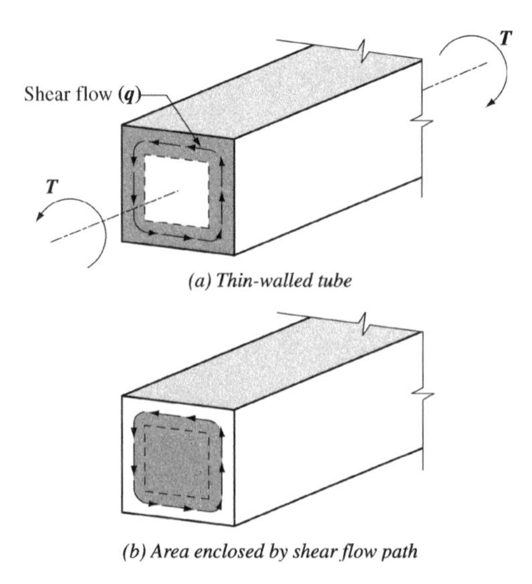

Fig. R11.6—(a) Thin-walled tube; (b) area enclosed by shear flow path

CODE

COMMENTARY

known as the shear flow, $q = \tau t$. The shear flow q due to torsion acts as shown in Fig. R11.6(a) and is constant at all points around the perimeter of the tube. The path along which it acts extends around the tube at midthickness of the walls of the tube. At any point along the perimeter of the tube the shear stress due to torsion is $\tau = T/(2A_o t)$ where A_o is the gross area enclosed by the shear flow path, shown shaded in Fig. R11.6(b), and t is the thickness of the wall at the point where τ is being computed. The shear flow follows the midthickness of the walls of the tube and A_o is the area enclosed by the path of the shearflow. For a hollow member with continuous walls, A_o includes the area of the hole.

In the 1995 code, the elliptical interaction between the nominal shear strength provided by the concrete, V_c, and the nominal torsion strength provided by the concrete was eliminated. V_c remains constant at the value it has when there is no torsion, and the torsion carried by the concrete is always taken as zero.

The design procedure is derived and compared with test results in References 11.29 and 11.30.

11.6.1 — Threshold torsion

It shall be permitted to neglect torsion effects if the factored torsional moment T_u is less than:

(a) For nonprestressed members:

$$\phi\sqrt{f_c'}\left(\frac{A_{cp}^2}{p_{cp}}\right)$$

(b) For prestressed members:

$$\phi\sqrt{f_c'}\left(\frac{A_{cp}^2}{p_{cp}}\right)\sqrt{1+\frac{f_{pc}}{4\sqrt{f_c'}}}$$

(c) For nonprestressed members subjected to an axial tensile or compressive force:

$$\phi\sqrt{f_c'}\left(\frac{A_{cp}^2}{p_{cp}}\right)\sqrt{1+\frac{N_u}{4A_g\sqrt{f_c'}}}$$

For members cast monolithically with a slab, the overhanging flange width used in computing A_{cp} and p_{cp} shall conform to 13.2.4. For a hollow section, A_g shall be used in place of A_{cp} in 11.6.1, and the outer boundaries of the section shall conform to 13.2.4.

11.6.1.1 — For isolated members with flanges and for members cast monolithically with a slab, the overhanging flange width used to compute A_{cp} and p_{cp} shall conform

R11.6.1 — Threshold torsion

Torques that do not exceed approximately one-quarter of the cracking torque T_{cr} will not cause a structurally significant reduction in either the flexural or shear strength and can be ignored. The cracking torsion under pure torsion T_{cr} is derived by replacing the actual section with an equivalent thin-walled tube with a wall thickness t prior to cracking of $0.75A_{cp}/p_{cp}$ and an area enclosed by the wall centerline A_o equal to $2A_{cp}/3$. Cracking is assumed to occur when the principal tensile stress reaches $4\sqrt{f_c'}$. In a nonprestressed beam loaded with torsion alone, the principal tensile stress is equal to the torsional shear stress, $\tau = T/(2A_o t)$. Thus, cracking occurs when τ reaches $4\sqrt{f_c'}$, giving the cracking torque T_{cr} as:

$$T_{cr} = 4\sqrt{f_c'}\left(\frac{A_{cp}^2}{p_{cp}}\right)$$

For solid members, the interaction between the cracking torsion and the inclined cracking shear is approximately circular or elliptical. For such a relationship, a torque of $0.25T_{cr}$, as used in 11.6.1, corresponds to a reduction of 3 percent in the inclined cracking shear. This reduction in the inclined cracking shear was considered negligible. The stress at cracking $4\sqrt{f_c'}$ has purposely been taken as a lower bound value.

For prestressed members, the torsional cracking load is increased by the prestress. A Mohr's Circle analysis based on average stresses indicates the torque required to cause a principal tensile stress equal to $4\sqrt{f_c'}$ is $\sqrt{1+f_{pc}/(4\sqrt{f_c'})}$ times the corresponding torque in a nonprestressed beam. A similar modification is made in part (c) of 11.6.1 for members subjected to axial load and torsion.

CODE

to 13.2.4, except that the overhanging flanges shall be neglected in cases where the parameter A_{cp}^2/p_{cp} calculated for a beam with flanges is less than that computed for the same beam ignoring the flanges.

11.6.2 — Calculation of factored torsional moment

11.6.2.1 — If the factored torsional moment, T_u, in a member is required to maintain equilibrium and exceeds the minimum value given in 11.6.1, the member shall be designed to carry T_u in accordance with 11.6.3 through 11.6.6.

11.6.2.2 — In a statically indeterminate structure where reduction of the torsional moment in a member can occur due to redistribution of internal forces upon cracking, the maximum T_u shall be permitted to be reduced to the values given in (a), (b), or (c), as applicable:

(a) For nonprestressed members, at the sections described in 11.6.2.4:

$$\phi 4\sqrt{f_c'}\left(\frac{A_{cp}^2}{p_{cp}}\right)$$

(b) For prestressed members, at the sections described in 11.6.2.5:

$$\phi 4\sqrt{f_c'}\left(\frac{A_{cp}^2}{p_{cp}}\right)\sqrt{1+\frac{f_{pc}}{4\sqrt{f_c'}}}$$

(c) For nonprestressed members subjected to an axial tensile or compressive force:

COMMENTARY

For torsion, a hollow member is defined as having one or more longitudinal voids, such as a single-cell or multiple-cell box girder. Small longitudinal voids, such as ungrouted post-tensioning ducts that result in A_g/A_{cp} greater than or equal to **0.95**, can be ignored when computing the threshold torque in 11.6.1. The interaction between torsional cracking and shear cracking for hollow sections is assumed to vary from the elliptical relationship for members with small voids, to a straight-line relationship for thin-walled sections with large voids. For a straight-line interaction, a torque of $0.25T_{cr}$ would cause a reduction in the inclined cracking shear of about 25 percent. This reduction was judged to be excessive.

In the 2002 code, two changes were made to modify 11.6.1 to apply to hollow sections. First, the minimum torque limits from the 1999 code were multiplied by (A_g/A_{cp}) because tests of solid and hollow beams[11.31] indicate that the cracking torque of a hollow section is approximately (A_g/A_{cp}) times the cracking torque of a solid section with the same outside dimensions. The second change was to multiply the cracking torque by (A_g/A_{cp}) a second time to reflect the transition from the circular interaction between the inclined cracking loads in shear and torsion for solid members, to the approximately linear interaction for thin-walled hollow sections.

R11.6.2 — Calculation of factored torsional moment

R11.6.2.1 and R11.6.2.2 — In designing for torsion in reinforced concrete structures, two conditions may be identified:[11.32,11.33]

(a) The torsional moment cannot be reduced by redistribution of internal forces (11.6.2.1). This is referred to as equilibrium torsion, since the torsional moment is required for the structure to be in equilibrium.

For this condition, illustrated in Fig. R11.6.2.1, torsion reinforcement designed according to 11.6.3 through 11.6.6 must be provided to resist the total design torsional moments.

(b) The torsional moment can be reduced by redistribution of internal forces after cracking (11.6.2.2) if the torsion arises from the member twisting to maintain compatibility of deformations. This type of torsion is referred to as compatibility torsion.

For this condition, illustrated in Fig. R11.6.2.2, the torsional stiffness before cracking corresponds to that of the uncracked section according to St. Venant's theory. At torsional cracking, however, a large twist occurs under an essentially constant torque, resulting in a large redistribution of forces in the structure.[11.32,11.33] The cracking torque under combined shear, flexure, and torsion corresponds to a principal tensile stress somewhat less than the $4\sqrt{f_c'}$ quoted in R11.6.1.

CODE

$$\phi 4\sqrt{f_c'}\left(\frac{A_{cp}^2}{p_{cp}}\right)\sqrt{1+\frac{N_u}{4A_g\sqrt{f_c'}}}$$

In (a), (b), or (c), the correspondingly redistributed bending moments and shears in the adjoining members shall be used in the design of these members. For hollow sections, A_{cp} shall not be replaced with A_g in 11.6.2.2.

COMMENTARY

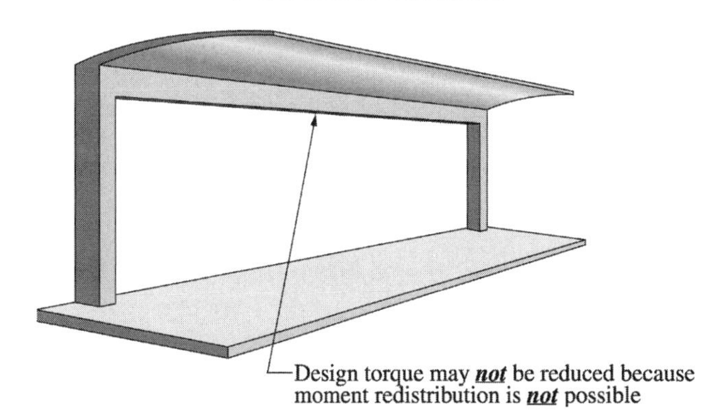

Fig. R11.6.2.1—Design torque may not be reduced (11.6.2.1)

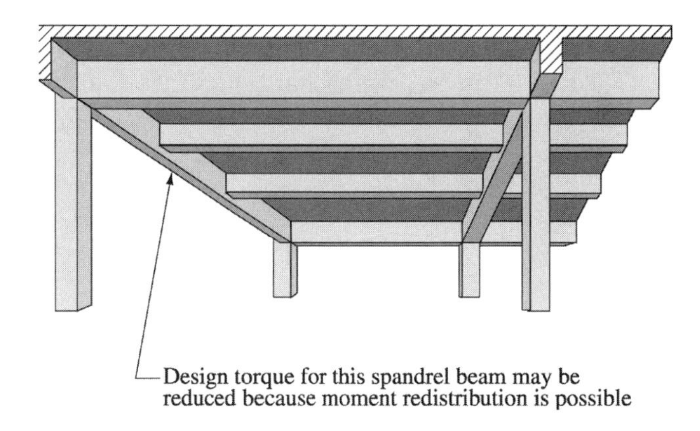

Fig. R11.6.2.2—Design torque may be reduced (11.6.2.2)

When the torsional moment exceeds the cracking torque, a maximum factored torsional moment equal to the cracking torque may be assumed to occur at the critical sections near the faces of the supports. This limit has been established to control the width of torsional cracks. The replacement of A_{cp} with A_g, as in the calculation of the threshold torque for hollow sections in 11.6.1, is not applied here. Thus, the torque after redistribution is larger and hence more conservative.

Section 11.6.2.2 applies to typical and regular framing conditions. With layouts that impose significant torsional rotations within a limited length of the member, such as a heavy torque loading located close to a stiff column, or a column that rotates in the reverse directions because of other loading, a more exact analysis is advisable.

When the factored torsional moment from an elastic analysis based on uncracked section properties is between the values in 11.6.1 and the values given in this section, torsion reinforcement should be designed to resist the computed torsional moments.

11.6.2.3 — Unless determined by a more exact analysis, it shall be permitted to take the torsional loading from a slab as uniformly distributed along the member.

CODE

11.6.2.4 — In nonprestressed members, sections located less than a distance d from the face of a support shall be designed for not less than T_u computed at a distance d. If a concentrated torque occurs within this distance, the critical section for design shall be at the face of the support.

11.6.2.5 — In prestressed members, sections located less than a distance $h/2$ from the face of a support shall be designed for not less than T_u computed at a distance $h/2$. If a concentrated torque occurs within this distance, the critical section for design shall be at the face of the support.

11.6.3 — Torsional moment strength

11.6.3.1 — The cross-sectional dimensions shall be such that:

(a) For solid sections:

$$\sqrt{\left(\frac{V_u}{b_w d}\right)^2 + \left(\frac{T_u p_h}{1.7A_{oh}^2}\right)^2} \leq \phi\left(\frac{V_c}{b_w d} + 8\sqrt{f_c'}\right) \qquad (11\text{-}18)$$

(b) For hollow sections:

$$\left(\frac{V_u}{b_w d}\right) + \left(\frac{T_u p_h}{1.7A_{oh}^2}\right) \leq \phi\left(\frac{V_c}{b_w d} + 8\sqrt{f_c'}\right) \qquad (11\text{-}19)$$

For prestressed members, d shall be determined in accordance with 11.5.3.

COMMENTARY

R11.6.2.4 and R11.6.2.5 — It is not uncommon for a beam to frame into one side of a girder near the support of the girder. In such a case a concentrated shear and torque are applied to the girder.

R11.6.3 — Torsional moment strength

R11.6.3.1 —The size of a cross section is limited for two reasons, first to reduce unsightly cracking and second to prevent crushing of the surface concrete due to inclined compressive stresses due to shear and torsion. In Eq. (11-18) and (11-19), the two terms on the left hand side are the shear stresses due to shear and torsion. The sum of these stresses may not exceed the stress causing shear cracking plus $8\sqrt{f_c'}$, similar to the limiting strength given in 11.5.7.9 for shear without torsion. The limit is expressed in terms of V_c to allow its use for nonprestressed or prestressed concrete. It was originally derived on the basis of crack control. It is not necessary to check against crushing of the web since this happens at higher shear stresses.

In a hollow section, the shear stresses due to shear and torsion both occur in the walls of the box as shown in Fig. 11.6.3.1(a)

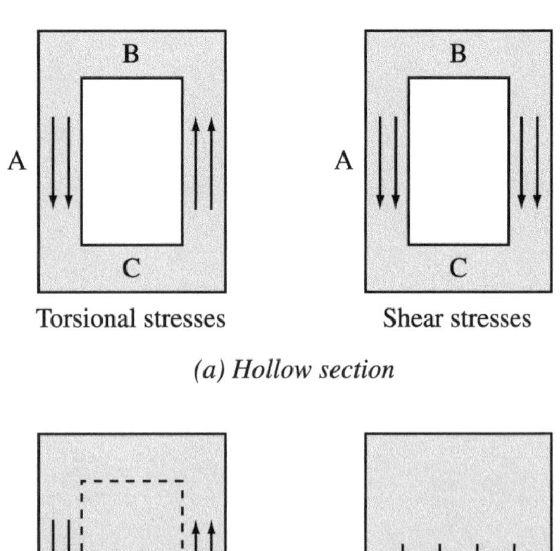

(a) Hollow section

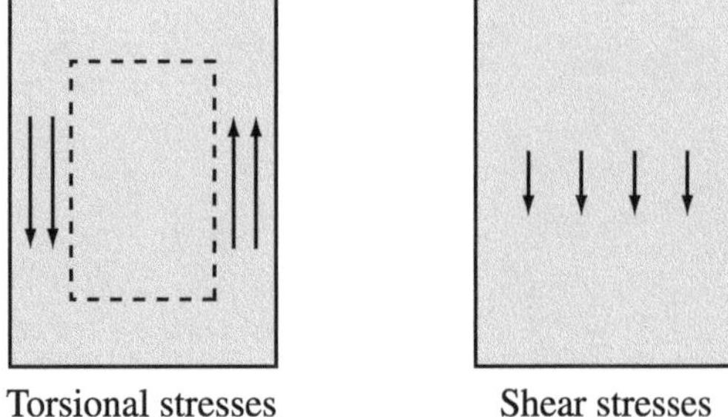

(b) Solid section

Fig. R11.6.3.1—Addition of torsional and shear stresses

CODE

COMMENTARY

and hence are directly additive at point A as given in Eq. (11-19). In a solid section the shear stresses due to torsion act in the "tubular" outside section while the shear stresses due to V_u are spread across the width of the section as shown in Fig. R11.6.3.1(b). For this reason stresses are combined in Eq. (11-18) using the square root of the sum of the squares rather than by direct addition.

11.6.3.2—If the wall thickness varies around the perimeter of a hollow section, Eq. (11-19) shall be evaluated at the location where the left-hand side of Eq. (11-19) is a maximum.

R11.6.3.2 — Generally, the maximum will be on the wall where the torsional and shearing stresses are additive [Point A in Fig. R11.6.3.1(a)]. If the top or bottom flanges are thinner than the vertical webs, it may be necessary to evaluate Eq. (11-19) at points B and C in Fig. R11.6.3.1(a). At these points the stresses due to the shear force are usually negligible.

11.6.3.3 — If the wall thickness is less than A_{oh}/p_h, the second term in Eq. (11-19) shall be taken as:

$$\left(\frac{T_u}{1.7A_{oh}t}\right)$$

where t is the thickness of the wall of the hollow section at the location where the stresses are being checked.

11.6.3.4 — The values of f_y and f_{yt} used for design of torsional reinforcement shall not exceed 60,000 psi.

R11.6.3.4 — Limiting the values of f_y and f_{yt} used in design of torsion reinforcement to 60,000 psi provides a control on diagonal crack width.

11.6.3.5 — Where T_u exceeds the threshold torsion, design of the cross section shall be based on:

$$\phi T_n \geq T_u \tag{11-20}$$

R11.6.3.5 — The factored torsional resistance ϕT_n must equal or exceed the torsion T_u due to the factored loads. In the calculation of T_n, all the torque is assumed to be resisted by stirrups and longitudinal steel with $T_c = 0$. At the same time, the nominal shear strength provided by concrete, V_c is assumed to be unchanged by the presence of torsion. For beams with V_u greater than about $0.8\phi V_c$ the resulting amount of combined shear and torsional reinforcement is essentially the same as required by the 1989 code. For smaller values of V_u, more shear and torsion reinforcement will be required.

11.6.3.6 — T_n shall be computed by:

$$T_n = \frac{2A_oA_tf_{yt}}{s}\cot\theta \tag{11-21}$$

where A_o shall be determined by analysis except that it shall be permitted to take A_o equal to $0.85A_{oh}$; θ shall not be taken smaller than 30 degrees nor larger than 60 degrees. It shall be permitted to take θ equal to:

(a) 45 degrees for nonprestressed members or members with less prestress than in (b); or

R11.6.3.6 — Eq. (11-21) is based on the space truss analogy shown in Fig. R11.6.3.6(a) with compression diagonals at an angle θ, assuming the concrete carries no tension and the reinforcement yields. After torsional cracking develops, the torsional resistance is provided mainly by closed stirrups, longitudinal bars, and compression diagonals. The concrete outside these stirrups is relatively ineffective. For this reason A_o, the gross area enclosed by the shear flow path around the perimeter of the tube, is defined after cracking in terms of A_{oh}, the area enclosed by the centerline of the outermost closed transverse torsional reinforcement. The area A_{oh} is shown in Fig. R11.6.3.6(b) for various cross sections. In an I-, T-, or L-shaped section, A_{oh} is taken as that area enclosed by the outermost legs of interlocking stirrups as shown in Fig. R11.6.3.6(b). The

CODE

(b) 37.5 degrees for prestressed members with an effective prestress force not less than 40 percent of the tensile strength of the longitudinal reinforcement.

11.6.3.7 — The additional area of longitudinal reinforcement to resist torsion, A_ℓ, shall not be less than:

$$A_\ell = \frac{A_t}{s} p_h \left(\frac{f_{yt}}{f_y} \right) \cot^2 \theta \quad \text{(11-22)}$$

where θ shall be the same value used in Eq. (11-21) and A_t/s shall be taken as the amount computed from Eq. (11-21) not modified in accordance with 11.6.5.2 or 11.6.5.3; f_{yt} refers to closed transverse torsional reinforcement, and f_y refers to longitudinal torsional reinforcement.

COMMENTARY

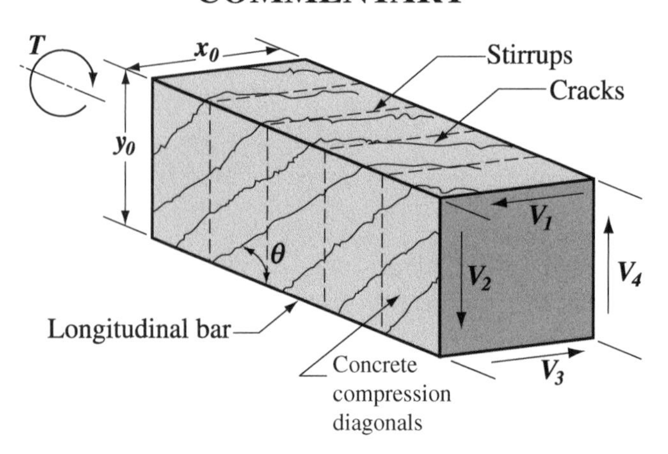

Fig. R11.6.3.6(a)—Space truss analogy

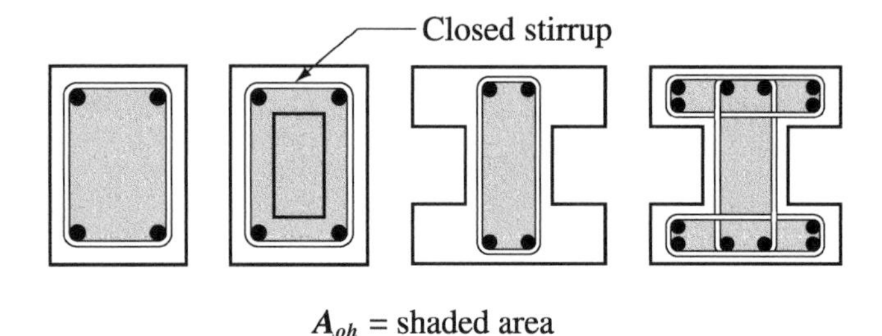

A_{oh} = shaded area

Fig. R11.6.3.6(b)—Definition of $\mathbf{A_{oh}}$

expression for A_o given by Hsu[11.34] may be used if greater accuracy is desired.

The shear flow q in the walls of the tube, discussed in R11.6, can be resolved into the shear forces V_1 to V_4 acting in the individual sides of the tube or space truss, as shown in Fig. R11.6.3.6(a).

The angle θ can be obtained by analysis[11.34] or may be taken to be equal to the values given in 11.6.3.6(a) or (b). The same value of θ should be used in both Eq. (11-21) and (11-22). As θ gets smaller, the amount of stirrups required by Eq. (11-21) decreases. At the same time the amount of longitudinal steel required by Eq. (11-22) increases.

R11.6.3.7 — Fig. R11.6.3.6(a) shows the shear forces V_1 to V_4 resulting from the shear flow around the walls of the tube. On a given wall of the tube, the shear flow V_i is resisted by a diagonal compression component, $D_i = V_i/\sin\theta$, in the concrete. An axial tension force, $N_i = V_i(\cot\theta)$, is needed in the longitudinal steel to complete the resolution of V_i.

Fig. R11.6.3.7 shows the diagonal compressive stresses and the axial tension force, N_i, acting on a short segment along one wall of the tube. Because the shear flow due to torsion is constant at all points around the perimeter of the tube, the resultants of D_i and N_i act through the midheight of side i. As a result, half of N_i can be assumed to be resisted by each of the top and bottom chords as shown. Longitudinal reinforcement

CODE

COMMENTARY

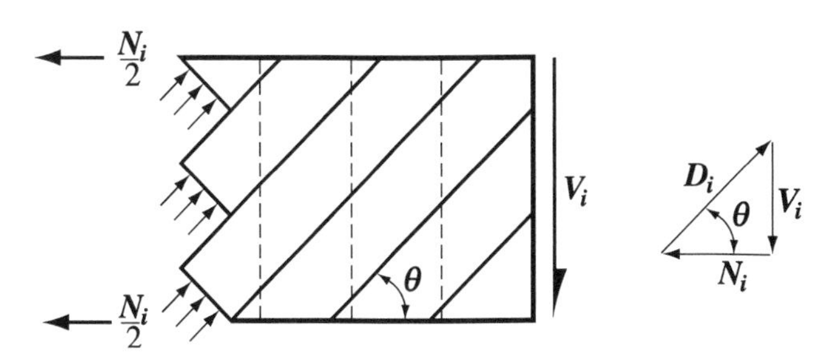

Fig. R11.6.3.7—Resolution of shear force $\mathbf{V_i}$ *into diagonal compression force* $\mathbf{D_i}$ *and axial tension force* $\mathbf{N_i}$ *in one wall of the tube*

with a capacity $A_\ell f_y$ should be provided to resist the sum of the N_i forces, ΣN_i, acting in all of the walls of the tube.

In the derivation of Eq. (11-22), axial tension forces are summed along the sides of the area A_o. These sides form a perimeter length, p_o, approximately equal to the length of the line joining the centers of the bars in the corners of the tube. For ease in computation this has been replaced with the perimeter of the closed stirrups, p_h.

Frequently, the maximum allowable stirrup spacing governs the amount of stirrups provided. Furthermore, when combined shear and torsion act, the total stirrup area is the sum of the amounts provided for shear and torsion. To avoid the need to provide excessive amounts of longitudinal reinforcement, 11.6.3.7 states that the A_t/s used in calculating A_ℓ at any given section should be taken as the A_t/s calculated at that section using Eq. (11-21).

11.6.3.8 — Reinforcement required for torsion shall be added to that required for the shear, moment and axial force that act in combination with the torsion. The most restrictive requirements for reinforcement spacing and placement shall be met.

R11.6.3.8 — The stirrup requirements for torsion and shear are added and stirrups are provided to supply at least the total amount required. Since the stirrup area A_v for shear is defined in terms of all the legs of a given stirrup while the stirrup area A_t for torsion is defined in terms of one leg only, the addition of stirrups is carried out as follows:

$$\text{Total}\left(\frac{A_{v+t}}{s}\right) = \frac{A_v}{s} + 2\frac{A_t}{s}$$

If a stirrup group had four legs for shear, only the legs adjacent to the sides of the beam would be included in this summation since the inner legs would be ineffective for torsion.

The longitudinal reinforcement required for torsion is added at each section to the longitudinal reinforcement required for bending moment that acts at the same time as the torsion. The longitudinal reinforcement is then chosen for this sum, but should not be less than the amount required for the maximum bending moment at that section if this exceeds the moment acting at the same time as the torsion. If the maximum bending moment occurs at one section, such as the midspan, while the maximum torsional moment occurs at another, such as the support, the total longitudinal steel required may be less than

CODE

COMMENTARY

that obtained by adding the maximum flexural steel plus the maximum torsional steel. In such a case the required longitudinal steel is evaluated at several locations.

The most restrictive requirements for spacing, cut-off points, and placement for flexural, shear, and torsional steel should be satisfied. The flexural steel should be extended a distance d, but not less than $12d_b$, past where it is no longer needed for flexure as required in 12.10.3.

11.6.3.9 — It shall be permitted to reduce the area of longitudinal torsion reinforcement in the flexural compression zone by an amount equal to $M_u/(0.9df_y)$, where M_u occurs at the section simultaneously with T_u, except that the reinforcement provided shall not be less than that required by 11.6.5.3 or 11.6.6.2.

R11.6.3.9 — The longitudinal tension due to torsion is offset in part by the compression in the flexural compression zone, allowing a reduction in the longitudinal torsion steel required in the compression zone.

11.6.3.10 — In prestressed beams:

(a) The total longitudinal reinforcement including prestressing steel at each section shall resist M_u at that section plus an additional concentric longitudinal tensile force equal to $A_\ell f_y$, based on T_u at that section;

(b) The spacing of the longitudinal reinforcement including tendons shall satisfy the requirements in 11.6.6.2.

R11.6.3.10 — As explained in R11.6.3.7, torsion causes an axial tension force. In a nonprestressed beam this force is resisted by longitudinal reinforcement having an axial tensile capacity of $A_\ell f_y$. This steel is in addition to the flexural reinforcement and is distributed uniformly around the sides of the perimeter so that the resultant of $A_\ell f_y$ acts along the axis of the member.

In a prestressed beam, the same technique (providing additional reinforcing bars with capacity $A_\ell f_y$) can be followed, or the designer can use any overcapacity of the prestressing steel to resist some of the axial force $A_\ell f_y$ as outlined in the next paragraph.

In a prestressed beam, the stress in the prestressing steel at nominal flexural strength at the section of the maximum moment is f_{ps}. At other sections, the stress in the prestressing steel at nominal flexural strength will be between f_{se} and f_{ps}. A portion of the $A_\ell f_y$ force acting on the sides of the perimeter where the prestressing steel is located can be resisted by a force $A_{ps}\Delta f_{pt}$ in the prestressing steel, where Δf_{pt} is f_{ps} at the section of maximum moment minus the stress in the prestressing steel due to prestressing and factored bending moments at the section under consideration. This can be taken as M_u at the section, divided by $(\phi 0.9 d_p A_{ps})$, but Δf_{pt} should not be more than 60 ksi. Longitudinal reinforcing bars will be required on the other sides of the member to provide the remainder of the $A_\ell f_y$ force, or to satisfy the spacing requirements given in 11.6.6.2, or both.

11.6.3.11 — In prestressed beams, it shall be permitted to reduce the area of longitudinal torsional reinforcement on the side of the member in compression due to flexure below that required by 11.6.3.10 in accordance with 11.6.3.9.

CODE

11.6.4 — Details of torsional reinforcement

11.6.4.1 — Torsion reinforcement shall consist of longitudinal bars or tendons and one or more of the following:

(a) Closed stirrups or closed ties, perpendicular to the axis of the member;

(b) A closed cage of welded wire reinforcement with transverse wires perpendicular to the axis of the member;

(c) In nonprestressed beams, spiral reinforcement.

11.6.4.2 — Transverse torsional reinforcement shall be anchored by one of the following:

(a) A 135 degree standard hook, or seismic hook as defined in 21.1, around a longitudinal bar;

(b) According to 12.13.2.1, 12.13.2.2, or 12.13.2.3 in regions where the concrete surrounding the anchorage is restrained against spalling by a flange or slab or similar member.

COMMENTARY

R11.6.4 — Details of torsional reinforcement

R11.6.4.1 — Both longitudinal and closed transverse reinforcement are required to resist the diagonal tension stresses due to torsion. The stirrups must be closed, since inclined cracking due to torsion may occur on all faces of a member.

In the case of sections subjected primarily to torsion, the concrete side cover over the stirrups spalls off at high torques.[11.35] This renders lapped-spliced stirrups ineffective, leading to a premature torsional failure.[11.36] In such cases, closed stirrups should not be made up of pairs of U-stirrups lapping one another.

R11.6.4.2 — When a rectangular beam fails in torsion, the corners of the beam tend to spall off due to the inclined compressive stresses in the concrete diagonals of the space truss changing direction at the corner as shown in Fig. R11.6.4.2(a). In tests,[11.35] closed stirrups anchored by 90 degree hooks failed when this occurred. For this reason, 135 degree standard hooks or seismic hooks are preferable for torsional stirrups in all cases. In regions where this spalling is prevented by an adjacent slab or flange, 11.6.4.2(b) relaxes this and allows 90 deg hooks.

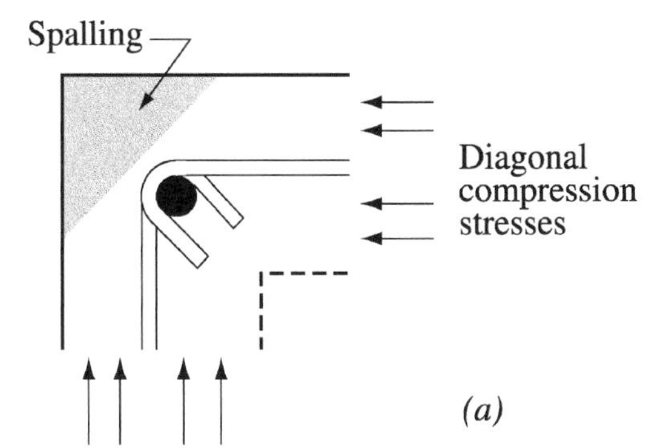

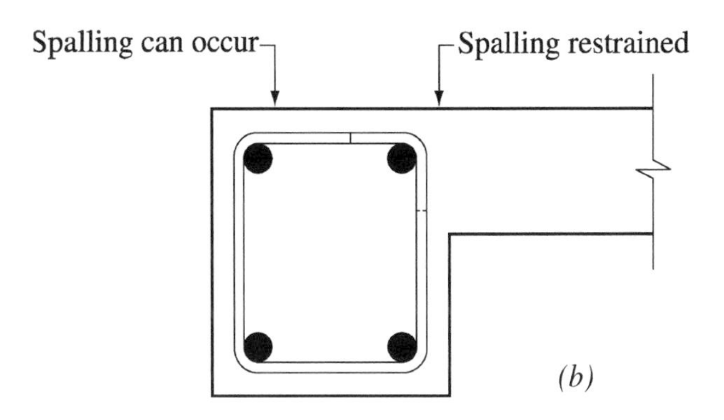

Fig. R11.6.4.2—Spalling of corners of beams loaded in torsion

CODE

11.6.4.3 — Longitudinal torsion reinforcement shall be developed at both ends.

11.6.4.4 — For hollow sections in torsion, the distance from the centerline of the transverse torsional reinforcement to the inside face of the wall of the hollow section shall not be less than $0.5A_{oh}/p_h$.

11.6.5 — Minimum torsion reinforcement

11.6.5.1 — A minimum area of torsional reinforcement shall be provided in all regions where T_u exceeds the threshold torsion given in 11.6.1.

11.6.5.2 — Where torsional reinforcement is required by 11.6.5.1, the minimum area of transverse closed stirrups shall be computed by:

$$(A_v + 2A_t) = 0.75\sqrt{f_c'}\frac{b_w s}{f_{yt}} \tag{11-23}$$

but shall not be less than $(50b_w s)/f_{yt}$.

11.6.5.3 — Where torsional reinforcement is required by 11.6.5.1, the minimum total area of longitudinal torsional reinforcement, $A_{\ell,min}$, shall be computed by:

$$A_{\ell,min} = \frac{5\sqrt{f_c'}A_{cp}}{f_y} - \left(\frac{A_t}{s}\right)p_h\frac{f_{yt}}{f_y} \tag{11-24}$$

where A_t/s shall not be taken less than $25b_w/f_{yt}$; f_{yt} refers to closed transverse torsional reinforcement, and f_y refers to longitudinal torsional reinforcement.

11.6.6 — Spacing of torsion reinforcement

11.6.6.1 — The spacing of transverse torsion reinforcement shall not exceed the smaller of $p_h/8$ or 12 in.

COMMENTARY

R11.6.4.3 — If high torsion acts near the end of a beam, the longitudinal torsion reinforcement should be adequately anchored. Sufficient development length should be provided outside the inner face of the support to develop the needed tension force in the bars or tendons. In the case of bars, this may require hooks or horizontal U-shaped bars lapped with the longitudinal torsion reinforcement.

R11.6.4.4 — The closed stirrups provided for torsion in a hollow section should be located in the outer half of the wall thickness effective for torsion where the wall thickness can be taken as A_{oh}/p_h.

R11.6.5 — Minimum torsion reinforcement

R11.6.5.1 and R11.6.5.2 — If a member is subject to a factored torsional moment T_u greater than the values specified in 11.6.1, the minimum amount of transverse web reinforcement for combined shear and torsion is $50b_w s/f_{yt}$. The differences in the definition of A_v and the symbol A_t should be noted; A_v is the area of two legs of a closed stirrup while A_t is the area of only one leg of a closed stirrup.

Tests[11.9] of high-strength reinforced concrete beams have indicated the need to increase the minimum area of shear reinforcement to prevent shear failures when inclined cracking occurs. Although there are a limited number of tests of high-strength concrete beams in torsion, the equation for the minimum area of transverse closed stirrups has been changed for consistency with calculations required for minimum shear reinforcement.

R11.6.5.3 — Reinforced concrete beam specimens with less than 1 percent torsional reinforcement by volume have failed in pure torsion at torsional cracking.[11.29] In the 1989 and prior codes, a relationship was presented that required about 1 percent torsional reinforcement in beams loaded in pure torsion and less in beams with combined shear and torsion, as a function of the ratio of shear stresses due to torsion and shear. Eq. (11-24) was simplified by assuming a single value of this reduction factor and results in a volumetric ratio of about 0.5 percent.

R11.6.6 — Spacing of torsion reinforcement

R11.6.6.1 — The spacing of the stirrups is limited to ensure the development of the ultimate torsional strength of the beam, to prevent excessive loss of torsional stiffness after cracking, and to control crack widths. For a square cross section the $p_h/8$ limitation requires stirrups at $d/2$, which corresponds to 11.5.5.1.

CODE

11.6.6.2 — The longitudinal reinforcement required for torsion shall be distributed around the perimeter of the closed stirrups with a maximum spacing of 12 in. The longitudinal bars or tendons shall be inside the stirrups. There shall be at least one longitudinal bar or tendon in each corner of the stirrups. Longitudinal bars shall have a diameter at least 0.042 times the stirrup spacing, but not less than 3/8 in.

11.6.6.3 — Torsional reinforcement shall be provided for a distance of at least $(b_t + d)$ beyond the point required by analysis.

11.6.7 — Alternative design for torsion

For torsion design of solid sections within the scope of this code with an aspect ratio, h/b_t, of three or greater, it shall be permitted to use another procedure, the adequacy of which has been shown by analysis and substantial agreement with results of comprehensive tests. Sections 11.6.4 and 11.6.6 shall apply.

11.7 — Shear-friction

11.7.1 — Provisions of 11.7 are to be applied where it is appropriate to consider shear transfer across a given plane, such as: an existing or potential crack, an interface between dissimilar materials, or an interface between two concretes cast at different times.

11.7.2 — Design of cross sections subject to shear transfer as described in 11.7.1 shall be based on Eq. (11-1), where V_n is calculated in accordance with provisions of 11.7.3 or 11.7.4.

11.7.3 — A crack shall be assumed to occur along the shear plane considered. The required area of shear-friction reinforcement A_{vf} across the shear plane shall be designed using either 11.7.4 or any other shear transfer design methods that result in prediction of strength in substantial agreement with results of comprehensive tests.

COMMENTARY

R11.6.6.2 — In R11.6.3.7 it was shown that longitudinal reinforcement is needed to resist the sum of the longitudinal tensile forces due to torsion in the walls of the thin-walled tube. Since the force acts along the centroidal axis of the section, the centroid of the additional longitudinal reinforcement for torsion should approximately coincide with the centroid of the section. The code accomplishes this by requiring the longitudinal torsional reinforcement to be distributed around the perimeter of the closed stirrups. Longitudinal bars or tendons are required in each corner of the stirrups to provide anchorage for the legs of the stirrups. Corner bars have also been found to be very effective in developing torsional strength and in controlling cracks.

R11.6.6.3 — The distance $(b_t + d)$ beyond the point theoretically required for torsional reinforcement is larger than that used for shear and flexural reinforcement because torsional diagonal tension cracks develop in a helical form.

R11.6.7 — Alternative design for torsion

Examples of such procedures are to be found in References 11.37 to 11.39, which have been extensively and successfully used for design of precast, prestressed concrete beams with ledges. The procedure described in References 11.37 and 11.38 is an extension to prestressed concrete sections of the torsion procedures of pre-1995 editions of ACI 318. The fourth edition of the *PCI Design Handbook*[11.40] describes the procedure of References 11.35 and 11.36. This procedure was experimentally verified by the tests described in Reference 11.41.

R11.7 — Shear-friction

R11.7.1 — With the exception of 11.7, virtually all provisions regarding shear are intended to prevent diagonal tension failures rather than direct shear transfer failures. The purpose of 11.7 is to provide design methods for conditions where shear transfer should be considered: an interface between concretes cast at different times, an interface between concrete and steel, reinforcement details for precast concrete structures, and other situations where it is considered appropriate to investigate shear transfer across a plane in structural concrete. (See References 11.42 and 11.43.)

R11.7.3 — Although uncracked concrete is relatively strong in direct shear there is always the possibility that a crack will form in an unfavorable location. The shear-friction concept assumes that such a crack will form, and that reinforcement must be provided across the crack to resist relative displacement along it. When shear acts along a crack, one crack face slips relative to the other. If the crack faces are rough and irregular, this slip is accompanied by separation of the crack faces. At ultimate,

CODE

11.7.3.1 — Provisions of 11.7.5 through 11.7.10 shall apply for all calculations of shear transfer strength.

COMMENTARY

the separation is sufficient to stress the reinforcement crossing the crack to its yield point. The reinforcement provides a clamping force $A_{vf}f_y$ across the crack faces. The applied shear is then resisted by friction between the crack faces, by resistance to the shearing off of protrusions on the crack faces, and by dowel action of the reinforcement crossing the crack. Successful application of 11.7 depends on proper selection of the location of an assumed crack.[11.18,11.42]

The relationship between shear-transfer strength and the reinforcement crossing the shear plane can be expressed in various ways. Eq. (11-25) and (11-26) of 11.7.4 are based on the shear-friction model. This gives a conservative prediction of shear-transfer strength. Other relationships that give a closer estimate of shear-transfer strength[11.18,11.44,11.45] can be used under the provisions of 11.7.3. For example, when the shear-friction reinforcement is perpendicular to the shear plane, the nominal shear strength V_n is given by[11.44,11.45]

$$V_n = 0.8A_{vf}f_y + A_cK_1$$

where A_c is the area of concrete section resisting shear transfer (in.^{2}) and $K_1 = 400$ psi for normalweight concrete, 200 psi for all-lightweight concrete, and 250 psi for sand-lightweight concrete. These values of K_1 apply to both monolithically cast concrete and to concrete cast against hardened concrete with a rough surface, as defined in 11.7.9.

In this equation, the first term represents the contribution of friction to shear-transfer resistance (0.8 representing the coefficient of friction). The second term represents the sum of the resistance to shearing of protrusions on the crack faces and the dowel action of the reinforcement.

When the shear-friction reinforcement is inclined to the shear plane, such that the shear force produces tension in that reinforcement, the nominal shear strength V_n is given by

$$V_n = A_{vf}f_y\,(0.8\sin\alpha + \cos\alpha) + A_cK_1\sin^2\alpha$$

where α is the angle between the shear-friction reinforcement and the shear plane, (that is, $0 < \alpha < 90$ **degrees**).

When using the modified shear-friction method, the terms $(A_{vf}f_y/A_c)$ or $(A_{vf}f_y \sin\alpha/A_c)$ should not be less than 200 psi for the design equations to be valid.

11.7.4 — Shear-friction design method

11.7.4.1 — Where shear-friction reinforcement is perpendicular to the shear plane, V_n shall be computed by

$$V_n = A_{vf}f_y\mu \qquad (11\text{-}25)$$

where μ is coefficient of friction in accordance with 11.7.4.3.

R11.7.4 — Shear-friction design method

R11.7.4.1 — The required area of shear-friction reinforcement A_{vf} is computed using

$$A_{vf} = \frac{V_u}{\phi f_y\mu}$$

The specified upper limit on shear strength should also be observed.

CODE

COMMENTARY

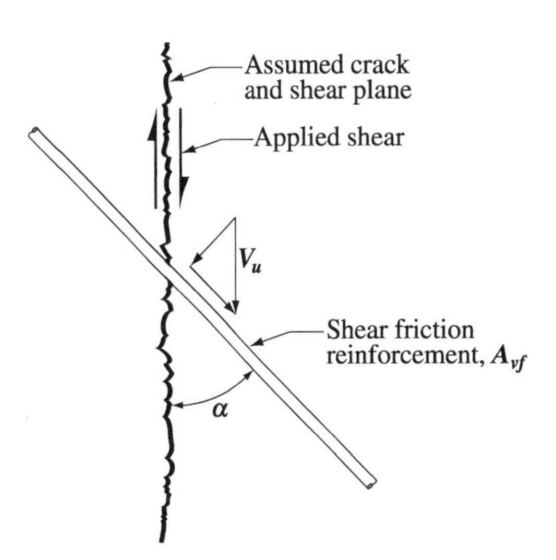

Fig. R11.7.4—Shear-friction reinforcement at an angle to assumed crack

11.7.4.2 — Where shear-friction reinforcement is inclined to the shear plane, such that the shear force produces tension in shear-friction reinforcement, V_n shall be computed by

$$V_n = A_{vf} f_y(\mu \sin \alpha + \cos \alpha) \quad (11\text{-}26)$$

where α is angle between shear-friction reinforcement and shear plane.

R11.7.4.2 — When the shear-friction reinforcement is inclined to the shear plane, such that the component of the shear force parallel to the reinforcement tends to produce tension in the reinforcement, as shown in Fig. R11.7.4, part of the shear is resisted by the component parallel to the shear plane of the tension force in the reinforcement.[11.45] Eq. (11-26) should be used only when the shear force component parallel to the reinforcement produces tension in the reinforcement, as shown in Fig. R11.7.4. When α is greater than 90 degrees, the relative movement of the surfaces tends to compress the bar and Eq. (11-26) is not valid.

11.7.4.3 — The coefficient of friction μ in Eq. (11-25) and Eq. (11-26) shall be taken as:

Concrete placed monolithically **1.4λ**

Concrete placed against hardened concrete with surface intentionally roughened as specified in 11.7.9 .... **1.0λ**

Concrete placed against hardened concrete not intentionally roughened **0.6λ**

Concrete anchored to as-rolled structural steel by headed studs or by reinforcing bars (see 11.7.10) **0.7λ**

where λ = **1.0** for normalweight concrete, 0.85 for sand-lightweight concrete and 0.75 for all lightweight concrete. Linear interpolation shall be permitted if partial sand replacement is used.

R11.7.4.3 — In the shear-friction method of calculation, it is assumed that all the shear resistance is due to the friction between the crack faces. It is, therefore, necessary to use artificially high values of the coefficient of friction in the shear-friction equations so that the calculated shear strength will be in reasonable agreement with test results. For concrete cast against hardened concrete not roughened in accordance with 11.7.9, shear resistance is primarily due to dowel action of the reinforcement and tests[11.46] indicate that reduced value of μ = **0.6λ** specified for this case is appropriate.

The value of μ for concrete placed against as-rolled structural steel relates to the design of connections between precast concrete members, or between structural steel members and structural concrete members. The shear-transfer reinforcement may be either reinforcing bars or headed stud shear connectors; also, field welding to steel plates after casting of concrete is common. The design of shear connectors for composite action of concrete slabs and steel beams is not covered by these provisions, but should be in accordance with Reference 11.47.

CODE

11.7.5 — V_n shall not be taken greater than the smaller of $0.2f_c'A_c$ and $800A_c$, where A_c is area of concrete section resisting shear transfer.

11.7.6 — The value of f_y used for design of shear-friction reinforcement shall not exceed 60,000 psi.

11.7.7 — Net tension across shear plane shall be resisted by additional reinforcement. Permanent net compression across shear plane shall be permitted to be taken as additive to $A_{vf}f_y$, the force in the shear-friction reinforcement, when calculating required A_{vf}.

11.7.8 — Shear-friction reinforcement shall be appropriately placed along the shear plane and shall be anchored to develop f_y on both sides by embedment, hooks, or welding to special devices.

COMMENTARY

R11.7.5 — This upper limit on shear strength is specified because Eq. (11-25) and (11-26) become unconservative if V_n has a greater value.

R11.7.7 — If a resultant tensile force acts across a shear plane, reinforcement to carry that tension should be provided in addition to that provided for shear transfer. Tension may be caused by restraint of deformations due to temperature change, creep, and shrinkage. Such tensile forces have caused failures, particularly in beam bearings.

When moment acts on a shear plane, the flexural tension stresses and flexural compression stresses are in equilibrium. There is no change in the resultant compression $A_{vf}f_y$ acting across the shear plane and the shear-transfer strength is not changed. It is therefore not necessary to provide additional reinforcement to resist the flexural tension stresses, unless the required flexural tension reinforcement exceeds the amount of shear-transfer reinforcement provided in the flexural tension zone. This has been demonstrated experimentally.[11.48]

It has also been demonstrated experimentally[11.43] that if a resultant compressive force acts across a shear plane, the shear-transfer strength is a function of the sum of the resultant compressive force and the force $A_{vf}f_y$ in the shear-friction reinforcement. In design, advantage should be taken of the existence of a compressive force across the shear plane to reduce the amount of shear-friction reinforcement required, only if it is certain that the compressive force is permanent.

R11.7.8 — If no moment acts across the shear plane, reinforcement should be uniformly distributed along the shear plane to minimize crack widths. If a moment acts across the shear plane, it is desirable to distribute the shear-transfer reinforcement primarily in the flexural tension zone.

Since the shear-friction reinforcement acts in tension, it should have full tensile anchorage on both sides of the shear plane. Further, the shear-friction reinforcement anchorage should engage the primary reinforcement, otherwise a potential crack may pass between the shear-friction reinforcement and the body of the concrete. This requirement applies particularly to welded headed studs used with steel inserts for connections in precast and cast-in-place concrete. Anchorage may be developed by bond, by a welded mechanical anchorage, or by threaded dowels and screw inserts. Space limitations often require a welded mechanical anchorage. For anchorage of headed studs in concrete see Reference 11.18.

CODE

11.7.9 — For the purpose of 11.7, when concrete is placed against previously hardened concrete, the interface for shear transfer shall be clean and free of laitance. If μ is assumed equal to $\mathbf{1.0\lambda}$, interface shall be roughened to a full amplitude of approximately 1/4 in.

11.7.10 — When shear is transferred between as-rolled steel and concrete using headed studs or welded reinforcing bars, steel shall be clean and free of paint.

11.8 — Deep beams

11.8.1 — The provisions of 11.8 shall apply to members with ℓ_n not exceeding four times the overall member depth or regions of beams with concentrated loads within twice the member depth from the support that are loaded on one face and supported on the opposite face so that compression struts can develop between the loads and supports. See also 12.10.6.

11.8.2 — Deep beams shall be designed using either nonlinear analysis as permitted in 10.7.1, or Appendix A.

11.8.3 — $\boldsymbol{V_n}$ for deep beams shall not exceed $\mathbf{10\sqrt{f_c'}\,b_w d}$.

11.8.4 — The area of shear reinforcement perpendicular to the flexural tension reinforcement, $\boldsymbol{A_v}$, shall not be less than $\mathbf{0.0025 b_w s}$, and $\boldsymbol{s}$ shall not exceed the smaller of $\boldsymbol{d/5}$ and 12 in.

11.8.5 — The area of shear reinforcement parallel to the flexural tension reinforcement, $\boldsymbol{A_{vh}}$, shall not be less than $\mathbf{0.0015 b_w s_2}$, and $\boldsymbol{s_2}$ shall not exceed the smaller of $\boldsymbol{d/5}$ and 12 in.

11.8.6 — It shall be permitted to provide reinforcement satisfying A.3.3 instead of the minimum horizontal and vertical reinforcement specified in 11.8.4 and 11.8.5.

COMMENTARY

R11.8 — Deep beams

R11.8.1 — The behavior of a deep beam is discussed in References 11.5 and 11.45. For a deep beam supporting gravity loads, this section applies if the loads are applied on the top of the beam and the beam is supported on its bottom face. If the loads are applied through the sides or bottom of such a member, the design for shear should be the same as for ordinary beams.

The longitudinal reinforcement in deep beams should be extended to the supports and adequately anchored by embedment, hooks, or welding to special devices. Bent-up bars are not recommended.

R11.8.2 — Deep beams can be designed using strut-and-tie models, regardless of how they are loaded and supported. Section 10.7.1 allows the use of nonlinear stress fields when proportioning deep beams. Such analyses should consider the effects of cracking on the stress distribution.

R11.8.3 — In the 1999 and earlier codes, a sliding maximum shear strength was specified. A re-examination of the test data suggests that this strength limit was derived from tests in which the beams failed due to crushing of support regions. This possibility is specifically addressed in the design process specified in this code.

R11.8.4 and **R11.8.5** — The relative amounts of horizontal and vertical shear reinforcement have been interchanged from those required in the 1999 and earlier codes because tests[11.49-11.51] have shown that vertical shear reinforcement is more effective than horizontal shear reinforcement. The maximum spacing of bars has been reduced from 18 in. to 12 in. because this steel is provided to restrain the width of the cracks.

CODE

11.9 — Special provisions for brackets and corbels

11.9.1 — Brackets and corbels with a shear span-to-depth ratio a_v/d less than 2 shall be permitted to be designed using Appendix A. Design shall be permitted using 11.9.3 and 11.9.4 for brackets and corbels with:

(a) a_v/d not greater than 1, and

(b) subject to factored horizontal tensile force, N_{uc}, not larger than V_u.

The requirements of 11.9.2, 11.9.3.2.1, 11.9.3.2.2, 11.9.5, 11.9.6, and 11.9.7 shall apply to design of brackets and corbels. Effective depth d shall be determined at the face of the support.

COMMENTARY

R11.9 — Special provisions for brackets and corbels

Brackets and corbels are cantilevers having shear span-to-depth ratios not greater than unity, which tend to act as simple trusses or deep beams, rather than flexural members designed for shear according to 11.3.

The corbel shown in Fig. R11.9.1 may fail by shearing along the interface between the column and the corbel, by yielding of the tension tie, by crushing or splitting of the compression strut, or by localized bearing or shearing failure under the loading plate. These failure modes are illustrated and are discussed more fully in Reference 11.1. The notation used in 11.9 is illustrated in Fig. R11.9.2.

R11.9.1 — An upper limit of 1.0 for a_v/d is imposed for design by 11.9.3 and 11.9.4 for two reasons. First, for a_v/d shear span-to-depth ratios exceeding unity, the diagonal tension cracks are less steeply inclined and the use of horizontal stirrups alone as specified in 11.9.4 is not appropriate. Second, this method of design has only been validated experimentally for a_v/d of unity or less. An upper limit is provided for N_{uc} because this method of design has only been validated experimentally for N_{uc} less than or equal to V_u, including N_{uc} equal to zero.

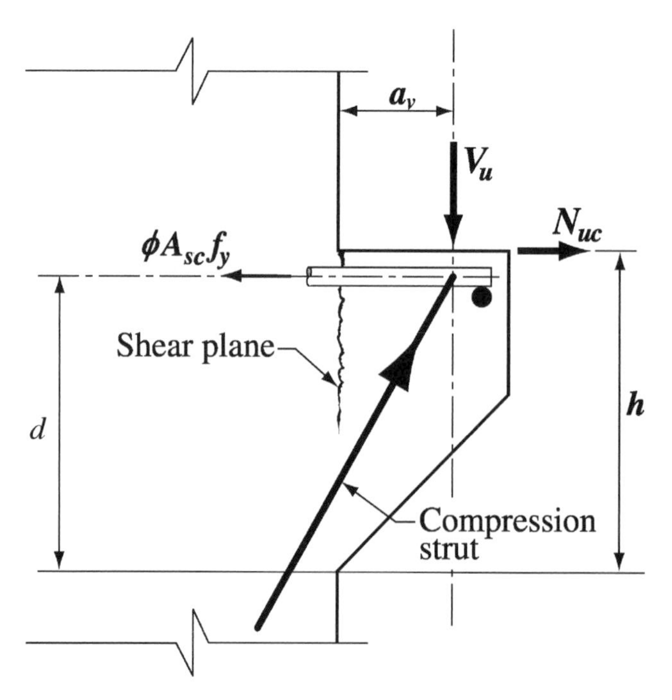

Fig. R11.9.1—Structural action of a corbel

CODE

COMMENTARY

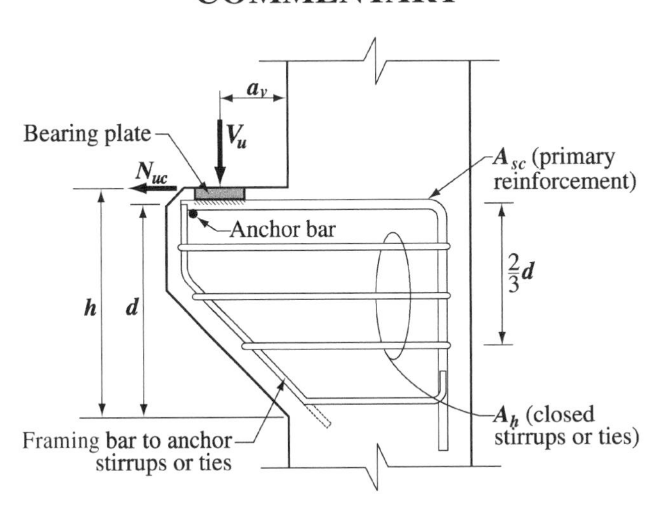

Fig. R11.9.2—Notation used in Section 11.9

11.9.2 — Depth at outside edge of bearing area shall not be less than $\mathbf{0.5}\boldsymbol{d}$.

R11.9.2 — A minimum depth is required at the outside edge of the bearing area so that a premature failure will not occur due to a major diagonal tension crack propagating from below the bearing area to the outer sloping face of the corbel or bracket. Failures of this type have been observed[11.52] in corbels having depths at the outside edge of the bearing area less than required in this section of the code.

11.9.3 — Section at face of support shall be designed to resist simultaneously V_u, a factored moment $[V_u a_v + N_{uc}(h - d)]$, and a factored horizontal tensile force, N_{uc}.

11.9.3.1 — In all design calculations in accordance with 11.9, ϕ shall be taken equal to 0.75

R11.9.3.1 — Corbel and bracket behavior is predominantly controlled by shear; therefore, a single value of $\phi = 0.75$ is required for all design conditions.

11.9.3.2 — Design of shear-friction reinforcement, A_{vf}, to resist V_u shall be in accordance with 11.7.

11.9.3.2.1 — For normalweight concrete, V_n shall not be taken greater than the smaller of $0.2 f_c' b_w d$ and $800 b_w d$.

11.9.3.2.2 — For all-lightweight or sand-lightweight concrete, V_n shall not be taken greater than the smaller of $(0.2 - 0.07 a_v / d) f_c' b_w d$ and $(800 - 280 a_v / d) b_w d$.

R11.9.3.2.2 — Tests[11.53] have shown that the maximum shear strength of lightweight concrete corbels or brackets is a function of both f_c' and a_v/d. No data are available for corbels or brackets made of sand-lightweight concrete. As a result, the same limitations have been placed on both all-lightweight and sand-lightweight brackets and corbels.

11.9.3.3 — Reinforcement A_f to resist factored moment $[V_u a_v + N_{uc}(h - d)]$ shall be computed in accordance with 10.2 and 10.3.

R11.9.3.3 — Reinforcement required to resist moment can be calculated using flexural theory. The factored moment is calculated by summing moments about the flexural reinforcement at the face of support.

CODE

11.9.3.4 — Reinforcement A_n to resist factored tensile force N_{uc} shall be determined from $\phi A_n f_y \geq N_{uc}$. Factored tensile force, N_{uc}, shall not be taken less than $0.2V_u$ unless special provisions are made to avoid tensile forces. N_{uc} shall be regarded as a live load even if tension results from restraint of creep, shrinkage, or temperature change.

11.9.3.5 — Area of primary tension reinforcement A_{sc} shall not be less than the larger of $(A_f + A_n)$ and $(2A_{vf}/3 + A_n)$.

11.9.4 — Total area, A_h, of closed stirrups or ties parallel to primary tension reinforcement shall not be less than $0.5\ (A_{sc} - A_n)$. Distribute A_h uniformly within $(2/3)d$ adjacent to primary tension reinforcement.

11.9.5 — A_{sc}/bd shall not be less than $0.04\ (f_c'/f_y)$.

11.9.6 — At front face of bracket or corbel, primary tension reinforcement shall be anchored by one of the following:

(a) By a structural weld to a transverse bar of at least equal size; weld to be designed to develop f_y of primary tension reinforcement;

(b) By bending primary tension reinforcement back to form a horizontal loop; or

(c) By some other means of positive anchorage.

COMMENTARY

R11.9.3.4 — Because the magnitude of horizontal forces acting on corbels or brackets cannot usually be determined with great accuracy, it is required that N_{uc} be regarded as a live load.

R11.9.3.5 — Tests[11.53] suggest that the total amount of reinforcement $(A_{sc} + A_h)$ required to cross the face of support should be the greater of:

(a) The sum of A_{vf} calculated according to 11.9.3.2 and A_n calculated according to 11.9.3.4;

(b) The sum of **1.5** times A_f calculated according to 11.9.3.3 and A_n calculated according to 11.9.3.4.

If (a) controls, $A_{sc} = (2A_{vf}/3 + A_n)$ is required as primary tensile reinforcement, and the remaining $A_{vf}/3$ should be provided as closed stirrups parallel to A_{sc} and distributed within $2d/3$, adjacent to A_{sc}. Section 11.9.4 satisfies this by requiring $A_h = 0.5(2A_{vf}/3)$.

If (b) controls, $A_{sc} = (A_f + A_n)$ is required as primary tension reinforcement, and the remaining $A_f/2$ should be provided as closed stirrups parallel to A_{sc} and distributed within $2d/3$, adjacent to A_{sc}. Again 11.9.4 satisfies this requirement.

R11.9.4 — Closed stirrups parallel to the primary tension reinforcement are necessary to prevent a premature diagonal tension failure of the corbel or bracket. The required area of closed stirrups $A_h = 0.5\ (A_{sc} - A_n)$ automatically yields the appropriate amounts, as discussed in R11.9.3.5 above.

R11.9.5 — A minimum amount of reinforcement is required to prevent the possibility of sudden failure should the bracket or corbel concrete crack under the action of flexural moment and outward tensile force N_{uc}.

R11.9.6 — Because the horizontal component of the inclined concrete compression strut (see Fig. R11.9.1) is transferred to the primary tension reinforcement at the location of the vertical load, the primary tension reinforcement is essentially uniformly stressed from the face of the support to the point where the vertical load is applied. It should, therefore, be anchored at its outer end and in the supporting column, so as to be able to develop its specified yield strength from the face of support to the vertical load. Satisfactory anchorage at the outer end can be obtained by bending the primary tension reinforcement bars in a horizontal loop as specified in (b), or by welding a bar of equal diameter or a suitably sized angle across the ends of the primary tension reinforcement bars. The welds should be designed to develop the yield strength of the primary tension rein-

CODE

COMMENTARY

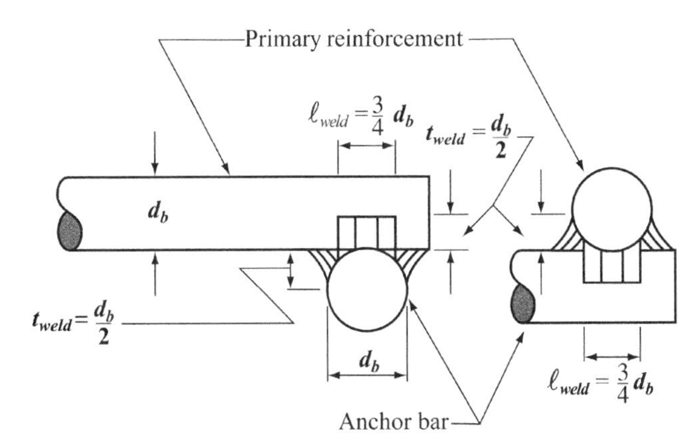

Fig. R11.9.6—Weld details used in tests of Reference 11.53.

forcement. The weld detail used successfully in the corbel tests reported in Reference 11.53 is shown in Fig. R11.9.6. The primary tension reinforcement should be anchored within the supporting column in accordance with the requirements of Chapter 12. See additional discussion on end anchorage in R12.10.6.

11.9.7 — Bearing area on bracket or corbel shall not project beyond straight portion of primary tension reinforcement, nor project beyond interior face of transverse anchor bar (if one is provided).

R11.9.7 — The restriction on the location of the bearing area is necessary to ensure development of the specified yield strength of the primary tension reinforcement near the load. When corbels are designed to resist horizontal forces, the bearing plate should be welded to the primary tension reinforcement.

11.10 — Special provisions for walls

R11.10 — Special provisions for walls

11.10.1 — Design for shear forces perpendicular to face of wall shall be in accordance with provisions for slabs in 11.12. Design for horizontal in-plane shear forces in a wall shall be in accordance with 11.10.2 through 11.10.9. Alternatively, it shall be permitted to design walls with a height not exceeding two times the length of the wall for horizontal shear forces in accordance with Appendix A and 11.10.9.2 through 11.10.9.5.

R11.10.1 — Shear in the plane of the wall is primarily of importance for shearwalls with a small height-to-length ratio. The design of higher walls, particularly walls with uniformly distributed reinforcement, will probably be controlled by flexural considerations.

11.10.2 — Design of horizontal section for shear in plane of wall shall be based on Eq. (11-1) and (11-2), where V_c shall be in accordance with 11.10.5 or 11.10.6 and V_s shall be in accordance with 11.10.9.

11.10.3 — V_n at any horizontal section for shear in plane of wall shall not be taken greater than $10\sqrt{f_c'}\,hd$, where h is thickness of wall, and d is defined in 11.10.4.

R11.10.3 — Although the width-to-depth ratio of shearwalls is less than that for ordinary beams, tests[11.54] on shearwalls with a thickness equal to $\ell_w/25$ have indicated that ultimate shear stresses in excess of $10\sqrt{f_c'}$ can be obtained.

CODE

11.10.4 — For design for horizontal shear forces in plane of wall, d shall be taken equal to $0.8\ell_w$. A larger value of d, equal to the distance from extreme compression fiber to center of force of all reinforcement in tension, shall be permitted to be used when determined by a strain compatibility analysis.

11.10.5 — Unless a more detailed calculation is made in accordance with 11.10.6, V_c shall not be taken greater than $2\sqrt{f_c'}hd$ for walls subject to axial compression, or V_c shall not be taken greater than the value given in 11.3.2.3 for walls subject to axial tension.

11.10.6 — V_c shall be permitted to be the lesser of the values computed from Eq. (11-29) and (11-30).

$$V_c = 3.3\sqrt{f_c'}hd + \frac{N_u d}{4\ell_w} \qquad (11\text{-}29)$$

or

$$V_c = \left[0.6\sqrt{f_c'} + \frac{\ell_w\left(1.25\sqrt{f_c'} + 0.2\frac{N_u}{\ell_w h}\right)}{\frac{M_u}{V_u} - \frac{\ell_w}{2}}\right]hd \qquad (11\text{-}30)$$

where ℓ_w is the overall length of the wall, and N_u is positive for compression and negative for tension. If $(M_u/V_u - \ell_w/2)$ is negative, Eq. (11-30) shall not apply.

11.10.7 — Sections located closer to wall base than a distance $\ell_w/2$ or one-half the wall height, whichever is less, shall be permitted to be designed for the same V_c as that computed at a distance $\ell_w/2$ or one-half the height.

11.10.8 — Where V_u is less than $0.5\phi V_c$, reinforcement shall be provided in accordance with 11.10.9 or in accordance with Chapter 14. Where V_u exceeds $0.5\phi V_c$, wall reinforcement for resisting shear shall be provided in accordance with 11.10.9.

11.10.9 — Design of shear reinforcement for walls

11.10.9.1 — Where V_u exceeds ϕV_c, horizontal shear reinforcement shall be provided to satisfy Eq. (11-1) and (11-2), where V_s shall be computed by

$$V_s = \frac{A_v f_y d}{s} \qquad (11\text{-}31)$$

COMMENTARY

R11.10.5 and R11.10.6 — Eq. (11-29) and (11-30) may be used to determine the inclined cracking strength at any section through a shearwall. Eq. (11-29) corresponds to the occurrence of a principal tensile stress of approximately $4\sqrt{f_c'}$ at the centroid of the shearwall cross section. Eq. (11-30) corresponds approximately to the occurrence of a flexural tensile stress of $6\sqrt{f_c'}$ at a section $\ell_w/2$ above the section being investigated. As the term

$$\left(\frac{M_u}{V_u} - \frac{\ell_w}{2}\right)$$

decreases, Eq. (11-29) will control before this term becomes negative. When this term becomes negative Eq. (11-29) should be used.

R11.10.7 — The values of V_c computed from Eq. (11-29) and (11-30) at a section located a lesser distance of $\ell_w/2$ or $h_w/2$ above the base apply to that and all sections between this section and the base. However, the maximum factored shear force V_u at any section, including the base of the wall, is limited to ϕV_n in accordance with 11.10.3.

R11.10.9 — Design of shear reinforcement for walls

Both horizontal and vertical shear reinforcement are required for all walls. The notation used to identify the direction of the distributed shear reinforcement in walls was updated in 2005 to eliminate conflicts between the notation used for ordinary structural walls in Chapters 11 and 14 and the notation used for special structural walls in Chapter 21. The distributed reinforcement is now identified as being oriented parallel to either the longitudinal or transverse axis of the wall. Therefore, for vertical wall

CODE

where A_v is area of horizontal shear reinforcement within spacing s, and d is determined in accordance with 11.10.4. Vertical shear reinforcement shall be provided in accordance with 11.10.9.4.

11.10.9.2 — Ratio of horizontal shear reinforcement area to gross concrete area of vertical section, ρ_t, shall not be less than 0.0025.

11.10.9.3 — Spacing of horizontal shear reinforcement shall not exceed the smallest of $\ell_w/5$, $3h$, and 18 in., where ℓ_w is the overall length of the wall.

11.10.9.4 — Ratio of vertical shear reinforcement area to gross concrete area of horizontal section, ρ_ℓ, shall not be less than the larger of

$$\rho_\ell = 0.0025 + 0.5\left(2.5 - \frac{h_w}{\ell_w}\right)(\rho_t - 0.0025) \quad (11\text{-}32)$$

and 0.0025, but need not be greater than ρ_t required by 11.10.9.1. In Eq. (11-32), ℓ_w is the overall length of the wall, and h_w is the overall height of the wall.

11.10.9.5 — Spacing of vertical shear reinforcement shall not exceed the smallest of $\ell_w/3$, $3h$, and 18 in., where ℓ_w is the overall length of the wall.

11.11 — Transfer of moments to columns

11.11.1 — When gravity load, wind, earthquake, or other lateral forces cause transfer of moment at connections of framing elements to columns, the shear resulting from moment transfer shall be considered in the design of lateral reinforcement in the columns.

11.11.2 — Except for connections not part of a primary seismic load-resisting system that are restrained on four sides by beams or slabs of approximately equal depth, connections shall have lateral reinforcement not less than that required by Eq. (11-13) within the column for a depth not less than that of the deepest connection of framing elements to the columns. See also 7.9.

11.12 — Special provisions for slabs and footings

11.12.1 — The shear strength of slabs and footings in the vicinity of columns, concentrated loads, or reactions is governed by the more severe of two conditions:

COMMENTARY

segments, the notation used to describe the horizontal distributed reinforcement ratio is ρ_t, and the notation used to describe the vertical distributed reinforcement ratio is ρ_ℓ.

For low walls, test data[11.55] indicate that horizontal shear reinforcement becomes less effective with vertical reinforcement becoming more effective. This change in effectiveness of the horizontal versus vertical reinforcement is recognized in Eq. (11-32); if h_w/ℓ_w is less than 0.5, the amount of vertical reinforcement is equal to the amount of horizontal reinforcement. If h_w/ℓ_w is greater than 2.5, only a minimum amount of vertical reinforcement is required **(0.0025 *sh*)**.

Eq. (11-31) is presented in terms of shear strength V_s provided by the horizontal shear reinforcement for direct application in Eq. (11-1) and (11-2).

Vertical shear reinforcement also should be provided in accordance with 11.10.9.4 within the spacing limitation of 11.10.9.5.

R11.11 — Transfer of moments to columns

R11.11.1 — Tests[11.56] have shown that the joint region of a beam-to-column connection in the interior of a building does not require shear reinforcement if the joint is confined on four sides by beams of approximately equal depth. However, joints without lateral confinement, such as at the exterior of a building, need shear reinforcement to prevent deterioration due to shear cracking.[11.57]

For regions where strong earthquakes may occur, joints may be required to withstand several reversals of loading that develop the flexural capacity of the adjoining beams. See Chapter 21 for special provisions for seismic design.

R11.12 — Special provisions for slabs and footings

R11.12.1 — Differentiation should be made between a long and narrow slab or footing acting as a beam, and a slab or footing subject to two-way action where failure may occur by punching along a truncated cone or pyramid around a concentrated load or reaction area.

CODE

11.12.1.1 — Beam action where each critical section to be investigated extends in a plane across the entire width. For beam action the slab or footing shall be designed in accordance with 11.1 through 11.5.

11.12.1.2 — For two-way action, each of the critical sections to be investigated shall be located so that its perimeter b_o is a minimum but need not approach closer than $d/2$ to

(a) Edges or corners of columns, concentrated loads, or reaction areas; and

(b) Changes in slab thickness such as edges of capitals or drop panels.

For two-way action the slab or footing shall be designed in accordance with 11.12.2 through 11.12.6.

11.12.1.3 — For square or rectangular columns, concentrated loads, or reaction areas, the critical sections with four straight sides shall be permitted.

11.12.2 — The design of a slab or footing for two-way action is based on Eq. (11-1) and (11-2). V_c shall be computed in accordance with 11.12.2.1, 11.12.2.2, or 11.12.3.1. V_s shall be computed in accordance with 11.12.3. For slabs with shearheads, V_n shall be in accordance with 11.12.4. When moment is transferred between a slab and a column, 11.12.6 shall apply.

11.12.2.1 — For nonprestressed slabs and footings, V_c shall be the smallest of (a), (b), and (c):

(a)
$$V_c = \left(2 + \frac{4}{\beta}\right)\sqrt{f_c'}\,b_o d \qquad (11\text{-}33)$$

where β is the ratio of long side to short side of the column, concentrated load or reaction area;

(b)
$$V_c = \left(\frac{\alpha_s d}{b_o} + 2\right)\sqrt{f_c'}\,b_o d \qquad (11\text{-}34)$$

where α_s is 40 for interior columns, 30 for edge columns, 20 for corner columns; and

COMMENTARY

R11.12.1.2 — The critical section for shear in slabs subjected to bending in two directions follows the perimeter at the edge of the loaded area.[11.3] The shear stress acting on this section at factored loads is a function of $\sqrt{f_c'}$ and the ratio of the side dimension of the column to the effective slab depth. A much simpler design equation results by assuming a pseudocritical section located at a distance $d/2$ from the periphery of the concentrated load. When this is done, the shear strength is almost independent of the ratio of column size to slab depth. For rectangular columns, this critical section was defined by straight lines drawn parallel to and at a distance $d/2$ from the edges of the loaded area. Section 11.12.1.3 allows the use of a rectangular critical section.

For slabs of uniform thickness, it is sufficient to check shear on one section. For slabs with changes in thickness, such as the edge of drop panels, it is necessary to check shear at several sections.

For edge columns at points where the slab cantilevers beyond the column, the critical perimeter will either be three-sided or four-sided.

R11.12.2.1 — For square columns, the shear stress due to ultimate loads in slabs subjected to bending in two directions is limited to $4\sqrt{f_c'}$. However, tests[11.58] have indicated that the value of $4\sqrt{f_c'}$ is unconservative when the ratio β of the lengths of the long and short sides of a rectangular column or loaded area is larger than 2.0. In such cases, the actual shear stress on the critical section at punching shear failure varies from a maximum of about $4\sqrt{f_c'}$ around the corners of the column or loaded area, down to $2\sqrt{f_c'}$ or less along the long sides between the two end sections. Other tests[11.59] indicate that v_c decreases as the ratio b_o/d increases. Eq. (11-33) and (11-34) were developed to account for these two effects. The words "interior," "edge," and "corner columns" in 11.12.2.1(b) refer to critical sections with 4, 3, and 2 sides, respectively.

CODE

(c) $$V_c = 4\sqrt{f_c'}\,b_o d \qquad (11\text{-}35)$$

11.12.2.2 — At columns of two-way prestressed slabs and footings that meet the requirements of 18.9.3

$$V_c = (\beta_p\sqrt{f_c'} + 0.3f_{pc})b_o d + V_p \qquad (11\text{-}36)$$

where β_p is the smaller of 3.5 and **($\alpha_s d/b_o$+ 1.5)**, α_s is 40 for interior columns, 30 for edge columns, and 20 for corner columns, b_o is perimeter of critical section defined in 11.12.1.2, f_{pc} is taken as the average value of f_{pc} for the two directions, and V_p is the vertical component of all effective prestress forces crossing the critical section. V_c shall be permitted to be computed by Eq. (11-36) if the following are satisfied; otherwise, 11.12.2.1 shall apply:

(a) No portion of the column cross section shall be closer to a discontinuous edge than 4 times the slab thickness;

(b) The value of $\sqrt{f_c'}$ used in Eq. (11-36) shall not be taken greater than 70 psi; and

(c) In each direction, f_{pc} shall not be less than 125 psi, nor be taken greater than 500 psi.

COMMENTARY

For shapes other than rectangular, β is taken to be the ratio of the longest overall dimension of the effective loaded area to the largest overall perpendicular dimension of the effective loaded area, as illustrated for an L-shaped reaction area in Fig. R11.12.2. The effective loaded area is that area totally enclosing the actual loaded area, for which the perimeter is a minimum.

R11.12.2.2 — For prestressed slabs and footings, a modified form of code Eq. (11-33) and (11-36) is specified for two-way action shear strength. Research[11.60,11.61] indicates that the shear strength of two-way prestressed slabs around interior columns is conservatively predicted by Eq. (11-36). V_c from Eq. (11-36) corresponds to a diagonal tension failure of the concrete initiating at the critical section defined in 11.12.1.2. The mode of failure differs from a punching shear failure of the concrete compression zone around the perimeter of the loaded area predicted by Eq. (11-33). Consequently, the term β does not enter into Eq. (11-36). Values for $\sqrt{f_c'}$ and f_{pc} are restricted in design due to limited test data available for higher values. When computing f_{pc}, loss of prestress due to restraint of the slab by shearwalls and other structural elements should be taken into account.

In a prestressed slab with distributed tendons, the V_p term in Eq. (11-36) contributes only a small amount to the shear strength; therefore, it may be conservatively taken as zero. If V_p is to be included, the tendon profile assumed in the calculations should be noted.

For an exterior column support where the distance from the outside of the column to the edge of the slab is less than four times the slab thickness, the prestress is not fully effective around b_o, the total perimeter of the critical section. Shear strength in this case is therefore conservatively taken the same as for a nonprestressed slab.

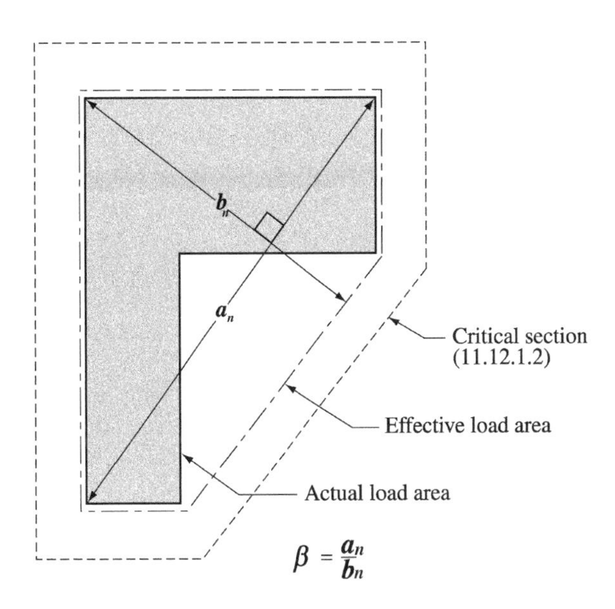

Fig. R11.12.2—Value of β for a nonrectangular loaded area

CODE

11.12.3 — Shear reinforcement consisting of bars or wires and single- or multiple-leg stirrups shall be permitted in slabs and footings with d greater than or equal to 6 in., but not less than 16 times the shear reinforcement bar diameter. Shear reinforcement shall be in accordance with 11.12.3.1 through 11.12.3.4.

11.12.3.1 — V_n shall be computed by Eq. (11-2), where V_c shall not be taken greater than $2\sqrt{f_c'}\,b_o d$, and V_s shall be calculated in accordance with 11.5. In Eq. (11-15), A_v shall be taken as the cross-sectional area of all legs of reinforcement on one peripheral line that is geometrically similar to the perimeter of the column section.

11.12.3.2 — V_n shall not be taken greater than $6\sqrt{f_c'}\,b_o d$.

11.12.3.3 — The distance between the column face and the first line of stirrup legs that surround the column shall not exceed $d/2$. The spacing between adjacent stirrup legs in the first line of shear reinforcement shall not exceed $2d$ measured in a direction parallel to the column face. The spacing between successive lines of shear reinforcement that surround the column shall not exceed $d/2$ measured in a direction perpendicular to the column face.

11.12.3.4 — Slab shear reinforcement shall satisfy the anchorage requirements of 12.13 and shall engage the longitudinal flexural reinforcement in the direction being considered.

11.12.4—Shear reinforcement consisting of structural steel I- or channel-shaped sections (shearheads) shall be permitted in slabs. The provisions of 11.12.4.1 through 11.12.4.9 shall apply where shear due to gravity load is transferred at interior column supports. Where moment is transferred to columns, 11.12.6.3 shall apply.

11.12.4.1 — Each shearhead shall consist of steel shapes fabricated by welding with a full penetration weld into identical arms at right angles. Shearhead arms shall not be interrupted within the column section.

11.12.4.2 — A shearhead shall not be deeper than 70 times the web thickness of the steel shape.

11.12.4.3 — The ends of each shearhead arm shall be permitted to be cut at angles not less than 30 deg with the horizontal, provided the plastic moment strength of the remaining tapered section is adequate to resist the shear force attributed to that arm of the shearhead.

COMMENTARY

R11.12.3 — Research[11.62-11.66] has shown that shear reinforcement consisting of properly anchored bars or wires and single- or multiple-leg stirrups, or closed stirrups, can increase the punching shear resistance of slabs. The spacing limits given in 11.12.3.3 correspond to slab shear reinforcement details that have been shown to be effective. Sections 12.13.2 and 12.13.3 give anchorage requirements for stirrup-type shear reinforcement that should also be applied for bars or wires used as slab shear reinforcement. It is essential that this shear reinforcement engage longitudinal reinforcement at both the top and bottom of the slab, as shown for typical details in Fig. R11.12.3(a) to (c). Anchorage of shear reinforcement according to the requirements of 12.13 is difficult in slabs thinner than 10 in. Shear reinforcement consisting of vertical bars mechanically anchored at each end by a plate or head capable of developing the yield strength of the bars has been used successfully.[11.66]

In a slab-column connection for which the moment transfer is negligible, the shear reinforcement should be symmetrical about the centroid of the critical section (Fig. R11.12.3(d)). Spacing limits defined in 11.12.3.3 are also shown in Fig. R11.12.3(d) and (e). At edge columns or for interior connections where moment transfer is significant, closed stirrups are recommended in a pattern as symmetrical as possible. Although the average shear stresses on faces *AD* and *BC* of the exterior column in Fig. R11.12.3(e) are lower than on face *AB*, the closed stirrups extending from faces *AD* and *BC* provide some torsional capacity along the edge of the slab.

R11.12.4—Based on reported test data,[11.67] design procedures are presented for shearhead reinforcement consisting of structural steel shapes. For a column connection transferring moment, the design of shearheads is given in 11.12.6.3.

Three basic criteria should be considered in the design of shearhead reinforcement for connections transferring shear due to gravity load. First, a minimum flexural strength should be provided to ensure that the required shear strength of the slab is reached before the flexural strength of the shearhead is exceeded. Second, the shear stress in the slab at the end of the shearhead reinforcement should be limited. Third, after these two requirements are satisfied, the designer can reduce the negative moment slab reinforcement in proportion to the moment contribution of the shearhead at the design section.

CODE

COMMENTARY

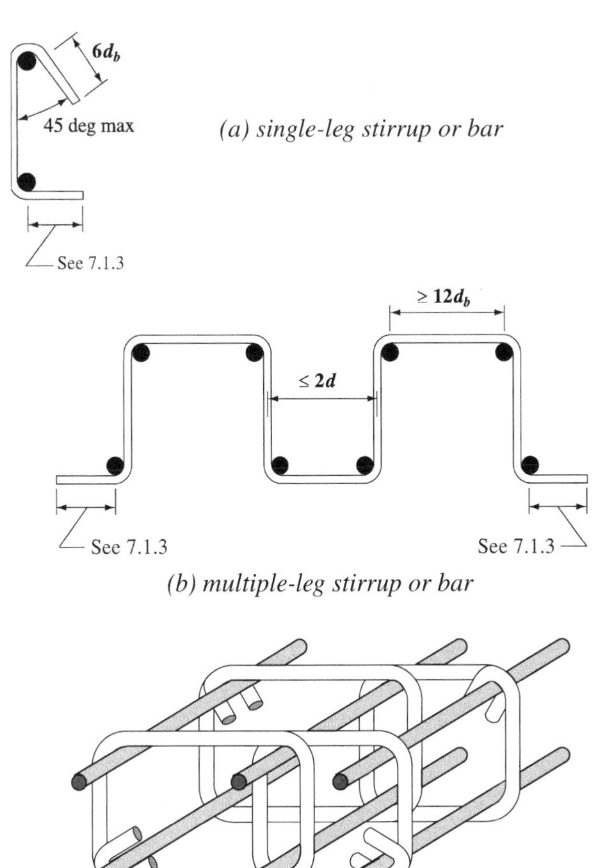

Fig. R11.12.3(a)-(c): Single- or multiple-leg stirrup-type slab shear reinforcement.

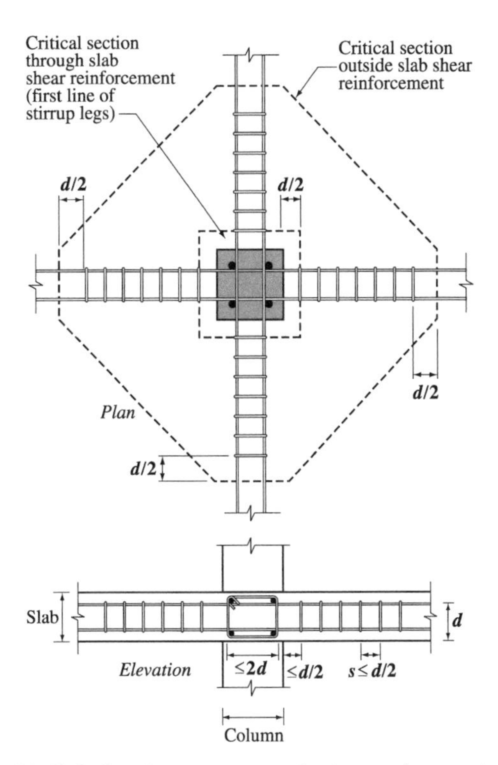

Fig. R11.12.3(d)—Arrangement of stirrup shear reinforcement, interior column.

CODE

COMMENTARY

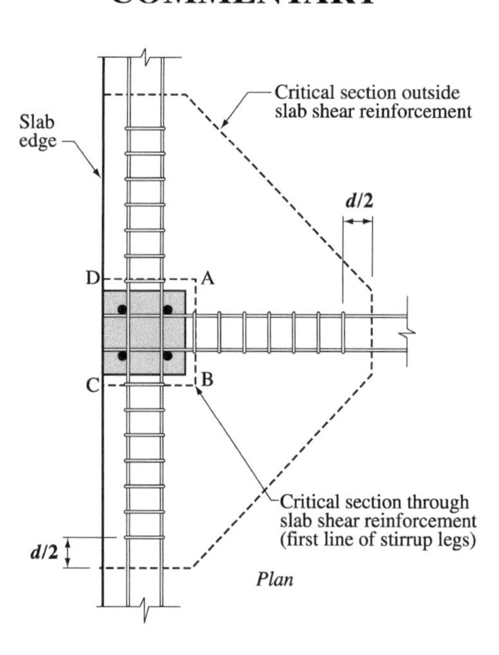

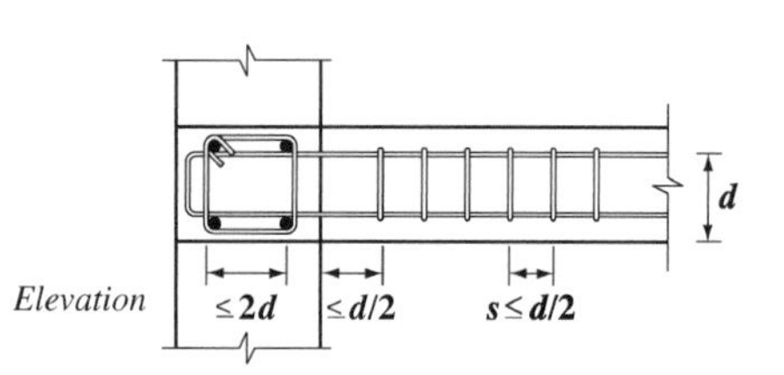

Fig. R11.12.3(e)—Arrangement of stirrup shear reinforcement, edge column

11.12.4.4 — All compression flanges of steel shapes shall be located within $0.3d$ of compression surface of slab.

11.12.4.5 — The ratio α_v between the flexural stiffness of each shearhead arm and that of the surrounding composite cracked slab section of width $(c_2 + d)$ shall not be less than 0.15.

11.12.4.6 — Plastic moment strength, M_p, required for each arm of the shearhead shall be computed by

$$M_p = \frac{V_u}{2\phi n}\left[h_v + \alpha_v\left(\ell_v - \frac{c_1}{2}\right)\right] \quad (11\text{-}37)$$

where ϕ is for tension-controlled members, n is number of shearhead arms, and ℓ_v is minimum length of each shearhead arm required to comply with requirements of 11.12.4.7 and 11.12.4.8.

R11.12.4.5 and R11.12.4.6 — The assumed idealized shear distribution along an arm of a shearhead at an interior column is shown in Fig. R11.12.4.5. The shear along each of the arms is taken as $\alpha_v V_c / n$. However, the peak shear at the face of the column is taken as the total shear considered per arm $V_u / \phi n$ minus the shear considered carried to the column by the concrete compression zone of the slab. The latter term is expressed as $(V_c / n)(1 - \alpha_v)$, so that it approaches zero for a heavy shearhead and approaches $V_u / \phi n$ when a light shearhead is used. Equation (11-37) then follows from the assumption that ϕV_c is about one-half the factored shear force V_u. In this equation, M_p is the required plastic moment strength of each shearhead arm necessary to ensure that V_u is attained as the moment strength of the shearhead is reached. The quantity ℓ_v is the length from the center of the column to the point at which the shearhead is no longer required, and the distance $c_1/2$ is one-half the dimension of the column in the direction considered.

CODE

COMMENTARY

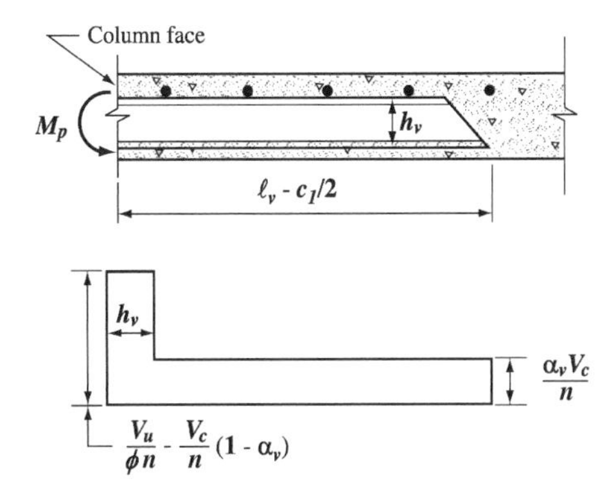

Fig. R11.12.4.5—Idealized shear acting on shearhead

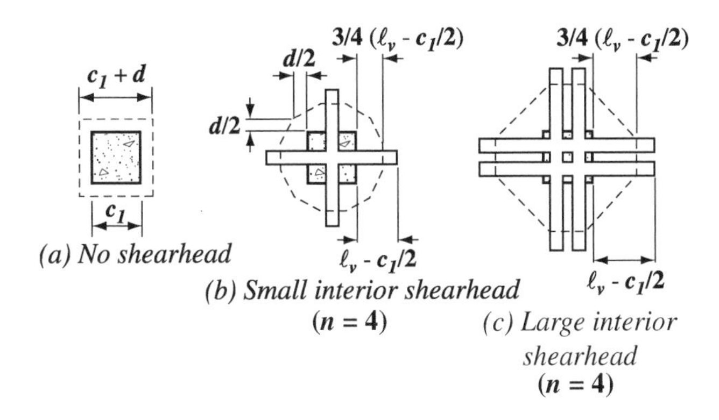

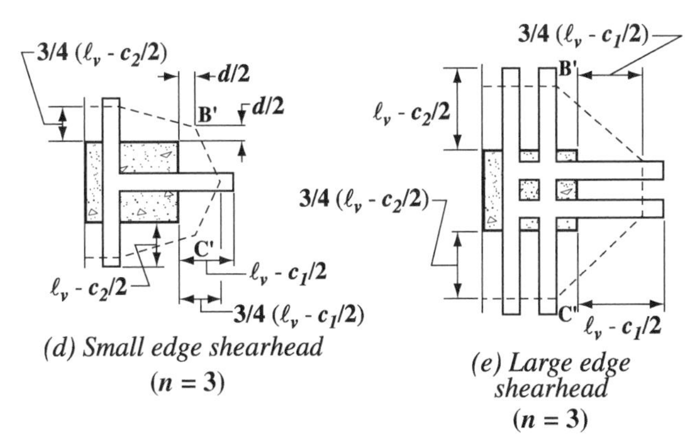

Fig. R11.12.4.7—Location of critical section defined in 11.12.4.7

11.12.4.7 — The critical slab section for shear shall be perpendicular to the plane of the slab and shall cross each shearhead arm at three-quarters the distance $[\ell_v - (c_1/2)]$ from the column face to the end of the shearhead arm. The critical section shall be located so that its perimeter b_o is a minimum, but need not be closer than the perimeter defined in 11.12.1.2(a).

R11.12.4.7 — The test results[11.67] indicated that slabs containing under-reinforcing shearheads failed at a shear stress on a critical section at the end of the shearhead reinforcement less than $4\sqrt{f_c'}$. Although the use of over-reinforcing shearheads brought the shear strength back to about the equivalent of $4\sqrt{f_c'}$, the limited test data suggest that a conservative design is desirable. Therefore, the shear strength is calculated as $4\sqrt{f_c'}$ on an assumed critical section located inside the end of the shearhead reinforcement.

The critical section is taken through the shearhead arms three-fourths of the distance $[\ell_v - (c_1/2)]$ from the face of the column to the end of the shearhead. However, this assumed critical section need not be taken closer than $d/2$ to the column. See Fig. R11.12.4.7.

CODE

11.12.4.8 — V_n shall not be taken greater than $4\sqrt{f_c'}\,b_o d$ on the critical section defined in 11.12.4.7. When shearhead reinforcement is provided, V_n shall not be taken greater than $7\sqrt{f_c'}\,b_o d$ on the critical section defined in 11.12.1.2(a).

11.12.4.9 — Moment resistance M_v contributed to each slab column strip by a shearhead shall not be taken greater than

$$M_v = \frac{\phi \alpha_v V_u}{2n}\left(\ell_v - \frac{c_1}{2}\right) \qquad (11\text{-}38)$$

where ϕ is for tension-controlled members, n is number of shearhead arms, and ℓ_v is length of each shearhead arm actually provided. However, M_v shall not be taken larger than the smallest of:

(a) 30 percent of the total factored moment required for each slab column strip;

(b) The change in column strip moment over the length ℓ_v;

(c) M_p computed by Eq. (11-37).

11.12.4.10 — When unbalanced moments are considered, the shearhead must have adequate anchorage to transmit M_p to the column.

11.12.5 — Openings in slabs

When openings in slabs are located at a distance less than 10 times the slab thickness from a concentrated load or reaction area, or when openings in flat slabs are located within column strips as defined in Chapter 13, the critical slab sections for shear defined in 11.12.1.2 and 11.12.4.7 shall be modified as follows:

11.12.5.1 — For slabs without shearheads, that part of the perimeter of the critical section that is enclosed by straight lines projecting from the centroid of the column, concentrated load, or reaction area and tangent to the boundaries of the openings shall be considered ineffective.

11.12.5.2 — For slabs with shearheads, the ineffective portion of the perimeter shall be one-half of that defined in 11.12.5.1.

COMMENTARY

R11.12.4.9 — If the peak shear at the face of the column is neglected, and ϕV_c is again assumed to be about one-half of V_u, the moment resistance contribution of the shearhead M_v can be conservatively computed from Eq. (11-38), in which ϕ is the factor for flexure.

R11.12.4.10 — See R11.12.6.3.

R11.12.5 — Openings in slabs

Provisions for design of openings in slabs (and footings) were developed in Reference 11.3. The locations of the effective portions of the critical section near typical openings and free edges are shown by the dashed lines in Fig. R11.12.5. Additional research[11.58] has confirmed that these provisions are conservative.

CODE

COMMENTARY

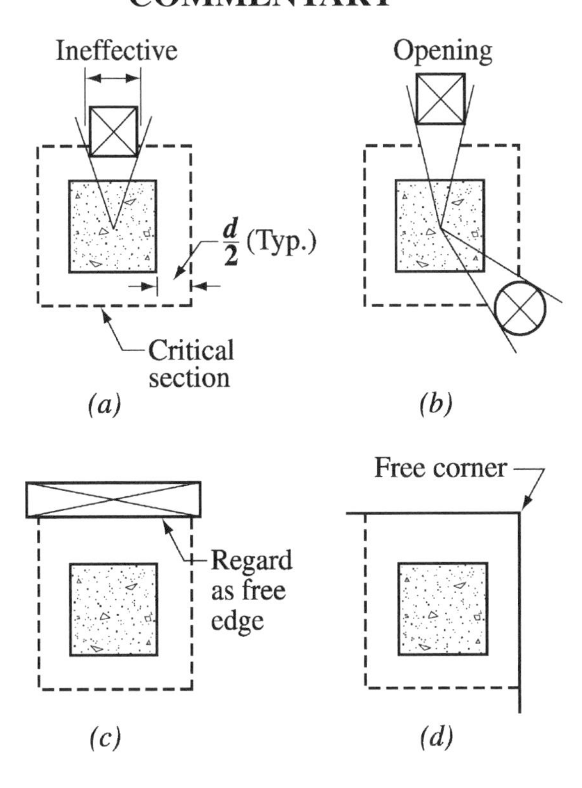

Fig. R11.12.5—Effect of openings and free edges (effective perimeter shown with dashed lines)

11.12.6 — Transfer of moment in slab-column connections

11.12.6.1 — Where gravity load, wind, earthquake, or other lateral forces cause transfer of unbalanced moment M_u between a slab and column, $\gamma_f M_u$ shall be transferred by flexure in accordance with 13.5.3. The remainder of the unbalanced moment, $\gamma_v M_u$, shall be considered to be transferred by eccentricity of shear about the centroid of the critical section defined in 11.12.1.2 where

$$\gamma_v = (1 - \gamma_f) \tag{11-39}$$

11.12.6.2 — The shear stress resulting from moment transfer by eccentricity of shear shall be assumed to vary linearly about the centroid of the critical sections defined in 11.12.1.2. The maximum shear stress due to V_u and M_u shall not exceed ϕv_n:

(a) For members without shear reinforcement:

$$\phi v_n = \phi V_c/(b_o d) \tag{11-40}$$

R11.12.6 — Transfer of moment in slab-column connections

R11.12.6.1 — In Reference 11.68 it was found that where moment is transferred between a column and a slab, 60 percent of the moment should be considered transferred by flexure across the perimeter of the critical section defined in 11.12.1.2, and 40 percent by eccentricity of the shear about the centroid of the critical section. For rectangular columns, the portion of the moment transferred by flexure increases as the width of the face of the critical section resisting the moment increases, as given by Eq. (13-1).

Most of the data in Reference 11.68 were obtained from tests of square columns, and little information is available for round columns. These can be approximated as square columns. Fig. R13.6.2.5 shows square supports having the same area as some nonrectangular members.

R11.12.6.2 — The stress distribution is assumed as illustrated in Fig. R11.12.6.2 for an interior or exterior column. The perimeter of the critical section, ***ABCD***, is determined in accordance with 11.12.1.2. The factored shear force V_u and unbalanced factored moment M_u are determined at the centroidal axis ***c-c*** of the critical section. The maximum factored shear stress may be calculated from:

$$v_{u(AB)} = \frac{V_u}{A_c} + \frac{\gamma_v M_u c_{AB}}{J_c}$$

CODE

where V_c is as defined in 11.12.2.1 or 11.12.2.2.

(b) For members with shear reinforcement other than shearheads:

$$\phi v_n = \phi(V_c + V_s)/(b_o d) \quad (11\text{-}41)$$

where V_c and V_s are defined in 11.12.3.1. The design shall take into account the variation of shear stress around the column. The shear stress due to factored shear force and moment shall not exceed $\phi(2\sqrt{f_c'})$ at the critical section located $d/2$ outside the outermost line of stirrup legs that surround the column.

11.12.6.3 — When shear reinforcement consisting of structural steel I- or channel-shaped sections (shearheads) is provided, the sum of the shear stresses due to vertical load acting on the critical section defined by 11.12.4.7 and the shear stresses resulting from moment transferred by eccentricity of shear about the centroid of the critical section defined in 11.12.1.2(a) and 11.12.1.3 shall not exceed $\phi 4\sqrt{f_c'}$.

COMMENTARY

or

$$v_{u(CD)} = \frac{V_u}{A_c} - \frac{\gamma_v M_u c_{CD}}{J_c}$$

where γ_v is given by Eq. (11-39). For an interior column, A_c and J_c may be calculated by

A_c = area of concrete of assumed critical section
$= 2d(c_1 + c_2 + 2d)$

J_c = property of assumed critical section analogous to polar moment of inertia

$$= \frac{d(c_1+d)^3}{6} + \frac{(c_1+d)d^3}{6} + \frac{d(c_2+d)(c_1+d)^2}{2}$$

Similar equations may be developed for A_c and J_c for columns located at the edge or corner of a slab.

The fraction of the unbalanced moment between slab and column not transferred by eccentricity of the shear should be transferred by flexure in accordance with 13.5.3. A conservative method assigns the fraction transferred by flexure over an effective slab width defined in 13.5.3.2. Often designers concentrate column strip reinforcement near the column to accommodate this unbalanced moment. Available test data[11.68] seem to indicate that this practice does not increase shear strength but may be desirable to increase the stiffness of the slab-column junction.

Test data[11.69] indicate that the moment transfer capacity of a prestressed slab to column connection can be calculated using the procedures of 11.12.6 and 13.5.3.

Where shear reinforcement has been used, the critical section beyond the shear reinforcement generally has a polygonal shape (Fig. R11.12.3(d) and (e)). Equations for calculating shear stresses on such sections are given in Reference 11.66.

R11.12.6.3 — Tests[11.70] indicate that the critical sections are defined in 11.12.1.2(a) and 11.12.1.3 and are appropriate for calculations of shear stresses caused by transfer of moments even when shearheads are used. Then, even though the critical sections for direct shear and shear due to moment transfer differ, they coincide or are in close proximity at the column corners where the failures initiate. Because a shearhead attracts most of the shear as it funnels toward the column, it is conservative to take the maximum shear stress as the sum of the two components.

Section 11.12.4.10 requires the moment M_p to be transferred to the column in shearhead connections transferring unbalanced moments. This may be done by bearing within the column or by mechanical anchorage.

COMMENTARY

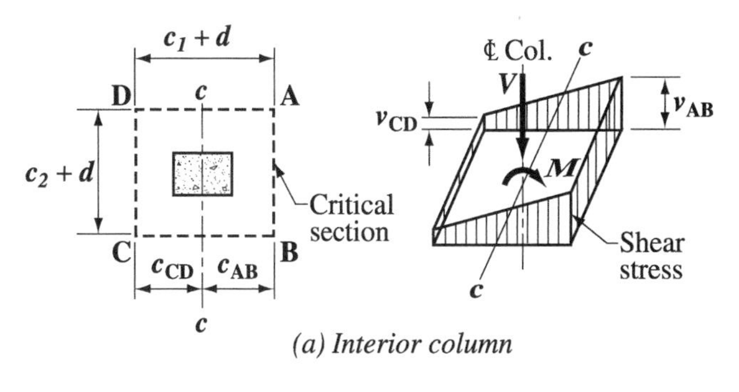

(a) Interior column

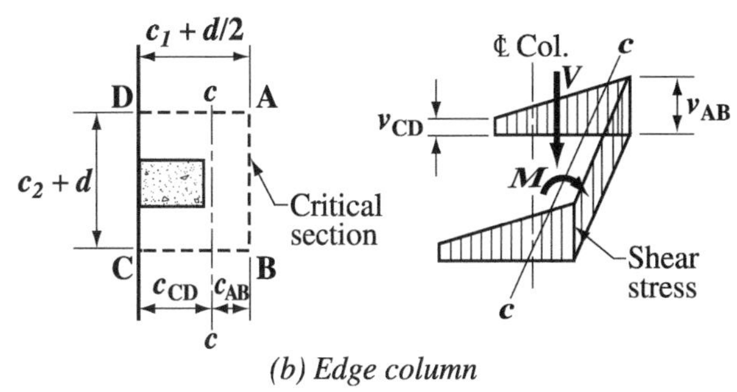

(b) Edge column

Fig. R11.12.6.2—Assumed distribution of shear stress

Notes

CHAPTER 12 — DEVELOPMENT AND SPLICES OF REINFORCEMENT

CODE

12.1 — Development of reinforcement — General

12.1.1 — Calculated tension or compression in reinforcement at each section of structural concrete members shall be developed on each side of that section by embedment length, hook or mechanical device, or a combination thereof. Hooks shall not be used to develop bars in compression.

12.1.2 — The values of $\sqrt{f_c'}$ used in this chapter shall not exceed 100 psi.

COMMENTARY

R12.1 — Development of reinforcement — General

The development length concept for anchorage of reinforcement was first introduced in the 1971 code, to replace the dual requirements for flexural bond and anchorage bond contained in earlier editions. It is no longer necessary to consider the flexural bond concept, which placed emphasis on the computation of nominal peak bond stresses. Consideration of an average bond resistance over a full development length of the reinforcement is more meaningful, partially because all bond tests consider an average bond resistance over a length of embedment of the reinforcement, and partially because uncalculated extreme variations in local bond stresses exist near flexural cracks.[12.1]

The development length concept is based on the attainable average bond stress over the length of embedment of the reinforcement. Development lengths are required because of the tendency of highly stressed bars to split relatively thin sections of restraining concrete. A single bar embedded in a mass of concrete should not require as great a development length; although a row of bars, even in mass concrete, can create a weakened plane with longitudinal splitting along the plane of the bars.

In application, the development length concept requires minimum lengths or extensions of reinforcement beyond all points of peak stress in the reinforcement. Such peak stresses generally occur at the points in 12.10.2.

The strength reduction factor ϕ is not used in the development length and lap splice equations. An allowance for strength reduction is already included in the expressions for determining development and splice lengths.

Units of measurement are given in the Notation to assist the user and are not intended to preclude the use of other correctly applied units for the same symbol, such as ft or kip.

From a point of peak stress in reinforcement, some length of reinforcement or anchorage is necessary to develop the stress. This development length or anchorage is necessary on both sides of such peak stress points. Often the reinforcement continues for a considerable distance on one side of a critical stress point so that calculations need involve only the other side, for example, the negative moment reinforcement continuing through a support to the middle of the next span.

CODE

12.2 — Development of deformed bars and deformed wire in tension

12.2.1 — Development length for deformed bars and deformed wire in tension, ℓ_d shall be determined from either 12.2.2 or 12.2.3, but shall not be less than 12 in.

12.2.2 — For deformed bars or deformed wire, ℓ_d shall be as follows:

	No. 6 and smaller bars and deformed wires	No. 7 and larger bars
Clear spacing of bars or wires being developed or spliced not less than d_b, clear cover not less than d_b, and stirrups or ties throughout ℓ_d not less than the code minimum or Clear spacing of bars or wires being developed or spliced not less than $2d_b$ and clear cover not less than d_b	$\left(\frac{f_y\psi_t\psi_e\lambda}{25\sqrt{f_c'}}\right)d_b$	$\left(\frac{f_y\psi_t\psi_e\lambda}{20\sqrt{f_c'}}\right)d_b$
Other cases	$\left(\frac{3f_y\psi_t\psi_e\lambda}{50\sqrt{f_c'}}\right)d_b$	$\left(\frac{3f_y\psi_t\psi_e\lambda}{40\sqrt{f_c'}}\right)d_b$

12.2.3 — For deformed bars or deformed wire, ℓ_d shall be:

$$\ell_d = \left(\frac{3}{40}\frac{f_y}{\sqrt{f_c'}}\frac{\psi_t\psi_e\psi_s\lambda}{\left(\frac{c_b + K_{tr}}{d_b}\right)}\right)d_b \quad (12\text{-}1)$$

in which the term $(c_b + K_{tr})/d_b$ shall not be taken greater than 2.5, and

$$K_{tr} = \frac{A_{tr}f_{yt}}{1500sn} \quad (12\text{-}2)$$

where n is the number of bars or wires being spliced or developed along the plane of splitting. It shall be permitted to use $K_{tr} = 0$ as a design simplification even if transverse reinforcement is present.

COMMENTARY

R12.2 — Development of deformed bars and deformed wire in tension

The general development length equation (Eq. (12-1)) is given in 12.2.3. The equation is based on the expression for development length previously endorsed by Committee 408.[12.2,12.3] In Eq. (12-1), c_b is a factor that represents the smallest of the side cover, the cover over the bar or wire (in both cases measured to the center of the bar or wire), or one-half the center-to-center spacing of the bars or wires. K_{tr} is a factor that represents the contribution of confining reinforcement across potential splitting planes. ψ_t is the traditional reinforcement location factor to reflect the adverse effects of the top reinforcement casting position. ψ_e is a coating factor reflecting the effects of epoxy coating. There is a limit on the product $\psi_t\psi_e$. The reinforcement size factor ψ_s reflects the more favorable performance of smaller diameter reinforcement. λ is a factor reflecting the lower tensile strength of lightweight concrete and the resulting reduction of the splitting resistance, which increases the development length in lightweight concrete. A limit of 2.5 is placed on the term $(c_b + K_{tr})/d_b$. When $(c_b + K_{tr})/d_b$ is less than 2.5, splitting failures are likely to occur. For values above 2.5, a pullout failure is expected and an increase in cover or transverse reinforcement is unlikely to increase the anchorage capacity.

Equation (12-1) allows the designer to see the effect of all variables controlling the development length. The designer is permitted to disregard terms when such omission results in longer and hence, more conservative, development lengths.

The provisions of 12.2.2 and 12.2.3 give a two-tier approach. The user can either calculate ℓ_d based on the actual $(c_b + K_{tr})/d_b$ (12.2.3) or calculate ℓ_d using 12.2.2, which is based on two preselected values of $(c_b + K_{tr})/d_b$.

Section 12.2.2 recognizes that many current practical construction cases utilize spacing and cover values along with confining reinforcement, such as stirrups or ties, that result in a value of $(c_b + K_{tr})/d_b$ of at least 1.5. Examples include a minimum clear cover of d_b along with either minimum clear spacing of $2d_b$, or a combination of minimum clear spacing of d_b and minimum ties or stirrups. For these frequently occurring cases, the development length for larger bars can be taken as $\ell_d = [f_y\psi_t\psi_e\lambda/(20\sqrt{f_c'})]d_b$. In the development of ACI 318-95, a comparison with past provisions and a check of a database of experimental results maintained by ACI Committee 408[12.2] indicated that for No. 6 deformed bars and smaller, as well as for deformed wire, the development lengths could be reduced 20 percent using ψ_s = **0.8**. This is the basis for the middle column of the table in 12.2.2. With less cover and in the absence of minimum ties or stirrups, the minimum clear spacing limits of 7.6.1 and the minimum concrete cover requirements of 7.7 result in minimum

CODE

COMMENTARY

values of c_b equal to d_b. Thus, for "other cases," the values are based on using $(c_b + K_{tr})/d_b = 1.0$ in Eq. (12-1).

The user may easily construct simple, useful expressions. For example, in all structures with normalweight concrete ($\lambda = 1.0$), uncoated reinforcement ($\psi_e = 1.0$), No. 7 or larger bottom bars ($\psi_t = 1.0$) with f_c' = 4000 psi and Grade 60 reinforcement, the equations reduce to

$$\ell_d = \frac{(60{,}000)(1.0)(1.0)(1.0)}{20\sqrt{4000}} d_b = 47 d_b$$

or

$$\ell_d = \frac{3(60{,}000)(1.0)(1.0)(1.0)}{40\sqrt{4000}} d_b = 71 d_b$$

Thus, as long as minimum cover of d_b is provided along with a minimum clear spacing of $2d_b$, or a minimum clear cover of d_b and a minimum clear spacing of d_b are provided along with minimum ties or stirrups, a designer knows that $\ell_d = 47d_b$. The penalty for spacing bars closer or providing less cover is the requirement that $\ell_d = 71d_b$.

Many practical combinations of side cover, clear cover, and confining reinforcement can be used with 12.2.3 to produce significantly shorter development lengths than allowed by 12.2.2. For example, bars or wires with minimum clear cover not less than $2d_b$ and minimum clear spacing not less than $4d_b$ and without any confining reinforcement would have a $(c_b + K_{tr})/d_b$ value of 2.5 and would require a development length of only $28d_b$ for the example above.

12.2.4 — The factors used in the expressions for development of deformed bars and deformed wires in tension in 12.2 are as follows:

(a) Where horizontal reinforcement is placed such that more than 12 in. of fresh concrete is cast below the development length or splice, ψ_t = **1.3**. For other situations, ψ_t = **1.0**.

(b) For epoxy-coated bars or wires with cover less than $3d_b$, or clear spacing less than $6d_b$, ψ_e = **1.5**. For all other epoxy-coated bars or wires, ψ_e = **1.2**. For uncoated reinforcement, ψ_e = **1.0**.

However, the product $\psi_t\psi_e$ need not be greater than 1.7.

(c) For No. 6 and smaller bars and deformed wires, ψ_s = **0.8**. For No. 7 and larger bars, ψ_s = **1.0**.

R12.2.4 — The reinforcement location factor ψ_t accounts for position of the reinforcement in freshly placed concrete. The factor was reduced to 1.3 in the 1989 code to reflect research.[12.4,12.5]

The factor λ for lightweight aggregate concrete was made the same for all types of lightweight aggregates in the 1989 code. Research on hooked bar anchorages did not support the variations in previous codes for all-lightweight and sand-lightweight concrete and a single value, 1.3, was selected. Section 12.2.4 allows a lower factor to be used when the splitting tensile strength of the lightweight concrete is specified. See 5.1.4.

Studies[12.6-12.8] of the anchorage of epoxy-coated bars show that bond strength is reduced because the coating prevents adhesion and friction between the bar and the concrete. The factors reflect the type of anchorage failure likely to occur. When the cover or spacing is small, a splitting failure can occur and the anchorage or bond strength is substantially reduced. If the cover and spacing between bars is large, a splitting failure is

CODE

(d) Where lightweight concrete is used, λ = **1.3**. However, when f_{ct} is specified, λ shall be permitted to be taken as $6.7\sqrt{f_c'}/f_{ct}$ but not less than 1.0. Where normalweight concrete is used, λ = **1.0**.

COMMENTARY

precluded and the effect of the epoxy coating on anchorage strength is not as large. Studies[12.9] have shown that although the cover or spacing may be small, the anchorage strength may be increased by adding transverse steel crossing the plane of splitting, and restraining the splitting crack.

Because the bond of epoxy-coated bars is already reduced due to the loss of adhesion between the bar and the concrete, an upper limit of 1.7 is established for the product of the top reinforcement and epoxy-coated reinforcement factors.

Although there is no requirement for transverse reinforcement along the tension development or splice length, recent research[12.10,12.11] indicates that in concrete with very high compressive strength, brittle anchorage failure occurred in bars with inadequate transverse reinforcement. In splice tests of No. 8 and No. 11 bars in concrete with an f_c' of approximately 15,000 psi, transverse reinforcement improved ductile anchorage behavior.

12.2.5 — Excess reinforcement

Reduction in ℓ_d shall be permitted where reinforcement in a flexural member is in excess of that required by analysis except where anchorage or development for f_y is specifically required or the reinforcement is designed under provisions of 21.2.1.4.......................... (A_s required)/(A_s provided).

R12.2.5 — Excess reinforcement

The reduction factor based on area is not to be used in those cases where anchorage development for full f_y is required. For example, the excess reinforcement factor does not apply for development of positive moment reinforcement at supports according to 12.11.2, for development of shrinkage and temperature reinforcement according to 7.12.2.3, or for development of reinforcement provided according to 7.13 and 13.3.8.5.

12.3 — Development of deformed bars and deformed wire in compression

12.3.1 — Development length for deformed bars and deformed wire in compression, ℓ_{dc}, shall be determined from 12.3.2 and applicable modification factors of 12.3.3, but ℓ_{dc} shall not be less than 8 in.

12.3.2 — For deformed bars and deformed wire, ℓ_{dc} shall be taken as the larger of $(0.02f_y/\sqrt{f_c'})d_b$ and $(0.0003f_y)d_b$, where the constant 0.0003 carries the unit of in.^{2}/lb.

12.3.3 — Length ℓ_{dc} in 12.3.2 shall be permitted to be multiplied by the applicable factors for:

(a) Reinforcement in excess of that required by analysis .... (A_s required)/(A_s provided)

(b) Reinforcement enclosed within spiral reinforcement not less than 1/4 in. diameter and not more than 4 in. pitch or within No. 4 ties in conformance with 7.10.5 and spaced at not more than 4 in. on center 0.75

R12.3 — Development of deformed bars and deformed wire in compression

The weakening effect of flexural tension cracks is not present for bars and wire in compression, and usually end bearing of the bars on the concrete is beneficial. Therefore, shorter development lengths are specified for compression than for tension. The development length may be reduced 25 percent when the reinforcement is enclosed within spirals or ties. A reduction in development length is also permitted if excess reinforcement is provided.

CODE

12.4 — Development of bundled bars

12.4.1 — Development length of individual bars within a bundle, in tension or compression, shall be that for the individual bar, increased 20 percent for three-bar bundle, and 33 percent for four-bar bundle.

12.4.2 — For determining the appropriate factors in 12.2, a unit of bundled bars shall be treated as a single bar of a diameter derived from the equivalent total area.

12.5 — Development of standard hooks in tension

12.5.1 — Development length for deformed bars in tension terminating in a standard hook (see 7.1), ℓ_{dh}, shall be determined from 12.5.2 and the applicable modification factors of 12.5.3, but ℓ_{dh} shall not be less than the larger of $8d_b$ and 6 in.

12.5.2 — For deformed bars, ℓ_{dh} shall be $(0.02\psi_e\lambda f_y/\sqrt{f_c'})d_b$ with ψ_e taken as 1.2 for epoxy-coated reinforcement, and λ taken as 1.3 for lightweight concrete. For other cases, ψ_e and λ shall be taken as 1.0.

12.5.3 — Length ℓ_{dh} in 12.5.2 shall be permitted to be multiplied by the following applicable factors:

(a) For No. 11 bar and smaller hooks with side cover (normal to plane of hook) not less than 2-1/2 in., and for 90 degree hook with cover on bar extension beyond hook not less than 2 in. 0.7

(b) For 90 degree hooks of No. 11 and smaller bars that are either enclosed within ties or stirrups perpendicular to the bar being developed, spaced not greater than $3d_b$ along ℓ_{dh}; or enclosed within ties or stirrups parallel to the bar being developed, spaced not greater than $3d_b$ along the length of the tail extension of the hook plus bend 0.8

COMMENTARY

R12.4 — Development of bundled bars

R12.4.1 — An increased development length for individual bars is required when three or four bars are bundled together. The extra extension is needed because the grouping makes it more difficult to mobilize bond resistance from the core between the bars.

The designer should also note 7.6.6.4 relating to the cutoff points of individual bars within a bundle and 12.14.2.2 relating to splices of bundled bars. The increases in development length of 12.4 do apply when computing splice lengths of bundled bars in accordance with 12.14.2.2. The development of bundled bars by a standard hook of the bundle is not covered by the provisions of 12.5.

R12.4.2 — Although splice and development lengths of bundled bars are based on the diameter of individual bars increased by 20 or 33 percent, as appropriate, it is necessary to use an equivalent diameter of the entire bundle derived from the equivalent total area of bars when determining factors in 12.2, which considers cover and clear spacing and represents the tendency of concrete to split.

R12.5 — Development of standard hooks in tension

The provisions for hooked bar anchorage were extensively revised in the 1983 code. Study of failures of hooked bars indicate that splitting of the concrete cover in the plane of the hook is the primary cause of failure and that splitting originates at the inside of the hook where the local stress concentrations are very high. Thus, hook development is a direct function of bar diameter d_b,which governs the magnitude of compressive stresses on the inside of the hook. Only standard hooks (see 7.1) are considered and the influence of larger radius of bend cannot be evaluated by 12.5.

The hooked bar anchorage provisions give the total hooked bar embedment length as shown in Fig. R12.5. The development length ℓ_{dh} is measured from the critical section to the outside end (or edge) of the hook.

The development length for standard hooks ℓ_{dh} of 12.5.2 can be reduced by all applicable modification factors of 12.5.3. As an example, if the conditions of both 12.5.3(a) and (c) are met, both factors may be applied.

The effects of bar yield strength, excess reinforcement, lightweight concrete, and factors to reflect the resistance to splitting provided from confinement by concrete and transverse ties or stirrups are based on recommendations from References 12.2 and 12.3.

CODE

(c) For 180 degree hooks of No. 11 and smaller bars that are enclosed within ties or stirrups perpendicular to the bar being developed, spaced not greater than $3d_b$ along ℓ_{dh} 0.8

(d) Where anchorage or development for f_y is not specifically required, reinforcement in excess of that required by analysis......................... (A_s required)/(A_s provided)

In 12.5.3(b) and 12.5.3(c), d_b is the diameter of the hooked bar, and the first tie or stirrup shall enclose the bent portion of the hook, within $2d_b$ of the outside of the bend.

COMMENTARY

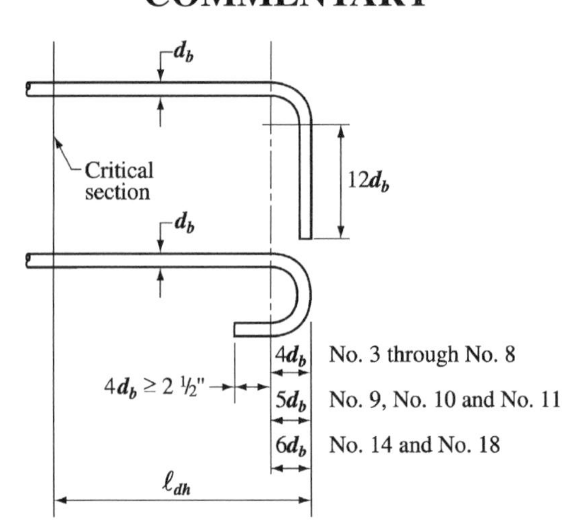

Fig. R12.5—Hooked bar details for development of standard hooks

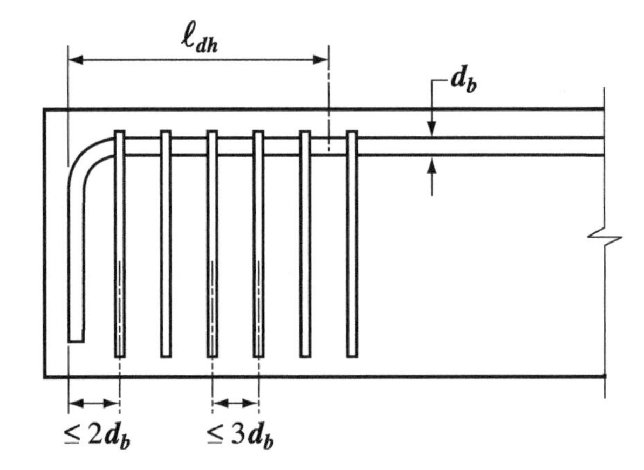

Fig. R12.5.3(a)—Ties or stirrups placed perpendicular to the bar being developed, spaced along the development length ℓ_{dh}.

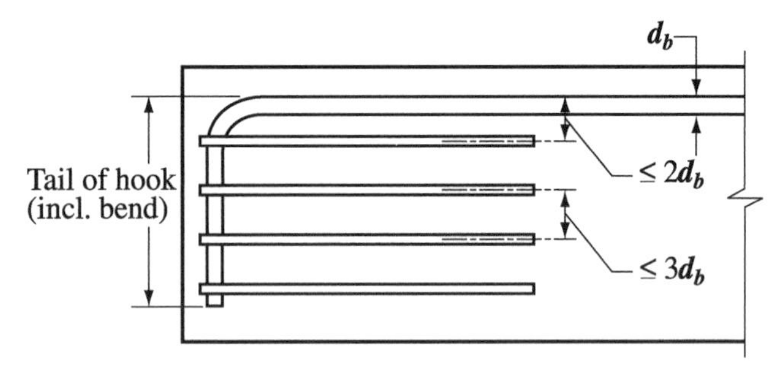

Fig. R12.5.3(b)—Ties or stirrups placed parallel to the bar being developed, spaced along the length of the tail extension of the hook plus bend.

Tests[12.12] indicate that closely spaced ties at or near the bend portion of a hooked bar are most effective in confining the hooked bar. For construction purposes, this is not always practicable. The cases where the modification factor of 12.5.3(b) may be used are illustrated in Fig. R12.5.3(a) and (b). Figure R12.5.3(a) shows placement of ties or stirrups perpendicular to the bar being developed, spaced along the development length, ℓ_{dh}, of the hook. Figure R12.5.3(b)

CODE

COMMENTARY

shows placement of ties or stirrups parallel to the bar being developed along the length of the tail extension of the hook plus bend. The latter configuration would be typical in a beam column joint.

The factor for excess reinforcement in 12.5.3(d) applies only where anchorage or development for full f_y is not specifically required. The λ factor for lightweight concrete is a simplification over the procedure in 12.2.3.3 of ACI 318-83 in which the increase varies from 18 to 33 percent, depending on the amount of lightweight aggregate used. Unlike straight bar development, no distinction is made between top bars and other bars; such a distinction is difficult for hooked bars in any case. A minimum value of ℓ_{dh} is specified to prevent failure by direct pullout in cases where a hook may be located very near the critical section. Hooks cannot be considered effective in compression.

Tests[12.13] indicate that the development length for hooked bars should be increased by 20 percent to account for reduced bond when reinforcement is epoxy coated.

12.5.4 — For bars being developed by a standard hook at discontinuous ends of members with both side cover and top (or bottom) cover over hook less than 2-1/2 in., the hooked bar shall be enclosed within ties or stirrups perpendicular to the bar being developed, spaced not greater than $3d_b$ along ℓ_{dh}. The first tie or stirrup shall enclose the bent portion of the hook, within $2d_b$ of the outside of the bend, where d_b is the diameter of the hooked bar. For this case, the factors of 12.5.3(b) and (c) shall not apply.

R12.5.4 — Bar hooks are especially susceptible to a concrete splitting failure if both side cover (normal to plane of hook) and top or bottom cover (in plane of hook) are small. See Fig. R12.5.4. With minimum confinement provided by concrete, additional confinement provided by ties or stirrups is essential, especially if full bar strength should be developed by a hooked bar with such small cover. Cases where hooks may require ties or stirrups for confinement are at ends of simply supported beams, at free end of cantilevers, and at ends of members framing into a joint where members do not extend beyond the joint. In contrast, if calculated bar stress is so low that the hook is not needed for bar anchorage, the ties or stirrups are not necessary. Also, provisions of 12.5.4 do not apply for hooked bars at discontinuous ends of slabs with confinement provided by the slab continuous on both sides normal to the plane of the hook.

12.5.5 — Hooks shall not be considered effective in developing bars in compression.

R12.5.5 — In compression, hooks are ineffective and may not be used as anchorage.

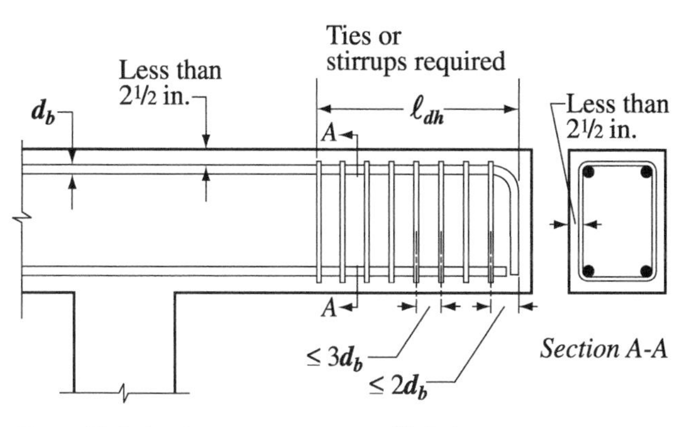

Fig. R12.5.4—Concrete cover per 12.5.4

CODE

12.6 — Mechanical anchorage

12.6.1 — Any mechanical device capable of developing the strength of reinforcement without damage to concrete is allowed as anchorage.

12.6.2 — Test results showing adequacy of such mechanical devices shall be presented to the building official.

12.6.3 — Development of reinforcement shall be permitted to consist of a combination of mechanical anchorage plus additional embedment length of reinforcement between the point of maximum bar stress and the mechanical anchorage.

12.7 — Development of welded deformed wire reinforcement in tension

12.7.1 — Development length in tension for welded deformed wire reinforcement, ℓ_d, measured from the point of critical section to the end of wire shall be computed as the product of ℓ_d, from 12.2.2 or 12.2.3, times a welded wire reinforcement factor from 12.7.2 or 12.7.3. It shall be permitted to reduce ℓ_d in accordance with 12.2.5 when applicable, but ℓ_d shall not be less than 8 in. except in computation of lap splices by 12.18. When using the welded wire reinforcement factor from 12.7.2, it shall be permitted to use an epoxy-coating factor ψ_e of 1.0 for epoxy-coated welded wire reinforcement in 12.2.2 and 12.2.3.

12.7.2 — For welded deformed wire reinforcement with at least one cross wire within ℓ_d and not less than 2 in. from the point of the critical section, the welded wire reinforcement factor shall be the greater of:

$$\left(\frac{f_y - 35{,}000}{f_y}\right)$$

and

$$\left(\frac{5d_b}{s}\right)$$

but not greater than 1.0, where s is the spacing between the wires to be developed.

12.7.3 — For welded deformed wire reinforcement with no cross wires within ℓ_d or with a single cross wire less than 2 in. from the point of the critical section, the welded wire reinforcement factor shall be taken as 1.0, and ℓ_d shall be determined as for deformed wire.

12.7.4—When any plain wires are present in the welded deformed wire reinforcement in the direction of the development length, the reinforcement shall be developed in accordance with 12.8.

COMMENTARY

R12.6 — Mechanical anchorage

R12.6.1 — Mechanical anchorage can be made adequate for strength both for tendons and for bar reinforcement.

R12.6.3 — Total development of a bar consists of the sum of all the parts that contribute to anchorage. When a mechanical anchorage is not capable of developing the required design strength of the reinforcement, additional embedment length of reinforcement should be provided between the mechanical anchorage and the critical section.

R12.7 — Development of welded deformed wire reinforcement in tension

Figure R12.7 shows the development requirements for welded deformed wire reinforcement with one cross wire within the development length. ASTM A 497 for welded deformed wire reinforcement requires the same strength of the weld as required for welded plain wire reinforcement (ASTM A 185). Some of the development is assigned to welds and some assigned to the length of deformed wire. The development computations are simplified from earlier code provisions for wire development by assuming that only one cross wire is contained in the development length. The factors in 12.7.2 are applied to the deformed wire development length computed from 12.2, but with an absolute minimum of 8 in. The explicit statement that the welded plain wire reinforcement multiplier not be taken greater than 1 corrects an oversight in earlier codes. The multipliers were derived using the general relationships between deformed welded wire reinforcement and deformed wires in the ℓ_{db} values of the 1983 code.

Tests[12.14] have indicated that epoxy-coated welded wire reinforcement has essentially the same development and splice strengths as uncoated welded wire reinforcement because the cross wires provide the primary anchorage for the wire. Therefore, an epoxy-coating factor of 1.0 is used for development and splice lengths of epoxy-coated welded wire reinforcement with cross wires within the splice or development length.

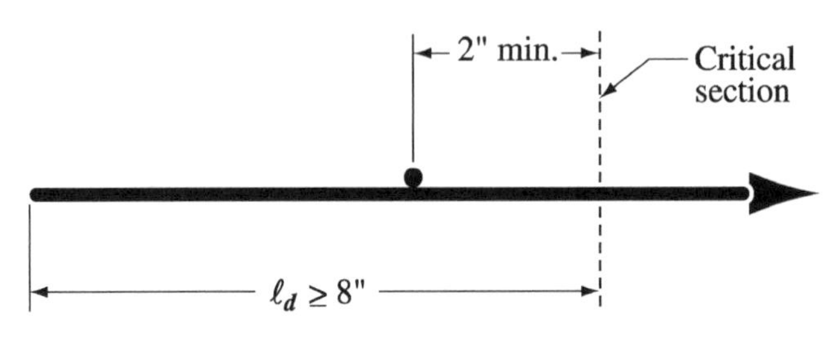

Fig. R12.7—Development of welded deformed wire reinforcement.

CODE

12.8—Development of welded plain wire reinforcement in tension

Yield strength of welded plain wire reinforcement shall be considered developed by embedment of two cross wires with the closer cross wire not less than 2 in. from the point of the critical section. However, ℓ_d shall not be less than

$$\ell_d = 0.27\frac{A_b}{s}\left(\frac{f_y}{\sqrt{f_c'}}\right)\lambda \qquad (12\text{-}3)$$

where ℓ_d is measured from the point of the critical section to the outermost crosswire, and s is the spacing between the wires to be developed. Where reinforcement provided is in excess of that required, ℓ_d may be reduced in accordance with 12.2.5. Length, ℓ_d, shall not be less than 6 in. except in computation of lap splices by 12.19.

12.9 — Development of prestressing strand

12.9.1 — Except as provided in 12.9.1.1, seven-wire strand shall be bonded beyond the critical section, a distance not less than

$$\ell_d = \left(\frac{f_{se}}{3000}\right)d_b + \left(\frac{f_{ps}-f_{se}}{1000}\right)d_b \qquad (12\text{-}4)$$

The expressions in parentheses are used as constants without units.

COMMENTARY

R12.8 — Development of welded plain wire reinforcement in tension

Fig. R12.8 shows the development requirements for welded plain wire reinforcement with development primarily dependent on the location of cross wires. For welded plain wire reinforcement made with the smaller wires, an embedment of at least two cross wires 2 in. or more beyond the point of critical section is adequate to develop the full yield strength of the anchored wires. However, for welded plain wire reinforcement made with larger closely spaced wires, a longer embedment is required and a minimum development length is provided for this reinforcement.

R12.9 — Development of prestressing strand

The development requirements for prestressing strand are intended to provide bond integrity for the strength of the member. The provisions are based on tests performed on normalweight concrete members with a minimum cover of 2 in. These tests may not represent the behavior of strand in low water-cementitious materials ratio, no-slump, concrete. Fabrication methods should ensure consolidation of concrete around the strand with complete contact between the steel and concrete. Extra precautions should be exercised when low water-cementitious material ratio, no-slump concrete is used.

The first term in Eq. (12-4) represents the transfer length of the strand, that is, the distance over which the strand should be bonded to the concrete to develop the effective prestress in the prestressing steel f_{se}. The second term represents the additional length over which the strand should be bonded so that a stress in the prestressing steel at nominal strength of the member, f_{ps}, may develop.

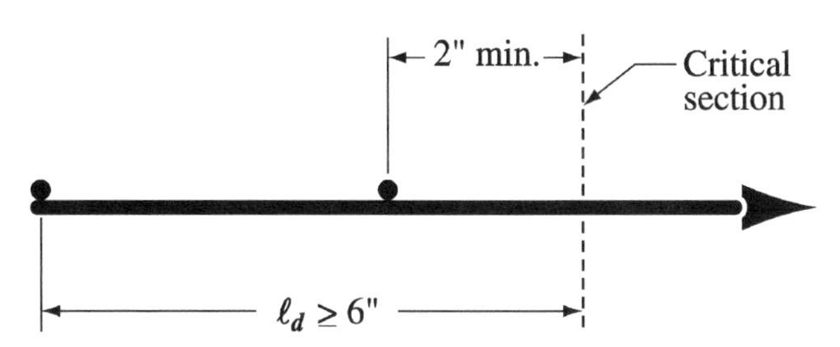

Fig. R12.8—Development of welded plain wire reinforcement

CODE

COMMENTARY

The bond of strand is a function of a number of factors, including the configuration and surface condition of the steel, the stress in the steel, the depth of concrete beneath the strand, and the method used to transfer the force in the strand to the concrete. For bonded applications, quality assurance procedures should be used to confirm that the strand is capable of adequate bond.[12.15,12.16] The precast concrete manufacturer may rely on certification from the strand manufacturer that the strand has bond characteristics that comply with this section. Strand with a slightly rusted surface can have an appreciably shorter transfer length than clean strand. Gentle release of the strand will permit a shorter transfer length than abruptly cutting the strands.

The provisions of 12.9 do not apply to plain wires or to end-anchored tendons. The length for smooth wire could be expected to be considerably greater due to the absence of mechanical interlock. Flexural bond failure would occur with plain wire when first slip occurred.

12.9.1.1 — Embedment less than ℓ_d shall be permitted at a section of a member provided the design strand stress at that section does not exceed values obtained from the bilinear relationship defined by Eq. (12-4).

R12.9.1.1 — Figure R12.9 shows the relationship between steel stress and the distance over which the strand is bonded to the concrete represented by Eq. (12-4). This idealized variation of strand stress may be used for analyzing sections within the development region.[12.17,12.18] The expressions for transfer length, and for the additional bonded length necessary to develop an increase in stress of $(f_{ps} - f_{se})$, are based on tests of members prestressed with clean, 1/4, 3/8, and 1/2 in. diameter strands for which the maximum value of f_{ps} was 275 kips/in.^{2} See References 12.19, 12.20, and 12.21.

12.9.2 — Limiting the investigation to cross sections nearest each end of the member that are required to develop full design strength under specified factored loads shall be permitted except where bonding of one or more strands does not extend to the end of the member, or where concentrated loads are applied within the strand development length.

R12.9.2 — Where bonding of one or more strands does not extend to the end of the member, critical sections may be at locations other than where full design strength is required to be developed, and detailed analysis may be required. References 12.17 and 12.18 show a method that may be used in

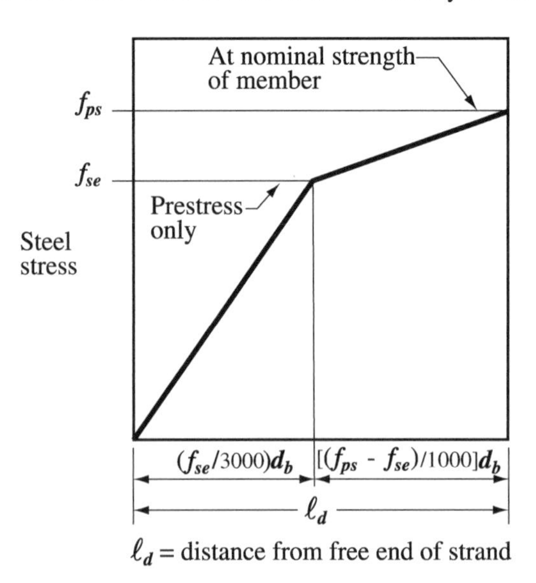

Fig. R12.9—Idealized bilinear relationship between steel stress and distance from the free end of strand.

CODE | COMMENTARY

the case of strands with different points of full development. Conservatively, only the strands that are fully developed at a section may be considered effective at that section. If critical sections occur in the transfer region, special considerations may be necessary. Some loading conditions, such as where heavy concentrated loads occur within the strand development length, may cause critical sections to occur away from the section that is required to develop full design strength.

12.9.3 — Where bonding of a strand does not extend to end of member, and design includes tension at service load in precompressed tensile zone as permitted by 18.4.2, ℓ_d specified in 12.9.1 shall be doubled.

R12.9.3 — Exploratory tests conducted in 1965[12.19] that study the effect of debonded strand (bond not permitted to extend to the ends of members) on performance of pretensioned girders indicated that the performance of these girders with embedment lengths twice those required by 12.9.1 closely matched the flexural performance of similar pretensioned girders with strand fully bonded to ends of girders. Accordingly, doubled development length is required for strand not bonded through to the end of a member. Subsequent tests[12.22] indicated that in pretensioned members designed for zero tension in the concrete under service load conditions (see 18.4.2), the development length for debonded strands need not be doubled. For analysis of sections with debonded strands at locations where strand is not fully developed, it is usually assumed that both the transfer length and development length are doubled.

12.10 — Development of flexural reinforcement — General

R12.10 — Development of flexural reinforcement — General

12.10.1 — Development of tension reinforcement by bending across the web to be anchored or made continuous with reinforcement on the opposite face of member shall be permitted.

12.10.2 — Critical sections for development of reinforcement in flexural members are at points of maximum stress and at points within the span where adjacent reinforcement terminates, or is bent. Provisions of 12.11.3 must be satisfied.

R12.10.2 — Critical sections for a typical continuous beam are indicated with a "c" or an "x" in Fig. R12.10.2. For uniform loading, the positive reinforcement extending into the support is more apt to be governed by the requirements of 12.11.3 rather than by development length measured from a point of maximum moment or bar cutoff.

12.10.3 — Reinforcement shall extend beyond the point at which it is no longer required to resist flexure for a distance equal to $\boldsymbol{d}$ or $\mathbf{12}\boldsymbol{d_b}$, whichever is greater, except at supports of simple spans and at free end of cantilevers.

R12.10.3 — The moment diagrams customarily used in design are approximate; some shifting of the location of maximum moments may occur due to changes in loading, settlement of supports, lateral loads, or other causes. A diagonal tension crack in a flexural member without stirrups may shift the location of the calculated tensile stress approximately a distance $\boldsymbol{d}$ towards a point of zero moment. When stirrups are provided, this effect is less severe, although still present to some extent.

To provide for shifts in the location of maximum moments, the code requires the extension of reinforcement a distance $\boldsymbol{d}$ or $\mathbf{12}\boldsymbol{d_b}$ beyond the point at which it is theoretically no longer required to resist flexure, except as noted.

CODE

12.10.4 — Continuing reinforcement shall have an embedment length not less than ℓ_d beyond the point where bent or terminated tension reinforcement is no longer required to resist flexure.

12.10.5 — Flexural reinforcement shall not be terminated in a tension zone unless 12.10.5.1, 12.10.5.2, or 12.10.5.3 is satisfied.

12.10.5.1 — V_u at the cutoff point does not exceed $(2/3)\phi V_n$.

12.10.5.2 — Stirrup area in excess of that required for shear and torsion is provided along each terminated bar or wire over a distance $(3/4)d$ from the termination point. Excess stirrup area shall be not less than $60b_w s/f_{yt}$. Spacing s shall not exceed $d/(8\beta_b)$.

12.10.5.3 — For No. 11 bars and smaller, continuing reinforcement provides double the area required for flexure at the cutoff point and V_u does not exceed $(3/4)\phi V_n$.

COMMENTARY

Cutoff points of bars to meet this requirement are illustrated in Fig. R12.10.2.

When bars of different sizes are used, the extension should be in accordance with the diameter of bar being terminated. A bar bent to the far face of a beam and continued there may logically be considered effective, in satisfying this section, to the point where the bar crosses the mid-depth of the member.

R12.10.4 — Peak stresses exist in the remaining bars wherever adjacent bars are cut off, or bent, in tension regions. In Fig. R12.10.2 an "x" is used to indicate the peak stress points remaining in continuing bars after part of the bars have been cut off. If bars are cut off as short as the moment diagrams allow, these peak stresses become the full f_y, which requires a full ℓ_d extension as indicated. This extension may exceed the length required for flexure.

R12.10.5 — Reduced shear strength and loss of ductility when bars are cut off in a tension zone, as in Fig. R12.10.2, have been reported. The code does not permit flexural reinforcement to be terminated in a tension zone unless special conditions are satisfied. Flexure cracks tend to open early wherever any reinforcement is terminated in a tension zone. If the steel stress in the continuing reinforcement and the shear strength are each near their limiting values, diagonal

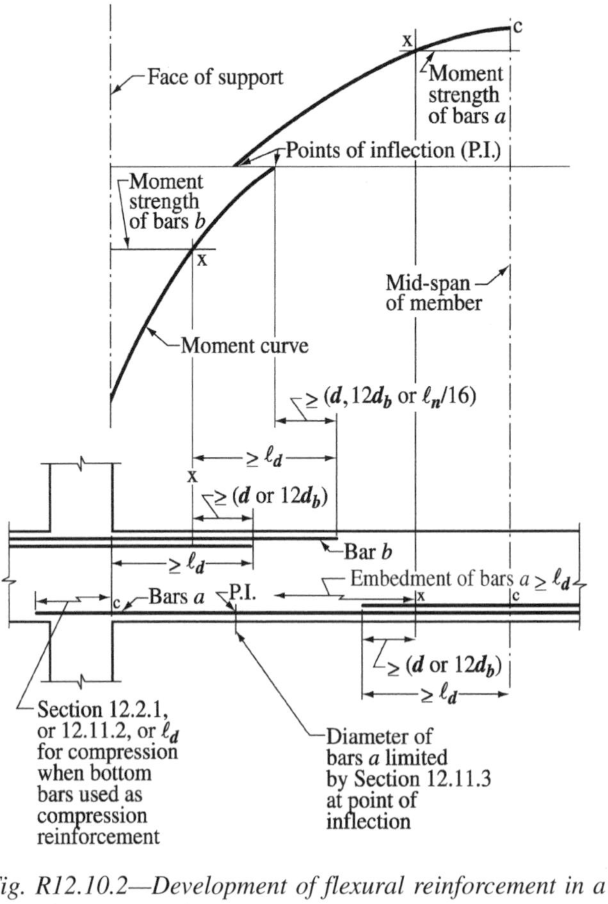

Fig. R12.10.2—Development of flexural reinforcement in a typical continuous beam

CODE | COMMENTARY

tension cracking tends to develop prematurely from these flexure cracks. Diagonal cracks are less likely to form where shear stress is low (see 12.10.5.1). Diagonal cracks can be restrained by closely spaced stirrups (see 12.10.5.2). A lower steel stress reduces the probability of such diagonal cracking (see 12.10.5.3). These requirements are not intended to apply to tension splices which are covered by 12.2, 12.13.5, and the related 12.15.

12.10.6 — Adequate anchorage shall be provided for tension reinforcement in flexural members where reinforcement stress is not directly proportional to moment, such as: sloped, stepped, or tapered footings; brackets; deep flexural members; or members in which tension reinforcement is not parallel to compression face. See 12.11.4 and 12.12.4 for deep flexural members.

R12.10.6 — Brackets, members of variable depth, and other members where f_s, calculated stress in reinforcement at service loads, does not decrease linearly in proportion to a decreasing moment require special consideration for proper development of the flexural reinforcement. For the bracket shown in Fig. R12.10.6, the stress at ultimate in the reinforcement is almost constant at approximately f_y from the face of support to the load point. In such a case, development of the flexural reinforcement depends largely on the end anchorage provided at the loaded end. Reference 12.1 suggests a welded cross bar of equal diameter as a means of providing effective end anchorage. An end hook in the vertical plane, with the minimum diameter bend, is not totally effective because an essentially plain concrete corner will exist near loads applied close to the corner. For wide brackets (perpendicular to the plane of the figure) and loads not applied close to the corners, U-shaped bars in a horizontal plane provide effective end hooks.

12.11 — Development of positive moment reinforcement

12.11.1 — At least one-third the positive moment reinforcement in simple members and one-fourth the positive moment reinforcement in continuous members shall extend along the same face of member into the support. In beams, such reinforcement shall extend into the support at least 6 in.

R12.11 — Development of positive moment reinforcement

R12.11.1 — Positive moment reinforcement is carried into the support to provide for some shifting of the moments due to changes in loading, settlement of supports, and lateral loads.

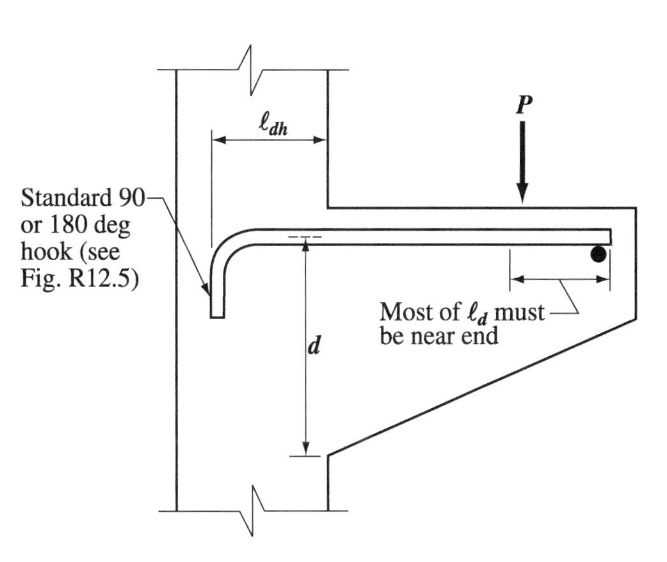

Fig. R12.10.6—Special member largely dependent on end anchorage

CODE

12.11.2 — When a flexural member is part of a primary lateral load resisting system, positive moment reinforcement required to be extended into the support by 12.11.1 shall be anchored to develop f_y in tension at the face of support.

12.11.3 — At simple supports and at points of inflection, positive moment tension reinforcement shall be limited to a diameter such that ℓ_d computed for f_y by 12.2 satisfies Eq. (12-5); except, Eq. (12-5) need not be satisfied for reinforcement terminating beyond centerline of simple supports by a standard hook, or a mechanical anchorage at least equivalent to a standard hook.

$$\ell_d \le \frac{M_n}{V_u} + \ell_a \qquad (12\text{-}5)$$

where:

M_n is calculated assuming all reinforcement at the section to be stressed to f_y;

V_u is calculated at the section;

ℓ_a at a support shall be the embedment length beyond center of support; or

ℓ_a at a point of inflection shall be limited to d or $12d_b$, whichever is greater.

An increase of 30 percent in the value of M_n/V_u shall be permitted when the ends of reinforcement are confined by a compressive reaction.

COMMENTARY

R12.11.2 — When a flexural member is part of a primary lateral load resisting system, loads greater than those anticipated in design may cause reversal of moment at supports; some positive reinforcement should be well anchored into the support. This anchorage is required to ensure ductility of response in the event of serious overstress, such as from blast or earthquake. It is not sufficient to use more reinforcement at lower stresses.

R12.11.3 — At simple supports and points of inflection such as "P.I." in Fig. R12.10.2, the diameter of the positive reinforcement should be small enough so that computed development length of the bar ℓ_d does not exceed $M_n/V_u + \ell_a$, or under favorable support conditions, $1.3M_n/V_u + \ell_a$. Fig. R12.11.3(a) illustrates the use of the provision.

At the point of inflection the value of ℓ_a should not exceed the actual bar extension used beyond the point of zero moment. The M_n/V_u portion of the available length is a theoretical quantity not generally associated with an obvious maximum stress point. M_n is the nominal flexural strength of the cross section without the ϕ-factor and is not the applied factored moment.

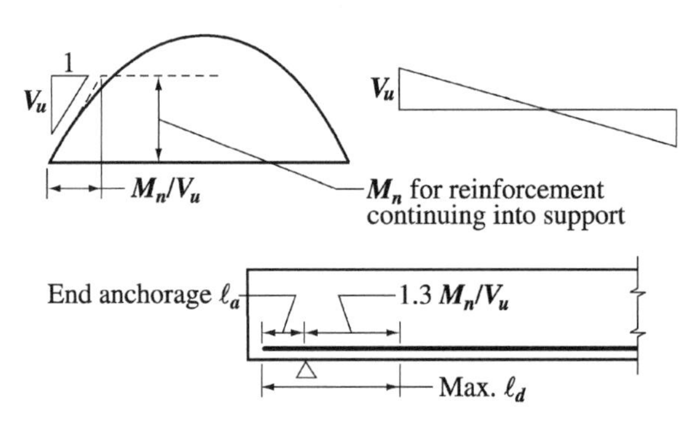

(a) Maximum size of bar at simple support

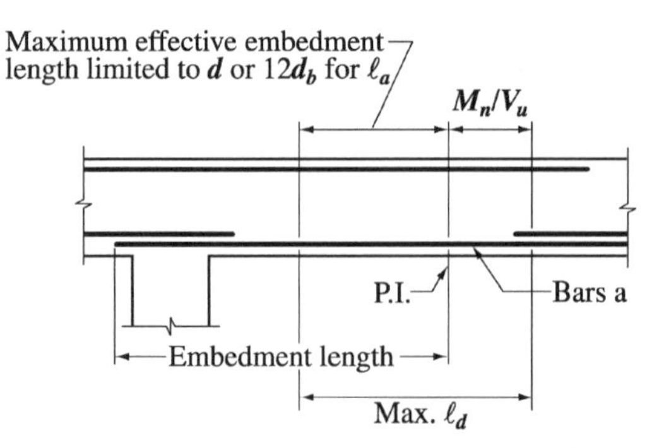

(b) Maximum size of bar "a" at point of inflection

Fig. R12.11.3—Concept for determining maximum bar size per 12.11.3

CODE | COMMENTARY

The length M_n/V_u corresponds to the development length for the maximum size bar obtained from the previously used flexural bond equation $\Sigma_o = V/ujd$, where u is bond stress, and jd is the moment arm. In the 1971 code, this anchorage requirement was relaxed from previous codes by crediting the available end anchorage length ℓ_a and by including a 30 percent increase for M_n/V_u when the ends of the reinforcement are confined by a compressive reaction.

For example, a bar size is provided at a simple support such that ℓ_d is computed in accordance with 12.2. The bar size provided is satisfactory only if computed ℓ_d does not exceed $1.3M_n/V_u + \ell_a$.

The ℓ_a to be used at points of inflection is limited to the effective depth of the member d or 12 bar diameters ($12d_b$), whichever is greater. Fig. R12.11.3(b) illustrates this provision at points of inflection. The ℓ_a limitation is added since test data are not available to show that a long end anchorage length will be fully effective in developing a bar that has only a short length between a point of inflection and a point of maximum stress.

12.11.4 — At simple supports of deep beams, positive moment tension reinforcement shall be anchored to develop f_y in tension at the face of the support except that if design is carried out using Appendix A, the positive moment tension reinforcement shall be anchored in accordance with A.4.3. At interior supports of deep beams, positive moment tension reinforcement shall be continuous or be spliced with that of the adjacent spans.

R12.11.4 — The use of the strut and tie model for the design of reinforced concrete deep flexural members clarifies that there is significant tension in the reinforcement at the face of the support. This requires the tension reinforcement to be continuous or be developed through and beyond the support.[12.23]

12.12 — Development of negative moment reinforcement

12.12.1 — Negative moment reinforcement in a continuous, restrained, or cantilever member, or in any member of a rigid frame, shall be anchored in or through the supporting member by embedment length, hooks, or mechanical anchorage.

12.12.2 — Negative moment reinforcement shall have an embedment length into the span as required by 12.1 and 12.10.3.

12.12.3 — At least one-third the total tension reinforcement provided for negative moment at a support shall have an embedment length beyond the point of inflection not less than d, $12d_b$, or $\ell_n/16$, whichever is greater.

12.12.4 — At interior supports of deep flexural members, negative moment tension reinforcement shall be continuous with that of the adjacent spans.

R12.12 — Development of negative moment reinforcement

Fig. R12.12 illustrates two methods of satisfying requirements for anchorage of tension reinforcement beyond the face of support. For anchorage of reinforcement with hooks, see R12.5.

Section 12.12.3 provides for possible shifting of the moment diagram at a point of inflection, as discussed under R12.10.3. This requirement may exceed that of 12.10.3, and the more restrictive of the two provisions governs.

CODE

COMMENTARY

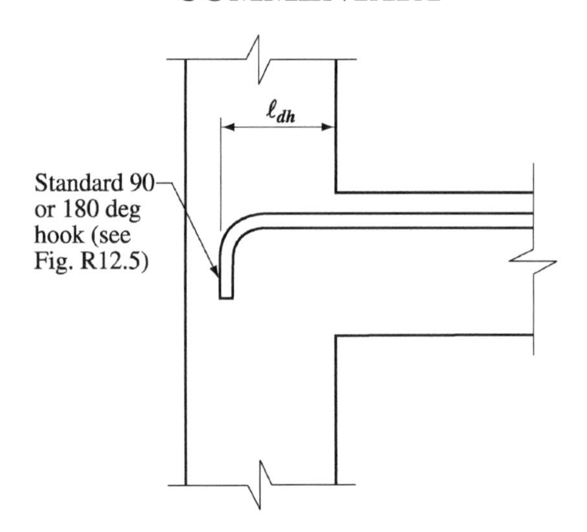

(a) Anchorage into exterior column

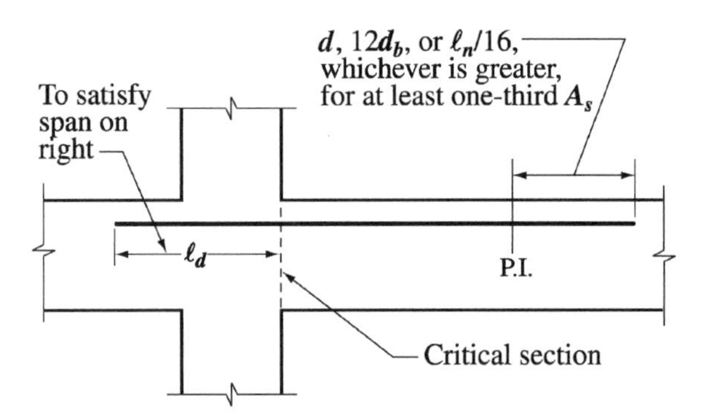

Note: Usually such anchorage becomes part of the adjacent beam reinforcement.

(b) Anchorage into adjacent beam

Fig. R12.12—Development of negative moment reinforcement

12.13 — Development of web reinforcement

12.13.1 — Web reinforcement shall be as close to the compression and tension surfaces of the member as cover requirements and proximity of other reinforcement permits.

12.13.2 — Ends of single leg, simple U-, or multiple U-stirrups shall be anchored as required by 12.13.2.1 through 12.13.2.5.

R12.13 — Development of web reinforcement

R12.13.1 — Stirrups should be carried as close to the compression face of the member as possible because near ultimate load the flexural tension cracks penetrate deeply.

R12.13.2 — The anchorage or development requirements for stirrups composed of bars or deformed wire were changed in the 1989 code to simplify the requirements. The straight anchorage was deleted as this stirrup is difficult to hold in place during concrete placement and the lack of a hook may make the stirrup ineffective as it crosses shear cracks near the end of the stirrup.

CODE

12.13.2.1 — For No. 5 bar and D31 wire, and smaller, and for No. 6, No. 7, and No. 8 bars with f_{yt} of 40,000 psi or less, a standard hook around longitudinal reinforcement.

12.13.2.2 — For No. 6, No. 7, and No. 8 stirrups with f_{yt} greater than 40,000 psi, a standard stirrup hook around a longitudinal bar plus an embedment between midheight of the member and the outside end of the hook equal to or greater than $0.014d_b f_{yt}/\sqrt{f_c'}$.

12.13.2.3 — For each leg of welded plain wire reinforcement forming simple U-stirrups, either:

(a) Two longitudinal wires spaced at a 2 in. spacing along the member at the top of the U; or

(b) One longitudinal wire located not more than $d/4$ from the compression face and a second wire closer to the compression face and spaced not less than 2 in. from the first wire. The second wire shall be permitted to be located on the stirrup leg beyond a bend, or on a bend with an inside diameter of bend not less than $8d_b$.

COMMENTARY

R12.13.2.1 — For a No. 5 bar or smaller, anchorage is provided by a standard stirrup hook, as defined in 7.1.3, hooked around a longitudinal bar. The 1989 code eliminated the need for a calculated straight embedment length in addition to the hook for these small bars, but 12.13.1 requires a full depth stirrup. Likewise, larger stirrups with f_{yt} equal to or less than 40,000 psi are sufficiently anchored with a standard stirrup hook around the longitudinal reinforcement.

R12.13.2.2 — Since it is not possible to bend a No. 6, No. 7, or No. 8 stirrup tightly around a longitudinal bar and due to the force in a bar with a design stress greater than 40,000 psi, stirrup anchorage depends on both the value of the hook and whatever development length is provided. A longitudinal bar within a stirrup hook limits the width of any flexural cracks, even in a tensile zone. Since such a stirrup hook cannot fail by splitting parallel to the plane of the hooked bar, the hook strength as utilized in 12.5.2 has been adjusted to reflect cover and confinement around the stirrup hook.

For stirrups with f_{yt} of only 40,000 psi, a standard stirrup hook provides sufficient anchorage and these bars are covered in 12.13.2.1. For bars with higher strength, the embedment should be checked. A 135 deg or 180 deg hook is preferred, but a 90 deg hook may be used provided the free end of the 90 deg hook is extended the full 12 bar diameters as required in 7.1.3.

R12.13.2.3—The requirements for anchorage of welded plain wire reinforcement stirrups are illustrated in Fig. R12.13.2.3.

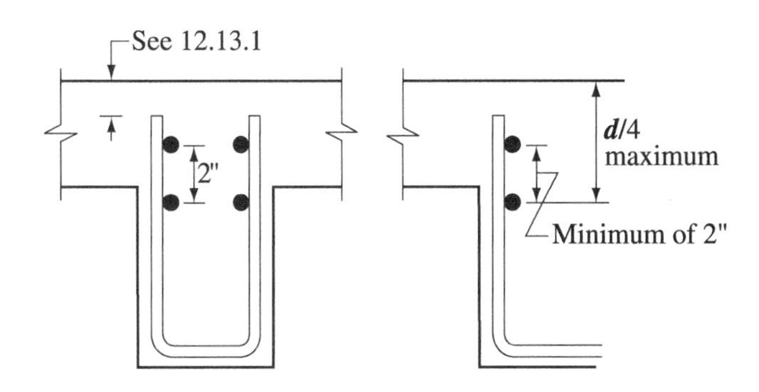

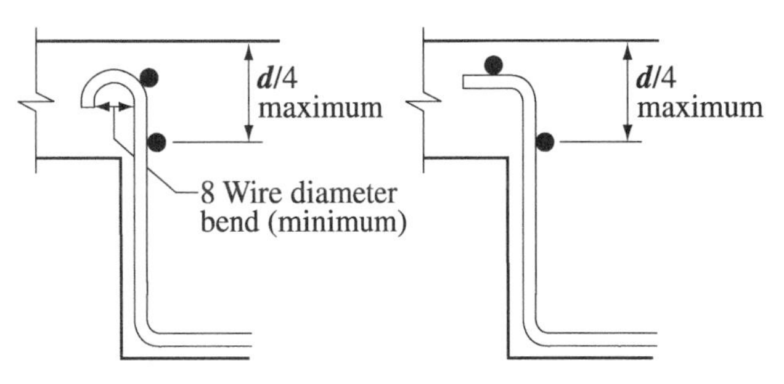

Fig. R12.13.2.3—Anchorage in compression zone of welded plain wire reinforcement U-stirrups

CODE

COMMENTARY

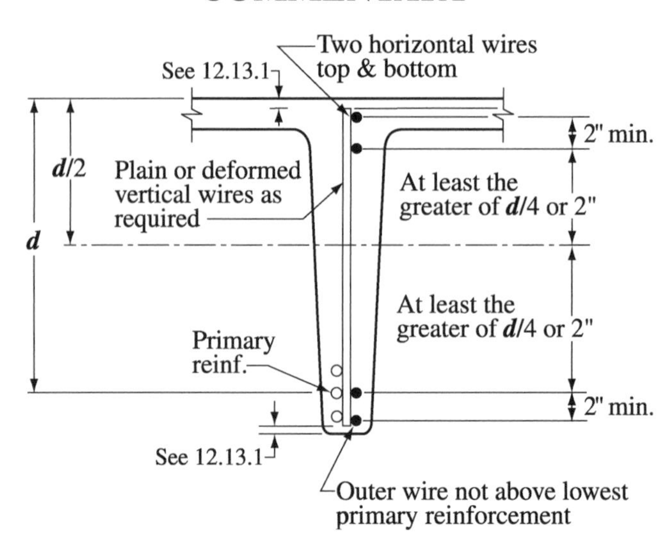

Fig. R12.13.2.4—Anchorage of single leg welded wire reinforcement shear reinforcement

12.13.2.4 — For each end of a single leg stirrup of welded wire reinforcement, two longitudinal wires at a minimum spacing of 2 in. and with the inner wire at least the greater of $d/4$ or 2 in. from $d/2$. Outer longitudinal wire at tension face shall not be farther from the face than the portion of primary flexural reinforcement closest to the face.

R12.13.2.4 — Use of welded wire reinforcement for shear reinforcement has become commonplace in the precast, prestressed concrete industry. The rationale for acceptance of straight sheets of welded wire reinforcement as shear reinforcement is presented in a report by a joint PCI/WRI Ad Hoc Committee on Welded Wire Fabric for Shear Reinforcement.[12.24]

The provisions for anchorage of single leg welded wire reinforcement in the tension face emphasize the location of the longitudinal wire at the same depth as the primary flexural reinforcement to avoid a splitting problem at the tension steel level. Fig. R12.13.2.4 illustrates the anchorage requirements for single leg, welded wire reinforcement. For anchorage of single leg, welded wire reinforcement, the code has permitted hooks and embedment length in the compression and tension faces of members (see 12.13.2.1 and 12.13.2.3), and embedment only in the compression face (see 12.13.2.2). Section 12.13.2.4 provides for anchorage of straight, single leg, welded wire reinforcement using longitudinal wire anchorage with adequate embedment length in compression and tension faces of members.

12.13.2.5 — In joist construction as defined in 8.11, for No. 4 bar and D20 wire and smaller, a standard hook.

R12.13.2.5 — In joists, a small bar or wire can be anchored by a standard hook not engaging longitudinal reinforcement, allowing a continuously bent bar to form a series of single-leg stirrups in the joist.

12.13.3 — Between anchored ends, each bend in the continuous portion of a simple U-stirrup or multiple U-stirrup shall enclose a longitudinal bar.

12.13.4 — Longitudinal bars bent to act as shear reinforcement, if extended into a region of tension, shall be continuous with longitudinal reinforcement and, if extended into a region of compression, shall be anchored beyond mid-depth $d/2$ as specified for development length in 12.2 for that part of f_{yt} required to satisfy Eq. (11-17).

CODE

12.13.5 — Pairs of U-stirrups or ties so placed as to form a closed unit shall be considered properly spliced when length of laps are $1.3\ell_d$. In members at least 18 in. deep, such splices with $A_b f_{yt}$ not more than 9000 lb per leg shall be considered adequate if stirrup legs extend the full available depth of member.

COMMENTARY

R12.13.5 — These requirements for lapping of double U-stirrups to form closed stirrups control over the provisions of 12.15.

12.14 — Splices of reinforcement — General

R12.14 — Splices of reinforcement — General

12.14.1 — Splices of reinforcement shall be made only as required or permitted on design drawings, or in specifications, or as authorized by the engineer.

Splices should, if possible, be located away from points of maximum tensile stress. The lap splice requirements of 12.15 encourage this practice.

12.14.2 — Lap splices

R12.14.2 — Lap splices

12.14.2.1 — Lap splices shall not be used for bars larger than No. 11 except as provided in 12.16.2 and 15.8.2.3.

R12.14.2.1 — Because of lack of adequate experimental data on lap splices of No. 14 and No. 18 bars in compression and in tension, lap splicing of these bar sizes is prohibited except as permitted in 12.16.2 and 15.8.2.3 for compression lap splices of No. 14 and No. 18 bars with smaller bars.

12.14.2.2 — Lap splices of bars in a bundle shall be based on the lap splice length required for individual bars within the bundle, increased in accordance with 12.4. Individual bar splices within a bundle shall not overlap. Entire bundles shall not be lap spliced.

R12.14.2.2 — The increased length of lap required for bars in bundles is based on the reduction in the exposed perimeter of the bars. Only individual bars are lap spliced along the bundle.

12.14.2.3 — Bars spliced by noncontact lap splices in flexural members shall not be spaced transversely farther apart than the smaller of one-fifth the required lap splice length, and 6 in.

R12.14.2.3 — If individual bars in noncontact lap splices are too widely spaced, an unreinforced section is created. Forcing a potential crack to follow a zigzag line (5 to 1 slope) is considered a minimum precaution. The 6 in. maximum spacing is added because most research available on the lap splicing of deformed bars was conducted with reinforcement within this spacing.

12.14.3 — Mechanical and welded splices

R12.14.3 — Mechanical and welded splices

12.14.3.1 — Mechanical and welded splices shall be permitted.

12.14.3.2 — A full mechanical splice shall develop in tension or compression, as required, at least $1.25f_y$ of the bar.

R12.14.3.2 — The maximum reinforcement stress used in design under the code is the specified yield strength. To ensure sufficient strength in splices so that yielding can be achieved in a member and thus brittle failure avoided, the 25 percent increase above the specified yield strength was selected as both an adequate minimum for safety and a practicable maximum for economy.

12.14.3.3 — Except as provided in this code, all welding shall conform to "Structural Welding Code—Reinforcing Steel" (ANSI/AWS D1.4).

R12.14.3.3 — See R3.5.2 for discussion on welding.

CODE

12.14.3.4 — A full welded splice shall develop at least $1.25f_y$ of the bar.

12.14.3.5 — Mechanical or welded splices not meeting requirements of 12.14.3.2 or 12.14.3.4 shall be permitted only for No. 5 bars and smaller and in accordance with 12.15.4.

12.15 — Splices of deformed bars and deformed wire in tension

12.15.1 — Minimum length of lap for tension lap splices shall be as required for Class A or B splice, but not less than 12 in., where:

Class A splice .... $1.0\ell_d$

Class B splice .... $1.3\ell_d$

where ℓ_d is calculated in accordance with 12.2 to develop f_y without the modification factor of 12.2.5.

12.15.2 — Lap splices of deformed bars and deformed wire in tension shall be Class B splices except that Class A splices are allowed when:

(a) the area of reinforcement provided is at least

COMMENTARY

R12.14.3.4 — A full welded splice is primarily intended for large bars (No. 6 and larger) in main members. The tensile strength requirement of 125 percent of specified yield strength is intended to provide sound welding that is also adequate for compression. See the discussion on strength in R12.14.3.2. The 1995 code eliminated a requirement that the bars be butted since indirect butt welds are permitted by ANSI/AWS D1.4, although ANSI/AWS D1.4 does indicate that wherever practical, direct butt splices are preferable for No. 7 bars and larger.

R12.14.3.5 — The use of mechanical or welded splices of less strength than 125 percent of specified yield strength is permitted if the minimum design criteria of 12.15.4 are met. Therefore, lap welds of reinforcing bars, either with or without backup material, welds to plate connections, and end-bearing splices are allowed under certain conditions. The 1995 code limited these lower strength welds and connections to No. 5 bars and smaller due to the potentially brittle nature of failure at these welds.

R12.15 — Splices of deformed bars and deformed wire in tension

R12.15.1 — Lap splices in tension are classified as Type A or B, with length of lap a multiple of the tensile development length ℓ_d. The development length ℓ_d used to obtain lap length should be based on f_y because the splice classifications already reflect any excess reinforcement at the splice location; therefore, the factor from 12.2.5 for excess A_s should not be used. When multiple bars located in the same plane are spliced at the same section, the clear spacing is the minimum clear distance between the adjacent splices. For splices in columns with offset bars, Fig. R12.15.1(a) illustrates the clear spacing to be used. For staggered splices, the clear spacing is taken as the minimum distance between adjacent splices [Fig. R12.15.1(b)].

The 1989 code contained several changes in development length in tension that eliminated many of the concerns regarding tension splices due to closely spaced bars with minimal cover. Thus, the Class C splice was eliminated although development lengths, on which splice lengths are based, have in some cases increased. Committee 318 considered suggestions from many sources, including ACI Committee 408, but has retained a two-level splice length primarily to encourage designers to splice bars at points of minimum stress and to stagger splices to improve behavior of critical details.

R12.15.2 — The tension lap splice requirements of 12.15.1 encourage the location of splices away from regions of high tensile stress to locations where the area of steel provided is at least twice that required by analysis. Table R12.15.2 presents the splice requirements in tabular form as presented in earlier code editions.

CODE

twice that required by analysis over the entire length of the splice; and

(b) one-half or less of the total reinforcement is spliced within the required lap length.

12.15.3 — Mechanical or welded splices used where area of reinforcement provided is less than twice that required by analysis shall meet requirements of 12.14.3.2 or 12.14.3.4.

12.15.4 — Mechanical or welded splices not meeting the requirements of 12.14.3.2 or 12.14.3.4 shall be permitted for No. 5 bars and smaller if the requirements of 12.15.4.1 through 12.15.4.3 are met:

12.15.4.1 — Splices shall be staggered at least 24 in.

12.15.4.2 — In computing the tensile forces that can be developed at each section, the spliced reinforcement stress shall be taken as the specified splice strength, but not greater than f_y. The stress in the unspliced reinforcement shall be taken as f_y times the ratio of the shortest length embedded beyond the section to ℓ_d, but not greater than f_y.

COMMENTARY

TABLE R12.15.2—TENSION LAP SPLICES

A_s provided* / A_s required	Maximum percent of A_s spliced within required lap length	
	50	100
Equal to or greater than 2	Class A	Class B
Less than 2	Class B	Class B

* Ratio of area of reinforcement provided to area of reinforcement required by analysis at splice locations.

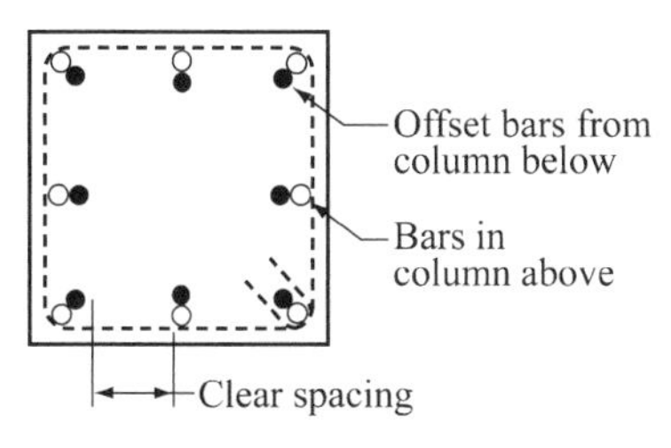

(a) Offset column bars

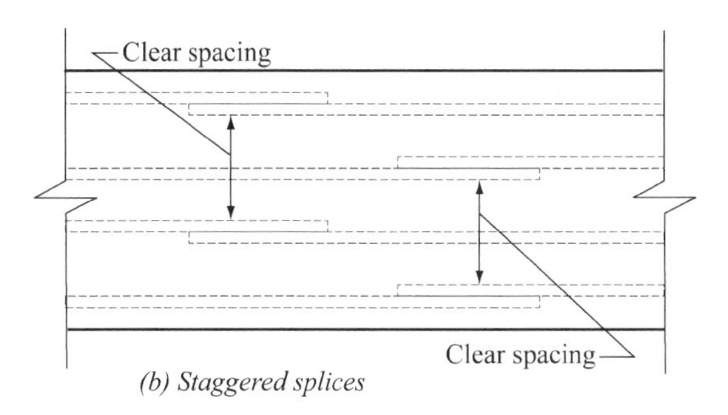

(b) Staggered splices

Fig. R12.15.1—Clear spacing of spliced bars

R12.15.3 — A mechanical or welded splice should develop at least 125 percent of the specified yield strength when located in regions of high tensile stress in the reinforcement. Such splices need not be staggered, although such staggering is encouraged where the area of reinforcement provided is less than twice that required by the analysis.

R12.15.4 — See R12.14.3.5. Section 12.15.4 concerns the situation where mechanical or welded splices of strength less than 125 percent of the specified yield strength of the reinforcement may be used. It provides a relaxation in the splice requirements where the splices are staggered and excess reinforcement area is available. The criterion of twice the computed tensile force is used to cover sections containing partial tensile splices with various percentages of total continuous steel. The usual partial tensile splice is a flare groove weld between bars or bar and structural steel piece.

To detail such welding, the length of weld should be specified. Such welds are rated at the product of total weld length times effective size of groove weld (established by bar size) times allowable stress permitted by **"Structural Welding Code—Reinforcing Steel"** (ANSI/AWS D1.4).

CODE

12.15.4.3 — The total tensile force that can be developed at each section must be at least twice that required by analysis, and at least 20,000 psi times the total area of reinforcement provided.

COMMENTARY

A full mechanical or welded splice conforming to 12.14.3.2 or 12.14.3.4 can be used without the stagger requirement in lieu of the lower strength mechanical or welded splice.

12.15.5 — Splices in tension tie members shall be made with a full mechanical or full welded splice in accordance with 12.14.3.2 or 12.14.3.4 and splices in adjacent bars shall be staggered at least 30 in.

R12.15.5 — A tension tie member, has the following characteristics: member having an axial tensile force sufficient to create tension over the cross section; a level of stress in the reinforcement such that every bar must be fully effective; and limited concrete cover on all sides. Examples of members that may be classified as tension ties are arch ties, hangers carrying load to an overhead supporting structure, and main tension elements in a truss.

In determining if a member should be classified as a tension tie, consideration should be given to the importance, function, proportions, and stress conditions of the member related to the above characteristics. For example, a usual large circular tank, with many bars and with splices well staggered and widely spaced should not be classified as a tension tie member, and Class B splices may be used.

12.16 — Splices of deformed bars in compression

R12.16 — Splices of deformed bars in compression

Bond research has been primarily related to bars in tension. Bond behavior of compression bars is not complicated by the problem of transverse tension cracking and thus compression splices do not require provisions as strict as those specified for tension splices. The minimum lengths for column splices contained originally in the 1956 code have been carried forward in later codes, and extended to compression bars in beams and to higher strength steels. No changes have been made in the provisions for compression splices since the 1971 code.

12.16.1 — Compression lap splice length shall be $0.0005 f_y d_b$, for f_y of 60,000 psi or less, or $(0.0009 f_y - 24) d_b$ for f_y greater than 60,000 psi, but not less than 12 in. For f_c' less than 3000 psi, length of lap shall be increased by one-third.

R12.16.1 — Essentially, lap requirements for compression splices have remained the same since the 1963 code.

The 1963 code values were modified in the 1971 code to recognize various degrees of confinement and to permit design with reinforcement having a specified yield strength up to 80,000 psi. Tests[12.1,12.25] have shown that splice strengths in compression depend considerably on end bearing and do not increase proportionally in strength when the splice length is doubled. Accordingly, for specified yield strengths above 60,000 psi, compression lap lengths are significantly increased, except where spiral enclosures are used (as in spiral columns) the where the increase is about 10 percent for an increase in specified yield strength from 60,000 to 75,000 psi.

12.16.2 — When bars of different size are lap spliced in compression, splice length shall be the larger of ℓ_{dc} of larger bar and splice length of smaller bar. Lap splices of No. 14 and No. 18 bars to No. 11 and smaller bars shall be permitted.

R12.16.2 — The lap splice length is to be computed based on the larger of the compression splice length of the smaller bar; or the compression development length of the larger bar. Lap splices are generally prohibited for No. 14 or No. 18 bars; however, for compression only, lap splices are permitted for No. 14 or No. 18 bars to No. 11 or smaller bars.

CODE

12.16.3 — Mechanical or welded splices used in compression shall meet requirements of 12.14.3.2 or 12.14.3.4.

12.16.4 — End-bearing splices

12.16.4.1 — In bars required for compression only, transmission of compressive stress by bearing of square cut ends held in concentric contact by a suitable device shall be permitted.

12.16.4.2 — Bar ends shall terminate in flat surfaces within 1.5 deg of a right angle to the axis of the bars and shall be fitted within 3 deg of full bearing after assembly.

12.16.4.3 — End-bearing splices shall be used only in members containing closed ties, closed stirrups, or spirals.

12.17 — Special splice requirements for columns

12.17.1 — Lap splices, mechanical splices, butt-welded splices, and end-bearing splices shall be used with the limitations of 12.17.2 through 12.17.4. A splice shall satisfy requirements for all load combinations for the column.

12.17.2 — Lap splices in columns

12.17.2.1 — Where the bar stress due to factored loads is compressive, lap splices shall conform to 12.16.1, 12.16.2, and, where applicable, to 12.17.2.4 or 12.17.2.5.

12.17.2.2 — Where the bar stress due to factored loads is tensile and does not exceed $\mathbf{0.5}f_y$ in tension, lap splices shall be Class B tension lap splices if more than one-half of the bars are spliced at any section, or Class A tension lap splices if half or fewer of the bars are spliced at any section and alternate lap splices are staggered by ℓ_d.

COMMENTARY

R12.16.4 — End-bearing splices

R12.16.4.1 — Experience with end-bearing splices has been almost exclusively with vertical bars in columns. If bars are significantly inclined from the vertical, special attention is required to ensure that adequate end-bearing contact can be achieved and maintained.

R12.16.4.2 — These tolerances were added in the 1971 code, representing practice based on tests of full-size members containing No. 18 bars.

R12.16.4.3 — This limitation was added in the 1971 code to ensure a minimum shear resistance in sections containing end-bearing splices.

R12.17 — Special splice requirements for columns

In columns subject to flexure and axial loads, tension stresses may occur on one face of the column for moderate and large eccentricities as shown in Fig. R12.17. When such tensions occur, 12.17 requires tension splices to be used or an adequate tensile resistance to be provided. Furthermore, a minimum tension capacity is required in each face of all columns even where analysis indicates compression only.

The 1989 code clarifies this section on the basis that a compressive lap splice has a tension capacity of at least one-quarter f_y, which simplifies the calculation requirements in previous codes.

Note that the column splice should satisfy requirements for all load combinations for the column. Frequently, the basic gravity load combination will govern the design of the column itself, but a load combination including wind or seismic loads may induce greater tension in some column bars, and the column splice should be designed for this tension.

R12.17.2 — Lap splices in columns

R12.17.2.1 — The 1989 code was simplified for column bars always in compression on the basis that a compressive lap splice is adequate for sufficient tension to preclude special requirements.

CODE

12.17.2.3 — Where the bar stress due to factored loads is greater than $0.5f_y$ in tension, lap splices shall be Class B tension lap splices.

12.17.2.4 — In tied reinforced compression members, where ties throughout the lap splice length have an effective area not less than $0.0015hs$, lap splice length shall be permitted to be multiplied by 0.83, but lap length shall not be less than 12 in. Tie legs perpendicular to dimension h shall be used in determining effective area.

12.17.2.5 — In spirally reinforced compression members, lap splice length of bars within a spiral shall be permitted to be multiplied by 0.75, but lap length shall not be less than 12 in.

12.17.3 — Mechanical or welded splices in columns

Mechanical or welded splices in columns shall meet the requirements of 12.14.3.2 or 12.14.3.4.

COMMENTARY

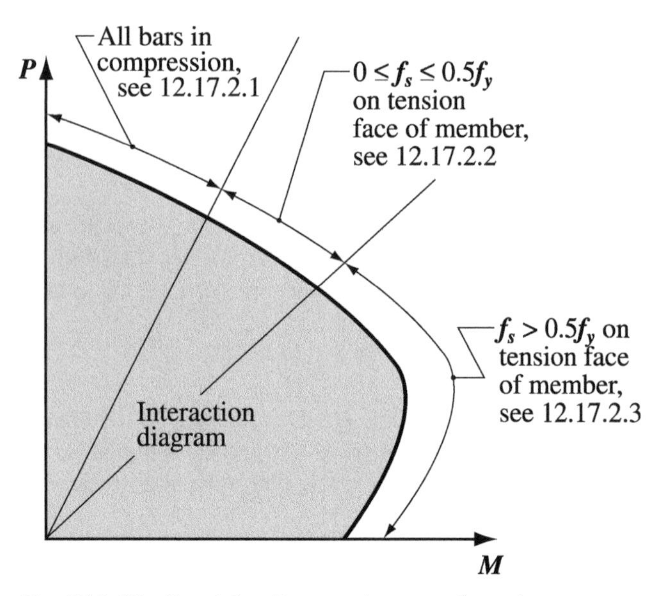

Fig. R12.17—Special splice requirements for columns

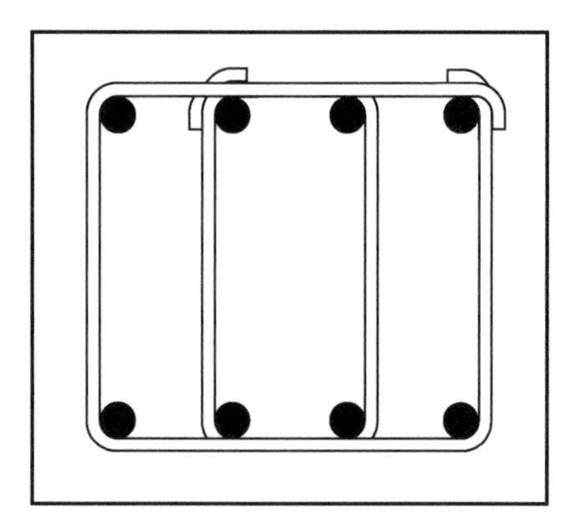

Fig. R.12.17.2—Tie legs which cross the axis of bending are used to compute effective area. In the case shown, four legs are effective

R12.17.2.4 — Reduced lap lengths are allowed when the splice is enclosed throughout its length by minimum ties.

The tie legs perpendicular to each direction are computed separately and the requirement must be satisfied in each direction. This is illustrated in Fig. R12.17.2, where four legs are effective in one direction and two legs in the other direction. This calculation is critical in one direction, which normally can be determined by inspection.

R12.17.2.5 — Compression lap lengths may be reduced when the lap splice is enclosed throughout its length by spirals because of increased splitting resistance. Spirals should meet requirements of 7.10.4 and 10.9.3.

R12.17.3 — Mechanical or welded splices in columns

Mechanical or welded splices are allowed for splices in columns but should be designed as a full mechanical splice or a

CODE

COMMENTARY

full welded splice developing 125 percent f_y as required by 12.14.3.2 or 12.14.3.4. Splice capacity is traditionally tested in tension and full strength is required to reflect the high compression loads possible in column reinforcement due to creep effects. If a mechanical splice developing less than a full mechanical splice is used, then the splice is required to conform to all requirements of end-bearing splices of 12.16.4 and 12.17.4.

12.17.4 — End-bearing splices in columns

End-bearing splices complying with 12.16.4 shall be permitted to be used for column bars stressed in compression provided the splices are staggered or additional bars are provided at splice locations. The continuing bars in each face of the column shall have a tensile strength, based on f_y, not less than $0.25f_y$ times the area of the vertical reinforcement in that face.

R12.17.4 — End-bearing splices in columns

End-bearing splices used to splice column bars always in compression should have a tension capacity of 25 percent of the specified yield strength of the steel area on each face of the column, either by staggering the end-bearing splices or by adding additional steel through the splice location. The end-bearing splice should conform to 12.16.4.

12.18 — Splices of welded deformed wire reinforcement in tension

12.18.1 — Minimum lap splice length of welded deformed wire reinforcement measured between the ends of each reinforcement sheet shall be not less than the larger of $1.3\ell_d$ and 8 in., and the overlap measured between outermost cross wires of each reinforcement sheet shall be not less than 2 in., where ℓ_d is calculated in accordance with 12.7 to develop f_y.

12.18.2 — Lap splices of welded deformed wire reinforcement, with no cross wires within the lap splice length, shall be determined as for deformed wire.

12.18.3—When any plain wires are present in the welded deformed wire reinforcement in the direction of the lap splice or when welded deformed wire reinforcement is lap spliced to welded plain wire reinforcement, the reinforcement shall be lap spliced in accordance with 12.19.

R12.18 — Splices of welded deformed wire reinforcement in tension

Splice provisions for welded deformed wire reinforcement are based on available tests.[12.26] The requirements were simplified (1976 code supplement) from provisions of the 1971 code by assuming that only one cross wire of each welded wire reinforcement sheet is overlapped and by computing the splice length as $1.3\ell_d$. The development length ℓ_d is that computed in accordance with the provisions of 12.7 without regard to the 8 in. minimum. The 8 in. applies to the overall splice length. See Fig. R12.18. If no cross wires are within the lap length, the provisions for deformed wire apply.

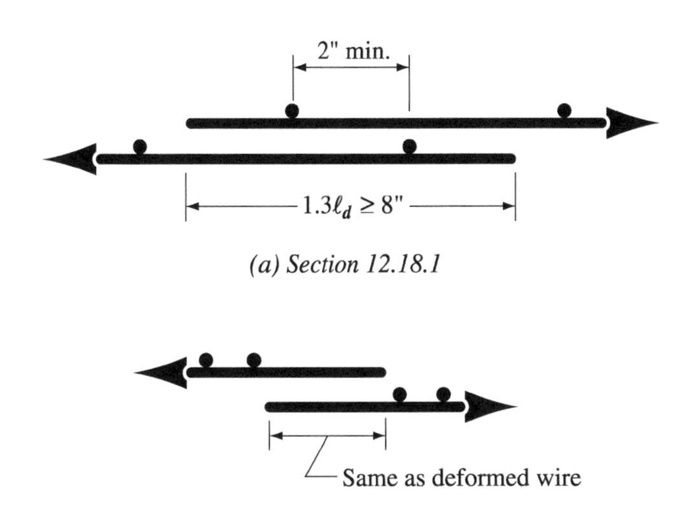

Fig. R12.18—Lap splices of welded deformed wire reinforcement

CODE

12.19 — Splices of welded plain wire reinforcement in tension

Minimum length of lap for lap splices of welded plain wire reinforcement shall be in accordance with 12.19.1 and 12.19.2.

12.19.1 — Where A_s provided is less than twice that required by analysis at splice location, length of overlap measured between outermost cross wires of each reinforcement sheet shall be not less than the largest of one spacing of cross wires plus 2 in., $1.5\ell_d$, and 6 in., where ℓ_d is calculated in accordance with 12.8 to develop f_y.

12.19.2 — Where A_s provided is at least twice that required by analysis at splice location, length of overlap measured between outermost cross wires of each reinforcement sheet shall not be less than the larger of $1.5\ell_d$, and 2 in., where ℓ_d is calculated in accordance with 12.8 to develop f_y.

COMMENTARY

R12.19 — Splices of welded plain wire reinforcement in tension

The strength of lap splices of welded plain wire reinforcement is dependent primarily on the anchorage obtained from the cross wires rather than on the length of wire in the splice. For this reason, the lap is specified in terms of overlap of cross wires rather than in wire diameters or inches. The 2 in. additional lap required is to assure overlapping of the cross wires and to provide space for satisfactory consolidation of the concrete between cross wires. Research[12.27] has shown an increased splice length is required when welded wire reinforcement of large, closely spaced wires is lapped and as a consequence additional splice length requirements are provided for this reinforcement, in addition to an absolute minimum of 6 in. The development length ℓ_d is that computed in accordance with the provisions of 12.8 without regard to the 6 in. minimum. Splice requirements are illustrated in Fig. R12.19.

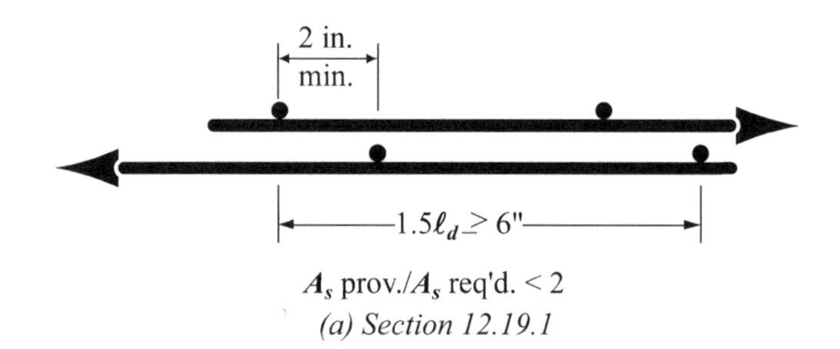

A_s prov./A_s req'd. < 2
(a) Section 12.19.1

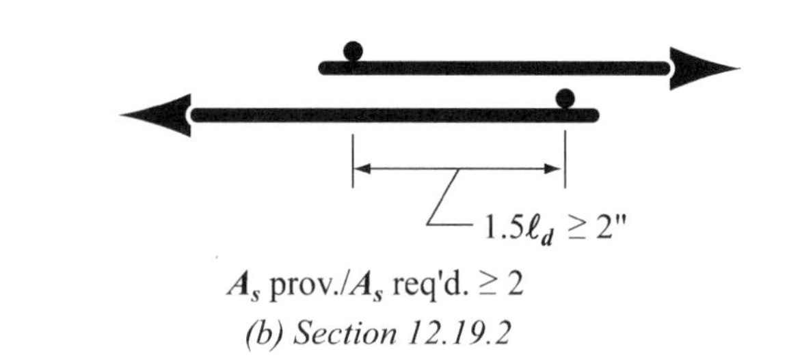

A_s prov./A_s req'd. ≥ 2
(b) Section 12.19.2

Fig. R12.19—Lap splices of plain welded wire reinforcement.

CHAPTER 13 — TWO-WAY SLAB SYSTEMS

CODE

13.1 — Scope

13.1.1 — Provisions of Chapter 13 shall apply for design of slab systems reinforced for flexure in more than one direction, with or without beams between supports.

13.1.2 — For a slab system supported by columns or walls, dimensions c_1, c_2, and ℓ_n shall be based on an effective support area defined by the intersection of the bottom surface of the slab, or of the drop panel if present, with the largest right circular cone, right pyramid, or tapered wedge whose surfaces are located within the column and the capital or bracket and are oriented no greater than 45 degrees to the axis of the column.

13.1.3 — Solid slabs and slabs with recesses or pockets made by permanent or removable fillers between ribs or joists in two directions are included within the scope of Chapter 13.

13.1.4 — Minimum thickness of slabs designed in accordance with Chapter 13 shall be as required by 9.5.3.

COMMENTARY

R13.1 — Scope

The design methods given in Chapter 13 are based on analysis of the results of an extensive series of tests[13.1-13.7] and the well established performance record of various slab systems. Much of Chapter 13 is concerned with the selection and distribution of flexural reinforcement. The designer is cautioned that the problem related to safety of a slab system is the transmission of load from the slab to the columns by flexure, torsion, and shear. Design criteria for shear and torsion in slabs are given in Chapter 11.

The fundamental design principles contained in Chapter 13 are applicable to all planar structural systems subjected to transverse loads. Some of the specific design rules, as well as historical precedents, limit the types of structures to which Chapter 13 applies. General characteristics of slab systems that may be designed according to Chapter 13 are described in this section. These systems include flat slabs, flat plates, two-way slabs, and waffle slabs. Slabs with paneled ceilings are two-way wide-band beam systems.

True one-way slabs, slabs reinforced to resist flexural stresses in only one direction, are excluded. Also excluded are soil-supported slabs, such as slabs on grade, that do not transmit vertical loads from other parts of the structure to the soil.

For slabs with beams, the explicit design procedures of Chapter 13 apply only when the beams are located at the edges of the panel and when the beams are supported by columns or other essentially nondeflecting supports at the corners of the panel. Two-way slabs with beams in one direction, with both slab and beams supported by girders in the other direction, may be designed under the general requirements of Chapter 13. Such designs should be based upon analysis compatible with the deflected position of the supporting beams and girders.

For slabs supported on walls, the explicit design procedures in this chapter treat the wall as a beam of infinite stiffness; therefore, each wall should support the entire length of an edge of the panel (see 13.2.3). Wall-like columns less than a full panel length can be treated as columns.

Design aids for use in the engineering analysis and design of two-way slab systems are given in the *ACI Design Handbook*.[13.8] Design aids are provided to simplify application of the direct design and equivalent frame methods of Chapter 13.

CODE / COMMENTARY

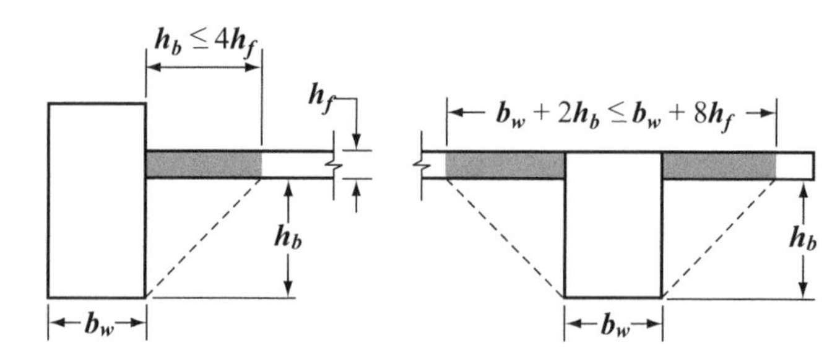

Fig. R13.2.4—Examples of the portion of slab to be included with the beam under 13.2.4

13.2 — Definitions

13.2.1 — Column strip is a design strip with a width on each side of a column centerline equal to $\mathbf{0.25\ell_2}$ or $\mathbf{0.25\ell_1}$, whichever is less. Column strip includes beams, if any.

13.2.2 — Middle strip is a design strip bounded by two column strips.

13.2.3 — A panel is bounded by column, beam, or wall centerlines on all sides.

13.2.4 — For monolithic or fully composite construction, a beam includes that portion of slab on each side of the beam extending a distance equal to the projection of the beam above or below the slab, whichever is greater, but not greater than four times the slab thickness.

13.2.5 — When used to reduce the amount of negative moment reinforcement over a column or minimum required slab thickness, a drop panel shall project below the slab at least one-quarter of the slab thickness beyond the drop and extend in each direction from the centerline of support a distance not less than one-sixth the span length measured from center-to-center of supports in that direction.

R13.2 — Definitions

R13.2.3 — A panel includes all flexural elements between column centerlines. Thus, the column strip includes the beam, if any.

R13.2.4 — For monolithic or fully composite construction, the beams include portions of the slab as flanges. Two examples of the rule are provided in Fig. R13.2.4.

R13.2.5 — Drop panel dimensions specified in 13.2.5 are necessary when reducing the amount of negative moment reinforcement following 13.3.7 or to satisfy some minimum slab thicknesses permitted in 9.5.3. Drop panels with dimensions less than those specified in 13.2.5 may be used to increase slab shear strength.

13.3 — Slab reinforcement

13.3.1 — Area of reinforcement in each direction for two-way slab systems shall be determined from moments at critical sections, but shall not be less than required by 7.12.

13.3.2 — Spacing of reinforcement at critical sections shall not exceed two times the slab thickness, except for portions of slab area of cellular or ribbed construction. In the slab over cellular spaces, reinforcement shall be provided as required by 7.12.

R13.3 — Slab reinforcement

R13.3.2 — The requirement that the center-to-center spacing of the reinforcement be not more than two times the slab thickness applies only to the reinforcement in solid slabs, and not to reinforcement joists or waffle slabs. This limitation is to ensure slab action, cracking, and provide for the possibility of loads concentrated on small areas of the slab. See also R10.6.

CODE

13.3.3 — Positive moment reinforcement perpendicular to a discontinuous edge shall extend to the edge of slab and have embedment, straight or hooked, at least 6 in. in spandrel beams, columns, or walls.

13.3.4 — Negative moment reinforcement perpendicular to a discontinuous edge shall be bent, hooked, or otherwise anchored in spandrel beams, columns, or walls, and shall be developed at face of support according to provisions of Chapter 12.

13.3.5 — Where a slab is not supported by a spandrel beam or wall at a discontinuous edge, or where a slab cantilevers beyond the support, anchorage of reinforcement shall be permitted within the slab.

13.3.6 — In slabs with beams between supports with a value of α_f greater than 1.0, special top and bottom slab reinforcement shall be provided at exterior corners in accordance with 13.3.6.1 through 13.3.6.4.

13.3.6.1 — The special reinforcement in both top and bottom of slab shall be sufficient to resist a moment per foot of width equal to the maximum positive moment in the slab.

13.3.6.2 — The moment shall be assumed to be about an axis perpendicular to the diagonal from the corner in the top of the slab and about an axis parallel to the diagonal from the corner in the bottom of the slab.

13.3.6.3 — The special reinforcement shall be provided for a distance in each direction from the corner equal to one-fifth the longer span.

13.3.6.4 — The special reinforcement shall be placed in a band parallel to the diagonal in the top of the slab and a band perpendicular to the diagonal in the bottom of the slab. Alternatively, the special reinforcement shall be placed in two layers parallel to the sides of the slab in both the top and bottom of the slab.

13.3.7 — When a drop panel is used to reduce the amount of negative moment reinforcement over the column of a flat slab, the dimensions of the drop panel shall be in accordance with 13.2.5. In computing required slab reinforcement, the thickness of the drop panel below the slab shall not be assumed to be greater than one-quarter the distance from the edge of drop panel to the face of column or column capital.

13.3.8 — Details of reinforcement in slabs without beams

13.3.8.1 — In addition to the other requirements of 13.3, reinforcement in slabs without beams shall have minimum extensions as prescribed in Fig. 13.3.8.

COMMENTARY

R13.3.3-R13.3.5 — Bending moments in slabs at spandrel beams can be subject to great variation. If spandrel beams are built solidly into walls, the slab approaches complete fixity. Without an integral wall, the slab could approach simply supported, depending on the torsional rigidity of the spandrel beam or slab edge. These requirements provide for unknown conditions that might normally occur in a structure.

R13.3.8 — Details of reinforcement in slabs without beams

In the 1989 code, bent bars were removed from Fig. 13.3.8. This was done because bent bars are seldom used and are difficult to place properly. Bent bars are permitted, however, if they comply with 13.3.8.3. Refer to 13.4.8 of the 1983 code.

CODE

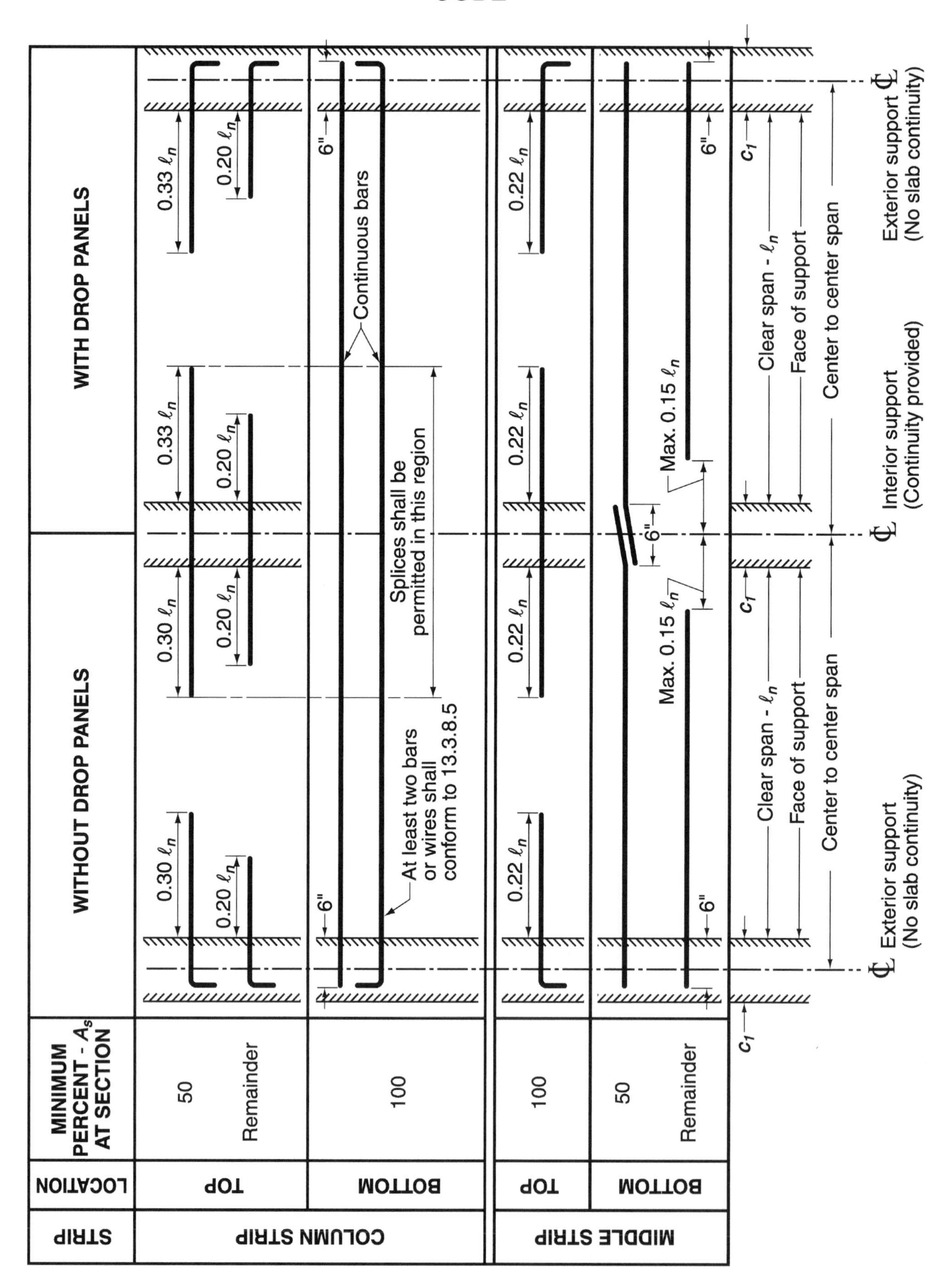

Fig. 13.3.8—Minimum extensions for reinforcement in slabs without beams. (See 12.11.1 for reinforcement extension into supports)

CODE

13.3.8.2 — Where adjacent spans are unequal, extensions of negative moment reinforcement beyond the face of support as prescribed in Fig. 13.3.8 shall be based on requirements of the longer span.

13.3.8.3 — Bent bars shall be permitted only when depth-span ratio permits use of bends of 45 degrees or less.

13.3.8.4 — In frames where two-way slabs act as primary members resisting lateral loads, lengths of reinforcement shall be determined by analysis but shall not be less than those prescribed in Fig. 13.3.8.

13.3.8.5 — All bottom bars or wires within the column strip, in each direction, shall be continuous or spliced with Class A tension splices or with mechanical or welded splices satisfying 12.14.3. Splices shall be located as shown in Fig. 13.3.8. At least two of the column strip bottom bars or wires in each direction shall pass within the column core and shall be anchored at exterior supports.

13.3.8.6 — In slabs with shearheads and in lift-slab construction where it is not practical to pass the bottom bars required by 13.3.8.5 through the column, at least two bonded bottom bars or wires in each direction shall pass through the shearhead or lifting collar as close to the column as practicable and be continuous or spliced with a Class A splice. At exterior columns, the reinforcement shall be anchored at the shearhead or lifting collar.

13.4 — Openings in slab systems

13.4.1 — Openings of any size shall be permitted in slab systems if shown by analysis that the design strength is at least equal to the required strength set forth in 9.2 and 9.3, and that all serviceability conditions, including the limits on deflections, are met.

13.4.2 — As an alternate to special analysis as required by 13.4.1, openings shall be permitted in slab systems without beams only in accordance with 13.4.2.1 through 13.4.2.4.

13.4.2.1 — Openings of any size shall be permitted in the area common to intersecting middle strips, provided total amount of reinforcement required for the panel without the opening is maintained.

13.4.2.2 — In the area common to intersecting column strips, not more than one-eighth the width of col-

COMMENTARY

R13.3.8.4 — For moments resulting from combined lateral and gravity loadings, the minimum lengths and extensions of bars in Fig. 13.3.8 may not be sufficient.

R13.3.8.5 — The continuous column strip bottom reinforcement provides the slab some residual ability to span to the adjacent supports should a single support be damaged. The two continuous column strip bottom bars or wires through the column may be termed integrity steel, and are provided to give the slab some residual capacity following a single punching shear failure at a single support.[13.9] In the 2002 code, mechanical and welded splices were explicitly recognized as alternative methods of splicing reinforcement.

R13.3.8.6 — In the 1992 code, this provision was added to require the same integrity steel as for other two-way slabs without beams in case of a punching shear failure at a support.

In some instances, there is sufficient clearance so that the bonded bottom bars can pass under shearheads and through the column. Where clearance under the shearhead is inadequate, the bottom bars should pass through holes in the shearhead arms or within the perimeter of the lifting collar. Shearheads should be kept as low as possible in the slab to increase their effectiveness.

R13.4 — Openings in slab systems

See R11.12.5.

CODE

COMMENTARY

umn strip in either span shall be interrupted by openings. An amount of reinforcement equivalent to that interrupted by an opening shall be added on the sides of the opening.

13.4.2.3 — In the area common to one column strip and one middle strip, not more than one-quarter of the reinforcement in either strip shall be interrupted by openings. An amount of reinforcement equivalent to that interrupted by an opening shall be added on the sides of the opening.

13.4.2.4 — Shear requirements of 11.12.5 shall be satisfied.

13.5 — Design procedures

13.5.1 — A slab system shall be designed by any procedure satisfying conditions of equilibrium and geometric compatibility, if shown that the design strength at every section is at least equal to the required strength set forth in 9.2 and 9.3, and that all serviceability conditions, including limits on deflections, are met.

13.5.1.1 — Design of a slab system for gravity loads, including the slab and beams (if any) between supports and supporting columns or walls forming orthogonal frames, by either the Direct Design Method of 13.6 or the Equivalent Frame Method of 13.7 shall be permitted.

13.5.1.2 — For lateral loads, analysis of frames shall take into account effects of cracking and reinforcement on stiffness of frame members.

R13.5 — Design procedures

R13.5.1 — This section permits a designer to base a design directly on fundamental principles of structural mechanics, provided it can be demonstrated explicitly that all safety and serviceability criteria are satisfied. The design of the slab may be achieved through the combined use of classic solutions based on a linearly elastic continuum, numerical solutions based on discrete elements, or yield-line analyses, including, in all cases, evaluation of the stress conditions around the supports in relation to shear and torsion as well as flexure. The designer should consider that the design of a slab system involves more than its analysis, and justify any deviations in physical dimensions of the slab from common practice on the basis of knowledge of the expected loads and the reliability of the calculated stresses and deformations of the structure.

R13.5.1.1 — For gravity load analysis of two-way slab systems, two analysis methods are given in 13.6 and 13.7. The specific provisions of both design methods are limited in application to orthogonal frames subject to gravity loads only. Both methods apply to two-way slabs with beams as well as to flat slabs and flat plates. In both methods, the distribution of moments to the critical sections of the slab reflects the effects of reduced stiffness of elements due to cracking and support geometry.

R13.5.1.2 — During the life of a structure, construction loads, ordinary occupancy loads, anticipated overloads, and volume changes will cause cracking of slabs. Cracking reduces stiffness of slab members, and increases lateral flexibility when lateral loads act on the structure. Cracking of slabs should be considered in stiffness assumptions so that drift caused by wind or earthquake is not grossly underestimated.

The designer may model the structure for lateral load analysis using any approach that is shown to satisfy equilibrium and geometric compatibility and to be in reasonable agreement with test data.[13.10,13.11] The selected approach should recognize effects of cracking as well as parameters such as ℓ_2/ℓ_1, c_1/ℓ_1, and c_2/c_1. Some of the available approaches are summarized in Reference 13.12, which includes a discus-

CODE

COMMENTARY

sion on the effects of cracking. Acceptable approaches include plate-bending finite-element models, the effective beam width model, and the equivalent frame model. In all cases, framing member stiffnesses should be reduced to account for cracking.

For nonprestressed slabs, it is normally appropriate to reduce slab bending stiffness to between one-half and one-quarter of the uncracked stiffness. For prestressed construction, stiffnesses greater than those of cracked, nonprestressed slabs may be appropriate. When the analysis is used to determine design drifts or moment magnification, lower-bound slab stiffnesses should be assumed. When the analysis is used to study interactions of the slab with other framing elements, such as structural walls, it may be appropriate to consider a range of slab stiffnesses so that the relative importance of the slab on those interactions can be assessed.

13.5.1.3 — Combining the results of the gravity load analysis with the results of the lateral load analysis shall be permitted.

13.5.2 — The slab and beams (if any) between supports shall be proportioned for factored moments prevailing at every section.

13.5.3 — When gravity load, wind, earthquake, or other lateral forces cause transfer of moment between slab and column, a fraction of the unbalanced moment shall be transferred by flexure in accordance with 13.5.3.2 and 13.5.3.3.

R13.5.3 — This section is concerned primarily with slab systems without beams. Tests and experience have shown that, unless special measures are taken to resist the torsional and shear stresses, all reinforcement resisting that part of the moment to be transferred to the column by flexure should be placed between lines that are one and one-half the slab or drop panel thickness, $1.5h$, on each side of the column. The calculated shear stresses in the slab around the column are required to conform to the requirements of 11.12.2. See R11.12.1.2 and R11.12.2.1 for more details on application of this section.

13.5.3.1 — The fraction of unbalanced moment not transferred by flexure shall be transferred by eccentricity of shear in accordance with 11.12.6.

13.5.3.2 — A fraction of the unbalanced moment given by $\gamma_f M_u$ shall be considered to be transferred by flexure within an effective slab width between lines that are one and one-half slab or drop panel thicknesses ($1.5h$) outside opposite faces of the column or capital, where M_u is the factored moment to be transferred and

$$\gamma_f = \frac{1}{1 + (2/3)\sqrt{b_1/b_2}} \qquad (13\text{-}1)$$

13.5.3.3 — For unbalanced moments about an axis parallel to the edge at exterior supports, the value of γ_f by Eq. (13-1) shall be permitted to be increased up to 1.0 provided that V_u at an edge support does not exceed $0.75\phi V_c$ or at a corner support does not exceed $0.5\phi V_c$, where V_c is calculated in accordance with 11.12.2.1. For unbalanced moments at interior supports, and for unbalanced moments about an axis transverse to the edge at exterior supports, the value of γ_f in Eq. (13-1) shall be permitted to be increased by up to 25 percent provided that V_u at the support does not exceed $0.4\phi V_c$. Reinforcement ratio ρ, within the

R13.5.3.3 — The 1989 code procedures remain unchanged, except that under certain conditions the designer is permitted to adjust the level of moment transferred by shear without revising member sizes. Tests indicate that some flexibility in distribution of unbalanced moments transferred by shear and flexure at both exterior and interior supports is possible. Interior, exterior, and corner supports refer to slab-column connections for which the critical perimeter for rectangular columns has 4, 3, or 2 sides, respectively. Changes in the 1995 code recognized, to some extent, design practices prior to the 1971 code.[13.13]

CODE

effective slab width defined in 13.5.3.2, shall not exceed $\mathbf{0.375}\rho_b$. No adjustments to γ_f shall be permitted for prestressed slab systems.

COMMENTARY

At exterior supports, for unbalanced moments about an axis parallel to the edge, the portion of moment transferred by eccentricity of shear $\gamma_v M_u$ may be reduced provided that the factored shear at the support (excluding the shear produced by moment transfer) does not exceed 75 percent of the shear capacity ϕV_c as defined in 11.12.2.1 for edge columns or 50 percent for corner columns. Tests[13.14,13.15] indicate that there is no significant interaction between shear and unbalanced moment at the exterior support in such cases. Note that as $\gamma_v M_u$ is decreased, $\gamma_f M_u$ is increased.

Evaluation of tests of interior supports indicate that some flexibility in distributing unbalanced moments transferred by shear and flexure is possible, but with more severe limitations than for exterior supports. For interior supports, the unbalanced moment transferred by flexure is permitted to be increased up to 25 percent provided that the factored shear (excluding the shear caused by the moment transfer) at the interior supports does not exceed 40 percent of the shear capacity ϕV_c as defined in 11.12.2.1.

Tests of slab-column connections indicate that a large degree of ductility is required because the interaction between shear and unbalanced moment is critical. When the factored shear is large, the column-slab joint cannot always develop all of the reinforcement provided in the effective width. The modifications for edge, corner, or interior slab-column connections in 13.5.3.3 are permitted only when the reinforcement ratio (within the effective width) required to develop the unbalanced moment $\gamma_f M_u$ does not exceed $\mathbf{0.375}\rho_b$. The use of Eq. (13-1) without the modification permitted in 13.5.3.3 will generally indicate overstress conditions on the joint. The provisions of 13.5.3.3 are intended to improve ductile behavior of the column-slab joint. When a reversal of moments occurs at opposite faces of an interior support, both top and bottom reinforcement should be concentrated within the effective width. A ratio of top to bottom reinforcement of about 2 has been observed to be appropriate.

13.5.3.4 — Concentration of reinforcement over the column by closer spacing or additional reinforcement shall be used to resist moment on the effective slab width defined in 13.5.3.2.

13.5.4 — Design for transfer of load from slabs to supporting columns or walls through shear and torsion shall be in accordance with Chapter 11.

13.6 — Direct design method

R13.6 — Direct design method

The direct design method consists of a set of rules for distributing moments to slab and beam sections to satisfy safety requirements and most serviceability requirements simultaneously. Three fundamental steps are involved as follows:

(1) Determination of the total factored static moment (see 13.6.2);

CODE

COMMENTARY

(2) Distribution of the total factored static moment to negative and positive sections (see 13.6.3);

(3) Distribution of the negative and positive factored moments to the column and middle strips and to the beams, if any (see 13.6.4 through 13.6.6). The distribution of moments to column and middle strips is also used in the equivalent frame method (see 13.7).

13.6.1 — Limitations

Design of slab systems within the limitations of 13.6.1.1 through 13.6.1.8 by the direct design method shall be permitted.

R13.6.1 — Limitations

The direct design method was developed from considerations of theoretical procedures for the determination of moments in slabs with and without beams, requirements for simple design and construction procedures, and precedents supplied by performance of slab systems. Consequently, the slab systems to be designed using the direct design method should conform to the limitations in this section.

13.6.1.1 — There shall be a minimum of three continuous spans in each direction.

R13.6.1.1 — The primary reason for the limitation in this section is the magnitude of the negative moments at the interior support in a structure with only two continuous spans. The rules given for the direct design method assume that the slab system at the first interior negative moment section is neither fixed against rotation nor discontinuous.

13.6.1.2 — Panels shall be rectangular, with a ratio of longer to shorter span center-to-center of supports within a panel not greater than 2.

R13.6.1.2 — If the ratio of the two spans (long span/short span) of a panel exceeds two, the slab resists the moment in the shorter span essentially as a one-way slab.

13.6.1.3 — Successive span lengths center-to-center of supports in each direction shall not differ by more than one-third the longer span.

R13.6.1.3 — The limitation in this section is related to the possibility of developing negative moments beyond the point where negative moment reinforcement is terminated, as prescribed in Fig. 13.3.8.

13.6.1.4 — Offset of columns by a maximum of 10 percent of the span (in direction of offset) from either axis between centerlines of successive columns shall be permitted.

R13.6.1.4 — Columns can be offset within specified limits from a regular rectangular array. A cumulative total offset of 20 percent of the span is established as the upper limit.

13.6.1.5 — All loads shall be due to gravity only and uniformly distributed over an entire panel. Live load shall not exceed two times dead load.

R13.6.1.5 — The direct design method is based on tests[13.16] for uniform gravity loads and resulting column reactions determined by statics. Lateral loads such as wind or seismic require a frame analysis. Inverted foundation mats designed as two-way slabs (see 15.10) involve application of known column loads. Therefore, even where the soil reaction is assumed to be uniform, a frame analysis should be performed.

In the 1995 code, the limit of applicability of the direct design method for ratios of live load to dead load was reduced from 3 to 2. In most slab systems, the live to dead load ratio will be less than 2 and it will not be necessary to check the effects of pattern loading.

13.6.1.6 — For a panel with beams between supports on all sides, Eq. (13-2) shall be satisfied for beams in the two perpendicular directions

R13.6.1.6 — The elastic distribution of moments will deviate significantly from those assumed in the direct design method unless the requirements for stiffness are satisfied.

CODE

$$0.2 \le \frac{\alpha_{f1}\ell_2^2}{\alpha_{f2}\ell_1^2} \le 5.0 \qquad (13\text{-}2)$$

where α_{f1} and α_{f2} are calculated in accordance with Eq. (13-3).

$$\alpha_f = \frac{E_{cb}I_b}{E_{cs}I_s} \qquad (13\text{-}3)$$

13.6.1.7 — Moment redistribution as permitted by 8.4 shall not be applied for slab systems designed by the Direct Design Method. See 13.6.7.

13.6.1.8 — Variations from the limitations of 13.6.1 shall be permitted if demonstrated by analysis that requirements of 13.5.1 are satisfied.

13.6.2 — Total factored static moment for a span

13.6.2.1 — Total factored static moment, M_o, for a span shall be determined in a strip bounded laterally by centerline of panel on each side of centerline of supports.

13.6.2.2 — Absolute sum of positive and average negative factored moments in each direction shall not be less than

$$M_o = \frac{q_u\ell_2\ell_n^2}{8} \qquad (13\text{-}4)$$

where ℓ_n is length of clear span in direction that moments are being determined.

13.6.2.3 — Where the transverse span of panels on either side of the centerline of supports varies, ℓ_2 in Eq. (13-4) shall be taken as the average of adjacent transverse spans.

13.6.2.4 — When the span adjacent and parallel to an edge is being considered, the distance from edge to panel centerline shall be substituted for ℓ_2 in Eq. (13-4).

COMMENTARY

R13.6.1.7 — Moment redistribution as permitted by 8.4 is not intended for use where approximate values for bending moments are used. For the direct design method, 10 percent modification is allowed by 13.6.7.

R13.6.1.8 — The designer is permitted to use the direct design method even if the structure does not fit the limitations in this section, provided it can be shown by analysis that the particular limitation does not apply to that structure. For a slab system carrying a nonmovable load (such as a water reservoir in which the load on all panels is expected to be the same), the designer need not satisfy the live load limitation of 13.6.1.5.

R13.6.2 — Total factored static moment for a span

R13.6.2.2 — Eq. (13-4) follows directly from Nichol's derivation[13.17] with the simplifying assumption that the reactions are concentrated along the faces of the support perpendicular to the span considered. In general, the designer will find it expedient to calculate static moments for two adjacent half panels that include a column strip with a half middle strip along each side.

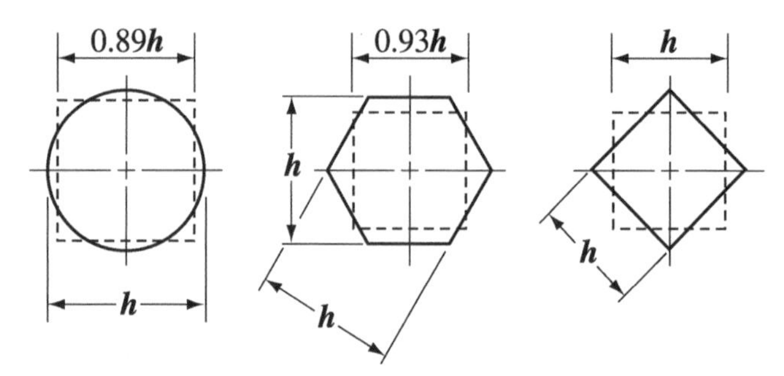

Fig. R13.6.2.5—Examples of equivalent square section for supporting members

CODE

13.6.2.5 — Clear span ℓ_n shall extend from face to face of columns, capitals, brackets, or walls. Value of ℓ_n used in Eq. (13-4) shall not be less than $\mathbf{0.65\ell_1}$. Circular or regular polygon shaped supports shall be treated as square supports with the same area.

COMMENTARY

R13.6.2.5 — If a supporting member does not have a rectangular cross section or if the sides of the rectangle are not parallel to the spans, it is to be treated as a square support having the same area, as illustrated in Fig. R13.6.2.5.

13.6.3 — Negative and positive factored moments

R13.6.3 — Negative and positive factored moments

13.6.3.1 — Negative factored moments shall be located at face of rectangular supports. Circular or regular polygon shaped supports shall be treated as square supports with the same area.

13.6.3.2 — In an interior span, total static moment, M_o, shall be distributed as follows:

Negative factored moment................................ 0.65

Positive factored moment................................. 0.35

13.6.3.3 — In an end span, total factored static moment, M_o, shall be distributed as follows:

	(1)	(2)	(3)	(4)	(5)
			Slab without beams between interior supports		
	Exterior edge unrestrained	Slab with beams between all supports	Without edge beam	With edge beam	Exterior edge fully restrained
Interior negative factored moment	0.75	0.70	0.70	0.70	0.65
Positive factored moment	0.63	0.57	0.52	0.50	0.35
Exterior negative factored moment	0	0.16	0.26	0.30	0.65

R13.6.3.3 — The moment coefficients for an end span are based on the equivalent column stiffness expressions from References 13.18, 13.19, and 13.20. The coefficients for an unrestrained edge would be used, for example, if the slab were simply supported on a masonry or concrete wall. Those for a fully restrained edge would apply if the slab were constructed integrally with a concrete wall having a flexural stiffness so large compared to that of the slab that little rotation occurs at the slab-to-wall connection.

For other than unrestrained or fully restrained edges, coefficients in the table were selected to be near the upper bound of the range for positive moments and interior negative moments. As a result, exterior negative moments were usually closer to a lower bound. The exterior negative moment capacity for most slab systems is governed by minimum reinforcement to control cracking. The final coefficients in the table have been adjusted so that the absolute sum of the positive and average moments equal M_o.

For two-way slab systems with beams between supports on all sides (two-way slabs), moment coefficients of column (2) of the table apply. For slab systems without beams between interior supports (flat plates and flat slabs), the moment coefficients of column (3) or (4) apply, without or with an edge (spandrel) beam, respectively.

In the 1977 code, distribution factors defined as a function of the stiffness ratio of the equivalent exterior support were used for proportioning the total static moment M_o in an end span. The approach may be used in place of values in 13.6.3.3.

13.6.3.4 — Negative moment sections shall be designed to resist the larger of the two interior negative factored moments determined for spans framing into a common support unless an analysis is made to distribute the unbalanced moment in accordance with stiffnesses of adjoining elements.

R13.6.3.4 — The differences in slab moment on either side of a column or other type of support should be accounted for in the design of the support. If an analysis is made to distribute unbalanced moments, flexural stiffness may be obtained on the basis of the gross concrete section of the members involved.

CODE

13.6.3.5 — Edge beams or edges of slab shall be proportioned to resist in torsion their share of exterior negative factored moments.

13.6.3.6 — The gravity load moment to be transferred between slab and edge column in accordance with 13.5.3.1 shall be $\mathbf{0.3}M_o$.

13.6.4 — Factored moments in column strips

13.6.4.1 — Column strips shall be proportioned to resist the following portions in percent of interior negative factored moments:

ℓ_2/ℓ_1	0.5	1.0	2.0
$(\alpha_{f1}\ell_2/\ell_1) = 0$	75	75	75
$(\alpha_{f1}\ell_2/\ell_1) \geq 1.0$	90	75	45

Linear interpolations shall be made between values shown.

13.6.4.2 — Column strips shall be proportioned to resist the following portions in percent of exterior negative factored moments:

ℓ_2/ℓ_1		0.5	1.0	2.0
$(\alpha_{f1}\ell_2/\ell_1) = 0$	$\beta_t = 0$	100	100	100
	$\beta_t \geq 2.5$	75	75	75
$(\alpha_{f1}\ell_2/\ell_1) \geq 1.0$	$\beta_t = 0$	100	100	100
	$\beta_t \geq 2.5$	90	75	45

Linear interpolations shall be made between values shown, where β_t is calculated in Eq. (13-5) and C is calculated in Eq. (13-6).

$$\beta_t = \frac{E_{cb}C}{2E_{cs}I_s} \qquad (13\text{-}5)$$

$$C = \sum\left(1 - 0.63\frac{x}{y}\right)\frac{x^3y}{3} \qquad (13\text{-}6)$$

The constant C for T- or L-sections shall be permitted to be evaluated by dividing the section into separate rectangular parts, as defined in 13.2.4, and summing the values of C for each part.

13.6.4.3 — Where supports consist of columns or walls extending for a distance equal to or greater than **(3/4)**ℓ_2 used to compute M_o, negative moments shall be considered to be uniformly distributed across ℓ_2.

COMMENTARY

R13.6.3.5 — Moments perpendicular to, and at the edge of, the slab structure should be transmitted to the supporting columns or walls. Torsional stresses caused by the moment assigned to the slab should be investigated.

R13.6.4, R13.6.5, and R13.6.6 — Factored moments in column strips, beams, and middle strips

The rules given for assigning moments to the column strips, beams, and middle strips are based on studies[13.21] of moments in linearly elastic slabs with different beam stiffness tempered by the moment coefficients that have been used successfully.

For the purpose of establishing moments in the half column strip adjacent to an edge supported by a wall, ℓ_n in Eq. (13-4) may be assumed equal to ℓ_n of the parallel adjacent column to column span, and the wall may be considered as a beam having a moment of inertia I_b equal to infinity.

R13.6.4.2 — The effect of the torsional stiffness parameter β_t is to assign all of the exterior negative factored moment to the column strip, and none to the middle strip, unless the beam torsional stiffness is high relative to the flexural stiffness of the supported slab. In the definition of β_t, the shear modulus has been taken as $E_{cb}/2$.

Where walls are used as supports along column lines, they can be regarded as very stiff beams with an $\alpha_{f1}\ell_2/\ell_1$ value greater than one. Where the exterior support consists of a wall perpendicular to the direction in which moments are being determined, β_t may be taken as zero if the wall is of masonry without torsional resistance, and β_t may be taken as 2.5 for a concrete wall with great torsional resistance that is monolithic with the slab.

CODE

13.6.4.4 — Column strips shall be proportioned to resist the following portions in percent of positive factored moments:

ℓ_2/ℓ_1	0.5	1.0	2.0
$(\alpha_{f1}\ell_2/\ell_1) = 0$	60	60	60
$(\alpha_{f1}\ell_2/\ell_1) \geq 1.0$	90	75	45

Linear interpolations shall be made between values shown.

13.6.4.5 — For slabs with beams between supports, the slab portion of column strips shall be proportioned to resist that portion of column strip moments not resisted by beams.

13.6.5 — Factored moments in beams

13.6.5.1 — Beams between supports shall be proportioned to resist 85 percent of column strip moments if $\alpha_{f1}\ell_2/\ell_1$ is equal to or greater than 1.0.

13.6.5.2 — For values of $\alpha_{f1}\ell_2/\ell_1$ between 1.0 and zero, proportion of column strip moments resisted by beams shall be obtained by linear interpolation between 85 and zero percent.

13.6.5.3 — In addition to moments calculated for uniform loads according to 13.6.2.2, 13.6.5.1, and 13.6.5.2, beams shall be proportioned to resist all moments caused by concentrated or linear loads applied directly to beams, including weight of projecting beam stem above or below the slab.

13.6.6 — Factored moments in middle strips

13.6.6.1 — That portion of negative and positive factored moments not resisted by column strips shall be proportionately assigned to corresponding half middle strips.

13.6.6.2 — Each middle strip shall be proportioned to resist the sum of the moments assigned to its two half middle strips.

13.6.6.3 — A middle strip adjacent to and parallel with a wall-supported edge shall be proportioned to resist twice the moment assigned to the half middle strip corresponding to the first row of interior supports.

13.6.7 — Modification of factored moments

Modification of negative and positive factored moments by 10 percent shall be permitted provided the total static moment for a panel, M_o, in the direction considered is not less than that required by Eq. (13-4).

COMMENTARY

R13.6.5 — Factored moments in beams

Loads assigned directly to beams are in addition to the uniform dead load of the slab; uniform superimposed dead loads, such as the ceiling, floor finish, or assumed equivalent partition loads; and uniform live loads. All of these loads are normally included with q_u in Eq. (13-4). Linear loads applied directly to beams include partition walls over or along beam centerlines and additional dead load of the projecting beam stem. Concentrated loads include posts above or hangers below the beams. For the purpose of assigning directly applied loads, only loads located within the width of the beam stem should be considered as directly applied to the beams. (The effective width of a beam as defined in 13.2.4 is solely for strength and relative stiffness calculations.) Line loads and concentrated loads located on the slab away from the beam stem require special consideration to determine their apportionment to slab and beams.

CODE

13.6.8 — Factored shear in slab systems with beams

13.6.8.1 — Beams with $\alpha_{f1}\ell_2/\ell_1$ equal to or greater than 1.0 shall be proportioned to resist shear caused by factored loads on tributary areas which are bounded by 45 degree lines drawn from the corners of the panels and the centerlines of the adjacent panels parallel to the long sides.

13.6.8.2 — In proportioning beams with $\alpha_{f1}\ell_2/\ell_1$ less than 1.0 to resist shear, linear interpolation, assuming beams carry no load at α_{f1} = **0**, shall be permitted.

13.6.8.3 — In addition to shears calculated according to 13.6.8.1 and 13.6.8.2, beams shall be proportioned to resist shears caused by factored loads applied directly on beams.

13.6.8.4 — Computation of slab shear strength on the assumption that load is distributed to supporting beams in accordance with 13.6.8.1 or 13.6.8.2 shall be permitted. Resistance to total shear occurring on a panel shall be provided.

13.6.8.5 — Shear strength shall satisfy the requirements of Chapter 11.

COMMENTARY

R13.6.8 — Factored shear in slab systems with beams

The tributary area for computing shear on an interior beam is shown shaded in Fig. R13.6.8. If the stiffness for the beam $\alpha_{f1}\ell_2/\ell_1$ is less than 1.0, the shear on the beam may be obtained by linear interpolation. In such cases, the beams framing into the column will not account for all of the shear force applied on the column. The remaining shear force will produce shear stresses in the slab around the column that should be checked in the same manner as for flat slabs, as required by 13.6.8.4. Sections 13.6.8.1 through 13.6.8.3 do not apply to the calculation of torsional moments on the beams. These moments should be based on the calculated flexural moments acting on the sides of the beam.

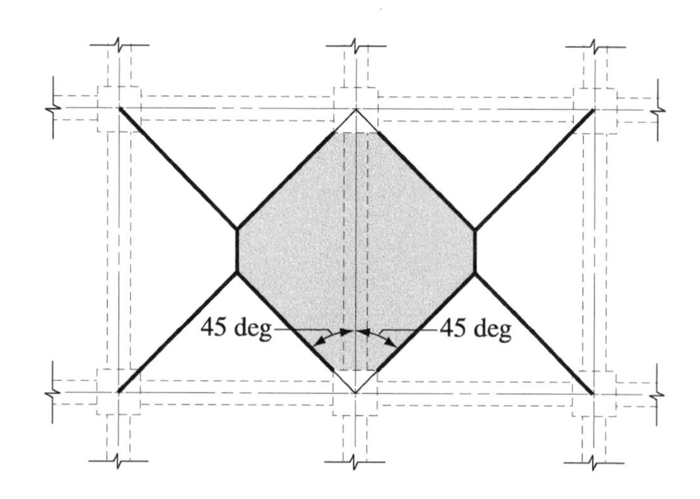

Fig. R13.6.8—Tributary area for shear on an interior beam

13.6.9 — Factored moments in columns and walls

13.6.9.1 — Columns and walls built integrally with a slab system shall resist moments caused by factored loads on the slab system.

13.6.9.2 — At an interior support, supporting elements above and below the slab shall resist the factored moment specified by Eq. (13-7) in direct proportion to their stiffnesses unless a general analysis is made.

$$M_u = 0.07[(q_{Du} + 0.5q_{Lu})\ell_2\ell_n^2 - q_{Du}'\ell_2'(\ell_n')^2] \quad (13\text{-}7)$$

where q_{Du}', ℓ_2', and ℓ_n' refer to shorter span.

R13.6.9 — Factored moments in columns and walls

Eq. (13-7) refers to two adjoining spans, with one span longer than the other, and with full dead load plus one-half live load applied on the longer span and only dead load applied on the shorter span.

Design and detailing of the reinforcement transferring the moment from the slab to the edge column is critical to both the performance and the safety of flat slabs or flat plates without edge beams or cantilever slabs. It is important that complete design details be shown on design drawings, such as concentration of reinforcement over the column by closer spacing or additional reinforcement.

CODE

13.7 — Equivalent frame method

13.7.1 — Design of slab systems by the equivalent frame method shall be based on assumptions given in 13.7.2 through 13.7.6, and all sections of slabs and supporting members shall be proportioned for moments and shears thus obtained.

13.7.1.1 — Where metal column capitals are used, it shall be permitted to take account of their contributions to stiffness and resistance to moment and to shear.

13.7.1.2 — It shall be permitted to neglect the change in length of columns and slabs due to direct stress, and deflections due to shear.

13.7.2 — Equivalent frame

13.7.2.1 — The structure shall be considered to be made up of equivalent frames on column lines taken longitudinally and transversely through the building.

13.7.2.2 — Each frame shall consist of a row of columns or supports and slab-beam strips, bounded laterally by the centerline of panel on each side of the centerline of columns or supports.

13.7.2.3 — Columns or supports shall be assumed to be attached to slab-beam strips by torsional members (see 13.7.5) transverse to the direction of the span for which moments are being determined and extending to bounding lateral panel centerlines on each side of a column.

13.7.2.4 — Frames adjacent and parallel to an edge shall be bounded by that edge and the centerline of adjacent panel.

13.7.2.5 — Analysis of each equivalent frame in its entirety shall be permitted. Alternatively, for gravity loading, a separate analysis of each floor or roof with far ends of columns considered fixed shall be permitted.

13.7.2.6 — Where slab-beams are analyzed separately, determination of moment at a given support assuming that the slab-beam is fixed at any support two panels distant therefrom, shall be permitted, provided the slab continues beyond that point.

COMMENTARY

R13.7 — Equivalent frame method

The equivalent frame method involves the representation of the three-dimensional slab system by a series of two-dimensional frames that are then analyzed for loads acting in the plane of the frames. The negative and positive moments so determined at the critical design sections of the frame are distributed to the slab sections in accordance with 13.6.4 (column strips), 13.6.5 (beams), and 13.6.6 (middle strips). The equivalent frame method is based on studies reported in References 13.18, 13.19, and 13.20. Many of the details of the equivalent frame method given in the Commentary in the 1989 code were removed in the 1995 code.

R13.7.2 — Equivalent frame

Application of the equivalent frame to a regular structure is illustrated in Fig. R13.7.2. The three-dimensional building is divided into a series of two-dimensional frame bents (equivalent frames) centered on column or support centerlines with each frame extending the full height of the building. The width of each equivalent frame is bounded by the centerlines of the adjacent panels. The complete analysis of a slab system for a building consists of analyzing a series of equivalent (interior and exterior) frames spanning longitudinally and transversely through the building.

The equivalent frame comprises three parts: (1) the horizontal slab strip, including any beams spanning in the direction of the frame, (2) the columns or other vertical supporting members, extending above and below the slab, and (3) the elements of the structure that provide moment transfer between the horizontal and vertical members.

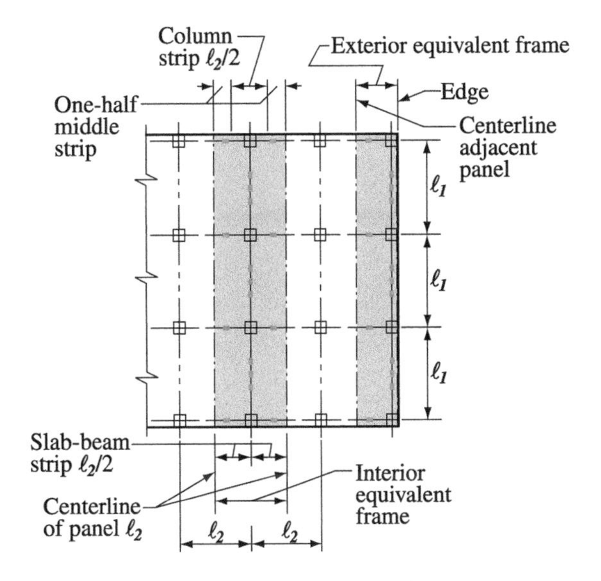

Fig. R13.7.2—Definitions of equivalent frame.

CODE

13.7.3 — Slab-beams

13.7.3.1—Determination of the moment of inertia of slab-beams at any cross section outside of joints or column capitals using the gross area of concrete shall be permitted.

13.7.3.2 — Variation in moment of inertia along axis of slab-beams shall be taken into account.

13.7.3.3 — Moment of inertia of slab-beams from center of column to face of column, bracket, or capital shall be assumed equal to the moment of inertia of the slab-beam at face of column, bracket, or capital divided by the quantity $(1 - c_2/\ell_2)^2$, where c_2 and ℓ_2 are measured transverse to the direction of the span for which moments are being determined.

13.7.4 — Columns

13.7.4.1 — Determination of the moment of inertia of columns at any cross section outside of joints or column capitals using the gross area of concrete shall be permitted.

13.7.4.2 — Variation in moment of inertia along axis of columns shall be taken into account.

13.7.4.3 — Moment of inertia of columns from top to bottom of the slab-beam at a joint shall be assumed to be infinite.

COMMENTARY

R13.7.3 — Slab-beams

R13.7.3.3 — A support is defined as a column, capital, bracket, or wall. A beam is not considered to be a support member for the equivalent frame.

R13.7.4 — Columns

Column stiffness is based on the length of the column from mid-depth of slab above to mid-depth of slab below. Column moment of inertia is computed on the basis of its cross section, taking into account the increase in stiffness provided by the capital, if any.

When slab-beams are analyzed separately for gravity loads, the concept of an equivalent column, combining the stiffness of the slab-beam and torsional member into a composite element, is used. The column flexibility is modified to account for the torsional flexibility of the slab-to-column connection that reduces its efficiency for transmission of moments. The equivalent column consists of the actual columns above and below the slab-beam, plus attached torsional members on each side of the columns extending to the centerline of the adjacent panels as shown in Fig. R13.7.4.

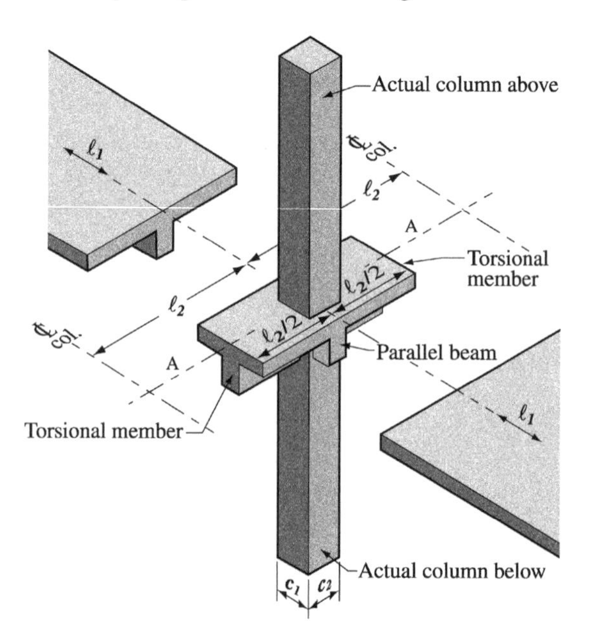

Fig. R13.7.4—Equivalent column (column plus torsional members)

CODE

13.7.5 — Torsional members

13.7.5.1 — Torsional members (see 13.7.2.3) shall be assumed to have a constant cross section throughout their length consisting of the largest of (a), (b), and (c):

(a) A portion of slab having a width equal to that of the column, bracket, or capital in the direction of the span for which moments are being determined;

(b) For monolithic or fully composite construction, the portion of slab specified in (a) plus that part of the transverse beam above and below the slab;

(c) The transverse beam as defined in 13.2.4.

13.7.5.2 — Where beams frame into columns in the direction of the span for which moments are being determined, the torsional stiffness shall be multiplied by the ratio of the moment of inertia of the slab with such a beam to the moment of inertia of the slab without such a beam.

COMMENTARY

R13.7.5 — Torsional members

Computation of the stiffness of the torsional member requires several simplifying assumptions. If no transverse-beam frames into the column, a portion of the slab equal to the width of the column or capital is assumed to be the torsional member. If a beam frames into the column, T-beam or L-beam action is assumed, with the flanges extending on each side of the beam a distance equal to the projection of the beam above or below the slab but not greater than four times the thickness of the slab. Furthermore, it is assumed that no torsional rotation occurs in the beam over the width of the support.

The member sections to be used for calculating the torsional stiffness are defined in 13.7.5.1. Up to the 1989 code, Eq. (13-6) specified the stiffness coefficient K_t of the torsional members. In 1995, the approximate expression for K_t was moved to the commentary.

Studies of three-dimensional analyses of various slab configurations suggest that a reasonable value of the torsional stiffness can be obtained by assuming a moment distribution along the torsional member that varies linearly from a maximum at the center of the column to zero at the middle of the panel. The assumed distribution of unit twisting moment along the column centerline is shown in Fig. R13.7.5.

An approximate expression for the stiffness of the torsional member, based on the results of three-dimensional analyses of various slab configurations (References 13.18, 13.19, and 13.20) is given below as

$$K_t = \sum \frac{9E_{cs}C}{\ell_2\left(1 - \frac{c_2}{\ell_2}\right)^3}$$

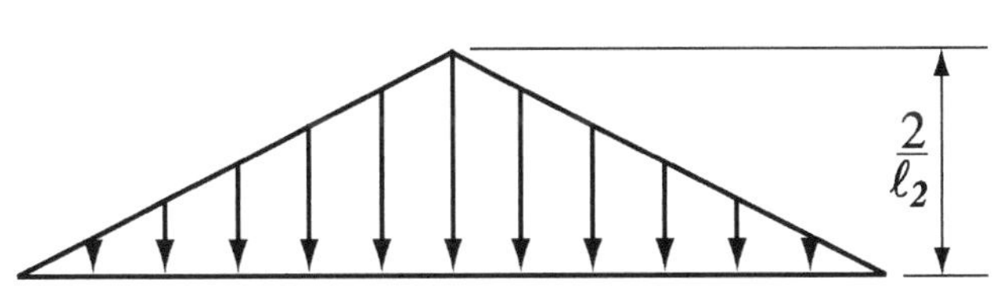

Fig. R13.7.5—Distribution of unit twisting moment along column centerline AA shown in Fig. R13.7.4

CODE

13.7.6 — Arrangement of live load

13.7.6.1 — When the loading pattern is known, the equivalent frame shall be analyzed for that load.

13.7.6.2 — When live load is variable but does not exceed three-quarters of the dead load, or the nature of live load is such that all panels will be loaded simultaneously, it shall be permitted to assume that maximum factored moments occur at all sections with full factored live load on entire slab system.

13.7.6.3 — For loading conditions other than those defined in 13.7.6.2, it shall be permitted to assume that maximum positive factored moment near midspan of a panel occurs with three-quarters of the full factored live load on the panel and on alternate panels; and it shall be permitted to assume that maximum negative factored moment in the slab at a support occurs with three-quarters of the full live load on adjacent panels only.

13.7.6.4 — Factored moments shall be taken not less than those occurring with full factored live load on all panels.

13.7.7 — Factored moments

13.7.7.1 — At interior supports, the critical section for negative factored moment (in both column and middle strips) shall be taken at face of rectilinear supports, but not farther away than $0.175\ell_1$ from the center of a column.

13.7.7.2 — At exterior supports with brackets or capitals, the critical section for negative factored moment in the span perpendicular to an edge shall be taken at a distance from face of supporting element not greater than one-half the projection of bracket or capital beyond face of supporting element.

13.7.7.3 — Circular or regular polygon shaped supports shall be treated as square supports with the same area for location of critical section for negative design moment.

13.7.7.4 — Where slab systems within limitations of 13.6.1 are analyzed by the equivalent frame method, it shall be permitted to reduce the resulting computed moments in such proportion that the absolute sum of the positive and average negative moments used in design need not exceed the value obtained from Eq. (13-4).

13.7.7.5 — Distribution of moments at critical sections across the slab-beam strip of each frame to column strips, beams, and middle strips as provided in 13.6.4, 13.6.5, and 13.6.6 shall be permitted if the requirement of 13.6.1.6 is satisfied.

COMMENTARY

R13.7.6 — Arrangement of live load

The use of only three-quarters of the full factored live load for maximum moment loading patterns is based on the fact that maximum negative and maximum positive live load moments cannot occur simultaneously and that redistribution of maximum moments is thus possible before failure occurs. This procedure, in effect, permits some local overstress under the full factored live load if it is distributed in the prescribed manner, but still ensures that the ultimate capacity of the slab system after redistribution of moment is not less than that required to carry the full factored dead and live loads on all panels.

R13.7.7 — Factored moments

R13.7.7.1-R13.7.7.3 — These code sections adjust the negative factored moments to the face of the supports. The adjustment is modified at an exterior support to limit reductions in the exterior negative moment. Fig. R13.6.2.5 illustrates several equivalent rectangular supports for use in establishing faces of supports for design with nonrectangular supports.

R13.7.7.4 — Previous codes have contained this section. It is based on the principle that if two different methods are prescribed to obtain a particular answer, the code should not require a value greater than the least acceptable value. Due to the long satisfactory experience with designs having total factored static moments not exceeding those given by Eq. (13-4), it is considered that these values are satisfactory for design when applicable limitations are met.

CHAPTER 14 — WALLS

CODE

14.1 — Scope

14.1.1 — Provisions of Chapter 14 shall apply for design of walls subjected to axial load, with or without flexure.

14.1.2 — Cantilever retaining walls are designed according to flexural design provisions of Chapter 10 with minimum horizontal reinforcement according to 14.3.3.

14.2 — General

14.2.1 — Walls shall be designed for eccentric loads and any lateral or other loads to which they are subjected.

14.2.2 — Walls subject to axial loads shall be designed in accordance with 14.2, 14.3, and either 14.4, 14.5, or 14.8.

14.2.3 — Design for shear shall be in accordance with 11.10.

14.2.4 — Unless otherwise demonstrated by an analysis, the horizontal length of wall considered as effective for each concentrated load shall not exceed center-to-center distance between loads, nor the bearing width plus four times the wall thickness.

14.2.5 — Compression members built integrally with walls shall conform to 10.8.2.

14.2.6 — Walls shall be anchored to intersecting elements, such as floors and roofs; or to columns, pilasters, buttresses, of intersecting walls; and to footings.

14.2.7 — Quantity of reinforcement and limits of thickness required by 14.3 and 14.5 shall be permitted to be waived where structural analysis shows adequate strength and stability.

14.2.8 — Transfer of force to footing at base of wall shall be in accordance with 15.8.

COMMENTARY

R14.1 — Scope

Chapter 14 applies generally to walls as vertical load carrying members. Cantilever retaining walls are designed according to the flexural design provisions of Chapter 10. Walls designed to resist shear forces, such as shearwalls, should be designed in accordance with Chapter 14 and 11.10 as applicable.

In the 1977 code, walls could be designed according to Chapter 14 or 10.15. In the 1983 code these two were combined in Chapter 14.

R14.2 — General

Walls should be designed to resist all loads to which they are subjected, including eccentric axial loads and lateral forces. Design is to be carried out in accordance with 14.4 unless the wall meets the requirements of 14.5.1.

CODE

14.3 — Minimum reinforcement

14.3.1 — Minimum vertical and horizontal reinforcement shall be in accordance with 14.3.2 and 14.3.3 unless a greater amount is required for shear by 11.10.8 and 11.10.9.

14.3.2 — Minimum ratio of vertical reinforcement area to gross concrete area, ρ_ℓ, shall be:

(a) 0.0012 for deformed bars not larger than No. 5 with f_y not less than 60,000 psi; or

(b) 0.0015 for other deformed bars; or

(c) 0.0012 for welded wire reinforcement not larger than W31 or D31.

14.3.3 — Minimum ratio of horizontal reinforcement area to gross concrete area, ρ_t, shall be:

(a) 0.0020 for deformed bars not larger than No. 5 with f_y not less than 60,000 psi; or

(b) 0.0025 for other deformed bars; or

(c) 0.0020 for welded wire reinforcement not larger than W31 or D31.

14.3.4 — Walls more than 10 in. thick, except basement walls, shall have reinforcement for each direction placed in two layers parallel with faces of wall in accordance with the following:

(a) One layer consisting of not less than one-half and not more than two-thirds of total reinforcement required for each direction shall be placed not less than 2 in. nor more than one-third the thickness of wall from the exterior surface;

(b) The other layer, consisting of the balance of required reinforcement in that direction, shall be placed not less than 3/4 in. nor more than one-third the thickness of wall from the interior surface.

14.3.5 — Vertical and horizontal reinforcement shall not be spaced farther apart than three times the wall thickness, nor farther apart than 18 in.

14.3.6 — Vertical reinforcement need not be enclosed by lateral ties if vertical reinforcement area is not greater than 0.01 times gross concrete area, or where vertical reinforcement is not required as compression reinforcement.

COMMENTARY

R14.3 — Minimum reinforcement

The requirements of 14.3 are similar to those in previous codes. These apply to walls designed according to 14.4, 14.5, or 14.8. For walls resisting horizontal shear forces in the plane of the wall, reinforcement designed according to 11.10.9.2 and 11.10.9.4 may exceed the minimum reinforcement in 14.3.

The notation used to identify the direction of the distributed reinforcement in walls was updated in 2005 to eliminate conflicts between the notation used for ordinary structural walls in Chapters 11 and 14 and the notation used for special structural walls in Chapter 21. The distributed reinforcement is now identified as being oriented parallel to either the longitudinal or transverse axis of the wall. Therefore, for vertical wall segments, the notation used to describe the horizontal distributed reinforcement ratio is ρ_t, and the notation used to describe the vertical distributed reinforcement ratio is ρ_ℓ.

CODE

14.3.7 — In addition to the minimum reinforcement required by 14.3.1, not less than two No. 5 bars shall be provided around all window and door openings. Such bars shall be extended to develop the bar beyond the corners of the openings but not less than 24 in.

14.4 — Walls designed as compression members

Except as provided in 14.5, walls subject to axial load or combined flexure and axial load shall be designed as compression members in accordance with provisions of 10.2, 10.3, 10.10, 10.11, 10.12, 10.13, 10.14, 10.17, 14.2, and 14.3.

14.5 — Empirical design method

14.5.1 — Walls of solid rectangular cross section shall be permitted to be designed by the empirical provisions of 14.5 if the resultant of all factored loads is located within the middle third of the overall thickness of the wall and all limits of 14.2, 14.3, and 14.5 are satisfied.

14.5.2 — Design axial strength ϕP_n of a wall satisfying limitations of 14.5.1 shall be computed by Eq. (14-1) unless designed in accordance with 14.4.

$$\phi P_n = 0.55\phi f_c' A_g\left[1-\left(\frac{k\ell_c}{32h}\right)^2\right] \qquad (14\text{-}1)$$

where ϕ shall correspond to compression-controlled sections in accordance with 9.3.2.2 and effective length factor k shall be:

For walls braced top and bottom against lateral translation and

(a) Restrained against rotation at one or both ends (top, bottom, or both) 0.8

(b) Unrestrained against rotation at both ends ... 1.0

For walls not braced against lateral translation....... 2.0

COMMENTARY

R14.5 — Empirical design method

The empirical design method applies only to solid rectangular cross sections. All other shapes should be designed according to 14.4.

Eccentric loads and lateral forces are used to determine the total eccentricity of the factored axial force P_u. When the resultant load for all applicable load combinations falls within the middle third of the wall thickness (eccentricity not greater than $h/6$) at all sections along the length of the undeformed wall, the empirical design method may be used. The design is then carried out considering P_u as the concentric load. The factored axial force P_u should be less than or equal to the design axial strength ϕP_n computed by Eq. (14-1), $P_u \leq \phi P_n$.

With the 1980 code supplement, (Eq. 14-1) was revised to reflect the general range of end conditions encountered in wall designs. The wall strength equation in the 1977 code was based on the assumption of a wall with top and bottom fixed against lateral movement, and with moment restraint at one end corresponding to an effective length factor between 0.8 and 0.9. Axial load strength values determined from the original equation were unconservative when compared to test results[14.1] for walls with pinned conditions at both ends, as occurs with some precast and tilt-up applications, or when the top of the wall is not effectively braced against translation, as occurs with free-standing walls or in large structures where significant roof diaphragm deflections occur due to wind and seismic loads. Eq. (14-1) gives the same results as the 1977 code for walls braced against translation and with reasonable base restraint against rotation.[14.2] Values of effective length factors k are given for commonly occurring wall end conditions. The end condition "restrained against rotation" required for a k of 0.8 implies attachment to a member having flexural stiffness EI/ℓ at least as large as that of the wall.

CODE

COMMENTARY

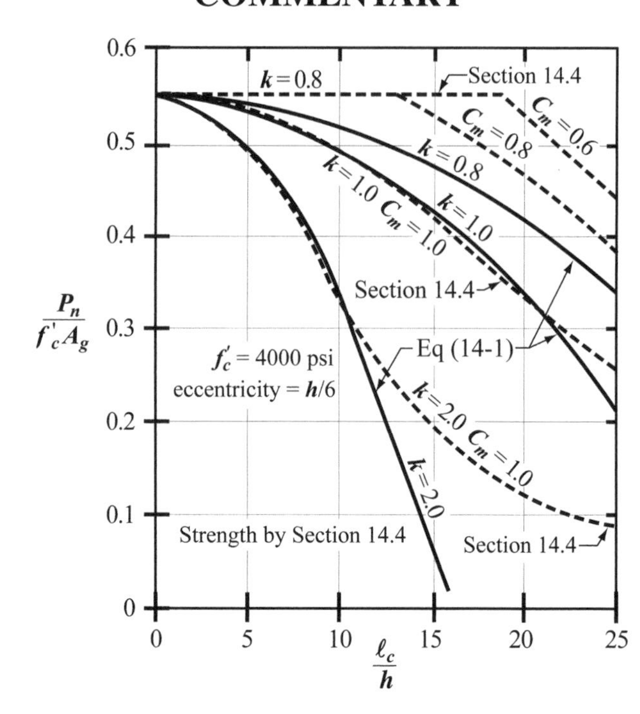

Fig. R14.5—Empirical design of walls, Eq. (14-1) versus 14.4.

The slenderness portion of Eq. (14-1) results in relatively comparable strengths by either 14.3 or 14.4 for members loaded at the middle third of the thickness with different braced and restrained end conditions. See Fig. R14.5.

14.5.3 — Minimum thickness of walls designed by empirical design method

14.5.3.1 — Thickness of bearing walls shall not be less than 1/25 the supported height or length, whichever is shorter, nor less than 4 in.

14.5.3.2 — Thickness of exterior basement walls and foundation walls shall not be less than 7-1/2 in.

R14.5.3 — Minimum thickness of walls designed by empirical design method

The minimum thickness requirements need not be applied to walls designed according to 14.4.

14.6 — Nonbearing walls

14.6.1 — Thickness of nonbearing walls shall not be less than 4 in., nor less than 1/30 the least distance between members that provide lateral support.

14.7 — Walls as grade beams

14.7.1 — Walls designed as grade beams shall have top and bottom reinforcement as required for moment in accordance with provisions of 10.2 through 10.7. Design for shear shall be in accordance with provisions of Chapter 11.

14.7.2 — Portions of grade beam walls exposed above grade shall also meet requirements of 14.3.

CODE

14.8 — Alternative design of slender walls

14.8.1 — When flexural tension controls the design of a wall, the requirements of 14.8 are considered to satisfy 10.10.

14.8.2 — Walls designed by the provisions of 14.8 shall satisfy 14.8.2.1 through 14.8.2.6.

14.8.2.1 — The wall panel shall be designed as a simply supported, axially loaded member subjected to an out-of-plane uniform lateral load, with maximum moments and deflections occurring at midspan.

14.8.2.2 — The cross section shall be constant over the height of the panel.

14.8.2.3 — The wall shall be tension-controlled.

14.8.2.4 — Reinforcement shall provide a design strength

$$\phi M_n \geq M_{cr} \quad (14\text{-}2)$$

where M_{cr} shall be obtained using the modulus of rupture, f_r, given by Eq. (9-10).

14.8.2.5 — Concentrated gravity loads applied to the wall above the design flexural section shall be assumed to be distributed over a width:

(a) Equal to the bearing width, plus a width on each side that increases at a slope of 2 vertical to 1 horizontal down to the design section; but

(b) Not greater than the spacing of the concentrated loads; and

(c) Not extending beyond the edges of the wall panel.

14.8.2.6 — Vertical stress P_u/A_g at the midheight section shall not exceed $0.06f_c'$.

14.8.3 — The design moment strength ϕM_n for combined flexure and axial loads at the midheight cross section shall be

$$\phi M_n \geq M_u \quad (14\text{-}3)$$

COMMENTARY

R14.8 — Alternative design of slender walls

Section 14.8 is based on the corresponding requirements in the Uniform Building Code (UBC)[14.3] and experimental research.[14.4] The procedure is presented as an alternative to the requirements of 10.10 for the out-of-plane design of precast wall panels, where the panels are restrained against overturning at the top.

The procedure, as prescribed in the UBC,[14.3] has been converted from working stress to factored load design.

Panels that have windows or other large openings are not considered to have constant cross section over the height of the panel. Such walls are to be designed taking into account the effects of openings.

Many aspects of the design of tilt-up walls and buildings are discussed in References 14.5 and 14.6.

R14.8.2.3 — This section was updated in the 2005 code to reflect the change in design approach that was introduced in 10.3 of the 2002 code. The previous requirement that the reinforcement ratio should not exceed $0.6\rho_{bal}$ was replaced by the requirement that the wall be tension-controlled, leading to approximately the same reinforcement ratio.

CODE

COMMENTARY

where:

$$M_u = M_{ua} + P_u \Delta_u \qquad (14\text{-}4)$$

M_{ua} is the moment at the midheight section of the wall due to factored lateral and eccentric vertical loads, and Δ_u is:

$$\Delta_u = \frac{5 M_u \ell_c^2}{(0.75)48 E_c I_{cr}} \qquad (14\text{-}5)$$

M_u shall be obtained by iteration of deflections, or by direct calculation using Eq. (14-6).

$$M_u = \frac{M_{ua}}{1 - \dfrac{5 P_u \ell_c^2}{(0.75)48 E_c I_{cr}}} \qquad (14\text{-}6)$$

where:

$$I_{cr} = \frac{E_s}{E_c}\left(A_s + \frac{P_u}{f_y}\right)(d - c)^2 + \frac{\ell_w c^3}{3} \qquad (14\text{-}7)$$

and the value of E_s/E_c shall not be taken less than 6.

14.8.4 — Δ_s, maximum deflection due to service loads, including $P\Delta$ effects, shall not exceed $\ell_c/150$. At midheight, Δ_s shall be calculated by:

$$\Delta_s = \frac{(5M)\ell_c^2}{48 E_c I_e} \qquad (14\text{-}8)$$

$$M = \frac{M_{sa}}{1 - \dfrac{5 P_s \ell_c^2}{48 E_c I_e}} \qquad (14\text{-}9)$$

I_e shall be calculated using the procedure of 9.5.2.3, substituting M for M_a, and I_{cr} shall be calculated using Eq. (14-7).

CHAPTER 15 — FOOTINGS

CODE

15.1 — Scope

15.1.1 — Provisions of Chapter 15 shall apply for design of isolated footings and, where applicable, to combined footings and mats.

15.1.2 — Additional requirements for design of combined footings and mats are given in 15.10.

15.2 — Loads and reactions

15.2.1 — Footings shall be proportioned to resist the factored loads and induced reactions, in accordance with the appropriate design requirements of this code and as provided in Chapter 15.

15.2.2 — Base area of footing or number and arrangement of piles shall be determined from unfactored forces and moments transmitted by footing to soil or piles and permissible soil pressure or permissible pile capacity determined through principles of soil mechanics.

15.2.3 — For footings on piles, computations for moments and shears shall be permitted to be based on the assumption that the reaction from any pile is concentrated at pile center.

COMMENTARY

R15.1 — Scope

While the provisions of Chapter 15 apply to isolated footings supporting a single column or wall, most of the provisions are generally applicable to combined footings and mats supporting several columns or walls or a combination thereof.[15.1,15.2]

R15.2 — Loads and reactions

Footings are required to be proportioned to sustain the applied factored loads and induced reactions which include axial loads, moments, and shears that have to be resisted at the base of the footing or pile cap.

After the permissible soil pressure or the permissible pile capacity has been determined by principles of soil mechanics and in accord with the general building code, the size of the base area of a footing on soil or the number and arrangement of the piles should be established on the basis of unfactored (service) loads such as D, L, W, and E in whatever combination that governs the design.

Only the computed end moments that exist at the base of a column (or pedestal) need to be transferred to the footing; the minimum moment requirement for slenderness considerations given in 10.12.3.2 need not be considered for transfer of forces and moments to footings.

In cases in which eccentric loads or moments are to be considered, the extreme soil pressure or pile reaction obtained from this loading should be within the permissible values. Similarly, the resultant reactions due to service loads combined with moments, shears, or both, caused by wind or earthquake loads should not exceed the increased values that may be permitted by the general building code.

To proportion a footing or pile cap for strength, the contact soil pressure or pile reaction due to the applied factored loading (see 8.1.1) should be determined. For a single concentrically loaded spread footing, the soil reaction q_s due to the factored loading is $q_s = U/A_f$, where U is the factored concentric load to be resisted by the footing, and A_f is the base area of the footing as determined by the principles stated in 15.2.2 using the unfactored loads and the permissible soil pressure.

q_s is a calculated reaction to the factored loading used to produce the same required strength conditions regarding

CODE

COMMENTARY

flexure, shear, and development of reinforcement in the footing or pile cap, as in any other member.

In the case of eccentric loading, load factors may cause eccentricities and reactions that are different from those obtained by unfactored loads.

15.3 — Footings supporting circular or regular polygon shaped columns or pedestals

For location of critical sections for moment, shear, and development of reinforcement in footings, it shall be permitted to treat circular or regular polygon shaped concrete columns or pedestals as square members with the same area.

15.4 — Moment in footings

R15.4 — Moment in footings

15.4.1 — External moment on any section of a footing shall be determined by passing a vertical plane through the footing, and computing the moment of the forces acting over entire area of footing on one side of that vertical plane.

15.4.2 — Maximum factored moment, M_u, for an isolated footing shall be computed as prescribed in 15.4.1 at critical sections located as follows:

(a) At face of column, pedestal, or wall, for footings supporting a concrete column, pedestal, or wall;

(b) Halfway between middle and edge of wall, for footings supporting a masonry wall;

(c) Halfway between face of column and edge of steel base plate, for footings supporting a column with steel base plate.

15.4.3 — In one-way footings and two-way square footings, reinforcement shall be distributed uniformly across entire width of footing.

15.4.4 — In two-way rectangular footings, reinforcement shall be distributed in accordance with 15.4.4.1 and 15.4.4.2.

15.4.4.1 — Reinforcement in long direction shall be distributed uniformly across entire width of footing.

15.4.4.2 — For reinforcement in short direction, a portion of the total reinforcement, $\gamma_s A_s$, shall be distributed uniformly over a band width (centered on cen-

R15.4.4 — In previous codes, the reinforcement in the short direction of rectangular footings should be distributed so that an area of steel given by Eq. (15-1) is provided in a band width equal to the length of the short side of the footing. The band width is centered about the column centerline.

The remaining reinforcement required in the short direction is to be distributed equally over the two segments outside the band width, one-half to each segment.

CODE

terline of column or pedestal) equal to the length of short side of footing. Remainder of reinforcement required in short direction, $(1 - \gamma_s)A_s$, shall be distributed uniformly outside center band width of footing.

$$\gamma_s = \frac{2}{(\beta + 1)} \qquad (15\text{-}1)$$

where β is ratio of long to short sides of footing.

15.5 — Shear in footings

15.5.1 — Shear strength of footings supported on soil or rock shall be in accordance with 11.12.

15.5.2 — Location of critical section for shear in accordance with Chapter 11 shall be measured from face of column, pedestal, or wall, for footings supporting a column, pedestal, or wall. For footings supporting a column or pedestal with steel base plates, the critical section shall be measured from location defined in 15.4.2(c).

15.5.3 — Where the distance between the axis of any pile to the axis of the column is more than two times the distance between the top of the pile cap and the top of the pile, the pile cap shall satisfy 11.12 and 15.5.4. Other pile caps shall satisfy either Appendix A, or both 11.12 and 15.5.4. If Appendix A is used, the effective concrete compression strength of the struts, f_{ce}, shall be determined using A.3.2.2(b).

COMMENTARY

R15.5 — Shear in footings

R15.5.1 and R15.5.2 — The shear strength of footings are determined for the more severe condition of 11.12.1.1 or 11.12.1.2. The critical section for shear is measured from the face of supported member (column, pedestal, or wall), except for supported members on steel base plates.

Computation of shear requires that the soil reaction q_s be obtained from the factored loads and the design be in accordance with the appropriate equations of Chapter 11.

Where necessary, shear around individual piles may be investigated in accordance with 11.12.1.2. If shear perimeters overlap, the modified critical perimeter b_o should be taken as that portion of the smallest envelope of individual shear perimeter that will actually resist the critical shear for the group under consideration. One such situation is illustrated in Fig. R15.5.

R15.5.3 — Pile caps supported on piles in more than one plane can be designed using three-dimensional strut-and-tie models satisfying Appendix A.[15.3] The effective concrete compressive strength is from A.3.2.2(b) because it is generally not feasible to provide confining reinforcement satisfying A.3.3.1 and A.3.3.2 in a pile cap.

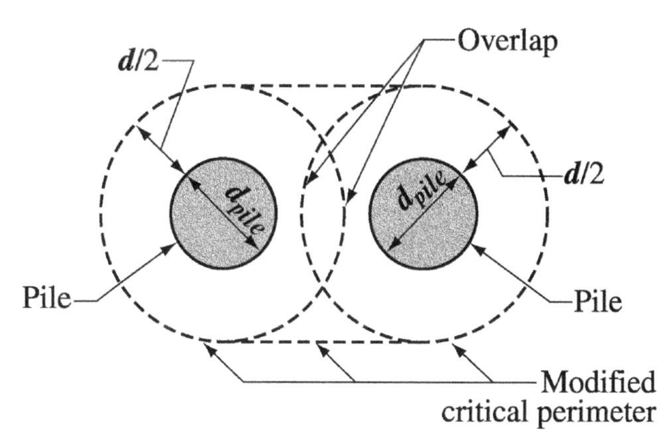

Fig. R15.5—Modified critical perimeter for shear with overlapping critical perimeters

CODE

15.5.4 —Computation of shear on any section through a footing supported on piles shall be in accordance with 15.5.4.1, 15.5.4.2, and 15.5.4.3.

15.5.4.1 — Entire reaction from any pile with its center located $d_{pile}/2$ or more outside the section shall be considered as producing shear on that section.

15.5.4.2 — Reaction from any pile with its center located $d_{pile}/2$ or more inside the section shall be considered as producing no shear on that section.

15.5.4.3 — For intermediate positions of pile center, the portion of the pile reaction to be considered as producing shear on the section shall be based on straight-line interpolation between full value at $d_{pile}/2$ outside the section and zero value at $d_{pile}/2$ inside the section.

COMMENTARY

R15.5.4 — When piles are located inside the critical sections d or $d/2$ from face of column, for one-way or two-way shear, respectively, an upper limit on the shear strength at a section adjacent to the face of the column should be considered. The *CRSI Handbook*[15.4] offers guidance for this situation.

15.6 — Development of reinforcement in footings

15.6.1 — Development of reinforcement in footings shall be in accordance with Chapter 12.

15.6.2 — Calculated tension or compression in reinforcement at each section shall be developed on each side of that section by embedment length, hook (tension only) or mechanical device, or a combination thereof.

15.6.3 — Critical sections for development of reinforcement shall be assumed at the same locations as defined in 15.4.2 for maximum factored moment, and at all other vertical planes where changes of section or reinforcement occur. See also 12.10.6.

15.7 — Minimum footing depth

Depth of footing above bottom reinforcement shall not be less than 6 in. for footings on soil, nor less than 12 in. for footings on piles.

15.8 — Transfer of force at base of column, wall, or reinforced pedestal

15.8.1 — Forces and moments at base of column, wall, or pedestal shall be transferred to supporting pedestal or footing by bearing on concrete and by reinforcement, dowels, and mechanical connectors.

15.8.1.1 — Bearing stress on concrete at contact surface between supported and supporting member shall not exceed concrete bearing strength for either surface as given by 10.17.

R15.8 — Transfer of force at base of column, wall, or reinforced pedestal

Section 15.8 provides the specific requirements for force transfer from a column, wall, or pedestal (supported member) to a pedestal or footing (supporting member). Force transfer should be by bearing on concrete (compressive force only) and by reinforcement (tensile or compressive force). Reinforcement may consist of extended longitudinal bars, dowels, anchor bolts, or suitable mechanical connectors.

The requirements of 15.8.1 apply to both cast-in-place construction and precast construction. Additional requirements

CODE

COMMENTARY

for cast-in-place construction are given in 15.8.2. Section 15.8.3 gives additional requirements for precast construction.

R15.8.1.1 — Compressive force may be transmitted to a supporting pedestal or footing by bearing on concrete. For strength design, allowable bearing stress on the loaded area is equal to $0.85\phi f_c'$, if the loaded area is equal to the area on which it is supported.

In the common case of a column bearing on a footing larger than the column, bearing strength should be checked at the base of the column and the top of the footing. Strength in the lower part of the column should be checked since the column reinforcement cannot be considered effective near the column base because the force in the reinforcement is not developed for some distance above the base, unless dowels are provided, or the column reinforcement is extended into the footing. The unit bearing stress on the column will normally be $0.85\phi f_c'$. The permissible bearing strength on the footing may be increased in accordance with 10.17 and will usually be two times $0.85\phi f_c'$. The compressive force that exceeds that developed by the permissible bearing strength at the base of the column or at the top of the footing should be carried by dowels or extended longitudinal bars.

15.8.1.2 — Reinforcement, dowels, or mechanical connectors between supported and supporting members shall be adequate to transfer:

(a) All compressive force that exceeds concrete bearing strength of either member;

(b) Any computed tensile force across interface.

In addition, reinforcement, dowels, or mechanical connectors shall satisfy 15.8.2 or 15.8.3.

R15.8.1.2 — All tensile forces, whether created by uplift, moment, or other means, should be transferred to supporting pedestal or footing entirely by reinforcement or suitable mechanical connectors. Generally, mechanical connectors would be used only in precast construction.

15.8.1.3 — If calculated moments are transferred to supporting pedestal or footing, then reinforcement, dowels, or mechanical connectors shall be adequate to satisfy 12.17.

R15.8.1.3 — If computed moments are transferred from the column to the footing, the concrete in the compression zone of the column will be stressed to $0.85f_c'$ under factored load conditions and, as a result, all the reinforcement will generally have to be doweled into the footing.

15.8.1.4 — Lateral forces shall be transferred to supporting pedestal or footing in accordance with shear-friction provisions of 11.7, or by other appropriate means.

R15.8.1.4 — The shear-friction method given in 11.7 may be used to check for transfer of lateral forces to supporting pedestal or footing. Shear keys may be used, provided that the reinforcement crossing the joint satisfies 15.8.2.1, 15.8.3.1, and the shear-friction requirements of 11.7. In precast construction, resistance to lateral forces may be provided by shear-friction, shear keys, or mechanical devices.

CODE

15.8.2 — In cast-in-place construction, reinforcement required to satisfy 15.8.1 shall be provided either by extending longitudinal bars into supporting pedestal or footing, or by dowels.

15.8.2.1 — For cast-in-place columns and pedestals, area of reinforcement across interface shall be not less than $\mathbf{0.005}\boldsymbol{A_g}$, where $\boldsymbol{A_g}$ is the gross area of the supported member.

15.8.2.2 — For cast-in-place walls, area of reinforcement across interface shall be not less than minimum vertical reinforcement given in 14.3.2.

15.8.2.3 — At footings, it shall be permitted to lap splice No. 14 and No. 18 longitudinal bars, in compression only, with dowels to provide reinforcement required to satisfy 15.8.1. Dowels shall not be larger than No. 11 bar and shall extend into supported member a distance not less than the development length, ℓ_{dc}, of No. 14 or No. 18 bars or the splice length of the dowels, whichever is greater, and into the footing a distance not less than the development length of the dowels.

15.8.2.4 — If a pinned or rocker connection is provided in cast-in-place construction, connection shall conform to 15.8.1 and 15.8.3.

15.8.3 — In precast construction, anchor bolts or suitable mechanical connectors shall be permitted for satisfying 15.8.1. Anchor bolts shall be designed in accordance with Appendix D.

15.8.3.1 — Connection between precast columns or pedestals and supporting members shall meet the requirements of 16.5.1.3(a).

15.8.3.2 — Connection between precast walls and supporting members shall meet the requirements of 16.5.1.3(b) and (c).

COMMENTARY

R15.8.2.1 and R15.8.2.2 — A minimum amount of reinforcement is required between all supported and supporting members to ensure ductile behavior. The code does not require that all bars in a column be extended through and be anchored into a footing. However, reinforcement with an area of 0.005 times the column area or an equal area of properly spliced dowels is required to extend into the footing with proper anchorage. This reinforcement is required to provide a degree of structural integrity during the construction stage and during the life of the structure.

R15.8.2.3 — Lap splices of No. 14 and No. 18 longitudinal bars in compression only to dowels from a footing are specifically permitted in 15.8.2.3. The dowel bars should be No. 11 or smaller in size. The dowel lap splice length should meet the larger of the two criteria: (a) be able to transfer the stress in the No. 14 and No. 18 bars, and (b) fully develop the stress in the dowels as a splice.

This provision is an exception to 12.14.2.1, which prohibits lap splicing of No. 14 and No. 18 bars. This exception results from many years of successful experience with the lap splicing of these large column bars with footing dowels of the smaller size. The reason for the restriction on dowel bar size is recognition of the anchorage length problem of the large bars, and to allow use of the smaller size dowels. A similar exception is allowed for compression splices between different size bars in 12.16.2.

R15.8.3.1 and R15.8.3.2 — For cast-in-place columns, 15.8.2.1 requires a minimum area of reinforcement equal to $\mathbf{0.005}\boldsymbol{A_g}$ across the column-footing interface to provide some degree of structural integrity. For precast columns this requirement is expressed in terms of an equivalent tensile force that should be transferred. Thus, across the joint, $\boldsymbol{A_s f_y} = \mathbf{200}\boldsymbol{A_g}$ [see 16.5.1.3(a)]. The minimum tensile strength required for precast wall-to-footing connection [see 16.5.1.3(b)] is somewhat less than that required for columns, since an overload would be distributed laterally and a sudden failure would be less likely. Since the tensile strength values of 16.5.1.3 have been arbitrarily chosen, it is not necessary to include a strength reduction factor ϕ for these calculations.

CODE

15.8.3.3 — Anchor bolts and mechanical connections shall be designed to reach their design strength before anchorage failure or failure of surrounding concrete. Anchor bolts shall be designed in accordance with Appendix D.

15.9 — Sloped or stepped footings

15.9.1 — In sloped or stepped footings, angle of slope or depth and location of steps shall be such that design requirements are satisfied at every section. (See also 12.10.6.)

15.9.2 — Sloped or stepped footings designed as a unit shall be constructed to ensure action as a unit.

15.10 — Combined footings and mats

15.10.1 — Footings supporting more than one column, pedestal, or wall (combined footings or mats) shall be proportioned to resist the factored loads and induced reactions, in accordance with appropriate design requirements of the code.

15.10.2 — The Direct Design Method of Chapter 13 shall not be used for design of combined footings and mats.

15.10.3 — Distribution of soil pressure under combined footings and mats shall be consistent with properties of the soil and the structure and with established principles of soil mechanics.

COMMENTARY

R15.10 — Combined footings and mats

R15.10.1 — Any reasonable assumption with respect to the distribution of soil pressure or pile reactions can be used as long as it is consistent with the type of structure and the properties of the soil, and conforms with established principles of soil mechanics (see 15.1). Similarly, as prescribed in 15.2.2 for isolated footings, the base area or pile arrangement of combined footings and mats should be determined using the unfactored forces, moments, or both, transmitted by the footing to the soil, considering permissible soil pressures and pile reactions.

Design methods using factored loads and strength reduction factors ϕ can be applied to combined footings or mats, regardless of the soil pressure distribution.

Detailed recommendations for design of combined footings and mats are reported by ACI Committee 336.[15.1] See also Reference 15.2.

Notes

CHAPTER 16 — PRECAST CONCRETE

CODE

16.1 — Scope

16.1.1 — All provisions of this code, not specifically excluded and not in conflict with the provisions of Chapter 16, shall apply to structures incorporating precast concrete structural members.

16.2 — General

16.2.1 — Design of precast members and connections shall include loading and restraint conditions from initial fabrication to end use in the structure, including form removal, storage, transportation, and erection.

16.2.2 — When precast members are incorporated into a structural system, the forces and deformations occurring in and adjacent to connections shall be included in the design.

16.2.3 — Tolerances for both precast members and interfacing members shall be specified. Design of precast members and connections shall include the effects of these tolerances.

COMMENTARY

R16.1 — Scope

R16.1.1 — See 2.2 for definition of precast concrete.

Design and construction requirements for precast concrete structural members differ in some respects from those for cast-in-place concrete structural members and these differences are addressed in this chapter. Where provisions for cast-in-place concrete applied to precast concrete, they have not been repeated. Similarly, items related to composite concrete in Chapter 17 and to prestressed concrete in Chapter 18 that apply to precast concrete are not restated.

More detailed recommendations concerning precast concrete are given in References 16.1 through 16.7. Tilt-up concrete construction is a form of precast concrete. It is recommended that Reference 16.8 be reviewed for tilt-up structures.

R16.2 — General

R16.2.1 — Stresses developed in precast members during the period from casting to final connection may be greater than the service load stresses. Handling procedures may cause undesirable deformations. Care should be given to the methods of storing, transporting, and erecting precast members so that performance at service loads and strength under factored loads meet code requirements.

R16.2.2 — The structural behavior of precast members may differ substantially from that of similar members that are cast-in-place. Design of connections to minimize or transmit forces due to shrinkage, creep, temperature change, elastic deformation, differential settlement, wind, and earthquake require special consideration in precast construction.

R16.2.3 — Design of precast members and connections is particularly sensitive to tolerances on the dimensions of individual members and on their location in the structure. To prevent misunderstanding, the tolerances used in design should be specified in the contract documents. The designer may specify the tolerance standard assumed in design. It is important to specify any deviations from accepted standards.

The tolerances required by 7.5 are considered to be a minimum acceptable standard for reinforcement in precast concrete. The designer should refer to publications of the Precast/Prestressed Concrete Institute (PCI) (References 16.9 through 16.11) for guidance on industry established standard product and erection tolerances. Added guidance is given in Reference 16.12.

CODE

16.2.4 — In addition to the requirements for drawings and specifications in 1.2, (a) and (b) shall be included in either the contract documents or shop drawings:

(a) Details of reinforcement, inserts and lifting devices required to resist temporary loads from handling, storage, transportation, and erection;

(b) Required concrete strength at stated ages or stages of construction.

16.3 — Distribution of forces among members

16.3.1 — Distribution of forces that are perpendicular to the plane of members shall be established by analysis or by test.

16.3.2 — Where the system behavior requires in-plane forces to be transferred between the members of a precast floor or wall system, 16.3.2.1 and 16.3.2.2 shall apply.

16.3.2.1 — In-plane force paths shall be continuous through both connections and members.

16.3.2.2 — Where tension forces occur, a continuous path of steel or steel reinforcement shall be provided.

16.4 — Member design

16.4.1 — In one-way precast floor and roof slabs and in one-way precast, prestressed wall panels, all not wider than 12 ft, and where members are not mechanically connected to cause restraint in the transverse direction, the shrinkage and temperature reinforcement requirements of 7.12 in the direction normal to

COMMENTARY

R16.2.4 — The additional requirements may be included in either contract documents or shop drawings, depending on the assignment of responsibility for design.

R16.3 — Distribution of forces among members

R16.3.1 — Concentrated point and line loads can be distributed among members provided they have sufficient torsional stiffness and that shear can be transferred across joints. Torsionally stiff members such as hollow-core or solid slabs have more favorable load distribution properties than do torsionally flexible members such as double tees with thin flanges. The actual distribution of the load depends on many factors discussed in detail in References 16.13 through 16.19. Large openings can cause significant changes in distribution of forces.

R16.3.2 — In-plane forces result primarily from diaphragm action in floors and roofs, causing tension or compression in the chords and shear in the body of the diaphragm. A continuous path of steel, steel reinforcement, or both, using lap splices, mechanical or welded splices, or mechanical connectors, should be provided to carry the tension, whereas the shear and compression may be carried by the net concrete section. A continuous path of steel through a connection includes bolts, weld plates, headed studs, or other steel devices. Tension forces in the connections are to be transferred to the primary reinforcement in the members.

In-plane forces in precast wall systems result primarily from diaphragm reactions and external lateral loads.

Connection details should provide for the forces and deformations due to shrinkage, creep, and thermal effects. Connection details may be selected to accommodate volume changes and rotations caused by temperature gradients and long-term deflections. When these effects are restrained, connections and members should be designed to provide adequate strength and ductility.

R16.4 — Member design

R16.4.1 — For prestressed concrete members not wider than 12 ft, such as hollow-core slabs, solid slabs, or slabs with closely spaced ribs, there is usually no need to provide transverse reinforcement to withstand shrinkage and temperature stresses in the short direction. This is generally true also for nonprestressed floor and roof slabs. The 12 ft width

CODE

the flexural reinforcement shall be permitted to be waived. This waiver shall not apply to members that require reinforcement to resist transverse flexural stresses.

COMMENTARY

is less than that in which shrinkage and temperature stresses can build up to a magnitude requiring transverse reinforcement. In addition, much of the shrinkage occurs before the members are tied into the structure. Once in the final structure, the members are usually not as rigidly connected transversely as monolithic concrete, thus the transverse restraint stresses due to both shrinkage and temperature change are significantly reduced.

The waiver does not apply to members such as single and double tees with thin, wide flanges.

16.4.2 — For precast, nonprestressed walls the reinforcement shall be designed in accordance with the provisions of Chapters 10 or 14, except that the area of horizontal and vertical reinforcement each shall be not less than $\mathbf{0.001}\boldsymbol{A_g}$, where $\boldsymbol{A_g}$ is the gross cross-sectional area of the wall panel. Spacing of reinforcement shall not exceed 5 times the wall thickness nor 30 in. for interior walls nor 18 in. for exterior walls.

R16.4.2 — This minimum area of wall reinforcement, instead of the minimum values in 14.3, has been used for many years and is recommended by the PCI [16.4] and the Canadian Building Code.[16.20] The provisions for reduced minimum reinforcement and greater spacing recognize that precast wall panels have very little restraint at their edges during early stages of curing and develop less shrinkage stress than comparable cast-in-place walls.

16.5 — Structural integrity

16.5.1 — Except where the provisions of 16.5.2 govern, the minimum provisions of 16.5.1.1 through 16.5.1.4 for structural integrity shall apply to all precast concrete structures.

16.5.1.1 — Longitudinal and transverse ties required by 7.13.3 shall connect members to a lateral load resisting system.

16.5.1.2 — Where precast elements form floor or roof diaphragms, the connections between diaphragm and those members being laterally supported shall have a nominal tensile strength capable of resisting not less than 300 lb per linear ft.

16.5.1.3 — Vertical tension tie requirements of 7.13.3 shall apply to all vertical structural members, except cladding, and shall be achieved by providing

R16.5 — Structural integrity

R16.5.1 — The provisions of 7.13.3 apply to all precast concrete structures. Sections 16.5.1 and 16.5.2 give minimum requirements to satisfy 7.13.3. It is not intended that these minimum requirements override other applicable provisions of the code for design of precast concrete structures.

The overall integrity of a structure can be substantially enhanced by minor changes in the amount, location, and detailing of member reinforcement and in the detailing of connection hardware.

R16.5.1.1 — Individual members may be connected into a lateral load resisting system by alternative methods. For example, a load-bearing spandrel could be connected to a diaphragm (part of the lateral load resisting system). Structural integrity could be achieved by connecting the spandrel into all or a portion of the deck members forming the diaphragm. Alternatively, the spandrel could be connected only to its supporting columns, which in turn is connected to the diaphragm.

R16.5.1.2 — Diaphragms are typically provided as part of the lateral load resisting system. The ties prescribed in 16.5.1.2 are the minimum required to attach members to the floor or roof diaphragms. The tie force is equivalent to the service load value of 200 lb/ft given in the Uniform Building Code.

R16.5.1.3 — Base connections and connections at horizontal joints in precast columns and wall panels, including shearwalls, are designed to transfer all design forces and

CODE

connections at horizontal joints in accordance with (a) through (c):

(a) Precast columns shall have a nominal strength in tension not less than **200** A_g, in lb. For columns with a larger cross section than required by consideration of loading, a reduced effective area A_g, based on cross section required but not less than one-half the total area, shall be permitted;

(b) Precast wall panels shall have a minimum of two ties per panel, with a nominal tensile strength not less than 10,000 lb per tie;

(c) When design forces result in no tension at the base, the ties required by 16.5.1.3(b) shall be permitted to be anchored into an appropriately reinforced concrete floor slab on grade.

16.5.1.4 — Connection details that rely solely on friction caused by gravity loads shall not be used.

16.5.2 — For precast concrete bearing wall structures three or more stories in height, the minimum provisions of 16.5.2.1 through 16.5.2.5 shall apply.

16.5.2.1 — Longitudinal and transverse ties shall be provided in floor and roof systems to provide a nominal strength of 1500 lb per foot of width or length. Ties shall be provided over interior wall supports and between members and exterior walls. Ties shall be positioned in or within 2 ft of the plane of the floor or roof system.

COMMENTARY

moments. The minimum tie requirements of 16.5.1.3 are not additive to these design requirements. Common practice is to place the wall ties symmetrically about the vertical centerline of the wall panel and within the outer quarters of the panel width, wherever possible.

R16.5.1.4 — In the event of damage to a beam, it is important that displacement of its supporting members be minimized, so that other members will not lose their load-carrying capacity. This situation shows why connection details that rely solely on friction caused by gravity loads are not used. An exception could be heavy modular unit structures (one or more cells in cell-type structures) where resistance to overturning or sliding in any direction has a large factor of safety. Acceptance of such systems should be based on the provisions of 1.4.

R16.5.2 — The structural integrity minimum tie provisions for bearing wall structures, often called large panel structures, are intended to provide catenary hanger supports in case of loss of a bearing wall support, as shown by test.[16.21] Forces induced by loading, temperature change, creep, and wind or seismic action may require a larger amount of tie force. It is intended that the general precast concrete provisions of 16.5.1 apply to bearing wall structures less than three stories in height.

Minimum ties in structures three or more stories in height, in accordance with 16.5.2.1, 16.5.2.2, 16.5.2.3, 16.5.2.4, and 16.5.2.5, are required for structural integrity (Fig. R16.5.2). These provisions are based on PCI's recommendations for design of precast concrete bearing wall buildings.[16.22] Tie capacity is based on yield strength.

R16.5.2.1 — Longitudinal ties may project from slabs and be lap spliced, welded, or mechanically connected, or they may be embedded in grout joints, with sufficient length and cover to develop the required force. Bond length for unstressed prestressing steel should be sufficient to develop the yield strength.[16.23] It is not uncommon to have ties positioned in the walls reasonably close to the plane of the floor or roof system.

CODE

COMMENTARY

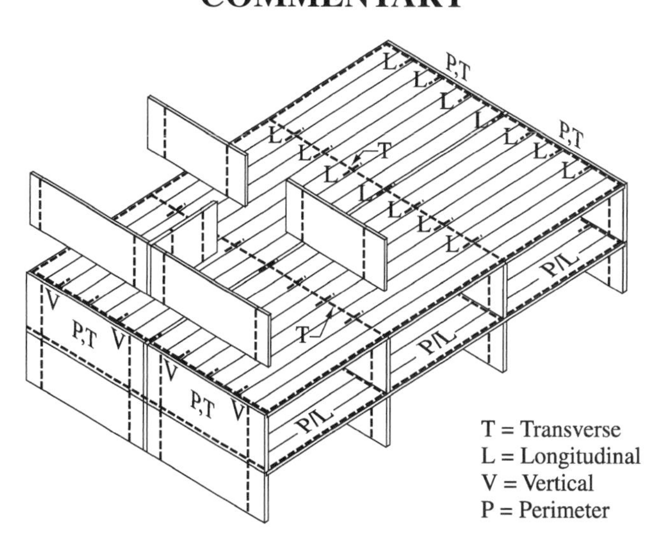

Fig. R16.5.2—Typical arrangement of tensile ties in large panel structures

16.5.2.2 — Longitudinal ties parallel to floor or roof slab spans shall be spaced not more than 10 ft on centers. Provisions shall be made to transfer forces around openings.

16.5.2.3 — Transverse ties perpendicular to floor or roof slab spans shall be spaced not greater than the bearing wall spacing.

R16.5.2.3 — Transverse ties may be uniformly spaced either encased in the panels or in a topping, or they may be concentrated at the transverse bearing walls.

16.5.2.4 — Ties around the perimeter of each floor and roof, within 4 ft of the edge, shall provide a nominal strength in tension not less than 16,000 lb.

R16.5.2.4 —The perimeter tie requirements need not be additive with the longitudinal and transverse tie requirements.

16.5.2.5 — Vertical tension ties shall be provided in all walls and shall be continuous over the height of the building. They shall provide a nominal tensile strength not less than 3000 lb per horizontal foot of wall. Not less than two ties shall be provided for each precast panel.

16.6 — Connection and bearing design

R16.6 — Connection and bearing design

16.6.1 — Forces shall be permitted to be transferred between members by grouted joints, shear keys, mechanical connectors, reinforcing steel connections, reinforced topping, or a combination of these means.

R16.6.1 — The code permits a variety of methods for connecting members. These are intended for transfer of forces both in-plane and perpendicular to the plane of the members.

16.6.1.1 — The adequacy of connections to transfer forces between members shall be determined by analysis or by test. Where shear is the primary result of imposed loading, it shall be permitted to use the provisions of 11.7 as applicable.

16.6.1.2 — When designing a connection using materials with different structural properties, their relative stiffnesses, strengths, and ductilities shall be considered.

R16.6.1.2 — Various components in a connection (such as bolts, welds, plates, and inserts) have different properties that can affect the overall behavior of the connection.

CODE

16.6.2 — Bearing for precast floor and roof members on simple supports shall satisfy 16.6.2.1 and 16.6.2.2.

16.6.2.1—The allowable bearing stress at the contact surface between supported and supporting members and between any intermediate bearing elements shall not exceed the bearing strength for either surface or the bearing element, or both. Concrete bearing strength shall be as given in 10.17.

16.6.2.2 — Unless shown by test or analysis that performance will not be impaired, (a) and (b) shall be met:

(a) Each member and its supporting system shall have design dimensions selected so that, after consideration of tolerances, the distance from the edge of the support to the end of the precast member in the direction of the span is at least ℓ_n**/180**, but not less than:

For solid or hollow-core slabs............................ 2 in.
For beams or stemmed members..................... 3 in.

(b) Bearing pads at unarmored edges shall be set back a minimum of 1/2 in. from the face of the support, or at least the chamfer dimension at chamfered edges.

16.6.2.3 — The requirements of 12.11.1 shall not apply to the positive bending moment reinforcement for statically determinate precast members, but at least one-third of such reinforcement shall extend to the center of the bearing length, taking into account permitted tolerances in 7.5.2.2 and 16.2.3.

COMMENTARY

R16.6.2.1 — When tensile forces occur in the plane of the bearing, it may be desirable to reduce the allowable bearing stress, provide confinement reinforcement, or both. Guidelines are provided in Reference 16.4.

R16.6.2.2 — This section differentiates between bearing length and length of the end of a precast member over the support (Fig. R16.6.2). Bearing pads distribute concentrated loads and reactions over the bearing area, and allow limited horizontal and rotational movements for stress relief. To prevent spalling under heavily loaded bearing areas, bearing pads should not extend to the edge of the support unless the edge is armored. Edges can be armored with anchored steel plates or angles. Section 11.9.7 gives requirements for bearing on brackets or corbels.

R16.6.2.3 — It is unnecessary to develop positive bending moment reinforcement beyond the ends of the precast element if the system is statically determinate. Tolerances need to be considered to avoid bearing on plain concrete where reinforcement has been discontinued.

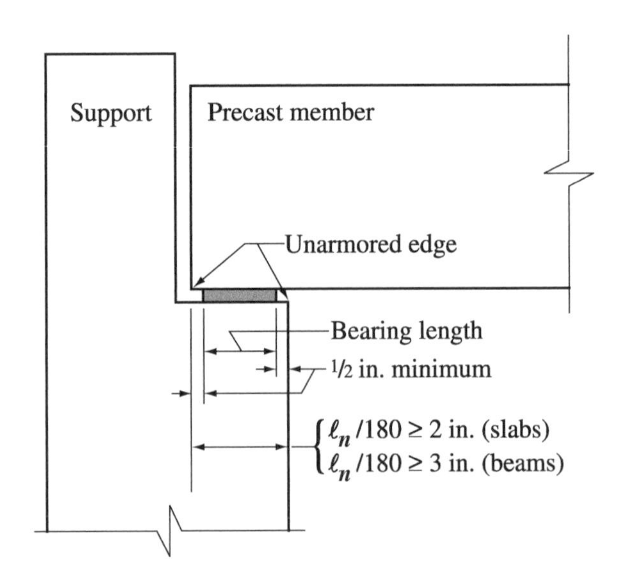

Fig. R16.6.2—Bearing length on support

CODE

COMMENTARY

16.7 — Items embedded after concrete placement

16.7.1 — When approved by the registered design professional, embedded items (such as dowels or inserts) that either protrude from the concrete or remain exposed for inspection shall be permitted to be embedded while the concrete is in a plastic state provided that 16.7.1.1, 16.7.1.2, and 16.7.1.3 are met.

16.7.1.1 — Embedded items are not required to be hooked or tied to reinforcement within the concrete.

16.7.1.2 — Embedded items are maintained in the correct position while the concrete remains plastic.

16.7.1.3 — The concrete is properly consolidated around the embedded item.

R16.7 — Items embedded after concrete placement

R16.7.1 — Section 16.7.1 is an exception to the provisions of 7.5.1. Many precast products are manufactured in such a way that it is difficult, if not impossible, to position reinforcement that protrudes from the concrete before the concrete is placed. Such items as ties for horizontal shear and inserts can be placed while the concrete is plastic, if proper precautions are taken. This exception is not applicable to reinforcement that is completely embedded, or to embedded items that will be hooked or tied to embedded reinforcement.

16.8 — Marking and identification

16.8.1 — Each precast member shall be marked to indicate its location and orientation in the structure and date of manufacture.

16.8.2 — Identification marks shall correspond to placing drawings.

16.9 — Handling

16.9.1 — Member design shall consider forces and distortions during curing, stripping, storage, transportation, and erection so that precast members are not overstressed or otherwise damaged.

16.9.2 — During erection, precast members and structures shall be adequately supported and braced to ensure proper alignment and structural integrity until permanent connections are completed.

R16.9 — Handling

R16.9.1 — The code requires acceptable performance at service loads and adequate strength under factored loads. However, handling loads should not produce permanent stresses, strains, cracking, or deflections inconsistent with the provisions of the code. A precast member should not be rejected for minor cracking or spalling where strength and durability are not affected. Guidance on assessing cracks is given in PCI reports on fabrication and shipment cracks.[16.24, 16.25]

R16.9.2 — All temporary erection connections, bracing, shoring as well as the sequencing of removal of these items are shown on contract or erection drawings.

16.10 — Strength evaluation of precast construction

16.10.1 — A precast element to be made composite with cast-in-place concrete shall be permitted to be tested in flexure as a precast element alone in accordance with 16.10.1.1 and 16.10.1.2.

16.10.1.1 — Test loads shall be applied only when calculations indicate the isolated precast element will not be critical in compression or buckling.

R16.10 — Strength evaluation of precast construction

The strength evaluation procedures of Chapter 20 are applicable to precast members.

CODE

16.10.1.2 — The test load shall be that load which, when applied to the precast member alone, induces the same total force in the tension reinforcement as would be induced by loading the composite member with the test load required by 20.3.2.

16.10.2 — The provisions of 20.5 shall be the basis for acceptance or rejection of the precast element.

COMMENTARY

CHAPTER 17 — COMPOSITE CONCRETE FLEXURAL MEMBERS

17.1 — Scope

17.1.1 — Provisions of Chapter 17 shall apply for design of composite concrete flexural members defined as precast concrete, cast-in-place concrete elements, or both, constructed in separate placements but so interconnected that all elements respond to loads as a unit.

17.1.2 — All provisions of the code shall apply to composite concrete flexural members, except as specifically modified in Chapter 17.

R17.1 — Scope

R17.1.1 — The scope of Chapter 17 is intended to include all types of composite concrete flexural members. In some cases with fully cast-in-place concrete, it may be necessary to design the interface of consecutive placements of concrete as required for composite members. Composite structural steel-concrete members are not covered in this chapter. Design provisions for such composite members are covered in Reference 17.1.

17.2 — General

17.2.1 — The use of an entire composite member or portions thereof for resisting shear and moment shall be permitted.

17.2.2 — Individual elements shall be investigated for all critical stages of loading.

17.2.3 — If the specified strength, unit weight, or other properties of the various elements are different, properties of the individual elements or the most critical values shall be used in design.

17.2.4 — In strength computations of composite members, no distinction shall be made between shored and unshored members.

17.2.5 — All elements shall be designed to support all loads introduced prior to full development of design strength of composite members.

17.2.6 — Reinforcement shall be provided as required to minimize cracking and to prevent separation of individual elements of composite members.

17.2.7 — Composite members shall meet requirements for control of deflections in accordance with 9.5.5.

R17.2 — General

R17.2.4 — Tests have indicated that the strength of a composite member is the same whether or not the first element cast is shored during casting and curing of the second element.

R17.2.6 — The extent of cracking is dependent on such factors as environment, aesthetics, and occupancy. In addition, composite action should not be impaired.

R17.2.7 — The premature loading of precast elements can cause excessive creep and shrinkage deflections. This is especially so at early ages when the moisture content is high and the strength low.

The transfer of shear by direct bond is important if excessive deflection from slippage is to be prevented. A shear key is an added mechanical factor of safety but it does not operate until slippage occurs.

CODE

17.3 — Shoring

When used, shoring shall not be removed until supported elements have developed design properties required to support all loads and limit deflections and cracking at time of shoring removal.

17.4 — Vertical shear strength

17.4.1 — Where an entire composite member is assumed to resist vertical shear, design shall be in accordance with requirements of Chapter 11 as for a monolithically cast member of the same cross-sectional shape.

17.4.2 — Shear reinforcement shall be fully anchored into interconnected elements in accordance with 12.13.

17.4.3 — Extended and anchored shear reinforcement shall be permitted to be included as ties for horizontal shear.

17.5 — Horizontal shear strength

17.5.1 — In a composite member, full transfer of horizontal shear forces shall be ensured at contact surfaces of interconnected elements.

17.5.2 — For the provisions of 17.5, d shall be taken as the distance from extreme compression fiber for entire composite section to centroid of prestressed and nonprestressed longitudinal tension reinforcement, if any, but need not be taken less than $0.80h$ for prestressed concrete members.

17.5.3 — Unless calculated in accordance with 17.5.4, design of cross sections subject to horizontal shear shall be based on

$$V_u \leq \phi V_{nh} \qquad (17\text{-}1)$$

where V_{nh} is nominal horizontal shear strength in accordance with 17.5.3.1 through 17.5.3.4.

17.5.3.1 — Where contact surfaces are clean, free of laitance, and intentionally roughened, V_{nh} shall not be taken greater than $80b_vd$.

17.5.3.2 — Where minimum ties are provided in accordance with 17.6, and contact surfaces are clean and free of laitance, but not intentionally roughened, V_{nh} shall not be taken greater than $80b_vd$.

COMMENTARY

R17.3 — Shoring

The provisions of 9.5.5 cover the requirements pertaining to deflections of shored and unshored members.

R17.5 — Horizontal shear strength

R17.5.1 — Full transfer of horizontal shear between segments of composite members should be ensured by horizontal shear strength at contact surfaces or properly anchored ties, or both.

R17.5.2 — Prestressed members used in composite construction may have variations in depth of tension reinforcement along member length due to draped or depressed tendons. Because of this variation, the definition of d used in Chapter 11 for determination of vertical shear strength is also appropriate when determining horizontal shear strength.

R17.5.3 — The nominal horizontal shear strengths V_{nh} apply when the design is based on the load factors and ϕ-factors of Chapter 9.

CODE

17.5.3.3 — Where ties are provided in accordance with 17.6, and contact surfaces are clean, free of laitance, and intentionally roughened to a full amplitude of approximately 1/4 in., V_{nh} shall be taken equal to $(260 + 0.6\rho_v f_y)\lambda b_v d$, but not greater than $500 b_v d$. Values for λ in 11.7.4.3 shall apply and ρ_v is $A_v/(b_v s)$.

17.5.3.4 — Where V_u at section considered exceeds $\phi(500 b_v d)$, design for horizontal shear shall be in accordance with 11.7.4.

17.5.4 — As an alternative to 17.5.3, horizontal shear shall be permitted to be determined by computing the actual change in compressive or tensile force in any segment, and provisions shall be made to transfer that force as horizontal shear to the supporting element. The factored horizontal shear force V_u shall not exceed horizontal shear strength ϕV_{nh} as given in 17.5.3.1 through 17.5.3.4, where area of contact surface shall be substituted for $b_v d$.

17.5.4.1 — Where ties provided to resist horizontal shear are designed to satisfy 17.5.4, the tie area to tie spacing ratio along the member shall approximately reflect the distribution of shear forces in the member.

17.5.5 — Where tension exists across any contact surface between interconnected elements, shear transfer by contact shall be permitted only when minimum ties are provided in accordance with 17.6.

17.6 — Ties for horizontal shear

17.6.1 — Where ties are provided to transfer horizontal shear, tie area shall not be less than that required by 11.5.6.3, and tie spacing shall not exceed four times the least dimension of supported element, nor exceed 24 in.

17.6.2 — Ties for horizontal shear shall consist of single bars or wire, multiple leg stirrups, or vertical legs of welded wire reinforcement.

17.6.3 — All ties shall be fully anchored into interconnected elements in accordance with 12.13.

COMMENTARY

R17.5.3.3 — The permitted horizontal shear strengths and the requirement of 1/4 in. amplitude for intentional roughness are based on tests discussed in References 17.2 through 17.4.

R17.5.4.1 — The distribution of horizontal shear stresses along the contact surface in a composite member will reflect the distribution of shear along the member. Horizontal shear failure will initiate where the horizontal shear stress is a maximum and will spread to regions of lower stress. Because the slip at peak horizontal shear resistance is small for a concrete-to-concrete contact surface, longitudinal redistribution of horizontal shear resistance is very limited. The spacing of the ties along the contact surface should, therefore, be such as to provide horizontal shear resistance distributed approximately as the shear acting on the member is distributed.

R17.5.5 — Proper anchorage of ties extending across interfaces is required to maintain contact of the interfaces.

R17.6 — Ties for horizontal shear

The minimum areas and maximum spacings are based on test data given in References 17.2 through 17.6.

Notes

CHAPTER 18 — PRESTRESSED CONCRETE

18.1 — Scope

18.1.1 — Provisions of Chapter 18 shall apply to members prestressed with wire, strands, or bars conforming to provisions for prestressing steel in 3.5.5.

18.1.2 — All provisions of this code not specifically excluded, and not in conflict with provisions of Chapter 18, shall apply to prestressed concrete.

18.1.3 — The following provisions of this code shall not apply to prestressed concrete, except as specifically noted: Sections 6.4.4, 7.6.5, 8.10.2, 8.10.3, 8.10.4, 8.11, 10.5, 10.6, 10.9.1, and 10.9.2; Chapter 13; and Sections 14.3, 14.5, and 14.6, except that certain sections of 10.6 apply as noted in 18.4.4.

COMMENTARY

R18.1 — Scope

R18.1.1 — The provisions of Chapter 18 were developed primarily for structural members such as slabs, beams, and columns that are commonly used in buildings. Many of the provisions may be applied to other types of construction, such as, pressure vessels, pavements, pipes, and crossties. Application of the provisions is left to the judgment of the engineer in cases not specifically cited in the code.

R18.1.3 — Some sections of the code are excluded from use in the design of prestressed concrete for specific reasons. The following discussion provides explanation for such exclusions:

Section 6.4.4 — Tendons of continuous post-tensioned beams and slabs are usually stressed at a point along the span where the tendon profile is at or near the centroid of the concrete cross section. Therefore, interior construction joints are usually located within the end thirds of the span, rather than the middle third of the span as required by 6.4.4. Construction joints located as described in continuous post-tensioned beams and slabs have a long history of satisfactory performance. Thus, 6.4.4 is excluded from application to prestressed concrete.

Section 7.6.5 — Section 7.6.5 of the code is excluded from application to prestressed concrete because the requirements for bonded reinforcement and unbonded tendons for cast-in-place members are provided in 18.9 and 18.12, respectively.

Sections 8.10.2, 8.10.3, and 8.10.4 — The empirical provisions of 8.10.2, 8.10.3, and 8.10.4 for T-beams were developed for nonprestressed reinforced concrete, and if applied to prestressed concrete would exclude many standard prestressed products in satisfactory use today. Hence, proof by experience permits variations.

By excluding 8.10.2, 8.10.3, and 8.10.4, no special requirements for prestressed concrete T-beams appear in the code. Instead, the determination of an effective width of flange is left to the experience and judgment of the engineer. Where possible, the flange widths in 8.10.2, 8.10.3, and 8.10.4 should be used unless experience has proven that variations are safe and satisfactory. It is not necessarily conservative in elastic analysis and design considerations to use the maximum flange width as permitted in 8.10.2.

CODE

COMMENTARY

Sections 8.10.1 and 8.10.5 provide general requirements for T-beams that are also applicable to prestressed concrete members. The spacing limitations for slab reinforcement are based on flange thickness, which for tapered flanges can be taken as the average thickness.

Section 8.11 — The empirical limits established for nonprestressed reinforced concrete joist floors are based on successful past performance of joist construction using standard joist forming systems. See R8.11. For prestressed joist construction, experience and judgment should be used. The provisions of 8.11 may be used as a guide.

Sections 10.5, 10.9.1, and 10.9.2 — For prestressed concrete, the limitations on reinforcement given in 10.5, 10.9.1, and 10.9.2 are replaced by those in 18.8.3, 18.9, and 18.11.2.

Section 10.6 —This section does not apply to prestressed members in its entirety. However, 10.6.4 and 10.6.7 are referenced in 18.4.4 pertaining to Class C prestressed flexural members.

Chapter 13 — The design of continuous prestressed concrete slabs requires recognition of secondary moments. Also, volume changes due to the prestressing force can create additional loads on the structure that are not adequately covered in Chapter 13. Because of these unique properties associated with prestressing, many of the design procedures of Chapter 13 are not appropriate for prestressed concrete structures and are replaced by the provisions of 18.12.

Sections 14.5 and 14.6 — The requirements for wall design in 14.5 and 14.6 are largely empirical, utilizing considerations not intended to apply to prestressed concrete.

18.2 — General

18.2.1 — Prestressed members shall meet the strength requirements of this code.

18.2.2 — Design of prestressed members shall be based on strength and on behavior at service conditions at all stages that will be critical during the life of the structure from the time prestress is first applied.

R18.2 — General

R18.2.1 and R18.2.2 — The design investigation should include all stages that may be significant. The three major stages are: (1) jacking stage, or prestress transfer stage—when the tensile force in the prestressing steel is transferred to the concrete and stress levels may be high relative to concrete strength; (2) service load stage—after long-term volume changes have occurred; and (3) the factored load stage—when the strength of the member is checked. There may be other load stages that require investigation. For example, if the cracking load is significant, this load stage may require study, or the handling and transporting stage may be critical.

From the standpoint of satisfactory behavior, the two stages of most importance are those for service load and factored load.

CODE

COMMENTARY

Service load stage refers to the loads defined in the general building code (without load factors), such as live load and dead load, while the factored load stage refers to loads multiplied by the appropriate load factors.

Section 18.3.2 provides assumptions that may be used for investigation at service loads and after transfer of the prestressing force.

18.2.3 — Stress concentrations due to prestressing shall be considered in design.

18.2.4 — Provisions shall be made for effects on adjoining construction of elastic and plastic deformations, deflections, changes in length, and rotations due to prestressing. Effects of temperature and shrinkage shall also be included.

18.2.5 — The possibility of buckling in a member between points where there is intermittent contact between the prestressing steel and an oversize duct, and buckling in thin webs and flanges shall be considered.

R18.2.5 — Section 18.2.5 refers to the type of post-tensioning where the prestressing steel makes intermittent contact with an oversize duct. Precautions should be taken to prevent buckling of such members.

If the prestressing steel is in complete contact with the member being prestressed, or is unbonded with the sheathing not excessively larger than the prestressing steel, it is not possible to buckle the member under the prestressing force being introduced.

18.2.6 — In computing section properties before bonding of prestressing steel, effect of loss of area due to open ducts shall be considered.

R18.2.6 — In considering the area of the open ducts, the critical sections should include those that have coupler sheaths that may be of a larger size than the duct containing the prestressing steel. Also, in some instances, the trumpet or transition piece from the conduit to the anchorage may be of such a size as to create a critical section. If the effect of the open duct area on design is deemed negligible, section properties may be based on total area.

In post-tensioned members after grouting and in pretensioned members, section properties may be based on effective sections using transformed areas of bonded prestressing steel and nonprestressed reinforcement gross sections, or net sections.

18.3 — Design assumptions

R18.3 — Design assumptions

18.3.1 — Strength design of prestressed members for flexure and axial loads shall be based on assumptions given in 10.2, except that 10.2.4 shall apply only to reinforcement conforming to 3.5.3.

18.3.2 — For investigation of stresses at transfer of prestress, at service loads, and at cracking loads, elastic theory shall be used with the assumptions of 18.3.2.1 and 18.3.2.2.

CODE

18.3.2.1 — Strains vary linearly with depth through the entire load range.

18.3.2.2 — At cracked sections, concrete resists no tension.

18.3.3 — Prestressed flexural members shall be classified as Class U, Class T, or Class C based on f_t, the computed extreme fiber stress in tension in the precompressed tensile zone calculated at service loads, as follows:

(a) Class U: $f_t \leq 7.5\sqrt{f_c'}$;
(b) Class T: $7.5\sqrt{f_c'} < f_t \leq 12\sqrt{f_c'}$;
(c) Class C: $f_t > 12\sqrt{f_c'}$;

Prestressed two-way slab systems shall be designed as Class U with $f_t \leq 6\sqrt{f_c'}$.

18.3.4 — For Class U and Class T flexural members, stresses at service loads shall be permitted to be calculated using the uncracked section. For Class C flexural members, stresses at service loads shall be calculated using the cracked transformed section.

18.3.5 — Deflections of prestressed flexural members shall be calculated in accordance with 9.5.4

18.4 — Serviceability requirements — Flexural members

18.4.1 — Stresses in concrete immediately after prestress transfer (before time-dependent prestress losses) shall not exceed the following:

(a) Extreme fiber stress in compression........**$0.60f_{ci}'$**

(b) Extreme fiber stress in tension except as permitted in (c).. **$3\sqrt{f_{ci}'}$**

(c) Extreme fiber stress in tension at ends of simply supported members...................................... **$6\sqrt{f_{ci}'}$**

COMMENTARY

R18.3.3 — This section defines three classes of behavior of prestressed flexural members. Class U members are assumed to behave as uncracked members. Class C members are assumed to behave as cracked members. The behavior of Class T members is assumed to be in transition between uncracked and cracked. The serviceability requirements for each class are summarized in Table R18.3.3. For comparison, Table R18.3.3 also shows corresponding requirements for nonprestressed members.

These classes apply to both bonded and unbonded prestressed flexural members, but prestressed two-way slab systems must be designed as Class U.

The precompressed tensile zone is that portion of a prestressed member where flexural tension, calculated using gross section properties, would occur under unfactored dead and live loads if the prestress force was not present. Prestressed concrete is usually designed so that the prestress force introduces compression into this zone, thus effectively reducing the magnitude of the tensile stress.

R18.3.4 — A method for computing stresses in a cracked section is given in Reference 18.1.

R18.3.5 — Reference 18.2 provides information on computing deflections of cracked members.

R18.4 — Serviceability requirements — Flexural members

Permissible stresses in concrete address serviceability. Permissible stresses do not ensure adequate structural strength, which should be checked in conformance with other code requirements.

R18.4.1 — The concrete stresses at this stage are caused by the force in the prestressing steel at transfer reduced by the losses due to elastic shortening of the concrete, relaxation of the prestressing steel, seating at transfer, and the stresses due to the weight of the member. Generally, shrinkage and creep effects are not included at this stage. These stresses apply to both pretensioned and post-tensioned concrete with proper modifications of the losses at transfer.

R18.4.1(b) and (c) — The tension stress limits of $3\sqrt{f_{ci}'}$ and $6\sqrt{f_{ci}'}$ refer to tensile stress at locations other than the precompressed tensile zone. Where tensile stresses exceed

CODE

COMMENTARY

TABLE R18.3.3 — SERVICEABILITY DESIGN REQUIREMENTS

	Prestressed			
	Class U	Class T	Class C	Nonprestressed
Assumed behavior	Uncracked	Transition between uncracked and cracked	Cracked	Cracked
Section properties for stress calculation at service loads	Gross section 18.3.4	Gross section 18.3.4	Cracked section 18.3.4	No requirement
Allowable stress at transfer	18.4.1	18.4.1	18.4.1	No requirement
Allowable compressive stress based on uncracked section properties	18.4.2	18.4.2	No requirement	No requirement
Tensile stress at service loads 18.3.3	$\leq 7.5\sqrt{f_c'}$	$7.5\sqrt{f_c'} < f_t \leq 12\sqrt{f_c'}$	No requirement	No requirement
Deflection calculation basis	9.5.4.1 Gross section	9.5.4.2 Cracked section, bilinear	9.5.4.2 Cracked section, bilinear	9.5.2, 9.5.3 Effective moment of inertia
Crack control	No requirement	No requirement	10.6.4 Modified by 18.4.4.1	10.6.4
Computation of Δf_{ps} or f_s for crack control	—	—	Cracked section analysis	$M/(A_s \times$ lever arm), or $0.6f_y$
Side skin reinforcement	No requirement	No requirement	10.6.7	10.6.7

Where computed tensile stresses, f_t, exceed the limits in (b) or (c), additional bonded reinforcement (nonprestressed or prestressed) shall be provided in the tensile zone to resist the total tensile force in concrete computed with the assumption of an uncracked section.

the permissible values, the total force in the tensile stress zone may be calculated and reinforcement proportioned on the basis of this force at a stress of $0.6f_y$, but not more than 30,000 psi. The effects of creep and shrinkage begin to reduce the tensile stress almost immediately; however, some tension remains in these areas after allowance is made for all prestress losses.

18.4.2 — For Class U and Class T prestressed flexural members, stresses in concrete at service loads (based on uncracked section properties, and after allowance for all prestress losses) shall not exceed the following:

(a) Extreme fiber stress in compression due to prestress plus sustained load.................... $0.45f_c'$

(b) Extreme fiber stress in compression due to prestress plus total load............................ $0.60f_c'$

R18.4.2(a) and (b) — The compression stress limit of $0.45f_c'$ was conservatively established to decrease the probability of failure of prestressed concrete members due to repeated loads. This limit seemed reasonable to preclude excessive creep deformation. At higher values of stress, creep strains tend to increase more rapidly as applied stress increases.

The change in allowable stress in the 1995 code recognized that fatigue tests of prestressed concrete beams have shown that concrete failures are not the controlling criterion. Designs with transient live loads that are large compared to sustained live and dead loads have been penalized by the previous single compression stress limit. Therefore, the stress limit of $0.60f_c'$ permits a one-third increase in allowable compression stress for members subject to transient loads.

Sustained live load is any portion of the service live load that will be sustained for a sufficient period to cause significant time-dependent deflections. Thus, when the sustained live and dead loads are a large percentage of total service load, the $0.45f_c'$ limit of 18.4.2(a) may control. On the other hand, when a large portion of the total service load consists of a transient or temporary service live load, the increased stress limit of 18.4.2(b) may apply.

The compression limit of $0.45f_c'$ for prestress plus sustained loads will continue to control the long-term behavior of prestressed members.

CODE

18.4.3 — Permissible stresses in 18.4.1 and 18.4.2 shall be permitted to be exceeded if shown by test or analysis that performance will not be impaired.

18.4.4 — For Class C prestressed flexural members not subject to fatigue or to aggressive exposure, the spacing of bonded reinforcement nearest the extreme tension face shall not exceed that given by 10.6.4.

For structures subject to fatigue or exposed to corrosive environments, special investigations and precautions are required.

18.4.4.1 — The spacing requirements shall be met by nonprestressed reinforcement and bonded tendons. The spacing of bonded tendons shall not exceed 2/3 of the maximum spacing permitted for nonprestressed reinforcement.

Where both reinforcement and bonded tendons are used to meet the spacing requirement, the spacing between a bar and a tendon shall not exceed 5/6 of that permitted by 10.6.4. See also 18.4.4.3.

18.4.4.2 — In applying Eq. (10-4) to prestressing tendons, Δf_{ps} shall be substituted for f_s, where Δf_{ps} shall be taken as the calculated stress in the prestressing steel at service loads based on a cracked section analysis minus the decompression stress f_{dc}. It shall be permitted to take f_{dc} equal to the effective stress in the prestressing steel f_{se}. See also 18.4.4.3.

18.4.4.3 — In applying Eq. (10-4) to prestressing tendons, the magnitude of Δf_{ps} shall not exceed 36,000 psi. When Δf_{ps} is less than or equal to 20,000 psi, the spacing requirements of 18.4.4.1 and 18.4.4.2 shall not apply.

18.4.4.4 — Where h of a beam exceeds 36 in., the area of longitudinal skin reinforcement consisting of reinforcement or bonded tendons shall be provided as required by 10.6.7.

COMMENTARY

R18.4.3 — This section provides a mechanism whereby development of new products, materials, and techniques in prestressed concrete construction need not be inhibited by code limits on stress. Approvals for the design should be in accordance with 1.4 of the code.

R18.4.4 — Spacing requirements for prestressed members with calculated tensile stress exceeding $12\sqrt{f_c'}$ were introduced in the 2002 edition of the code.

For conditions of corrosive environments, defined as an environment in which chemical attack (such as seawater, corrosive industrial atmosphere, or sewer gas) is encountered, cover greater than that required by 7.7.2 should be used, and tension stresses in the concrete reduced to eliminate possible cracking at service loads. The engineer should use judgment to determine the amount of increased cover and whether reduced tension stresses are required.

R18.4.4.1 — Only tension steel nearest the tension face need be considered in selecting the value of c_c used in computing spacing requirements. To account for prestressing steel, such as strand, having bond characteristics less effective than deformed reinforcement, a 2/3 effectiveness factor is used.

For post-tensioned members designed as cracked members, it will usually be advantageous to provide crack control by the use of deformed reinforcement, for which the provisions of 10.6 may be used directly. Bonded reinforcement required by other provisions of this code may also be used as crack control reinforcement.

R18.4.4.2 — It is conservative to take the decompression stress f_{dc} equal to f_{se}, the effective stress in the prestressing steel.

R18.4.4.3 — The maximum limitation of 36,000 psi for Δf_{ps} and the exemption for members with Δf_{ps} less than 20,000 psi are intended to be similar to the code requirements before the 2002 edition.

CODE

COMMENTARY

R18.4.4.4 — The steel area of reinforcement, bonded tendons, or a combination of both may be used to satisfy this requirement.

18.5 — Permissible stresses in prestressing steel

R18.5 — Permissible stresses in prestressing steel

The code does not distinguish between temporary and effective prestressing steel stresses. Only one limit on prestressing steel stress is provided because the initial prestressing steel stress (immediately after transfer) can prevail for a considerable time, even after the structure has been put into service. This stress, therefore, should have an adequate safety factor under service conditions and cannot be considered as a temporary stress. Any subsequent decrease in prestressing steel stress due to losses can only improve conditions and no limit on such stress decrease is provided in the code.

18.5.1 — Tensile stress in prestressing steel shall not exceed the following:

(a) Due to prestressing steel jacking force ... $0.94f_{py}$

but not greater than the lesser of $0.80f_{pu}$ and the maximum value recommended by the manufacturer of prestressing steel or anchorage devices.

(b) Immediately after prestress transfer........ $0.82f_{py}$

but not greater than $0.74f_{pu}$.

(c) Post-tensioning tendons, at anchorage devices and couplers, immediately after force transfer........ $0.70f_{pu}$

R18.5.1 — With the 1983 code, permissible stresses in prestressing steel were revised to recognize the higher yield strength of low-relaxation wire and strand meeting the requirements of ASTM A 421 and A 416. For such prestressing steel, it is more appropriate to specify permissible stresses in terms of specified minimum ASTM yield strength rather than specified minimum ASTM tensile strength. For the low-relaxation wire and strands, with f_{py} equal to $0.90f_{pu}$, the $0.94f_{py}$ and $0.82f_{py}$ limits are equivalent to $0.85f_{pu}$ and $0.74f_{pu}$, respectively. In the 1986 supplement and in the 1989 code, the maximum jacking stress for low-relaxation prestressing steel was reduced to $0.80f_{pu}$ to ensure closer compatibility with the maximum prestressing steel stress value of $0.74f_{pu}$ immediately after prestress transfer. The higher yield strength of the low-relaxation prestressing steel does not change the effectiveness of tendon anchorage devices; thus, the permissible stress at post-tensioning anchorage devices and couplers is not increased above the previously permitted value of $0.70f_{pu}$. For ordinary prestressing steel (wire, strands, and bars) with f_{py} equal to $0.85f_{pu}$, the $0.94f_{py}$ and $0.82f_{py}$ limits are equivalent to $0.80f_{pu}$ and $0.70f_{pu}$, respectively, the same as permitted in the 1977 code. For bar prestressing steel with f_{py} equal to $0.80f_{pu}$, the same limits are equivalent to $0.75f_{pu}$ and $0.66f_{pu}$, respectively.

Because of the higher allowable initial prestressing steel stresses permitted since the 1983 code, final stresses can be greater. Designers should be concerned with setting a limit on final stress when the structure is subject to corrosive conditions or repeated loadings.

18.6 — Loss of prestress

R18.6 — Loss of prestress

18.6.1 — To determine effective stress in the prestressing steel, f_{se}, allowance for the following sources of loss of prestress shall be considered:

R18.6.1 — For an explanation of how to compute prestress losses, see References 18.3 through 18.6. Lump sum values of prestress losses for both pretensioned and post-tensioned members that were indicated before the 1983 commentary

CODE

(a) Prestressing steel seating at transfer;

(b) Elastic shortening of concrete;

(c) Creep of concrete;

(d) Shrinkage of concrete;

(e) Relaxation of prestressing steel stress;

(f) Friction loss due to intended or unintended curvature in post-tensioning tendons.

18.6.2 — Friction loss in post-tensioning tendons

18.6.2.1 — P_{px}, force in post-tensioning tendons a distance ℓ_{px} from the jacking end shall be computed by

$$P_{px} = P_{pj} e^{-(K\ell_{px} + \mu_p \alpha_{px})} \quad (18\text{-}1)$$

Where $(K\ell_{px} + \mu_p\alpha_{px})$ is not greater than 0.3, P_{px} shall be permitted to be computed by

$$P_{px} = P_{pj}(1 + K\ell_{px} + \mu_p\alpha_{px})^{-1} \quad (18\text{-}2)$$

18.6.2.2—Friction loss shall be based on experimentally determined wobble K and curvature μ_p friction coefficients, and shall be verified during tendon stressing operations.

COMMENTARY

are considered obsolete. Reasonably accurate estimates of prestress losses can be calculated in accordance with the recommendations in Reference 18.6, which include consideration of initial stress level ($0.7f_{pu}$ or higher), type of steel (stress-relieved or low-relaxation wire, strand, or bar), exposure conditions, and type of construction (pretensioned, bonded post-tensioned, or unbonded post-tensioned).

Actual losses, greater or smaller than the computed values, have little effect on the design strength of the member, but affect service load behavior (deflections, camber, cracking load) and connections. At service loads, overestimation of prestress losses can be almost as detrimental as underestimation, since the former can result in excessive camber and horizontal movement.

R18.6.2 — Friction loss in post-tensioning tendons

The coefficients tabulated in Table R18.6.2 give a range that generally can be expected. Due to the many types of prestressing steel ducts and sheathing available, these values can only serve as a guide. Where rigid conduit is used, the wobble coefficient K can be considered as zero. For large-diameter prestressing steel in semirigid type conduit, the wobble factor can also be considered zero. Values of the coefficients to be used for the particular types of prestressing steel and particular types of ducts should be obtained from the manufacturers of the tendons. An unrealistically low evaluation of the friction loss can lead to improper camber of the member and inadequate prestress. Overestimation of the friction may result in extra prestressing force. This could lead to excessive camber and excessive shortening of a member. If the friction factors are determined to be less than those assumed in the design, the tendon stressing should be adjusted to give only that prestressing force in the critical portions of the structure required by the design.

TABLE R18.6.2—FRICTION COEFFICIENTS FOR POST-TENSIONED TENDONS FOR USE IN EQ. (18-1) OR (18-2)

			Wobble coefficient, K per foot	Curvature coefficient, μ_p per radian
Grouted tendons in metal sheathing		Wire tendons	0.0010-0.0015	0.15-0.25
		High-strength bars	0.0001-0.0006	0.08-0.30
		7-wire strand	0.0005-0.0020	0.15-0.25
Unbonded tendons	Mastic coated	Wire tendons	0.0010-0.0020	0.05-0.15
		7-wire strand	0.0010-0.0020	0.05-0.15
	Pre-greased	Wire tendons	0.0003-0.0020	0.05-0.15
		7-wire strand	0.0003-0.0020	0.05-0.15

CODE

18.6.2.3 — Values of K and μ_p used in design shall be shown on design drawings.

18.6.3—Where loss of prestress in a member occurs due to connection of the member to adjoining construction, such loss of prestress shall be allowed for in design.

18.7 — Flexural strength

18.7.1 — Design moment strength of flexural members shall be computed by the strength design methods of the code. For prestressing steel, f_{ps} shall be substituted for f_y in strength computations.

18.7.2 — As an alternative to a more accurate determination of f_{ps} based on strain compatibility, the following approximate values of f_{ps} shall be permitted to be used if f_{se} is not less than $0.5f_{pu}$.

(a) For members with bonded tendons:

$$f_{ps} = f_{pu}\left\{1 - \frac{\gamma_p}{\beta_1}\left[\rho_p\frac{f_{pu}}{f_c'} + \frac{d}{d_p}(\omega - \omega')\right]\right\} \quad (18\text{-}3)$$

where ω is $\rho f_y/f_c'$, ω' is $\rho' f_y/f_c'$, and γ_p is 0.55 for f_{py}/f_{pu} not less than 0.80; 0.40 for f_{py}/f_{pu} not less than 0.85; and 0.28 for f_{py}/f_{pu} not less than 0.90.

If any compression reinforcement is taken into account when calculating f_{ps} by Eq. (18-3), the term

$$\left[\rho_p\frac{f_{pu}}{f_c'} + \frac{d}{d_p}(\omega - \omega')\right]$$

shall be taken not less than 0.17 and d' shall be no greater than $0.15d_p$.

COMMENTARY

R18.6.2.3 — When the safety or serviceability of the structure may be involved, the acceptable range of prestressing steel jacking forces or other limiting requirements should either be given or approved by the structural engineer in conformance with the permissible stresses of 18.4 and 18.5.

R18.7 — Flexural strength

R18.7.1 — Design moment strength of prestressed flexural members may be computed using strength equations similar to those for nonprestressed concrete members. The 1983 code provided strength equations for rectangular and flanged sections, with tension reinforcement only and with tension and compression reinforcement. When part of the prestressing steel is in the compression zone, a method based on applicable conditions of equilibrium and compatibility of strains at a factored load condition should be used.

For other cross sections, the design moment strength ϕM_n is computed by an analysis based on stress and strain compatibility, using the stress-strain properties of the prestressing steel and the assumptions given in 10.2.

R18.7.2 — Eq. (18-3) may underestimate the strength of beams with high percentages of reinforcement and, for more accurate evaluations of their strength, the strain compatibility and equilibrium method should be used. Use of Eq. (18-3) is appropriate when all of the prestressed reinforcement is in the tension zone. When part of the prestressed reinforcement is in the compression zone, a strain compatibility and equilibrium method should be used.

By inclusion of the ω' term, Eq. (18-3) reflects the increased value of f_{ps} obtained when compression reinforcement is provided in a beam with a large reinforcement index. When the term $[\rho_p f_{pu}/f_c' + (d/d_p)(\omega - \omega')]$ in Eq. (18-3) is small, the neutral axis depth is small, the compressive reinforcement does not develop its yield strength, and Eq. (18-3) becomes unconservative. This is the reason why the term $[\rho_p f_{pu}/f_c' + (d/d_p)(\omega - \omega')]$ in Eq. (18-3) may not be taken less than 0.17 if compression reinforcement is taken into account when computing f_{ps}. If the compression reinforcement is neglected when using Eq. (18-3), ω' is taken as zero, then the term $[\rho_p f_{pu}/f_c' + (d/d_p)\omega]$ may be less than 0.17 and an increased and correct value of f_{ps} is obtained.

When d' is large, the strain in compression reinforcement can be considerably less than its yield strain. In such a case, the compression reinforcement does not influence f_{ps} as favorably as implied by Eq. (18-3). For this reason, the

CODE

(b) For members with unbonded tendons and with a span-to-depth ratio of 35 or less:

$$f_{ps} = f_{se} + 10{,}000 + \frac{f_c'}{100\rho_p} \qquad (18\text{-}4)$$

but f_{ps} in Eq. (18-4) shall not be taken greater than the lesser of f_{py} and $(f_{se} + 60{,}000)$.

(c) For members with unbonded tendons and with a span-to-depth ratio greater than 35:

$$f_{ps} = f_{se} + 10{,}000 + \frac{f_c'}{300\rho_p} \qquad (18\text{-}5)$$

but f_{ps} in Eq. (18-5) shall not be taken greater than the lesser of f_{py} and $(f_{se} + 30{,}000)$.

18.7.3 — Nonprestressed reinforcement conforming to 3.5.3, if used with prestressing steel, shall be permitted to be considered to contribute to the tensile force and to be included in moment strength computations at a stress equal to f_y. Other nonprestressed reinforcement shall be permitted to be included in strength computations only if a strain compatibility analysis is performed to determine stresses in such reinforcement.

COMMENTARY

applicability of Eq. (18-3) is limited to beams in which d' is less than or equal to $0.15d_p$.

The term $[\rho_p f_{pu}/f_c' + (d/d_p)(\omega - \omega')]$ in Eq. (18-3) may also be written $[\rho_p f_{pu}/f_c' + A_s f_y/(bd_p f_c') - A_s' f_y/(bd_p f_c')]$. This form may be more convenient, such as when there is no unprestressed tension reinforcement.

Eq. (18-5) reflects results of tests on members with unbonded tendons and span-to-depth ratios greater than 35 (one-way slabs, flat plates, and flat slabs).[18.7] These tests also indicate that Eq. (18-4), formerly used for all span-depth ratios, overestimates the amount of stress increase in such members. Although these same tests indicate that the moment strength of those shallow members designed using Eq. (18-4) meets the factored load strength requirements, this reflects the effect of the code requirements for minimum bonded reinforcement, as well as the limitation on concrete tensile stress that often controls the amount of prestressing force provided.

18.8 — Limits for reinforcement of flexural members

18.8.1 — Prestressed concrete sections shall be classified as either tension-controlled, transition, or compression-controlled sections, in accordance with 10.3.3 and 10.3.4. The appropriate strength reduction factors, ϕ, from 9.3.2 shall apply.

18.8.2 — Total amount of prestressed and nonprestressed reinforcement shall be adequate to develop a factored load at least 1.2 times the cracking load computed on the basis of the modulus of rupture f_r specified in 9.5.2.3. This provision shall be permitted to be waived for:

(a) Two-way, unbonded post-tensioned slabs; and

(b) Flexural members with shear and flexural strength at least twice that required by 9.2.

R18.8 — Limits for reinforcement of flexural members

R18.8.1 — The net tensile strain limits for compression- and tension-controlled sections given in 10.3.3 and 10.3.4 apply to prestressed sections. These provisions take the place of maximum reinforcement limits used in the 1999 code.

The net tensile strain limits for tension-controlled sections given in 10.3.4 may also be stated in terms of ω_p as defined in the 1999 and earlier editions of the code. The net tensile strain limit of 0.005 corresponds to $\omega_p = 0.32\beta_1$ for prestressed rectangular sections.

R18.8.2 — This provision is a precaution against abrupt flexural failure developing immediately after cracking. A flexural member designed according to code provisions requires considerable additional load beyond cracking to reach its flexural strength. Thus, considerable deflection would warn that the member strength is approaching. If the flexural strength were reached shortly after cracking, the warning deflection would not occur.

CODE

18.8.3 — Part or all of the bonded reinforcement consisting of bars or tendons shall be provided as close as practicable to the tension face in prestressed flexural members. In members prestressed with unbonded tendons, the minimum bonded reinforcement consisting of bars or tendons shall be as required by 18.9.

COMMENTARY

R18.8.3 — Some bonded steel is required to be placed near the tension face of prestressed flexural members. The purpose of this bonded steel is to control cracking under full service loads or overloads.

18.9 — Minimum bonded reinforcement

18.9.1 — A minimum area of bonded reinforcement shall be provided in all flexural members with unbonded tendons as required by 18.9.2 and 18.9.3.

18.9.2 — Except as provided in 18.9.3, minimum area of bonded reinforcement shall be computed by

$$A_s = 0.004A_{ct} \quad (18\text{-}6)$$

where A_{ct} is area of that part of cross section between the flexural tension face and center of gravity of gross section.

18.9.2.1 — Bonded reinforcement required by Eq. (18-6) shall be uniformly distributed over precompressed tensile zone as close as practicable to extreme tension fiber.

18.9.2.2 — Bonded reinforcement shall be required regardless of service load stress conditions.

18.9.3 — For two-way flat slab systems, minimum area and distribution of bonded reinforcement shall be as required in 18.9.3.1, 18.9.3.2, and 18.9.3.3.

R18.9 — Minimum bonded reinforcement

R18.9.1 — Some bonded reinforcement is required by the code in members prestressed with unbonded tendons to ensure flexural performance at ultimate member strength, rather than as a tied arch, and to limit crack width and spacing at service load when concrete tensile stresses exceed the modulus of rupture. Providing the minimum bonded reinforcement as stipulated in 18.9 helps to ensure adequate performance.

Research has shown that unbonded post-tensioned members do not inherently provide large capacity for energy dissipation under severe earthquake loadings because the member response is primarily elastic. For this reason, unbonded post-tensioned structural elements reinforced in accordance with the provisions of this section should be assumed to carry only vertical loads and to act as horizontal diaphragms between energy dissipating elements under earthquake loadings of the magnitude defined in 21.2.1.1. The minimum bonded reinforcement areas required by Eq. (18-6) and (18-8) are absolute minimum areas independent of grade of steel or design yield strength.

R18.9.2 — The minimum amount of bonded reinforcement for members other than two-way flat slab systems is based on research comparing the behavior of bonded and unbonded post-tensioned beams.[18.8] Based on this research, it is advisable to apply the provisions of 18.9.2 also to one-way slab systems.

R18.9.3 — The minimum amount of bonded reinforcement in two-way flat slab systems is based on reports by ACI-ASCE Committee 423.[18.3,18.9] Limited research available for two-way flat slabs with drop panels[18.10] indicates that behavior of these particular systems is similar to the behavior of flat plates. Reference 18.9 was revised by Committee 423 in 1983 to clarify that Section 18.9.3 applies to two-way flat slab systems.

CODE

18.9.3.1 — Bonded reinforcement shall not be required in positive moment areas where f_t, the extreme fiber stress in tension in the precompressed tensile zone at service loads, (after allowance for all prestress losses) does not exceed $2\sqrt{f_c'}$.

18.9.3.2 — In positive moment areas where computed tensile stress in concrete at service load exceeds $2\sqrt{f_c'}$, minimum area of bonded reinforcement shall be computed by

$$A_s = \frac{N_c}{0.5 f_y} \qquad \text{(18-7)}$$

where the value of f_y used in Eq. (18-7) shall not exceed 60,000 psi. Bonded reinforcement shall be uniformly distributed over precompressed tensile zone as close as practicable to the extreme tension fiber.

18.9.3.3 — In negative moment areas at column supports, the minimum area of bonded reinforcement A_s in the top of the slab in each direction shall be computed by

$$A_s = 0.00075 A_{cf} \qquad \text{(18-8)}$$

where A_{cf} is the larger gross cross-sectional area of the slab-beam strips in two orthogonal equivalent frames intersecting at a column in a two-way slab.

Bonded reinforcement required by Eq. (18-8) shall be distributed between lines that are $1.5h$ outside opposite faces of the column support. At least four bars or wires shall be provided in each direction. Spacing of bonded reinforcement shall not exceed 12 in.

18.9.4 — Minimum length of bonded reinforcement required by 18.9.2 and 18.9.3 shall be as required in 18.9.4.1, 18.9.4.2, and 18.9.4.3.

18.9.4.1 — In positive moment areas, minimum length of bonded reinforcement shall be one-third the clear span length, ℓ_n, and centered in positive moment area.

18.9.4.2 — In negative moment areas, bonded reinforcement shall extend one-sixth the clear span, ℓ_n, on each side of support.

COMMENTARY

R18.9.3.1 — For usual loads and span lengths, flat plate tests summarized in the Committee 423 report[18.3] and experience since the 1963 code was adopted indicate satisfactory performance without bonded reinforcement in the areas described in 18.9.3.1.

R18.9.3.2 — In positive moment areas, where the concrete tensile stresses are between $2\sqrt{f_c'}$ and $6\sqrt{f_c'}$, a minimum bonded reinforcement area proportioned according to Eq. (18-7) is required.The tensile force N_c is computed at service load on the basis of an uncracked, homogeneous section.

R18.9.3.3 — Research on unbonded post-tensioned two-way flat slab systems evaluated by ACI-ASCE Committee 423[18.1,18.3,18.9,18.10] shows that bonded reinforcement in negative moment regions, proportioned on the basis of 0.075 percent of the cross-sectional area of the slab-beam strip, provides sufficient ductility and reduces crack width and spacing. To account for different adjacent tributary spans, Eq. (18-8) is given on the basis of the equivalent frame as defined in 13.7.2 and pictured in Fig. R13.7.2. For rectangular slab panels, Eq. (18-8) is conservatively based upon the larger of the cross-sectional areas of the two intersecting equivalent frame slab-beam strips at the column. This ensures that the minimum percentage of steel recommended by research is provided in both directions. Concentration of this reinforcement in the top of the slab directly over and immediately adjacent to the column is important. Research also shows that where low tensile stresses occur at service loads, satisfactory behavior has been achieved at factored loads without bonded reinforcement. However, the code requires minimum bonded reinforcement regardless of service load stress levels to help ensure flexural continuity and ductility, and to limit crack widths and spacing due to overload, temperature, or shrinkage. Research on post-tensioned flat plate-to-column connections is reported in References 18.11 through 18.15.

R18.9.4 — Bonded reinforcement should be adequately anchored to develop factored load forces. The requirements of Chapter 12 will ensure that bonded reinforcement required for flexural strength under factored loads in accordance with 18.7.3, or for tensile stress conditions at service load in accordance with 18.9.3.2, will be adequately anchored to develop tension or compression forces. The minimum lengths apply for bonded reinforcement required by 18.9.2 or 18.9.3.3, but not required for flexural strength in accordance with 18.7.3. Research[18.10] on continuous spans shows that these minimum lengths provide adequate behavior under service load and factored load conditions.

CODE

18.9.4.3 — Where bonded reinforcement is provided for ϕM_n in accordance with 18.7.3, or for tensile stress conditions in accordance with 18.9.3.2, minimum length also shall conform to provisions of Chapter 12.

18.10 — Statically indeterminate structures

18.10.1 — Frames and continuous construction of prestressed concrete shall be designed for satisfactory performance at service load conditions and for adequate strength.

18.10.2 — Performance at service load conditions shall be determined by elastic analysis, considering reactions, moments, shears, and axial forces induced by prestressing, creep, shrinkage, temperature change, axial deformation, restraint of attached structural elements, and foundation settlement.

18.10.3 — Moments used to compute required strength shall be the sum of the moments due to reactions induced by prestressing (with a load factor of 1.0) and the moments due to factored loads. Adjustment of the sum of these moments shall be permitted as allowed in 18.10.4.

18.10.4 — Redistribution of negative moments in continuous prestressed flexural members

18.10.4.1 — Where bonded reinforcement is provided at supports in accordance with 18.9, it shall be permitted to increase or decrease negative moments calculated by elastic theory for any assumed loading, in accordance with 8.4.

18.10.4.2 — The modified negative moments shall be used for calculating moments at sections within spans for the same loading arrangement.

COMMENTARY

R18.10 — Statically indeterminate structures

R18.10.3 — For statically indeterminate structures, the moments due to reactions induced by prestressing forces, referred to as secondary moments, are significant in both the elastic and inelastic states (see References 18.16 through 18.18). The elastic deformations caused by a nonconcordant tendon change the amount of inelastic rotation required to obtain a given amount of moment redistribution. Conversely, for a beam with a given inelastic rotational capacity, the amount by which the moment at the support may be varied is changed by an amount equal to the secondary moment at the support due to prestressing. Thus, the code requires that secondary moments be included in determining design moments.

To determine the moments used in design, the order of calculation should be: (a) determine moments due to dead and live load; (b) modify by algebraic addition of secondary moments; (c) redistribute as permitted. A positive secondary moment at the support caused by a tendon transformed downward from a concordant profile will reduce the negative moments near the supports and increase the positive moments in the midspan regions. A tendon that is transformed upward will have the reverse effect.

R18.10.4 — Redistribution of negative moments in continuous prestressed flexural members

The provisions for redistribution of negative moments given in 8.4 apply equally to prestressed members. See Reference 18.19 for a comparison of research results and to Section 18.10.4 of the 1999 ACI Code.

For the moment redistribution principles of 18.10.4 to be applicable to beams with unbonded tendons, it is necessary that such beams contain sufficient bonded reinforcement to ensure they will act as beams after cracking and not as a series of tied arches. The minimum bonded reinforcement requirements of 18.9 will serve this purpose.

CODE

COMMENTARY

18.11 — Compression members — Combined flexure and axial loads

18.11.1 — Prestressed concrete members subject to combined flexure and axial load, with or without nonprestressed reinforcement, shall be proportioned by the strength design methods of this code. Effects of prestress, creep, shrinkage, and temperature change shall be included.

18.11.2 — Limits for reinforcement of prestressed compression members

18.11.2.1 — Members with average compressive stress in concrete due to effective prestress force only less than 225 psi shall have minimum reinforcement in accordance with 7.10, 10.9.1 and 10.9.2 for columns, or 14.3 for walls.

18.11.2.2—Except for walls, members with average compressive stress in concrete due to effective prestress force only equal to or greater than 225 psi shall have all tendons enclosed by spirals or lateral ties in accordance with (a) through (d):

(a) Spirals shall conform to 7.10.4;

(b) Lateral ties shall be at least No. 3 in size or welded wire reinforcement of equivalent area, and shall be spaced vertically not to exceed 48 tie bar or wire diameters, or the least dimension of the compression member;

(c) Ties shall be located vertically not more than half a tie spacing above top of footing or slab in any story, and not more than half a tie spacing below the lowest horizontal reinforcement in members supported above;

(d) Where beams or brackets frame into all sides of a column, ties shall be terminated not more than 3 in. below lowest reinforcement in such beams or brackets.

18.11.2.3 — For walls with average compressive stress in concrete due to effective prestress force only equal to or greater than 225 psi, minimum reinforcement required by 14.3 shall not apply where structural analysis shows adequate strength and stability.

18.12 — Slab systems

18.12.1 — Factored moments and shears in prestressed slab systems reinforced for flexure in more than one direction shall be determined in accordance with provisions of 13.7 (excluding 13.7.7.4 and 13.7.7.5), or by more detailed design procedures.

R18.11 — Compression members — Combined flexure and axial loads

R18.11.2 — Limits for reinforcement of prestressed compression members

R18.11.2.3 — The minimum amounts of reinforcement in 14.3 need not apply to prestressed concrete walls, provided the average compressive stress in concrete due to effective prestress force only is 225 psi or greater and a structural analysis is performed to show adequate strength and stability with lower amounts of reinforcement.

R18.12 — Slab systems

R18.12.1 — Use of the equivalent frame method of analysis (see 13.7) or more precise analysis procedures is required for determination of both service and factored moments and shears for prestressed slab systems. The equivalent frame method of analysis has been shown by tests of large struc-

CODE

COMMENTARY

tural models to satisfactorily predict factored moments and shears in prestressed slab systems. (See References 18.11 through 18.13, and 18.20 through 18.22.) The referenced research also shows that analysis using prismatic sections or other approximations of stiffness may provide erroneous results on the unsafe side. Section 13.7.7.4 is excluded from application to prestressed slab systems because it relates to reinforced slabs designed by the direct design method, and because moment redistribution for prestressed slabs is covered in 18.10.4. Section 13.7.7.5 does not apply to prestressed slab systems because the distribution of moments between column strips and middle strips required by 13.7.7.5 is based on tests for nonprestressed concrete slabs. Simplified methods of analysis using average coefficients do not apply to prestressed concrete slab systems.

18.12.2 — ϕM_n of prestressed slabs required by 9.3 at every section shall be greater than or equal to M_u considering 9.2, 18.10.3, and 18.10.4. ϕV_n of prestressed slabs at columns required by 9.3 shall be greater than or equal to V_u considering 9.2, 11.1, 11.12.2, and 11.12.6.2.

R18.12.2 — Tests indicate that the moment and shear strength of prestressed slabs is controlled by total prestessing steel strength and by the amount and location of nonprestressed reinforcement, rather than by tendon distribution. (See References 18.11 through 18.13, and 18.20 through 18.22.)

18.12.3 — At service load conditions, all serviceability limitations, including limits on deflections, shall be met, with appropriate consideration of the factors listed in 18.10.2.

R18.12.3 — For prestressed flat slabs continuous over two or more spans in each direction, the span-thickness ratio generally should not exceed 42 for floors and 48 for roofs; these limits may be increased to 48 and 52, respectively, if calculations verify that both short- and long-term deflection, camber, and vibration frequency and amplitude are not objectionable.

Short- and long-term deflection and camber should be computed and checked against the requirements of serviceability of the structure.

The maximum length of a slab between construction joints is generally limited to 100 to 150 ft to minimize the effects of slab shortening, and to avoid excessive loss of prestress due to friction.

18.12.4 — For uniformly distributed loads, spacing of tendons or groups of tendons in one direction shall not exceed the smaller of eight times the slab thickness and 5 ft. Spacing of tendons also shall provide a minimum average effective prestress of 125 psi on the slab section tributary to the tendon or tendon group. A minimum of two tendons shall be provided in each direction through the critical shear section over columns. Special consideration of tendon spacing shall be provided for slabs with concentrated loads.

R18.12.4 — This section provides specific guidance concerning tendon distribution that will permit the use of banded tendon distributions in one direction. This method of tendon distribution has been shown to provide satisfactory performance by structural research.

18.12.5 — In slabs with unbonded tendons, bonded reinforcement shall be provided in accordance with 18.9.3 and 18.9.4.

18.12.6 — In lift slabs, bonded bottom reinforcement shall be detailed in accordance with 13.3.8.6.

CODE

18.13 — Post-tensioned tendon anchorage zones

18.13.1 — Anchorage zone

The anchorage zone shall be considered as composed of two zones:

(a) The local zone is the rectangular prism (or equivalent rectangular prism for circular or oval anchorages) of concrete immediately surrounding the anchorage device and any confining reinforcement;

(b) The general zone is the anchorage zone as defined in 2.2 and includes the local zone.

18.13.2 — Local zone

18.13.2.1 — Design of local zones shall be based upon the factored prestressing force, P_{pu}, and the requirements of 9.2.5 and 9.3.2.5.

18.13.2.2 — Local-zone reinforcement shall be provided where required for proper functioning of the anchorage device.

18.13.2.3 — Local-zone requirements of 18.13.2.2 are satisfied by 18.14.1 or 18.15.1 and 18.15.2.

COMMENTARY

R18.13 — Post-tensioned tendon anchorage zones

Section 18.13 was extensively revised in the 1999 code and was made compatible with the 1996 AASHTO "Standard Specifications for Highway Bridges"[18.23] and the recommendations of NCHRP Report 356.[18.24]

Following the adoption by AASHTO 1994 of comprehensive provisions for post-tensioned anchorage zones, ACI Committee 318 revised the code to be generally consistent with the AASHTO requirements. Thus, the highly detailed AASHTO provisions for analysis and reinforcement detailing are deemed to satisfy the more general ACI 318 requirements. In the specific areas of anchorage device evaluation and acceptance testing, ACI 318 incorporates the detailed AASHTO provisions by reference.

R18.13.1 — Anchorage zone

Based on the Principle of Saint-Venant, the extent of the anchorage zone may be estimated as approximately equal to the largest dimension of the cross section. Local zone and general zone are shown in Fig. R18.13.1(a). When anchorage devices located away from the end of the member are tensioned, large tensile stresses exist locally behind and ahead of the device. These tensile stresses are induced by incompatibility of deformations ahead of [as shown in Fig. R.18.13.1(b)] and behind the anchorage device. The entire shaded region should be considered, as shown in Fig. R18.13.1(b).

R18.13.2 — Local zone

The local zone resists the very high local stresses introduced by the anchorage device and transfers them to the remainder of the anchorage zone. The behavior of the local zone is strongly influenced by the specific characteristics of the anchorage device and its confining reinforcement, and less influenced by the geometry and loading of the overall structure. Local-zone design sometimes cannot be completed until specific anchorage devices are determined at the shop drawing stage. When special anchorage devices are used, the anchorage device supplier should furnish the test information to show the device is satisfactory under AASHTO "Standard Specifications for Highway Bridges," Division II, Article 10.3.2.3 and provide information regarding necessary conditions for use of the device. The main considerations in local-zone design are the effects of the high bearing pressure and the adequacy of any confining reinforcement provided to increase the capacity of the concrete resisting bearing stresses.

The factored prestressing force P_{pu} is the product of the load factor (1.2 from Section 9.2.5) and the maximum pre-

CODE

COMMENTARY

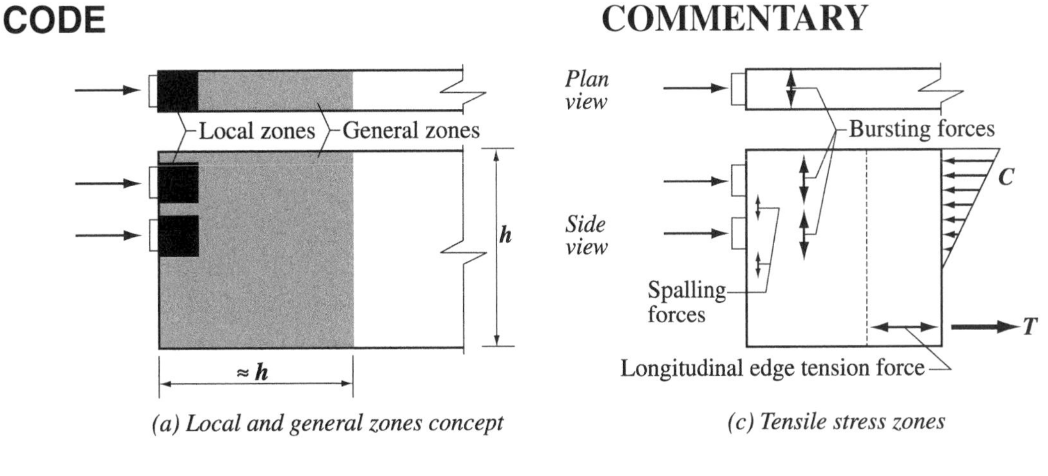

(a) Local and general zones concept

(c) Tensile stress zones

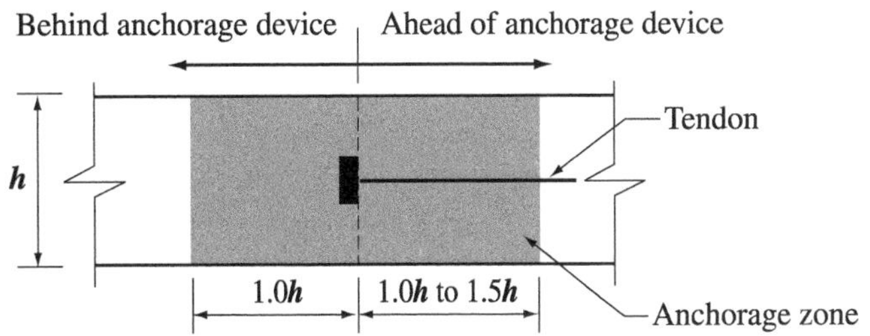

(b) General zone for anchorage device located away from the end of a member

Fig. R18.13.1—Anchorage zones

stressing force allowed. Under 18.5.1, this is usually overstressing due to $0.94f_{py}$, but not greater than $0.8f_{pu}$, which is permitted for short periods of time.

$$P_{pu} = (1.2)(0.80)f_{pu}A_{ps} = 0.96f_{pu}A_{ps}$$

18.13.3 — General zone

18.13.3.1 — Design of general zones shall be based upon the factored prestressing force, P_{pu}, and the requirements of 9.2.5 and 9.3.2.5.

18.13.3.2 — General-zone reinforcement shall be provided where required to resist bursting, spalling, and longitudinal edge tension forces induced by anchorage devices. Effects of abrupt change in section shall be considered.

18.13.3.3 — The general-zone requirements of 18.13.3.2 are satisfied by 18.13.4, 18.13.5, 18.13.6 and whichever one of 18.14.2 or 18.14.3 or 18.15.3 is applicable.

R18.13.3 — General zone

Within the general zone the usual assumption of beam theory that plane sections remain plane is not valid.

Design should consider all regions of tensile stresses that can be caused by the tendon anchorage device, including bursting, spalling, and edge tension as shown in Fig. R18.13.1(c). Also, the compressive stresses immediately ahead [as shown in Fig. R18.13.1(b)] of the local zone should be checked. Sometimes reinforcement requirements cannot be determined until specific tendon and anchorage device layouts are determined at the shop-drawing stage. Design and approval responsibilities should be clearly assigned in the project drawings and specifications.

Abrupt changes in section can cause substantial deviation in force paths. These deviations can greatly increase tension forces as shown in Fig. R18.13.3.

CODE

COMMENTARY

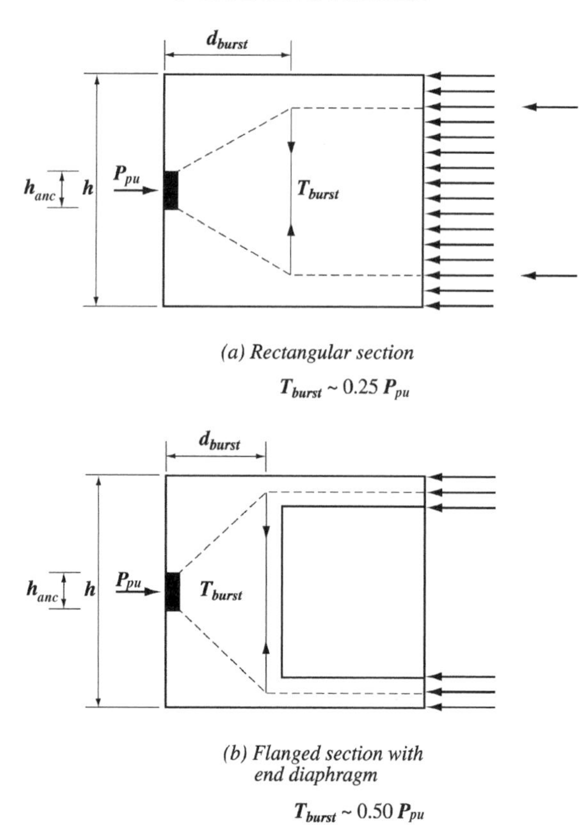

Fig. R18.13.3—Effect of cross section change.

18.13.4 — Nominal material strengths

18.13.4.1 — Nominal tensile stress of bonded reinforcement is limited to f_y for nonprestressed reinforcement and to f_{py} for prestressed reinforcement. Nominal tensile stress of unbonded prestressed reinforcement for resisting tensile forces in the anchorage zone shall be limited to $f_{ps} = f_{se} + 10{,}000$.

18.13.4.2 — Except for concrete confined within spirals or hoops providing confinement equivalent to that corresponding to Eq. (10-5), nominal compressive strength of concrete in the general zone shall be limited to $0.7\lambda f'_{ci}$.

18.13.4.3 — Compressive strength of concrete at time of post-tensioning shall be specified on the design drawings. Unless oversize anchorage devices sized to compensate for the lower compressive strength are used or the prestressing steel is stressed to no more than 50 percent of the final prestressing force, prestressing steel shall not be stressed until compressive strength of concrete as indicated by tests consistent with the curing of the member, is at least 4000 psi for multistrand tendons or at least 2500 psi for single-strand or bar tendons.

R18.13.4 — Nominal material strengths

Some inelastic deformation of concrete is expected because anchorage zone design is based on a strength approach. The low value for the nominal compressive strength for unconfined concrete reflects this possibility. For well-confined concrete, the effective compressive strength could be increased (See Reference 18.24). The value for nominal tensile strength of bonded prestressing steel is limited to the yield strength of the prestressing steel because Eq. (18-3) may not apply to these nonflexural applications. The value for unbonded prestressing steel is based on the values of 18.7.2 (b) and (c), but is somewhat limited for these short-length, nonflexural applications. Test results given in Reference 18.24 indicate that the compressive stress introduced by auxiliary prestressing applied perpendicular to the axis of the main tendons is effective in increasing the anchorage zone capacity. The inclusion of the λ factor for lightweight concrete reflects its lower tensile strength, which is an indirect factor in limiting compressive stresses, as well as the wide scatter and brittleness exhibited in some lightweight concrete anchorage zone tests.

The designer is required to specify concrete strength at the time of stressing in the project drawings and specifications. To limit early shrinkage cracking, monostrand tendons are sometimes stressed at concrete strengths less than 2500 psi. In such cases, either oversized monostrand anchorages are used, or the strands are stressed in stages, often to levels 1/3 to 1/2 the final prestressing force.

CODE

18.13.5 — Design methods

18.13.5.1 — The following methods shall be permitted for the design of general zones provided that the specific procedures used result in prediction of strength in substantial agreement with results of comprehensive tests:

(a) Equilibrium based plasticity models (strut-and-tie models);

(b) Linear stress analysis (including finite element analysis or equivalent); or

(c) Simplified equations where applicable.

18.13.5.2 — Simplified equations shall not be used where member cross sections are nonrectangular, where discontinuities in or near the general zone cause deviations in the force flow path, where minimum edge distance is less than 1-1/2 times the anchorage device lateral dimension in that direction, or where multiple anchorage devices are used in other than one closely spaced group.

COMMENTARY

R18.13.5 — Design methods

The list of design methods in 18.13.5.1 includes those procedures for which fairly specific guidelines have been given in References 18.23 and 18.24. These procedures have been shown to be conservative predictors of strength when compared to test results.[18.24] The use of strut-and-tie models is especially helpful for general zone design.[18.24] In many anchorage applications, where substantial or massive concrete regions surround the anchorages, simplified equations can be used except in the cases noted in 18.13.5.2.

For many cases, simplified equations based on References 18.23 and 18.24 can be used. Values for the magnitude of the bursting force, T_{burst}, and for its centroidal distance from the major bearing surface of the anchorage, d_{burst}, may be estimated from Eq. (R18-1) and (R18-2), respectively. The terms of Eq. (R18-1) and (R18-2) are shown in Fig. R18.13.5 for a prestressing force with small eccentricity. In the applications of Eq. (R18-1) and (R18-2), the specified stressing sequence should be considered if more than one tendon is present.

$$T_{burst} = 0.25\Sigma P_{pu}\left(1 - \frac{h_{anc}}{h}\right) \quad \text{(R18-1)}$$

$$d_{burst} = 0.5(h - 2e_{anc}) \quad \text{(R18-2)}$$

where:

ΣP_{pu} = the sum of the P_{pu} forces from the individual tendons, lb;

h_{anc} = the depth of anchorage device or single group of closely spaced devices in the direction considered, in.;

e_{anc} = the eccentricity (always taken as positive) of the anchorage device or group of closely spaced devices with respect to the centroid of the cross section, in.;

h = the depth of the cross section in the direction considered, in.

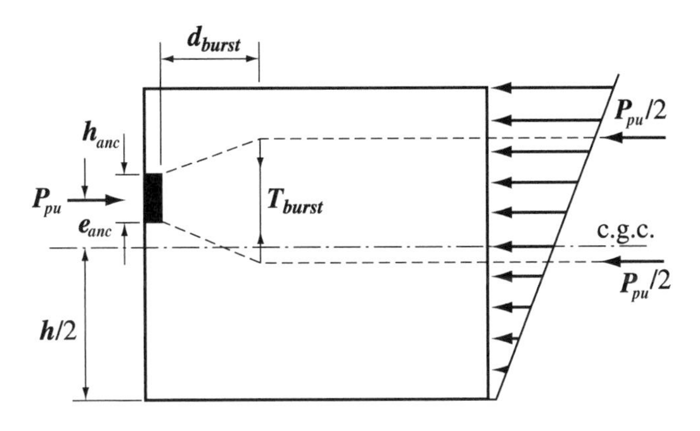

Fig. R18.13.5—Strut-and-tie model example

CODE | COMMENTARY

Anchorage devices should be treated as closely spaced if their center-to-center spacing does not exceed 1.5 times the width of the anchorage device in the direction considered.

The spalling force for tendons for which the centroid lies within the kern of the section may be estimated as 2 percent of the total factored prestressing force, except for multiple anchorage devices with center-to-center spacing greater than 0.4 times the depth of the section. For large spacings and for cases where the centroid of the tendons is located outside the kern, a detailed analysis is required. In addition, in the post-tensioning of thin sections, or flanged sections, or irregular sections, or when the tendons have appreciable curvature within the general zone, more general procedures such as those of AASHTO Articles 9.21.4 and 9.21.5 will be required. Detailed recommendations for design principles that apply to all design methods are given in Article 9.21.3.4 of Reference 18.23.

18.13.5.3 — The stressing sequence shall be specified on the design drawings and considered in the design.

R18.13.5.3 — The sequence of anchorage device stressing can have a significant effect on the general zone stresses. Therefore, it is important to consider not only the final stage of a stressing sequence with all tendons stressed, but also intermediate stages during construction. The most critical bursting forces caused by each of the sequentially post-tensioned tendon combinations, as well as that of the entire group of tendons, should be taken into account.

18.13.5.4 — Three-dimensional effects shall be considered in design and analyzed using three-dimensional procedures or approximated by considering the summation of effects for two orthogonal planes.

R18.13.5.4 — The provision for three-dimensional effects was included to alert the designer to effects perpendicular to the main plane of the member, such as bursting forces in the thin direction of webs or slabs. In many cases these effects can be determined independently for each direction, but some applications require a fully three-dimensional analysis (for example diaphragms for the anchorage of external tendons).

18.13.5.5 — For anchorage devices located away from the end of the member, bonded reinforcement shall be provided to transfer at least $0.35P_{pu}$ into the concrete section behind the anchor. Such reinforcement shall be placed symmetrically around the anchorage devices and shall be fully developed both behind and ahead of the anchorage devices.

R18.13.5.5 — Where anchorages are located away from the end of a member, local tensile stresses are generated behind these anchorages [see Fig. R18.13.1(b)] due to compatibility requirements for deformations ahead of and behind the anchorages. Bonded tie-back reinforcement is required in the immediate vicinity of the anchorage to limit the extent of cracking behind the anchorage. The requirement of $0.35P_{pu}$ was developed using 25 percent of the unfactored prestressing force being resisted by reinforcement at $0.6f_y$.

18.13.5.6 — Where tendons are curved in the general zone, except for monostrand tendons in slabs or where analysis shows reinforcement is not required, bonded reinforcement shall be provided to resist radial and splitting forces.

CODE

18.13.5.7 — Except for monostrand tendons in slabs or where analysis shows reinforcement is not required, minimum reinforcement with a nominal tensile strength equal to 2 percent of each factored prestressing force shall be provided in orthogonal directions parallel to the back face of all anchorage zones to limit spalling.

18.13.5.8 — Tensile strength of concrete shall be neglected in calculations of reinforcement requirements.

18.13.6 — Detailing requirements

Selection of reinforcement sizes, spacings, cover, and other details for anchorage zones shall make allowances for tolerances on the bending, fabrication, and placement of reinforcement, for the size of aggregate, and for adequate placement and consolidation of the concrete.

18.14 — Design of anchorage zones for monostrand or single 5/8 in. diameter bar tendons

18.14.1 — Local zone design

Monostrand or single 5/8 in. or smaller diameter bar anchorage devices and local zone reinforcement shall meet the requirements of ACI 423.6 or the special anchorage device requirements of 18.15.2.

18.14.2 — General-zone design for slab tendons

18.14.2.1 — For anchorage devices for 0.5 in. or smaller diameter strands in normalweight concrete slabs, minimum reinforcement meeting the requirements of 18.14.2.2 and 18.14.2.3 shall be provided unless a detailed analysis satisfying 18.13.5 shows such reinforcement is not required.

18.14.2.2 — Two horizontal bars at least No. 4 in size shall be provided parallel to the slab edge. They shall be permitted to be in contact with the front face of the anchorage device and shall be within a distance of $1/2h$ ahead of each device. Those bars shall extend at least 6 in. either side of the outer edges of each device.

18.14.2.3 — If the center-to-center spacing of anchorage devices is 12 in. or less, the anchorage devices shall be considered as a group. For each group of six or more anchorage devices, $n + 1$ hairpin bars or closed stirrups at least No. 3 in size shall be provided, where n is the number of anchorage devices. One hairpin bar or stirrup shall be placed between each anchorage device and one on each side

COMMENTARY

R18.14 — Design of anchorage zones for monostrand or single 5/8 in. diameter bar tendons

R18.14.2 — General-zone design for slab tendons

For monostrand slab tendons, the general-zone minimum reinforcement requirements are based on the recommendations of ACI-ASCE Committee 423,[18.25] which shows typical details. The horizontal bars parallel to the edge required by 18.14.2.2 should be continuous where possible.

The tests on which the recommendations of Reference 18.24 were based were limited to anchorage devices for 1/2 in. diameter, 270 ksi strand, unbonded tendons in normalweight concrete. Thus, for larger strand anchorage devices and for all use in lightweight concrete slabs, ACI-ASCE Committee 423 recommended that the amount and spacing of reinforcement should be conservatively adjusted to provide for the larger anchorage force and smaller splitting tensile strength of lightweight concrete.[18.25]

Both References 18.24 and 18.25 recommend that hairpin bars also be furnished for anchorages located within 12 in. of slab corners to resist edge tension forces. The words “ahead of” in 18.14.2.3 have the meaning shown in Fig. R18.13.1.

In those cases where multistrand anchorage devices are used for slab tendons, 18.15 is applicable.

CODE

of the group. The hairpin bars or stirrups shall be placed with the legs extending into the slab perpendicular to the edge. The center portion of the hairpin bars or stirrups shall be placed perpendicular to the plane of the slab from $3h/8$ to $h/2$ ahead of the anchorage devices.

18.14.2.4 — For anchorage devices not conforming to 18.14.2.1, minimum reinforcement shall be based upon a detailed analysis satisfying 18.13.5.

18.14.3 — General-zone design for groups of monostrand tendons in beams and girders

Design of general zones for groups of monostrand tendons in beams and girders shall meet the requirements of 18.13.3 through 18.13.5.

18.15 — Design of anchorage zones for multistrand tendons

18.15.1 — Local zone design

Basic multistrand anchorage devices and local zone reinforcement shall meet the requirements of AASHTO "Standard Specification for Highway Bridges," Division I, Articles 9.21.7.2.2 through 9.21.7.2.4.

Special anchorage devices shall satisfy the tests required in AASHTO "Standard Specification for Highway Bridges," Division I, Article 9.21.7.3 and described in AASHTO "Standard Specification for Highway Bridges," Division II, Article 10.3.2.3.

18.15.2 — Use of special anchorage devices

Where special anchorage devices are to be used, supplemental skin reinforcement shall be furnished in the corresponding regions of the anchorage zone, in addition to the confining reinforcement specified for the anchorage device. This supplemental reinforcement shall be similar in configuration and at least equivalent in volumetric ratio to any supplementary skin reinforcement used in the qualifying acceptance tests of the anchorage device.

COMMENTARY

The bursting reinforcement perpendicular to the plane of the slab required by 18.14.2.3 for groups of relatively closely spaced tendons should also be provided in the case of widely spaced tendons if an anchorage device failure could cause more than local damage.

R18.14.3 — General-zone design for groups of monostrand tendons in beams and girders

Groups of monostrand tendons with individual monostrand anchorage devices are often used in beams and girders. Anchorage devices can be treated as closely spaced if their center-to-center spacing does not exceed 1.5 times the width of the anchorage device in the direction considered. If a beam or girder has a single anchorage device or a single group of closely spaced anchorage devices, the use of simplified equations such as those given in R18.13.5 is allowed, unless 18.13.5.2 governs. More complex conditions can be designed using strut-and-tie models. Detailed recommendations for use of such models are given in References 18.25 and 18.26 as well as in R18.13.5.

R18.15 — Design of anchorage zones for multistrand tendons

R18.15.1 — Local zone design

See R18.13.2.

R18.15.2 — Use of special anchorage devices

Skin reinforcement is reinforcement placed near the outer faces in the anchorage zone to limit local crack width and spacing. Reinforcement in the general zone for other actions (flexure, shear, shrinkage, temperature and similar) may be used in satisfying the supplementary skin reinforcement requirement. Determination of the supplementary skin reinforcement depends on the anchorage device hardware used and frequently cannot be determined until the shop-drawing stage.

CODE

18.15.3 — General-zone design

Design for general zones for multistrand tendons shall meet the requirements of 18.13.3 through 18.13.5.

18.16 — Corrosion protection for unbonded tendons

18.16.1 — Unbonded prestressing steel shall be encased with sheathing. The prestressing steel shall be completely coated and the sheathing around the prestressing steel filled with suitable material to inhibit corrosion.

18.16.2 — Sheathing shall be watertight and continuous over entire length to be unbonded.

18.16.3 — For applications in corrosive environments, the sheathing shall be connected to all stressing, intermediate and fixed anchorages in a watertight fashion.

18.16.4 — Unbonded single strand tendons shall be protected against corrosion in accordance with ACI's "Specification for Unbonded Single-Strand Tendons (ACI 423.6)."

18.17 — Post-tensioning ducts

18.17.1 — Ducts for grouted tendons shall be mortar-tight and nonreactive with concrete, prestressing steel, grout, and corrosion inhibitor.

18.17.2 — Ducts for grouted single wire, single strand, or single bar tendons shall have an inside diameter at least 1/4 in. larger than the prestressing steel diameter.

18.17.3 — Ducts for grouted multiple wire, multiple strand, or multiple bar tendons shall have an inside cross-sectional area at least two times the cross-sectional area of the prestressing steel.

18.17.4 — Ducts shall be maintained free of ponded water if members to be grouted are exposed to temperatures below freezing prior to grouting.

18.18 — Grout for bonded tendons

18.18.1 — Grout shall consist of portland cement and water; or portland cement, sand, and water.

COMMENTARY

R18.16 — Corrosion protection for unbonded tendons

R18.16.1 — Suitable material for corrosion protection of unbonded prestressing steel should have the properties identified in Section 5.1 of Reference 18.26.

R18.16.2 — Typically, sheathing is a continuous, seamless, high-density polythylene material that is extruded directly onto the coated prestressing steel.

R18.16.4 — In the 1989 code, corrosion protection requirements for unbonded single strand tendons were added in accordance with the Post-Tensioning Institute's "Specification for Unbonded Single Strand Tendons." In the 2002 code, the reference changed to ACI 423.6.

R18.17 — Post-tensioning ducts

R18.17.4 — Water in ducts may cause distress to the surrounding concrete upon freezing. When strands are present, ponded water in ducts should also be avoided. A corrosion inhibitor should be used to provide temporary corrosion protection if prestressing steel is exposed to prolonged periods of moisture in the ducts before grouting.[18.27]

R18.18 — Grout for bonded tendons

Proper grout and grouting procedures are critical to post-tensioned construction.[18.28,18.29] Grout provides bond between the prestressing steel and the duct, and provides corrosion protection to the prestressing steel.

CODE

COMMENTARY

Past success with grout for bonded tendons has been with portland cement. A blanket endorsement of all cementitious materials (defined in 2.1) for use with this grout is deemed inappropriate because of a lack of experience or tests with cementitious materials other than portland cement and a concern that some cementitious materials might introduce chemicals listed as harmful to tendons in R18.18.2. Thus, portland cement in 18.18.1 and water-cement ratio in 18.18.3.3 are retained in the code.

18.18.2 — Materials for grout shall conform to 18.18.2.1 through 18.18.2.4.

18.18.2.1 — Portland cement shall conform to 3.2.

18.18.2.2 — Water shall conform to 3.4.

18.18.2.3 — Sand, if used, shall conform to "Standard Specification for Aggregate for Masonry Mortar" (ASTM C 144) except that gradation shall be permitted to be modified as necessary to obtain satisfactory workability.

18.18.2.4 — Admixtures conforming to 3.6 and known to have no injurious effects on grout, steel, or concrete shall be permitted. Calcium chloride shall not be used.

R18.18.2 — The limitations on admixtures in 3.6 apply to grout. Substances known to be harmful to tendons, grout, or concrete are chlorides, fluorides, sulfites, and nitrates. Aluminum powder or other expansive admixtures, when approved, should produce an unconfined expansion of 5 to 10 percent. Neat cement grout is used in almost all building construction. Use of finely graded sand in the grout should only be considered with large ducts having large void areas.

18.18.3 — Selection of grout proportions

18.18.3.1 — Proportions of materials for grout shall be based on either (a) or (b):

(a) Results of tests on fresh and hardened grout prior to beginning grouting operations; or

(b) Prior documented experience with similar materials and equipment and under comparable field conditions.

18.18.3.2 — Cement used in the work shall correspond to that on which selection of grout proportions was based.

18.18.3.3 — Water content shall be minimum necessary for proper pumping of grout; however, water-cement ratio shall not exceed 0.45 by weight.

18.18.3.4 — Water shall not be added to increase grout flowability that has been decreased by delayed use of the grout.

R18.18.3 — Selection of grout proportions

Grout proportioned in accordance with these provisions will generally lead to 7-day compressive strength on standard 2 in. cubes in excess of 2500 psi and 28-day strengths of about 4000 psi. The handling and placing properties of grout are usually given more consideration than strength when designing grout mixtures.

18.18.4 — Mixing and pumping grout

18.18.4.1 — Grout shall be mixed in equipment capable of continuous mechanical mixing and agitation that will produce uniform distribution of materials, passed through screens, and pumped in a manner that will completely fill the ducts.

R18.18.4 — Mixing and pumping grout

In an ambient temperature of 35 F, grout with an initial minimum temperature of 60 F may require as much as 5 days to reach 800 psi. A minimum grout temperature of 60 F is suggested because it is consistent with the recommended minimum temperature for concrete placed at an ambient temperature of 35 F. Quickset grouts, when approved, may require shorter periods of protection and the recommenda-

CODE

18.18.4.2 — Temperature of members at time of grouting shall be above 35 F and shall be maintained above 35 F until field-cured 2 in. cubes of grout reach a minimum compressive strength of 800 psi.

18.18.4.3 — Grout temperatures shall not be above 90 F during mixing and pumping.

18.19 — Protection for prestressing steel

Burning or welding operations in the vicinity of prestressing steel shall be performed so that prestressing steel is not subject to excessive temperatures, welding sparks, or ground currents.

18.20 — Application and measurement of prestressing force

18.20.1 — Prestressing force shall be determined by both of (a) and (b):

(a) Measurement of steel elongation. Required elongation shall be determined from average load-elongation curves for the prestressing steel used;

(b) Observation of jacking force on a calibrated gage or load cell or by use of a calibrated dynamometer.

Cause of any difference in force determination between (a) and (b) that exceeds 5 percent for pretensioned elements or 7 percent for post-tensioned construction shall be ascertained and corrected.

18.20.2 — Where the transfer of force from the bulkheads of pretensioning bed to the concrete is accomplished by flame cutting prestressing steel, cutting points and cutting sequence shall be predetermined to avoid undesired temporary stresses.

18.20.3 — Long lengths of exposed pretensioned strand shall be cut near the member to minimize shock to concrete.

COMMENTARY

tions of the suppliers should be followed. Test cubes should be cured under temperature and moisture conditions as close as possible to those of the grout in the member. Grout temperatures in excess of 90 F will lead to difficulties in pumping.

R18.20 — Application and measurement of prestressing force

R18.20.1 — Elongation measurements for prestressed elements should be in accordance with the procedures outlined in the "Manual for Quality Control for Plants and Production of Precast and Prestressed Concrete Products," published by the Precast/Prestressed Concrete Institute.[18.30]

Section 18.18.1 of the 1989 code was revised to permit 7 percent tolerance in prestressing steel force determined by gage pressure and elongation measurements for post-tensioned construction. Elongation measurements for post-tensioned construction are affected by several factors that are less significant, or that do not exist, for pretensioned elements. The friction along prestressing steel in post-tensioning applications may be affected to varying degrees by placing tolerances and small irregularities in tendon profile due to concrete placement. The friction coefficients between the prestressing steel and the duct are also subject to variation. The 5 percent tolerance that has appeared since the 1963 code was proposed by ACI-ASCE Committee 423 in 1958,[18.3] and primarily reflected experience with production of pretensioned concrete elements. Because the tendons for pretensioned elements are usually stressed in the air with minimal friction effects, the 5 percent tolerance for such elements was retained.

CODE

18.20.4 — Total loss of prestress due to unreplaced broken prestressing steel shall not exceed 2 percent of total prestress.

18.21 — Post-tensioning anchorages and couplers

18.21.1 — Anchorages and couplers for bonded and unbonded tendons shall develop at least 95 percent of the specified breaking strength of the prestressing steel, when tested in an unbonded condition, without exceeding anticipated set. For bonded tendons, anchorages and couplers shall be located so that 100 percent of the specified breaking strength of the prestressing steel shall be developed at critical sections after the prestressing steel is bonded in the member.

18.21.2 — Couplers shall be placed in areas approved by the engineer and enclosed in housing long enough to permit necessary movements.

18.21.3 — In unbonded construction subject to repetitive loads, special attention shall be given to the possibility of fatigue in anchorages and couplers.

18.21.4 — Anchorages, couplers, and end fittings shall be permanently protected against corrosion.

COMMENTARY

R18.20.4 — This provision applies to all prestressed concrete members. For cast-in-place post-tensioned slab systems, a member should be that portion considered as an element in the design, such as the joist and effective slab width in one-way joist systems, or the column strip or middle strip in two-way flat plate systems.

R18.21 — Post-tensioning anchorages and couplers

R18.21.1 — In the 1986 interim code, the separate provisions for strength of unbonded and bonded tendon anchorages and couplers presented in 18.19.1 and 18.19.2 of the 1983 code were combined into a single revised 18.19.1 covering anchorages and couplers for both unbonded and bonded tendons. Since the 1989 code, the required strength of the tendon-anchorage or tendon-coupler assemblies for both unbonded and bonded tendons, when tested in an unbonded state, is based on 95 percent of the specified breaking strength of the prestressing steel in the test. The prestressing steel material should comply with the minimum provisions of the applicable ASTM specifications as outlined in 3.5.5. The specified strength of anchorages and couplers exceeds the maximum design strength of the prestressing steel by a substantial margin, and, at the same time, recognizes the stress-riser effects associated with most available post-tensioning anchorages and couplers. Anchorage and coupler strength should be attained with a minimum amount of permanent deformation and successive set, recognizing that some deformation and set will occur when testing to failure. Tendon assemblies should conform to the 2 percent elongation requirements in ACI 301[18.31] and industry recommendations.[18.14] Anchorages and couplers for bonded tendons that develop less than 100 percent of the specified breaking strength of the prestressing steel should be used only where the bond transfer length between the anchorage or coupler and critical sections equals or exceeds that required to develop the prestressing steel strength. This bond length may be calculated by the results of tests of bond characteristics of untensioned prestressing strand,[18.32] or by bond tests on other prestressing steel materials, as appropriate.

R18.21.3 — For discussion on fatigue loading, see Reference 18.33.

For detailed recommendations on tests for static and cyclic loading conditions for tendons and anchorage fittings of unbonded tendons, see Section 4.1.3 of Reference 18.9, and Section 15.2.2 of Reference 18.31.

R18.21.4 — For recommendations regarding protection see Sections 4.2 and 4.3 of Reference 18.9, and Sections 3.4, 3.6, 5, 6, and 8.3 of Reference 18.26.

CODE

18.22 — External post-tensioning

18.22.1 — Post-tensioning tendons shall be permitted to be external to any concrete section of a member. The strength and serviceability design methods of this code shall be used in evaluating the effects of external tendon forces on the concrete structure.

18.22.2 — External tendons shall be considered as unbonded tendons when computing flexural strength unless provisions are made to effectively bond the external tendons to the concrete section along its entire length.

18.22.3 — External tendons shall be attached to the concrete member in a manner that maintains the desired eccentricity between the tendons and the concrete centroid throughout the full range of anticipated member deflection.

18.22.4 — External tendons and tendon anchorage regions shall be protected against corrosion, and the details of the protection method shall be indicated on the drawings or in the project specifications.

COMMENTARY

R18.22 — External post-tensioning

External attachment of tendons is a versatile method of providing additional strength, or improving serviceability, or both, in existing structures. It is well suited to repair or upgrade existing structures and permits a wide variety of tendon arrangements.

Additional information on external post-tensioning is given in Reference 18.34.

R18.22.3 — External tendons are often attached to the concrete member at various locations between anchorages (such as midspan, quarter points, or third points) for desired load balancing effects, for tendon alignment, or to address tendon vibration concerns. Consideration should be given to the effects caused by the tendon profile shifting in relationship to the concrete centroid as the member deforms under effects of post-tensioning and applied load.

R18.22.4 — Permanent corrosion protection can be achieved by a variety of methods. The corrosion protection provided should be suitable to the environment in which the tendons are located. Some conditions will require that the prestressing steel be protected by concrete cover or by cement grout in polyethylene or metal tubing; other conditions will permit the protection provided by coatings such as paint or grease. Corrosion protection methods should meet the fire protection requirements of the general building code, unless the installation of external post-tensioning is to only improve serviceability.

Notes

CHAPTER 19 — SHELLS AND FOLDED PLATE MEMBERS

CODE

19.1 — Scope and definitions

19.1.1 — Provisions of Chapter 19 shall apply to thin shell and folded plate concrete structures, including ribs and edge members.

19.1.2 — All provisions of this code not specifically excluded, and not in conflict with provisions of Chapter 19, shall apply to thin-shell structures.

19.1.3 — ***Thin shells*** — Three-dimensional spatial structures made up of one or more curved slabs or folded plates whose thicknesses are small compared to their other dimensions. Thin shells are characterized by their three-dimensional load-carrying behavior, which is determined by the geometry of their forms, by the manner in which they are supported, and by the nature of the applied load.

19.1.4 — ***Folded plates*** — A special class of shell structure formed by joining flat, thin slabs along their edges to create a three-dimensional spatial structure.

COMMENTARY

R19.1 — Scope and definitions

The code and commentary provide information on the design, analysis, and construction of concrete thin shells and folded plates. The process began in 1964 with the publication of a practice and commentary by ACI Committee 334,[19.1] and continued with the inclusion of Chapter 19 in the 1971 code. The 1982 revision of ACI 334R.1 reflected additional experience in design, analysis, and construction and was influenced by the publication of the **"Recommendations for Reinforced Concrete Shells and Folded Plates"** of the International Association for Shell and Spatial Structures (IASS) in 1979.[19.2]

Since Chapter 19 applies to concrete thin shells and folded plates of all shapes, extensive discussion of their design, analysis, and construction in the commentary is not possible. Additional information can be obtained from the references. Performance of shells and folded plates requires special attention to detail.[19.3]

R19.1.1 — Discussion of the application of thin shells in special structures such as cooling towers and circular prestressed concrete tanks may be found in the reports of ACI Committee 334[19.4] and ACI Committee 373.[19.5]

R19.1.3 — Common types of thin shells are domes (surfaces of revolution),[19.6,19.7] cylindrical shells,[19.7] barrel vaults,[19.8] conoids,[19.8] elliptical paraboloids,[19.8] hyperbolic paraboloids,[19.9] and groined vaults.[19.9]

R19.1.4 — Folded plates may be prismatic,[19.6,19.7] nonprismatic,[19.7] or faceted. The first two types consist generally of planar thin slabs joined along their longitudinal edges to form a beam-like structure spanning between supports. Faceted folded plates are made up of triangular or polygonal planar thin slabs joined along their edges to form three-dimensional spatial structures.

CODE

19.1.5 — *Ribbed shells* — Spatial structures with material placed primarily along certain preferred rib lines, with the area between the ribs filled with thin slabs or left open.

19.1.6 — *Auxiliary members* — Ribs or edge beams that serve to strengthen, stiffen, or support the shell; usually, auxiliary members act jointly with the shell.

19.1.7 — *Elastic analysis* — An analysis of deformations and internal forces based on equilibrium, compatibility of strains, and assumed elastic behavior, and representing to a suitable approximation the three-dimensional action of the shell together with its auxiliary members.

19.1.8 — *Inelastic analysis* — an analysis of deformations and internal forces based on equilibrium, nonlinear stress-strain relations for concrete and reinforcement, consideration of cracking and time-dependent effects, and compatibility of strains. The analysis shall represent to a suitable approximation three-dimensional action of the shell together with its auxiliary members.

19.1.9 — *Experimental analysis* — An analysis procedure based on the measurement of deformations or strains, or both, of the structure or its model; experimental analysis is based on either elastic or inelastic behavior.

COMMENTARY

R19.1.5 — Ribbed shells[19.8,19.9] generally have been used for larger spans where the increased thickness of the curved slab alone becomes excessive or uneconomical. Ribbed shells are also used because of the construction techniques employed and to enhance the aesthetic impact of the completed structure.

R19.1.6 — Most thin shell structures require ribs or edge beams at their boundaries to carry the shell boundary forces, to assist in transmitting them to the supporting structure, and to accommodate the increased amount of reinforcement in these areas.

R19.1.7 — Elastic analysis of thin shells and folded plates can be performed using any method of structural analysis based on assumptions that provide suitable approximations to the three-dimensional behavior of the structure. The method should determine the internal forces and displacements needed in the design of the shell proper, the rib or edge members, and the supporting structure. Equilibrium of internal forces and external loads and compatibility of deformations should be satisfied.

Methods of elastic analysis based on classical shell theory, simplified mathematical or analytical models, or numerical solutions using finite element,[19.10] finite differences,[19.8] or numerical integration techniques,[19.8,19.11] are described in the cited references.

The choice of the method of analysis and the degree of accuracy required depends on certain critical factors. These include: the size of the structure, the geometry of the thin shell or folded plate, the manner in which the structure is supported, the nature of the applied load, and the extent of personal or documented experience regarding the reliability of the given method of analysis in predicting the behavior of the specific type of shell[19.8] or folded plate.[19.7]

R19.1.8—Inelastic analysis of thin shells and folded plates can be performed using a refined method of analysis based on the specific nonlinear material properties, nonlinear behavior due to the cracking of concrete, and time-dependent effects such as creep, shrinkage, temperature, and load history. These effects are incorporated in order to trace the response and crack propagation of a reinforced concrete shell through the elastic, inelastic, and ultimate ranges. Such analyses usually require incremental loading and iterative procedures to converge on solutions that satisfy both equilibrium and strain compatibility.[19.12,19.13]

CODE

19.2 — Analysis and design

19.2.1 — Elastic behavior shall be an accepted basis for determining internal forces and displacements of thin shells. This behavior shall be permitted to be established by computations based on an analysis of the uncracked concrete structure in which the material is assumed linearly elastic, homogeneous, and isotropic. Poisson's ratio of concrete shall be permitted to be taken equal to zero.

19.2.2 — Inelastic analyses shall be permitted to be used where it can be shown that such methods provide a safe basis for design.

19.2.3 — Equilibrium checks of internal resistances and external loads shall be made to ensure consistency of results.

19.2.4 — Experimental or numerical analysis procedures shall be permitted where it can be shown that such procedures provide a safe basis for design.

19.2.5 — Approximate methods of analysis shall be permitted where it can be shown that such methods provide a safe basis for design.

19.2.6 — In prestressed shells, the analysis shall also consider behavior under loads induced during prestressing, at cracking load, and at factored load. Where tendons are draped within a shell, design shall take into account force components on the shell resulting from the tendon profile not lying in one plane.

COMMENTARY

R19.2 — Analysis and design

R19.2.1 — For types of shell structures where experience, tests, and analyses have shown that the structure can sustain reasonable overloads without undergoing brittle failure, elastic analysis is an acceptable procedure. The designer may assume that reinforced concrete is ideally elastic, homogeneous, and isotropic, having identical properties in all directions. An analysis should be performed for the shell considering service load conditions. The analysis of shells of unusual size, shape, or complexity should consider behavior through the elastic, cracking, and inelastic stages.

R19.2.2 — Several inelastic analysis procedures contain possible solution methods.[19.12,19.13]

R19.2.4 — Experimental analysis of elastic models[19.14] has been used as a substitute for an analytical solution of a complex shell structure. Experimental analysis of reinforced microconcrete models through the elastic, cracking, inelastic, and ultimate stages should be considered for important shells of unusual size, shape, or complexity.

For model analysis, only those portions of the structure that significantly affect the items under study need be simulated. Every attempt should be made to ensure that the experiments reveal the quantitative behavior of the prototype structure.

Wind tunnel tests of a scaled-down model do not necessarily provide usable results and should be conducted by a recognized expert in wind tunnel testing of structural models.

R19.2.5 — Solutions that include both membrane and bending effects and satisfy conditions of compatibility and equilibrium are encouraged. Approximate solutions that satisfy statics but not the compatibility of strains may be used only when extensive experience has proved that safe designs have resulted from their use. Such methods include beam-type analysis for barrel shells and folded plates having large ratios of span to either width or radius of curvature, simple membrane analysis for shells of revolution, and others in which the equations of equilibrium are satisfied, while the strain compatibility equations are not.

R19.2.6 — If the shell is prestressed, the analysis should include its strength at factored loads as well as its adequacy under service loads, under the load that causes cracking, and under loads induced during prestressing. Axial forces due to draped tendons may not lie in one plane and due consideration should be given to the resulting force components. The effects of post-tensioning of shell supporting members should be taken into account.

CODE

19.2.7 — The thickness of a shell and its reinforcement shall be proportioned for the required strength and serviceability, using either the strength design method of 8.1.1 or the design method of 8.1.2.

19.2.8 — Shell instability shall be investigated and shown by design to be precluded.

19.2.9 — Auxiliary members shall be designed according to the applicable provisions of the code. It shall be permitted to assume that a portion of the shell equal to the flange width, as specified in 8.10, acts with the auxiliary member. In such portions of the shell, the reinforcement perpendicular to the auxiliary member shall be at least equal to that required for the flange of a T-beam by 8.10.5.

19.2.10 — Strength design of shell slabs for membrane and bending forces shall be based on the distribution of stresses and strains as determined from either an elastic or an inelastic analysis.

COMMENTARY

R19.2.7 — The thin shell's thickness and reinforcement are required to be proportioned to satisfy the strength provisions of this code, and to resist internal forces obtained from an analysis, an experimental model study, or a combination thereof. Reinforcement sufficient to minimize cracking under service load conditions should be provided. The thickness of the shell is often dictated by the required reinforcement and the construction constraints, by 19.2.8, or by the code minimum thickness requirements.

R19.2.8 — Thin shells, like other structures that experience in-plane membrane compressive forces, are subject to buckling when the applied load reaches a critical value. Because of the surface-like geometry of shells, the problem of calculating buckling load is complex. If one of the principal membrane forces is tensile, the shell is less likely to buckle than if both principal membrane forces are compressive. The kinds of membrane forces that develop in a shell depend on its initial shape and the manner in which the shell is supported and loaded. In some types of shells, post-buckling behavior should be considered in determining safety against instability.[19.2]

Investigation of thin shells for stability should consider the effect of (1) anticipated deviation of the geometry of the shell surface as-built from the idealized, geometry, (2) large deflections, (3) creep and shrinkage of concrete, (4) inelastic properties of materials, (5) cracking of concrete, (6) location, amount, and orientation of reinforcement, and (7) possible deformation of supporting elements.

Measures successfully used to improve resistance to buckling include the provision of two mats of reinforcement—one near each outer surface of the shell, a local increase of shell curvatures, the use of ribbed shells, and the use of concrete with high tensile strength and low creep.

A procedure for determining critical buckling loads of shells is given in the IASS recommendations.[19.2] Some recommendations for buckling design of domes used in industrial applications are given in References 19.5 and 19.15.

R19.2.10 — The stresses and strains in the shell slab used for design are those determined by analysis (elastic or inelastic) multiplied by appropriate load factors. Because of detrimental effects of membrane cracking, the computed tensile strain in the reinforcement under factored loads should be limited.

CODE

19.2.11 — In a region where membrane cracking is predicted, the nominal compressive strength parallel to the cracks shall be taken as $0.4f_c'$.

19.3 — Design strength of materials

19.3.1 — Specified compressive strength of concrete f_c' at 28 days shall not be less than 3000 psi.

19.3.2 — Specified yield strength of nonprestressed reinforcement f_y shall not exceed 60,000 psi.

19.4 — Shell reinforcement

19.4.1 — Shell reinforcement shall be provided to resist tensile stresses from internal membrane forces, to resist tension from bending and twisting moments, to limit shrinkage and temperature crack width and spacing, and as special reinforcement at shell boundaries, load attachments, and shell openings.

19.4.2 — Tensile reinforcement shall be provided in two or more directions and shall be proportioned such that its resistance in any direction equals or exceeds the component of internal forces in that direction.

Alternatively, reinforcement for the membrane forces in the slab shall be calculated as the reinforcement required to resist axial tensile forces plus the tensile force due to shear-friction required to transfer shear across any cross section of the membrane. The assumed coefficient of friction, μ, shall not exceed 1.0λ where $\lambda = 1.0$ for normalweight concrete, **0.85** for sand-lightweight concrete, and **0.75** for all-lightweight concrete. Linear interpolation shall be permitted when partial sand replacement is used.

COMMENTARY

R19.2.11 — When principal tensile stress produces membrane cracking in the shell, experiments indicate the attainable compressive strength in the direction parallel to the cracks is reduced.[19.16,19.17]

R19.4 — Shell reinforcement

R19.4.1 — At any point in a shell, two different kinds of internal forces may occur simultaneously: those associated with membrane action, and those associated with bending of the shell. The membrane forces are assumed to act in the tangential plane midway between the surfaces of the shell, and are the two axial forces and the membrane shears. Flexural effects include bending moments, twisting moments, and the associated transverse shears. Limiting membrane crack width and spacing due to shrinkage, temperature, and service load conditions is a major design consideration.

R19.4.2 — The requirement of ensuring strength in all directions is based on safety considerations. Any method that ensures sufficient strength consistent with equilibrium is acceptable. The direction of the principal membrane tensile force at any point may vary depending on the direction, magnitudes, and combinations of the various applied loads.

The magnitude of the internal membrane forces, acting at any point due to a specific load, is generally calculated on the basis of an elastic theory in which the shell is assumed as uncracked. The computation of the required amount of reinforcement to resist the internal membrane forces has been traditionally based on the assumption that concrete does not resist tension. The associated deflections, and the possibility of cracking, should be investigated in the serviceability phase of the design. Achieving this may require a working stress design for steel selection.

Where reinforcement is not placed in the direction of the principal tensile forces and where cracks at the service load level are objectionable, the computation of reinforcement may have to be based on a more refined approach[19.16,19.18,19.19] that considers the existence of cracks. In the cracked state, the concrete is assumed to be unable to resist either tension or shear. Thus, equilibrium is attained by equating tensile resisting forces in reinforcement and compressive resisting forces in concrete.

The alternative method to calculate orthogonal reinforcement is the shear-friction method. It is based on the assumption that shear integrity of a shell should be maintained at

CODE | COMMENTARY

factored loads. It is not necessary to calculate principal stresses if the alternative approach is used.

19.4.3 — The area of shell reinforcement at any section as measured in two orthogonal directions shall not be less than the slab shrinkage or temperature reinforcement required by 7.12.

R19.4.3 — Minimum membrane reinforcement corresponding to slab shrinkage and temperature reinforcement are to be provided in at least two approximately orthogonal directions even if the calculated membrane forces are compressive in one or more directions.

19.4.4 — Reinforcement for shear and bending moments about axes in the plane of the shell slab shall be calculated in accordance with Chapters 10, 11, and 13.

19.4.5 — The area of shell tension reinforcement shall be limited so that the reinforcement will yield before either crushing of concrete in compression or shell buckling can take place.

R19.4.5 — The requirement that the tensile reinforcement yields before the concrete crushes anywhere is consistent with 10.3.3. Such crushing can also occur in regions near supports and for some shells where the principal membrane forces are approximately equal and opposite in sign.

19.4.6 — In regions of high tension, membrane reinforcement shall, if practical, be placed in the general directions of the principal tensile membrane forces. Where this is not practical, it shall be permitted to place membrane reinforcement in two or more component directions.

R19.4.6 — Generally, for all shells, and particularly in regions of substantial tension, the orientation of reinforcement should approximate the directions of the principal tensile membrane forces. However, in some structures it is not possible to detail the reinforcement to follow the stress trajectories. For such cases, orthogonal component reinforcement is allowed.

19.4.7 — If the direction of reinforcement varies more than 10 deg from the direction of principal tensile membrane force, the amount of reinforcement shall be reviewed in relation to cracking at service loads.

R19.4.7 — When the directions of reinforcement deviate significantly (more than 10 degrees) from the directions of the principal membrane forces, higher strains in the shell occur to develop the capacity of reinforcement. This might lead to the development of unacceptable wide cracks. The crack width should be estimated and limited if necessary.

Permissible crack widths for service loads under different environmental conditions are given in the report of ACI Committee 224.[19.20] Crack width can be limited by an increase in the amount of reinforcement used, by reducing the stress at the service load level, by providing reinforcement in three or more directions in the plane of the shell, or by using closer spacing of smaller diameter bars.

19.4.8 — Where the magnitude of the principal tensile membrane stress within the shell varies greatly over the area of the shell surface, reinforcement resisting the total tension shall be permitted to be concentrated in the regions of largest tensile stress where it can be shown that this provides a safe basis for design. However, the ratio of shell reinforcement in any portion of the tensile zone shall be not less than 0.0035 based on the overall thickness of the shell.

R19.4.8 — The practice of concentrating tensile reinforcement in the regions of maximum tensile stress has led to a number of successful and economical designs, primarily for long folded plates, long barrel vault shells, and for domes. The requirement of providing the minimum reinforcement in the remaining tensile zone is intended to limit crack width and spacing.

CODE

19.4.9 — Reinforcement required to resist shell bending moments shall be proportioned with due regard to the simultaneous action of membrane axial forces at the same location. Where shell reinforcement is required in only one face to resist bending moments, equal amounts shall be placed near both surfaces of the shell even though a reversal of bending moments is not indicated by the analysis.

19.4.10 — Shell reinforcement in any direction shall not be spaced farther apart than 18 in. nor farther apart than five times the shell thickness. Where the principal membrane tensile stress on the gross concrete area due to factored loads exceeds $4\phi\sqrt{f_c'}$, reinforcement shall not be spaced farther apart than three times the shell thickness.

19.4.11 — Shell reinforcement at the junction of the shell and supporting members or edge members shall be anchored in or extended through such members in accordance with the requirements of Chapter 12, except that the minimum development length shall be $1.2\ell_d$ but not less than 18 in.

19.4.12 — Splice lengths of shell reinforcement shall be governed by the provisions of Chapter 12, except that the minimum splice length of tension bars shall be 1.2 times the value required by Chapter 12 but not less than 18 in. The number of splices in principal tensile reinforcement shall be kept to a practical minimum. Where splices are necessary they shall be staggered at least ℓ_d with not more than one-third of the reinforcement spliced at any section.

19.5 — Construction

19.5.1 — When removal of formwork is based on a specific modulus of elasticity of concrete because of stability or deflection considerations, the value of the modulus of elasticity, E_c, used shall be determined from flexural tests of field-cured beam specimens. The number of test specimens, the dimensions of test beam specimens, and test procedures shall be specified by the registered design professional.

19.5.2 — The engineer shall specify the tolerances for the shape of the shell. If construction results in deviations from the shape greater than the specified tolerances, an analysis of the effect of the deviations shall be made and any required remedial actions shall be taken to ensure safe behavior.

COMMENTARY

R19.4.9 — The design method should ensure that the concrete sections, including consideration of the reinforcement, are capable of developing the internal forces required by the equations of equilibrium.[19.21] The sign of bending moments may change rapidly from point to point of a shell. For this reason, reinforcement to resist bending, where required, is to be placed near both outer surfaces of the shell. In many cases, the thickness required to provide proper cover and spacing for the multiple layers of reinforcement may govern the design of the shell thickness.

R19.4.10 — The value of ϕ to be used is that prescribed in 9.3.2.1 for axial tension.

R19.4.11 and R19.4.12 — On curved shell surfaces it is difficult to control the alignment of precut reinforcement. This should be considered to avoid insufficient splice and development lengths. Sections 19.4.11 and 19.4.12 require extra reinforcement length to maintain the minimum lengths on curved surfaces.

R19.5 — Construction

R19.5.1 — When early removal of forms is necessary, the magnitude of the modulus of elasticity at the time of proposed form removal should be investigated to ensure safety of the shell with respect to buckling, and to restrict deflections.[19.3,19.22] The value of the modulus of elasticity E_c should be obtained from a flexural test of field-cured specimens. It is not sufficient to determine the modulus from the formula in 8.5.1, even if the compressive strength of concrete is determined for the field-cured specimen.

R19.5.2 — In some types of shells, small local deviations from the theoretical geometry of the shell can cause relatively large changes in local stresses and in overall safety against instability. These changes can result in local cracking and yielding that may make the structure unsafe or can greatly affect the critical load producing instability. The effect of such deviations should be evaluated and any necessary remedial actions should be taken. Special attention is needed when using air supported form systems.[19.23]

Notes

CHAPTER 20 — STRENGTH EVALUATION OF EXISTING STRUCTURES

CODE

20.1 — Strength evaluation — General

20.1.1 — If there is doubt that a part or all of a structure meets the safety requirements of this code, a strength evaluation shall be carried out as required by the engineer or building official.

20.1.2 — If the effect of the strength deficiency is well understood and if it is feasible to measure the dimensions and material properties required for analysis, analytical evaluations of strength based on those measurements shall suffice. Required data shall be determined in accordance with 20.2.

20.1.3 — If the effect of the strength deficiency is not well understood or if it is not feasible to establish the required dimensions and material properties by measurement, a load test shall be required if the structure is to remain in service.

COMMENTARY

R20.1 — Strength evaluation — General

Chapter 20 does not cover load testing for the approval of new design or construction methods. (See 16.10 for recommendations on strength evaluation of precast concrete members.) Provisions of Chapter 20 may be used to evaluate whether a structure or a portion of a structure satisfies the safety requirements of this code. A strength evaluation may be required if the materials are considered to be deficient in quality, if there is evidence indicating faulty construction, if a structure has deteriorated, if a building will be used for a new function, or if, for any reason, a structure or a portion of it does not appear to satisfy the requirements of the code. In such cases, Chapter 20 provides guidance for investigating the safety of the structure.

If the safety concerns are related to an assembly of elements or an entire structure, it is not feasible to load test every element and section to the maximum. In such cases, it is appropriate that an investigation plan be developed to address the specific safety concerns. If a load test is described as part of the strength evaluation process, it is desirable for all parties involved to come to an agreement about the region to be loaded, the magnitude of the load, the load test procedure, and acceptance criteria before any load tests are conducted.

R20.1.2 — Strength considerations related to axial load, flexure, and combined axial load and flexure are well understood. There are reliable theories relating strength and short-term displacement to load in terms of dimensional and material data for the structure.

To determine the strength of the structure by analysis, calculations should be based on data gathered on the actual dimensions of the structure, properties of the materials in place, and all pertinent details. Requirements for data collection are in 20.2.

R20.1.3 — If the shear or bond strength of an element is critical in relation to the doubt expressed about safety, a test may be the most efficient solution to eliminate or confirm the doubt. A test may also be appropriate if it is not feasible to determine the material and dimensional properties required for analysis, even if the cause of the concern relates to flexure or axial load.

Wherever possible and appropriate, support the results of the load test by analysis.

CODE

20.1.4 — If the doubt about safety of a part or all of a structure involves deterioration, and if the observed response during the load test satisfies the acceptance criteria, the structure or part of the structure shall be permitted to remain in service for a specified time period. If deemed necessary by the engineer, periodic reevaluations shall be conducted.

COMMENTARY

R20.1.4 — For a deteriorating structure, the acceptance provided by the load test may not be assumed to be without limits in terms of time. In such cases, a periodic inspection program is useful. A program that involves physical tests and periodic inspection can justify a longer period in service. Another option for maintaining the structure in service, while the periodic inspection program continues, is to limit the live load to a level determined to be appropriate.

The length of the specified time period should be based on consideration of (a) the nature of the problem, (b) environmental and load effects, (c) service history of the structure, and (d) scope of the periodic inspection program. At the end of a specified time period, further strength evaluation is required if the structure is to remain in service.

With the agreement of all concerned parties, special procedures may be devised for periodic testing that do not necessarily conform to the loading and acceptance criteria specified in Chapter 20.

20.2 — Determination of required dimensions and material properties

R20.2 — Determination of required dimensions and material properties

This section applies if it is decided to make an analytical evaluation (see 20.1.2).

20.2.1 — Dimensions of the structural elements shall be established at critical sections.

R20.2.1 — Critical sections are where each type of stress calculated for the load in question reaches its maximum value.

20.2.2 — Locations and sizes of the reinforcing bars, welded wire reinforcement, or tendons shall be determined by measurement. It shall be permitted to base reinforcement locations on available drawings if spot checks are made confirming the information on the drawings.

R20.2.2 — For individual elements, amount, size, arrangement, and location should be determined at the critical sections for reinforcement or tendons, or both, designed to resist applied load. Nondestructive investigation methods are acceptable. In large structures, determination of these data for approximately 5 percent of the reinforcement or tendons in critical regions may suffice if these measurements confirm the data provided in the construction drawings.

20.2.3 — If required, concrete strength shall be based on results of cylinder tests or tests of cores removed from the part of the structure where the strength is in doubt. Concrete strengths shall be determined as specified in 5.6.5.

R20.2.3 — The number of tests may depend on the size of the structure and the sensitivity of structural safety to concrete strength. In cases where the potential problem involves flexure only, investigation of concrete strength can be minimal for a lightly reinforced section ($\rho f_y / f_c' \leq \mathbf{0.15}$ for rectangular section).

20.2.4 — If required, reinforcement or prestressing steel strength shall be based on tensile tests of representative samples of the material in the structure in question.

R20.2.4 — The number of tests required depends on the uniformity of the material and is best determined by the engineer for the specific application.

20.2.5 — If the required dimensions and material properties are determined through measurements and testing, and if calculations can be made in accordance with 20.1.2, it shall be permitted to increase ϕ from those specified in 9.3, but ϕ shall not be more than:

R20.2.5 — Strength reduction factors given in 20.2.5 are larger than those specified in Chapter 9. These increased values are justified by the use of accurate field-obtained material properties, actual in-place dimensions, and well-understood methods of analysis.

CODE

Tension-controlled sections, as defined in 10.3.4 .... 1.0

Compression-controlled sections, as defined in 10.3.3:

Members with spiral reinforcement conforming to 10.9.3 .... 0.85

Other reinforced members 0.8

Shear and/or torsion 0.8

Bearing on concrete 0.8

COMMENTARY

The strength reduction factors in 20.2.5 were changed for the 2002 edition to be compatible with the load combinations and strength reduction factors of Chapter 9, which were revised at that time.

20.3 — Load test procedure

20.3.1 — Load arrangement

The number and arrangement of spans or panels loaded shall be selected to maximize the deflection and stresses in the critical regions of the structural elements of which strength is in doubt. More than one test load arrangement shall be used if a single arrangement will not simultaneously result in maximum values of the effects (such as deflection, rotation, or stress) necessary to demonstrate the adequacy of the structure.

20.3.2 — Load intensity

The total test load (including dead load already in place) shall not be less than **0.85 (1.4*D* + 1.7*L*)**. It shall be permitted to reduce L in accordance with the requirements of the applicable general building code.

20.3.3 — A load test shall not be made until that portion of the structure to be subjected to load is at least 56 days old. If the owner of the structure, the contractor, and all involved parties agree, it shall be permitted to make the test at an earlier age.

R20.3 — Load test procedure

R20.3.1—Load arrangement

It is important to apply the load at locations so that its effects on the suspected defect are a maximum and the probability of unloaded members sharing the applied load is a minimum. In cases where it is shown by analysis that adjoining unloaded elements will help carry some of the load, the load should be placed to develop effects consistent with the intent of the load factor.

R20.3.2 — Load intensity

The required load intensity follows previous load test practice. The live load L may be reduced as permitted by the general building code governing safety considerations for the structure. The live load should be increased to compensate for resistance provided by unloaded portions of the structure in questions. The increase in live load is determined from analysis of the loading conditions in relation to the selected pass/fail criterion for the test.

Although the load combinations and strength reduction factors of Chapter 9 were revised for the 2002 edition, the test load intensity remained the same. It is considered appropriate for designs using the load combinations and strength reduction factors of Chapter 9 or Appendix C.

20.4 — Loading criteria

20.4.1 — The initial value for all applicable response measurements (such as deflection, rotation, strain, slip, crack widths) shall be obtained not more than 1

R20.4 — Loading criteria

CODE

hour before application of the first load increment. Measurements shall be made at locations where maximum response is expected. Additional measurements shall be made if required.

20.4.2 — Test load shall be applied in not less than four approximately equal increments.

20.4.3 — Uniform test load shall be applied in a manner to ensure uniform distribution of the load transmitted to the structure or portion of the structure being tested. Arching of the applied load shall be avoided.

20.4.4 — A set of response measurements shall be made after each load increment is applied and after the total load has been applied on the structure for at least 24 hours.

20.4.5 — Total test load shall be removed immediately after all response measurements defined in 20.4.4 are made.

20.4.6 — A set of final response measurements shall be made 24 hours after the test load is removed.

20.5 — Acceptance criteria

20.5.1 — The portion of the structure tested shall show no evidence of failure. Spalling and crushing of compressed concrete shall be considered an indication of failure.

COMMENTARY

R20.4.2 — Inspecting the structure after each load increment is advisable.

R20.4.3 — Arching refers to the tendency for the load to be transmitted nonuniformly to the flexural element being tested. For example, if a slab is loaded by a uniform arrangement of bricks with the bricks in contact, arching would results in reduction of the load on the slab near the midspan of the slab.

R20.5 — Acceptance criteria

R20.5.1 — A general acceptance criterion for the behavior of a structure under the test load is that it does not show evidence of failure. Evidence of failure includes cracking, spalling, or deflection of such magnitude and extent that the observed result is obviously excessive and incompatible with the safety requirements of the structure. No simple rules have been developed for application to all types of structures and conditions. If sufficient damage has occurred so that the structure is considered to have failed that test, retesting is not permitted since it is considered that damaged members should not be put into service even at a lower load rating.

Local spalling or flaking of the compressed concrete in flexural elements related to casting imperfections need not indicate overall structural distress. Crack widths are good indicators of the state of the structure and should be observed to help determine whether the structure is satisfactory. However, exact prediction or measurement of crack widths in reinforced concrete elements is not likely to be achieved under field conditions. Establish criteria before the test, relative to the types of cracks anticipated; where the cracks will be measured; how they will be measured; and approximate limits or criteria to evaluate new cracks or limits for the changes in crack width.

CODE

20.5.2 — Measured deflections shall satisfy one of the following conditions:

$$\Delta_1 \leq \frac{\ell_t^2}{20{,}000h} \qquad (20\text{-}1)$$

$$\Delta_r \leq \frac{\Delta_1}{4} \qquad (20\text{-}2)$$

If the measured maximum and residual deflections, Δ_1 and Δ_r, do not satisfy Eq. (20-1) or (20-2), it shall be permitted to repeat the load test.

The repeat test shall be conducted not earlier than 72 hours after removal of the first test load. The portion of the structure tested in the repeat test shall be considered acceptable if deflection recovery Δ_r satisfies the condition:

$$\Delta_r \leq \frac{\Delta_2}{5} \qquad (20\text{-}3)$$

where Δ_2 is the maximum deflection measured during the second test relative to the position of the structure at the beginning of the second test.

COMMENTARY

R20.5.2 — The deflection limits and the retest option follow past practice. If the structure shows no evidence of failure, recovery of deflection after removal of the test load is used to determine whether the strength of the structure is satisfactory. In the case of a very stiff structure, however, the errors in measurements under field conditions may be of the same order as the actual deflections and recovery. To avoid penalizing a satisfactory structure in such a case, recovery measurements are waived if the maximum deflection is less than $\ell_t^2/(20{,}000h)$. The residual deflection Δ_r is the difference between the initial and final (after load removal) deflections for the load test or the repeat load test.

20.5.3 — Structural members tested shall not have cracks indicating the imminence of shear failure.

R20.5.3 — Forces are transmitted across a shear crack plane by a combination of aggregate interlock at the interface of the crack that is enhanced by clamping action of transverse stirrup reinforcing and by dowel action of stirrups crossing the crack. As crack lengths increase to approach a horizontal projected length equal to the depth of the member and concurrently widen to the extent that aggregate interlock cannot occur, and as transverse stirrups if present begin to yield or display loss of anchorage so as to threaten their integrity, the member is assumed to be approaching imminent shear failure.

20.5.4 — In regions of structural members without transverse reinforcement, appearance of structural cracks inclined to the longitudinal axis and having a horizontal projection longer than the depth of the member at midpoint of the crack shall be evaluated.

R20.5.4 — The intent of 20.5.4 is to make the professionals in charge of the test pay attention to the structural implication of observed inclined cracks that may lead to brittle collapse in members without transverse reinforcement.

20.5.5 — In regions of anchorage and lap splices, the appearance along the line of reinforcement of a series of short inclined cracks or horizontal cracks shall be evaluated.

R20.5.5 — Cracking along the axis of the reinforcement in anchorage zones may be related to high stresses associated with the transfer of forces between the reinforcement and the concrete. These cracks may be indicators of pending brittle failure of the element if they are associated with the main reinforcement. It is important that their causes and consequences be evaluated.

CODE

20.6 — Provision for lower load rating

If the structure under investigation does not satisfy conditions or criteria of 20.1.2, 20.5.2, or 20.5.3, the structure shall be permitted for use at a lower load rating based on the results of the load test or analysis, if approved by the building official.

20.7 — Safety

20.7.1 — Load tests shall be conducted in such a manner as to provide for safety of life and structure during the test.

20.7.2 — No safety measures shall interfere with load test procedures or affect results.

COMMENTARY

R20.6 — Provision for lower load rating

Except for load tested members that have failed under a test (see 20.5), the building official may permit the use of a structure or member at a lower load rating that is judged to be safe and appropriate on the basis of the test results.

CHAPTER 21 — SPECIAL PROVISIONS FOR SEISMIC DESIGN

CODE

21.1 — Definitions

Base of structure — Level at which earthquake motions are assumed to be imparted to a building. This level does not necessarily coincide with the ground level.

Boundary elements — Portions along structural wall and structural diaphragm edges strengthened by longitudinal and transverse reinforcement. Boundary elements do not necessarily require an increase in the thickness of the wall or diaphragm. Edges of openings within walls and diaphragms shall be provided with boundary elements as required by 21.7.6 or 21.9.5.3.

Collector elements — Elements that serve to transmit the inertial forces within structural diaphragms to members of the lateral-force-resisting systems.

Connection — A region that joins two or more members, of which one or more is precast.

> ***Ductile connection*** — Connection that experiences yielding as a result of the design displacements.
>
> ***Strong connection*** — Connection that remains elastic while adjoining members experience yielding as a result of the design displacements.

Crosstie — A continuous reinforcing bar having a seismic hook at one end and a hook not less than 90 degrees with at least a six-diameter extension at the other end. The hooks shall engage peripheral longitudinal bars. The 90 degree hooks of two successive crossties engaging the same longitudinal bars shall be alternated end for end.

Design displacement — Total lateral displacement expected for the design-basis earthquake, as required by the governing code for earthquake-resistant design.

Design load combinations — Combinations of factored loads and forces in 9.2.

Design story drift ratio — Relative difference of design displacement between the top and bottom of a story, divided by the story height.

Development length for a bar with a standard hook — The shortest distance from the critical section (where the strength of the bar is to be developed) to the outside end of the 90 degree hook.

COMMENTARY

R21.1 — Definitions

The design displacement is an index of the maximum lateral displacement expected in design for the design-basis earthquake. In documents such as the National Earthquake Hazards Reduction Provisions (NEHRP),[21.1] ASCE 7-95, the Uniform Building Code (UBC),[21.2] the BOCA/National Building Code (BOCA)[21.3] published by Building Officials and Code Administrators International, or the Standard Building Code (SBC)[21.4] published by Southern Building Code Congress International, the design-basis earthquake has approximately a 90 percent probability of nonexceedance in 50 years. In those documents, the design displacement is calculated using static or dynamic linear elastic analysis under code-specified actions considering effects of cracked sections, effects of torsion, effects of vertical forces

CODE

Factored loads and forces — Loads and forces multiplied by appropriate load factors in 9.2.

Hoop — A closed tie or continuously wound tie. A closed tie can be made up of several reinforcement elements each having seismic hooks at both ends. A continuously wound tie shall have a seismic hook at both ends.

Joint — Portion of structure common to intersecting members. The effective cross-sectional area of the joint, A_j, for shear strength computations is defined in 21.5.3.1.

Lateral-force resisting system — That portion of the structure composed of members proportioned to resist forces related to earthquake effects.

Lightweight aggregate concrete — All-lightweight or sand-lightweight aggregate concrete made with lightweight aggregates conforming to 3.3.

Moment frame — Frame in which members and joints resist forces through flexure, shear, and axial force. Moment frames shall be categorized as follows:

Intermediate moment frame — A cast-in-place frame complying with the requirements of 21.2.2.3 and 21.12 in addition to the requirements for ordinary moment frames.

Ordinary moment frame — A cast-in-place or precast concrete frame complying with the requirements of Chapters 1 through 18.

Special moment frame — A cast-in-place frame complying with the requirements of 21.2.2.3, 21.2.3 through 21.2.7, and 21.3 through 21.5 or a precast frame complying with the requirements of 21.2.2.3, 21.2.3 through 21.2.7, and 21.3 through 21.6. In addition, the requirements for ordinary moment frames shall be satisfied.

Plastic hinge region — Length of frame element over which flexural yielding is intended to occur due to design displacements, extending not less than a distance h from the critical section where flexural yielding initiates.

Seismic hook — A hook on a stirrup, hoop, or crosstie having a bend not less than 135 degrees, except that circular hoops shall have a bend not less than 90 degrees. Hooks shall have a six-diameter (but not less than 3 in.) extension that engages the longitudinal reinforcement and projects into the interior of the stirrup or hoop.

COMMENTARY

acting through lateral displacements, and modification factors to account for expected inelastic response. The design displacement generally is larger than the displacement calculated from design-level forces applied to a linear-elastic model of the building.

The provisions of 21.6 are intended to result in a special moment frame constructed using precast concrete having minimum strength and toughness equivalent to that for a special moment frame of cast-in-place concrete.

CODE

Special boundary elements — Boundary elements required by 21.7.6.2 or 21.7.6.3.

Specified lateral forces — Lateral forces corresponding to the appropriate distribution of the design base shear force prescribed by the governing code for earthquake-resistant design.

Structural diaphragms — Structural members, such as floor and roof slabs, that transmit inertial forces to lateral-force resisting members.

Structural trusses — Assemblages of reinforced concrete members subjected primarily to axial forces.

Structural walls — Walls proportioned to resist combinations of shears, moments, and axial forces induced by earthquake motions. A shearwall is a structural wall. Structural walls shall be categorized as follows:

Intermediate precast structural wall — A wall complying with all applicable requirements of Chapters 1 through 18 in addition to 21.13.

Ordinary reinforced concrete structural wall — A wall complying with the requirements of Chapters 1 through 18.

Ordinary structural plain concrete wall — A wall complying with the requirements of Chapter 22.

Special precast structural wall — A precast wall complying with the requirements of 21.8. In addition, the requirements for ordinary reinforced concrete structural walls and the requirements of 21.2.2.3, 21.2.3 through 21.2.7, and 21. 7 shall be satisfied.

Special reinforced concrete structural wall — A cast-in-place wall complying with the requirements of 21.2.2.3, 21.2.3 through 21.2.7, and 21. 7 in addition to the requirements for ordinary reinforced concrete structural walls.

Strut — An element of a structural diaphragm used to provide continuity around an opening in the diaphragm.

Tie elements — Elements that serve to transmit inertia forces and prevent separation of building components such as footings and walls.

21.2 — General requirements

21.2.1 — Scope

21.2.1.1 — Chapter 21 contains special requirements for design and construction of reinforced con-

COMMENTARY

The provisions of 21.13 are intended to result in an intermediate precast structural wall having minimum strength and toughness equivalent to that for an ordinary reinforced concrete structural wall of cast-in-place concrete. A precast concrete wall satisfying only the requirements of Chapters 1 through 18 and not the additional requirements of 21.13 or 21.8 is considered to have ductility and structural integrity less than that for an intermediate precast structural wall.

The provisions of 21.8 are intended to result in a special precast structural wall having minimum strength and toughness equivalent to that for a special reinforced concrete structural wall of cast-in-place concrete.

R21.2 — General requirements

R21.2.1 — Scope

Chapter 21 contains provisions considered to be the minimum requirements for a cast-in-place or precast concrete

CODE

crete members of a structure for which the design forces, related to earthquake motions, have been determined on the basis of energy dissipation in the nonlinear range of response. For applicable specified concrete strengths, see 1.1.1 and 21.2.4.1

21.2.1.2 — In regions of low seismic risk or for structures assigned to low seismic performance or design categories, the provisions of Chapters 1 through 18 and 22 shall apply. Where the design seismic loads are computed using provisions for intermediate or special concrete systems, the requirements of Chapter 21 for intermediate or special systems, as applicable, shall be satisfied.

21.2.1.3 — In regions of moderate seismic risk or for structures assigned to intermediate seismic performance or design categories, intermediate or special moment frames, or ordinary, intermediate, or special structural walls shall be used to resist forces induced by earthquake motions. Where the design seismic loads are computed using provisions for special concrete systems, the requirements of Chapter 21 for special systems, as applicable, shall be satisfied.

21.2.1.4 — In regions of high seismic risk or for structures assigned to high seismic performance or design categories, special moment frames, special structural walls, and diaphragms and trusses complying with 21.2.2 through 21.2.8 and 21.3 through 21.10 shall be used to resist forces induced by earthquake motions. Members not proportioned to resist earthquake forces shall comply with 21.11.

21.2.1.5 — A reinforced concrete structural system not satisfying the requirements of this chapter shall be permitted if it is demonstrated by experimental evidence and analysis that the proposed system will have strength and toughness equal to or exceeding those provided by a comparable monolithic reinforced concrete structure satisfying this chapter.

COMMENTARY

structure capable of sustaining a series of oscillations into the inelastic range of response without critical deterioration in strength. The integrity of the structure in the inelastic range of response should be maintained because the design forces defined in documents such as the IBC,[21.5] the UBC,[21.2] and the NEHRP[21.1] provisions are considered less than those corresponding to linear response at the anticipated earthquake intensity.[21.1,21.6-21.8]

As a properly detailed cast-in-place or precast concrete structure responds to strong ground motion, its effective stiffness decreases and its energy dissipation increases. These changes tend to reduce the response accelerations and lateral inertia forces relative to values that would occur were the structure to remain linearly elastic and lightly damped.[21.8] Thus, the use of design forces representing earthquake effects such as those in Reference 21.2 requires that the lateral-force resisting system retain a substantial portion of its strength into the inelastic range under displacement reversals.

The provisions of Chapter 21 relate detailing requirements to type of structural framing, earthquake risk level at the site, level of inelastic deformation intended in structural design, and use and occupancy of the structure. Earthquake risk levels traditionally have been classified as low, moderate, and high. The seismic risk level of a region or the seismic performance or design category of a structure is regulated by the legally adopted general building code or determined by local authority (see 1.1.8.3, R1.1.8.3, and Table R1.1.8.3). The 2000 IBC[21.5] and the 2000 NEHRP[21.9] provisions use the same terminology as the 1997 NEHRP[21.1] provisions.

The design and detailing requirements should be compatible with the level of energy dissipation (or toughness) assumed in the computation of the design seismic loads. The terms ordinary, intermediate, and special are specifically used to facilitate this compatibility. The degree of required toughness, and therefore the level of required detailing, increases for structures progressing from ordinary through intermediate to special categories. It is essential that structures in higher seismic zones or assigned to higher seismic performance or design categories possess a higher degree of toughness. It is permitted, however, to design for higher toughness in the lower seismic zones or design categories and take advantage of the lower design force levels.

The provisions of Chapters 1 through 18 and 22 are intended to provide adequate toughness for structures in regions of low seismic risk, or assigned to ordinary categories. Therefore, it is not required to apply the provisions of Chapter 21 to lateral-force resisting systems consisting of ordinary structural walls.

Chapter 21 requires special details for reinforced concrete structures in regions of moderate seismic risk, or assigned to intermediate seismic performance or design categories. These requirements are contained in 21.2.1.3, 21.12, and 21.13.

CODE

COMMENTARY

Although new provisions are provided in 21.13 for design of intermediate precast structural walls, general building codes that address seismic performance or design categories currently do not include intermediate structural walls.

Structures in regions of high seismic risk, or assigned to high seismic performance or design categories, may be subjected to strong ground shaking. Structures designed using seismic forces based upon response modification factors for special moment frames or special reinforced concrete structural walls are likely to experience multiple cycles of lateral displacements well beyond the point where reinforcement yields should the design earthquake ground shaking occur. The provisions of 21.2.2 through 21.2.8 and 21.3 through 21.11 have been developed to provide the structure with adequate toughness for this special response.

The requirements of Chapter 21 and 22 as they apply to various components of structures in regions of intermediate or high seismic risk, or assigned to intermediate or high seismic performance or design categories, are summarized in Table R21.2.1.

The special proportioning and detailing requirements in Chapter 21 are based predominantly on field and laboratory experience with monolithic reinforced concrete building structures and precast concrete building structures designed and detailed to behave like monolithic building structures. Extrapolation of these requirements to other types of cast-in-place or precast concrete structures should be based on evidence provided by field experience, tests, or analysis. ACI T1.1-01, "Acceptance Criteria for Moment Frames Based on Structural Testing," can be used in conjunction with Chapter 21 to demonstrate that the strength and toughness of a proposed frame system equals or exceeds that provided by a comparable monolithic concrete system.

The toughness requirements in 21.2.1.5 refer to the concern for the structural integrity of the entire lateral-force resisting

TABLE R21.2.1— SECTIONS OF CHAPTERS 21 AND 22 TO BE SATISFIED*

	Level of seismic risk or assigned seismic performance or design categories (as defined in code section)	
Component resisting earthquake effect, unless otherwise noted	Intermediate (21.2.1.3)	High (21.2.1.4)
Frame members	21.12	21.2, 21.3, 21.4, 21.5
Structural walls and coupling beams	None	21.2, 21.7
Precast structural walls	21.13	21.2, 21.8
Structural diaphragms and trusses	None	21.2, 21.9
Foundations	None	21.2, 21.10
Frame members not proportioned to resist forces induced by earthquake motions	None	2.11
Plain concrete	22.4	22.4, 22.10.1

*In addition to requirements of Chapters 1 through 18, except as modified by Chapter 21.

CODE

COMMENTARY

system at lateral displacements anticipated for ground motions corresponding to the design earthquake. Depending on the energy-dissipation characteristics of the structural system used, such displacements may be larger than for a monolithic reinforced concrete structure.

21.2.2 — Analysis and proportioning of structural members

21.2.2.1 — The interaction of all structural and nonstructural members that materially affect the linear and nonlinear response of the structure to earthquake motions shall be considered in the analysis.

21.2.2.2 — Rigid members assumed not to be a part of the lateral-force resisting system shall be permitted provided their effect on the response of the system is considered and accommodated in the structural design. Consequences of failure of structural and nonstructural members, which are not a part of the lateral-force resisting system, shall also be considered.

21.2.2.3 — Structural members below base of structure that are required to transmit to the foundation forces resulting from earthquake effects shall also comply with the requirements of Chapter 21.

21.2.2.4 — All structural members assumed not to be part of the lateral-force resisting system shall conform to 21.11.

R21.2.2 — Analysis and proportioning of structural members

It is assumed that the distribution of required strength to the various components of a lateral-force resisting system will be guided by the analysis of a linearly elastic model of the system acted upon by the factored forces required by the governing code. If nonlinear response history analyses are to be used, base motions should be selected after a detailed study of the site conditions and local seismic history.

Because the design basis admits nonlinear response, it is necessary to investigate the stability of the lateral-force resisting system as well as its interaction with other structural and nonstructural members at displacements larger than those indicated by linear analysis. To handle this without having to resort to nonlinear response analysis, one option is to multiply by a factor of at least two the displacements from linear analysis by using the factored lateral forces, unless the governing code specifies the factors to be used as in References 21.1 and 21.2. For lateral displacement calculations, assuming all the horizontal structural members to be fully cracked is likely to lead to better estimates of the possible drift than using uncracked stiffness for all members.

The main concern of Chapter 21 is the safety of the structure. The intent of 21.2.2.1 and 21.2.2.2 is to draw attention to the influence of nonstructural members on structural response and to hazards from falling objects.

Section 21.2.2.3 alerts the designer that the base of the structure as defined in analysis may not necessarily correspond to the foundation or ground level.

In selecting member sizes for earthquake-resistant structures, it is important to consider problems related to congestion of reinforcement. The designer should ensure that all reinforcement can be assembled and placed and that concrete can be cast and consolidated properly. Use of upper limits of reinforcement ratios permitted is likely to lead to insurmountable construction problems especially at frame joints.

21.2.3 — Strength reduction factors

Strength reduction factors shall be as given in 9.3.4.

CODE

21.2.4 — Concrete in members resisting earthquake-induced forces

21.2.4.1 — Specified compressive strength of concrete, f_c', shall be not less than 3000 psi.

21.2.4.2 — Specified compressive strength of lightweight concrete, f_c', shall not exceed 5000 psi unless demonstrated by experimental evidence that structural members made with that lightweight concrete provide strength and toughness equal to or exceeding those of comparable members made with normalweight concrete of the same strength.

21.2.5 — Reinforcement in members resisting earthquake-induced forces

Reinforcement resisting earthquake-induced flexural and axial forces in frame members and in structural wall boundary elements shall comply with ASTM A 706. ASTM A 615 Grades 40 and 60 reinforcement shall be permitted in these members if:

(a) The actual yield strength based on mill tests does not exceed f_y by more than 18,000 psi (retests shall not exceed this value by more than an additional 3000 psi); and

(b) The ratio of the actual tensile strength to the actual yield strength is not less than 1.25.

The value of f_{yt} for transverse reinforcement including spiral reinforcement shall not exceed 60,000 psi.

21.2.6 — Mechanical splices

21.2.6.1 — Mechanical splices shall be classified as either Type 1 or Type 2 mechanical splices, as follows:

(a) Type 1 mechanical splices shall conform to 12.14.3.2;

(b) Type 2 mechanical splices shall conform to 12.14.3.2 and shall develop the specified tensile strength of the spliced bar.

COMMENTARY

R21.2.4 — Concrete in members resisting earthquake-induced forces

Requirements of this section refer to concrete quality in frames, trusses, or walls proportioned to resist earthquake-induced forces. The maximum specified compressive strength of lightweight concrete to be used in structural design calculations is limited to 5000 psi, primarily because of paucity of experimental and field data on the behavior of members made with lightweight aggregate concrete subjected to displacement reversals in the nonlinear range. If convincing evidence is developed for a specific application, the limit on maximum specified compressive strength of lightweight concrete may be increased to a level justified by the evidence.

R21.2.5 — Reinforcement in members resisting earthquake-induced forces

Use of longitudinal reinforcement with strength substantially higher than that assumed in design will lead to higher shear and bond stresses at the time of development of yield moments. These conditions may lead to brittle failures in shear or bond and should be avoided even if such failures may occur at higher loads than those anticipated in design. Therefore, a ceiling is placed on the actual yield strength of the steel [see 21.2.5(a)].

The requirement for a tensile strength larger than the yield strength of the reinforcement [21.2.5(b)] is based on the assumption that the capability of a structural member to develop inelastic rotation capacity is a function of the length of the yield region along the axis of the member. In interpreting experimental results, the length of the yield region has been related to the relative magnitudes of ultimate and yield moments.[21.10] According to this interpretation, the larger the ratio of ultimate to yield moment, the longer the yield region. Chapter 21 requires that the ratio of actual tensile strength to actual yield strength is not less than 1.25. Members with reinforcement not satisfying this condition can also develop inelastic rotation, but their behavior is sufficiently different to exclude them from direct consideration on the basis of rules derived from experience with members reinforced with strain-hardening steel.

R21.2.6 — Mechanical splices

In a structure undergoing inelastic deformations during an earthquake, the tensile stresses in reinforcement may approach the tensile strength of the reinforcement. The requirements for Type 2 mechanical splices are intended to avoid a splice failure when the reinforcement is subjected to expected stress levels in yielding regions. Type 1 splices are not required to satisfy the more stringent requirements for Type 2 splices, and may not be capable of resisting the stress levels expected in yielding regions. The locations of

CODE

COMMENTARY

21.2.6.2 — Type 1 mechanical splices shall not be used within a distance equal to twice the member depth from the column or beam face for special moment frames or from sections where yielding of the reinforcement is likely to occur as a result of inelastic lateral displacements. Type 2 mechanical splices shall be permitted to be used at any location.

Type 1 splices are restricted because tensile stresses in reinforcement in yielding regions can exceed the strength requirements of 12.14.3.2.

Recommended detailing practice would preclude the use of splices in regions of potential yield in members resisting earthquake effects. If use of mechanical splices in regions of potential yielding cannot be avoided, the designer should have documentation on the actual strength characteristics of the bars to be spliced, on the force-deformation characteristics of the spliced bar, and on the ability of the Type 2 splice to be used to meet the specified performance requirements.

21.2.7 — Welded splices

21.2.7.1 — Welded splices in reinforcement resisting earthquake-induced forces shall conform to 12.14.3.4 and shall not be used within a distance equal to twice the member depth from the column or beam face for special moment frames or from sections where yielding of the reinforcement is likely to occur as a result of inelastic lateral displacements.

21.2.7.2 — Welding of stirrups, ties, inserts, or other similar elements to longitudinal reinforcement that is required by design shall not be permitted.

R21.2.7 — Welded splices

R21.2.7.1 — Welding of reinforcement should be according to ANSI/AWS D1.4 as required in Chapter 3. The locations of welded splices are restricted because reinforcement tension stresses in yielding regions can exceed the strength requirements of 12.14.3.4.

R21.2.7.2 — Welding of crossing reinforcing bars can lead to local embrittlement of the steel. If welding of crossing bars is used to facilitate fabrication or placement of reinforcement, it should be done only on bars added for such purposes. The prohibition of welding crossing reinforcing bars does not apply to bars that are welded with welding operations under continuous, competent control as in the manufacture of welded wire reinforcement.

21.2.8 — Anchoring to concrete

21.2.8.1 — Anchors resisting earthquake-induced forces in structures in regions of moderate or high seismic risk, or assigned to intermediate or high seismic performance or design categories shall conform to the additional requirements of D.3.3 of Appendix D.

21.3 — Flexural members of special moment frames

21.3.1 — Scope

Requirements of 21.3 apply to special moment frame members (a) resisting earthquake-induced forces and (b) proportioned primarily to resist flexure. These frame members shall also satisfy the conditions of 21.3.1.1 through 21.3.1.4.

21.3.1.1 — Factored axial compressive force on the member, P_u, shall not exceed $A_g f_c'/10$.

21.3.1.2 — Clear span for member, ℓ_n, shall not be less than four times its effective depth.

R21.3 — Flexural members of special moment frames

R21.3.1 — Scope

This section refers to beams of special moment frames resisting lateral loads induced by earthquake motions. Any frame member subjected to a factored axial compressive force exceeding $(A_g f_c'/10)$ is to be proportioned and detailed as described in 21.4.

Experimental evidence[21.11] indicates that, under reversals of displacement into the nonlinear range, behavior of continuous members having length-to-depth ratios of less than four is significantly different from the behavior of relatively slender members. Design rules derived from experience

CODE

21.3.1.3 — Width of member, b_w, shall not be less than the smaller of $0.3h$ and 10 in.

21.3.1.4 — Width of member, b_w, shall not exceed width of supporting member (measured on a plane perpendicular to the longitudinal axis of flexural member) plus distances on each side of supporting member not exceeding three-fourths of the depth of flexural member.

21.3.2 — Longitudinal reinforcement

21.3.2.1 — At any section of a flexural member, except as provided in 10.5.3, for top as well as for bottom reinforcement, the amount of reinforcement shall not be less than that given by Eq. (10-3) but not less than $200b_wd/f_y$, and the reinforcement ratio, ρ, shall not exceed 0.025. At least two bars shall be provided continuously both top and bottom.

21.3.2.2 — Positive moment strength at joint face shall be not less than one-half of the negative moment strength provided at that face of the joint. Neither the negative nor the positive moment strength at any section along member length shall be less than one-fourth the maximum moment strength provided at face of either joint.

21.3.2.3 — Lap splices of flexural reinforcement shall be permitted only if hoop or spiral reinforcement is provided over the lap length. Spacing of the transverse reinforcement enclosing the lapped bars shall not exceed the smaller of $d/4$ and 4 in. Lap splices shall not be used

(a) within the joints;

(b) within a distance of twice the member depth from the face of the joint; and

COMMENTARY

with relatively slender members do not apply directly to members with length-to-depth ratios less than four, especially with respect to shear strength.

Geometric constraints indicated in 21.3.1.3 and 21.3.1.4 were derived from practice with reinforced concrete frames resisting earthquake-induced forces.[21.12]

R21.3.2 — Longitudinal reinforcement

Section 10.3.5 limits the net tensile strain, ε_t, thereby indirectly limiting the tensile reinforcement ratio in a flexural member to a fraction of the amount that would produce balanced conditions. For a section subjected to bending only and loaded monotonically to yielding, this approach is feasible because the likelihood of compressive failure can be estimated reliably with the behavioral model assumed for determining the reinforcement ratio corresponding to balanced failure. The same behavioral model (because of incorrect assumptions such as linear strain distribution, well-defined yield point for the steel, limiting compressive strain in the concrete of 0.003, and compressive stresses in the shell concrete) fails to describe the conditions in a flexural member subjected to reversals of displacements well into the inelastic range. Thus, there is little rationale for continuing to refer to balanced conditions in earthquake-resistant design of reinforced concrete structures.

R21.3.2.1 — The limiting reinforcement ratio of 0.025 is based primarily on considerations of steel congestion and, indirectly, on limiting shear stresses in girders of typical proportions. The requirement of at least two bars, top and bottom, refers again to construction rather than behavioral requirements.

R21.3.2.3 — Lap splices of reinforcement are prohibited at regions where flexural yielding is anticipated because such splices are not reliable under conditions of cyclic loading into the inelastic range. Transverse reinforcement for lap splices at any location is mandatory because of the likelihood of loss of shell concrete.

CODE

COMMENTARY

(c) where analysis indicates flexural yielding is caused by inelastic lateral displacements of the frame.

21.3.2.4 — Mechanical splices shall conform to 21.2.6 and welded splices shall conform to 21.2.7.

21.3.3 — Transverse reinforcement

21.3.3.1 — Hoops shall be provided in the following regions of frame members:

(a) Over a length equal to twice the member depth measured from the face of the supporting member toward midspan, at both ends of the flexural member;

(b) Over lengths equal to twice the member depth on both sides of a section where flexural yielding is likely to occur in connection with inelastic lateral displacements of the frame.

21.3.3.2 — The first hoop shall be located not more than 2 in. from the face of a supporting member. Spacing of the hoops shall not exceed the smallest of (a), (b), (c) and (d):

(a) $d/4$;

(b) eight times the diameter of the smallest longitudinal bars;

(c) 24 times the diameter of the hoop bars; and

(d) 12 in.

21.3.3.3 — Where hoops are required, longitudinal bars on the perimeter shall have lateral support conforming to 7.10.5.3.

21.3.3.4 — Where hoops are not required, stirrups with seismic hooks at both ends shall be spaced at a distance not more than $d/2$ throughout the length of the member.

21.3.3.5 — Stirrups or ties required to resist shear shall be hoops over lengths of members in 21.3.3, 21.4.4, and 21.5.2.

21.3.3.6 — Hoops in flexural members shall be permitted to be made up of two pieces of reinforcement: a stirrup having seismic hooks at both ends and closed by a crosstie. Consecutive crossties engaging the same longitudinal bar shall have their 90 degree hooks at opposite sides of the flexural member. If the longitudinal reinforcing bars secured by the crossties are confined by a slab on only one side of the flexural frame member, the 90 degree hooks of the crossties shall be placed on that side.

R21.3.3 — Transverse reinforcement

Transverse reinforcement is required primarily to confine the concrete and maintain lateral support for the reinforcing bars in regions where yielding is expected. Examples of hoops suitable for flexural members of frames are shown in Fig. R21.3.3.

In the case of members with varying strength along the span or members for which the permanent load represents a large proportion of the total design load, concentrations of inelastic rotation may occur within the span. If such a condition is anticipated, transverse reinforcement also should be provided in regions where yielding is expected.

Because spalling of the concrete shell is anticipated during strong motion, especially at and near regions of flexural yielding, all web reinforcement should be provided in the form of closed hoops as defined in 21.3.3.5.

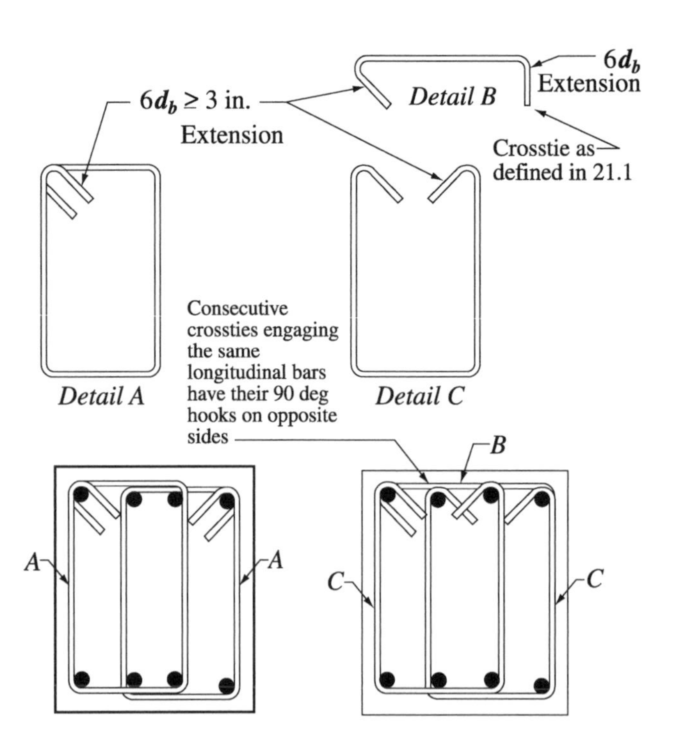

Fig. R21.3.3—Examples of overlapping hoops

CODE

21.3.4 — Shear strength requirements

21.3.4.1 — Design forces

The design shear force, V_e, shall be determined from consideration of the statical forces on the portion of the member between faces of the joints. It shall be assumed that moments of opposite sign corresponding to probable flexural moment strength, M_{pr}, act at the joint faces and that the member is loaded with the factored tributary gravity load along its span.

21.3.4.2 — Transverse reinforcement

Transverse reinforcement over the lengths identified in 21.3.3.1 shall be proportioned to resist shear assuming $V_c = 0$ when both (a) and (b) occur:

(a) The earthquake-induced shear force calculated in accordance with 21.3.4.1 represents one-half or more of the maximum required shear strength within those lengths;

(b) The factored axial compressive force, P_u, including earthquake effects is less than $A_g f_c'/20$.

21.4 — Special moment frame members subjected to bending and axial load

21.4.1 — Scope

The requirements of this section apply to special moment frame members (a) resisting earthquake-induced forces and (b) having a factored axial compressive force P_u exceeding $A_g f_c'/10$. These frame

COMMENTARY

R21.3.4 — Shear strength requirements

R21.3.4.1 — Design forces

In determining the equivalent lateral forces representing earthquake effects for the type of frames considered, it is assumed that frame members will dissipate energy in the nonlinear range of response. Unless a frame member possesses a strength that is a multiple on the order of 3 or 4 of the design forces, it should be assumed that it will yield in the event of a major earthquake. The design shear force should be a good approximation of the maximum shear that may develop in a member. Therefore, required shear strength for frame members is related to flexural strengths of the designed member rather than to factored shear forces indicated by lateral load analysis. The conditions described by 21.3.4.1 are illustrated in Fig. R21.3.4.

Because the actual yield strength of the longitudinal reinforcement may exceed the specified yield strength and because strain hardening of the reinforcement is likely to take place at a joint subjected to large rotations, required shear strengths are determined using a stress of at least $1.25f_y$ in the longitudinal reinforcement.

R21.3.4.2 — Transverse reinforcement

Experimental studies[21.13, 21.14] of reinforced concrete members subjected to cyclic loading have demonstrated that more shear reinforcement is required to ensure a flexural failure if the member is subjected to alternating nonlinear displacements than if the member is loaded in only one direction: the necessary increase of shear reinforcement being higher in the case of no axial load. This observation is reflected in the code (see 21.3.4.2) by eliminating the term representing the contribution of concrete to shear strength. The added conservatism on shear is deemed necessary in locations where potential flexural hinging may occur. However, this stratagem, chosen for its relative simplicity, should not be interpreted to mean that no concrete is required to resist shear. On the contrary, it may be argued that the concrete core resists all of the shear with the shear (transverse) reinforcement confining and strengthening the concrete. The confined concrete core plays an important role in the behavior of the beam and should not be reduced to a minimum just because the design expression does not explicitly recognize it.

R21.4 — Special moment frame members subjected to bending and axial load

R21.4.1 — Scope

Section 21.4.1 is intended primarily for columns of special moment frames. Frame members, other than columns, that do not satisfy 21.3.1 are to be proportioned and detailed according to this section.

CODE

members shall also satisfy the conditions of 21.4.1.1 and 21.4.1.2.

21.4.1.1 — The shortest cross-sectional dimension, measured on a straight line passing through the geometric centroid, shall not be less than 12 in.

21.4.1.2 — The ratio of the shortest cross-sectional dimension to the perpendicular dimension shall not be less than 0.4.

COMMENTARY

The geometric constraints in 21.4.1.1 and 21.4.1.2 follow from previous practice.[21.12]

Notes on Fig. R21.3.4:

1. Direction of shear force V_e depends on relative magnitudes of gravity loads and shear generated by end moments.

2. End moments M_{pr} based on steel tensile stress of $1.25f_y$, where f_y is specified yield strength. (Both end moments should be considered in both directions, clockwise and counter-clockwise).

3. End moment M_{pr} for columns need not be greater than moments generated by the M_{pr} of the beams framing into the beam-column joints. V_e should not be less than that required by analysis of the structure.

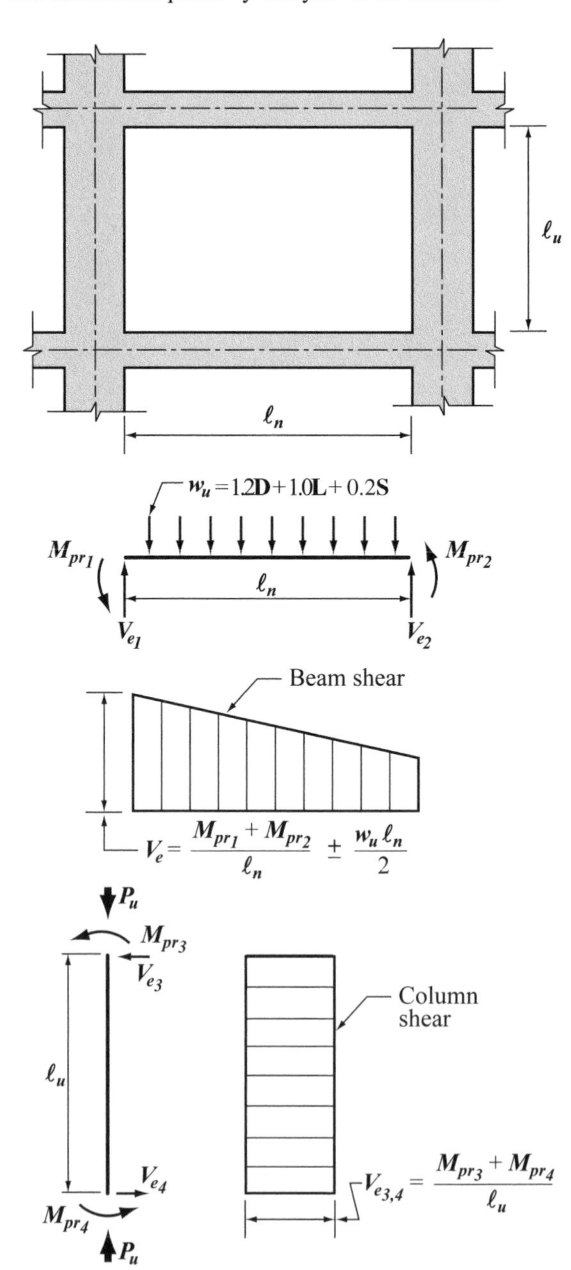

Fig. R21.3.4—Design shears for girders and columns

CODE

21.4.2 — Minimum flexural strength of columns

21.4.2.1 — Flexural strength of any column proportioned to resist P_u exceeding $A_g f_c'/10$ shall satisfy 21.4.2.2 or 21.4.2.3.

Lateral strength and stiffness of columns not satisfying 21.4.2.2 shall be ignored when determining the calculated strength and stiffness of the structure, but such columns shall conform to 21.11.

21.4.2.2 — The flexural strengths of the columns shall satisfy Eq. (21-1)

$$\Sigma M_{nc} \geq (6/5)\, \Sigma M_{nb} \qquad (21\text{-}1)$$

ΣM_{nc} = sum of nominal flexural strengths of columns framing into the joint, evaluated at the faces of the joint. Column flexural strength shall be calculated for the factored axial force, consistent with the direction of the lateral forces considered, resulting in the lowest flexural strength.

ΣM_{nb} = sum of nominal flexural strengths of the beams framing into the joint, evaluated at the faces of the joint. In T-beam construction, where the slab is in tension under moments at the face of the joint, slab reinforcement within an effective slab width defined in 8.10 shall be assumed to contribute to M_{nb} if the slab reinforcement is developed at the critical section for flexure.

Flexural strengths shall be summed such that the column moments oppose the beam moments. Eq. (21-1) shall be satisfied for beam moments acting in both directions in the vertical plane of the frame considered.

21.4.2.3 — If 21.4.2.2 is not satisfied at a joint, columns supporting reactions from that joint shall be provided with transverse reinforcement as specified in 21.4.4.1 through 21.4.4.3 over their full height.

21.4.3 — Longitudinal reinforcement

21.4.3.1 — Area of longitudinal reinforcement, A_{st}, shall not be less than $0.01A_g$ or more than $0.06A_g$.

21.4.3.2 — Mechanical splices shall conform to 21.2.6 and welded splices shall conform to 21.2.7. Lap splices shall be permitted only within the center half of the member length, shall be designed as tension lap splices, and shall be enclosed within transverse reinforcement conforming to 21.4.4.2 and 21.4.4.3.

COMMENTARY

R21.4.2 — Minimum flexural strength of columns

The intent of 21.4.2.2 is to reduce the likelihood of yielding in columns that are considered as part of the lateral force resisting system. If columns are not stronger than beams framing into a joint, there is likelihood of inelastic action. In the worst case of weak columns, flexural yielding can occur at both ends of all columns in a given story, resulting in a column failure mechanism that can lead to collapse.

In 21.4.2.2, the nominal strengths of the girders and columns are calculated at the joint faces, and those strengths are compared directly using Eq. (21-1). The 1995 code required design strengths to be compared at the center of the joint, which typically produced similar results but with added computational effort.

When determining the nominal flexural strength of a girder section in negative bending (top in tension), longitudinal reinforcement contained within an effective flange width of a top slab that acts monolithically with the girder increases the girder strength. Research[21.15] on beam-column subassemblies under lateral loading indicates that using the effective flange widths defined in 8.10 gives reasonable estimates of girder negative bending strengths of interior connections at interstory displacement levels approaching 2 percent of story height. This effective width is conservative where the slab terminates in a weak spandrel.

If 21.4.2.2 cannot be satisfied at a joint, any positive contribution of the column or columns involved to the lateral strength and stiffness of the structure is to be ignored. Negative contributions of the column or columns should not be ignored. For example, ignoring the stiffness of the columns ought not be used as a justification for reducing the design base shear. If inclusion of those columns in the analytical model of the building results in an increase in torsional effects, the increase should be considered as required by the governing code.

R21.4.3 — Longitudinal reinforcement

The lower limit of the area of longitudinal reinforcement is to control time-dependent deformations and to have the yield moment exceed the cracking moment. The upper limit of the section reflects concern for steel congestion, load transfer from floor elements to column especially in low-rise construction, and the development of high shear stresses.

Spalling of the shell concrete, which is likely to occur near the ends of the column in frames of typical configuration, makes lap splices in these locations vulnerable. If lap splices are to be used at all, they should be located near the midheight where stress reversal is likely to be limited to a smaller stress range than at locations near the joints. Special

CODE

21.4.4 — Transverse reinforcement

21.4.4.1 — Transverse reinforcement required in (a) through (e) shall be provided unless a larger amount is required by 21.4.3.2 or 21.4.5.

(a) The volumetric ratio of spiral or circular hoop reinforcement, ρ_s, shall not be less than required by Eq. (21-2).

$$\rho_s = 0.12 f_c'/f_{yt} \qquad (21\text{-}2)$$

and shall not be less than required by Eq. (10-5).

(b) The total cross-sectional area of rectangular hoop reinforcement, A_{sh}, shall not be less than required by Eq. (21-3) and (21-4).

$$A_{sh} = 0.3(sb_c f_c'/f_{yt})[(A_g/A_{ch}) - 1] \qquad (21\text{-}3)$$

$$A_{sh} = 0.09 sb_c f_c'/f_{yt} \qquad (21\text{-}4)$$

(c) Transverse reinforcement shall be provided by either single or overlapping hoops. Crossties of the same bar size and spacing as the hoops shall be permitted. Each end of the crosstie shall engage a peripheral longitudinal reinforcing bar. Consecutive crossties shall be alternated end for end along the longitudinal reinforcement.

(d) If the design strength of member core satisfies the requirement of the design loading combinations including E, load effects for earthquake effect, Eq. (21-3) and (10-5) need not be satisfied.

(e) If the thickness of the concrete outside the confining transverse reinforcement exceeds 4 in., additional transverse reinforcement shall be provided at a spacing not exceeding 12 in. Concrete cover on the additional reinforcement shall not exceed 4 in.

21.4.4.2 — Spacing of transverse reinforcement shall not exceed the smallest of (a), (b), and (c):

(a) one-quarter of the minimum member dimension;

(b) six times the diameter of the longitudinal reinforcement; and

(c) s_o, as defined by Eq. (21-5)

COMMENTARY

transverse reinforcement is required along the lap-splice length because of the uncertainty in moment distributions along the height and the need for confinement of lap splices subjected to stress reversals.[21.16]

R21.4.4 — Transverse reinforcement

Requirements of this section are concerned with confining the concrete and providing lateral support to the longitudinal reinforcement.

The effect of helical (spiral) reinforcement and adequately configured rectangular hoop reinforcement on strength and ductility of columns is well established.[21.17] While analytical procedures exist for calculation of strength and ductility capacity of columns under axial and moment reversals,[21.18] the axial load and deformation demands required during earthquake loading are not known with sufficient accuracy to justify calculation of required transverse reinforcement as a function of design earthquake demands. Instead, Eq. (10-5) and (21-3) are required, with the intent that spalling of shell concrete will not result in a loss of axial load strength of the column. Eq. (21-2) and (21-4) govern for large-diameter columns, and are intended to ensure adequate flexural curvature capacity in yielding regions.

Fig. R21.4.4 shows an example of transverse reinforcement provided by one hoop and three crossties. Crossties with a 90 degree hook are not as effective as either crossties with 135 degree hooks or hoops in providing confinement. Tests show that if crosstie ends with 90 degree hooks are alternated, confinement will be sufficient.

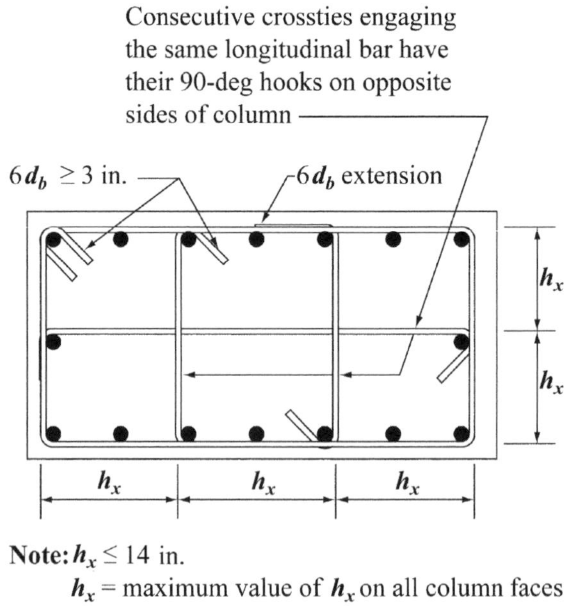

Consecutive crossties engaging the same longitudinal bar have their 90-deg hooks on opposite sides of column

$6d_b \geq 3$ in.

$6d_b$ extension

h_x h_x h_x h_x h_x

Note: $h_x \leq 14$ in.

h_x = maximum value of h_x on all column faces

Fig. R21.4.4—Example of transverse reinforcement in columns

CODE

$$s_o = 4 + \left(\frac{14 - h_x}{3}\right) \quad (21\text{-}5)$$

The value of s_o shall not exceed 6 in. and need not be taken less than 4 in.

21.4.4.3 — Horizontal spacing of crossties or legs of overlapping hoops, h_x, shall not exceed 14 in. on center.

21.4.4.4 — Transverse reinforcement as specified in 21.4.4.1 through 21.4.4.3 shall be provided over a length ℓ_o from each joint face and on both sides of any section where flexural yielding is likely to occur as a result of inelastic lateral displacements of the frame. Length ℓ_o shall not be less than the largest of (a), (b), and (c):

(a) the depth of the member at the joint face or at the section where flexural yielding is likely to occur;

(b) one-sixth of the clear span of the member; and

(c) 18 in.

21.4.4.5 — Columns supporting reactions from discontinued stiff members, such as walls, shall be provided with transverse reinforcement as required in 21.4.4.1 through 21.4.4.3 over their full height beneath the level at which the discontinuity occurs if the factored axial compressive force in these members, related to earthquake effect, exceeds $A_g f_c'/10$. Transverse reinforcement as required in 21.4.4.1 through 21.4.4.3 shall extend at least the development length in tension, ℓ_d, into discontinued member, where ℓ_d is determined in accordance of 21.5.4 using the largest longitudinal reinforcement in the column. If the lower end of the column terminates on a wall, transverse reinforcement as required in 21.4.4.1 through 21.4.4.3 shall extend into wall at least ℓ_d of the largest longitudinal column bar at the point of termination. If the column terminates on a footing or mat, transverse reinforcement as required in 21.4.4.1 through 21.4.4.3 shall extend at least 12 in. into the footing or mat.

21.4.4.6 — Where transverse reinforcement, as specified in 21.4.4.1 through 21.4.4.3, is not provided throughout the full length of the column, the remainder of the column length shall contain spiral or hoop reinforcement with center-to-center spacing, s, not exceeding the smaller of six times the diameter of the longitudinal column bars and 6 in.

COMMENTARY

Sections 21.4.4.2 and 21.4.4.3 are interrelated requirements for configuration of rectangular hoop reinforcement. The requirement that spacing not exceed one-quarter of the minimum member dimension is to obtain adequate concrete confinement. The requirement that spacing not exceed six bar diameters is intended to restrain longitudinal reinforcement buckling after spalling. The 4 in. spacing is for concrete confinement; 21.4.4.2 permits this limit to be relaxed to a maximum of 6 in. if the spacing of crossties or legs of overlapping hoops is limited to 8 in.

The unreinforced shell may spall as the column deforms to resist earthquake effects. Separation of portions of the shell from the core caused by local spalling creates a falling hazard. The additional reinforcement is required to reduce the risk of portions of the shell falling away from the column.

Section 21.4.4.4 stipulates a minimum length over which to provide closely-spaced transverse reinforcement at the member ends, where flexural yielding normally occurs. Research results indicate that the length should be increased by 50 percent or more in locations, such as the base of the building, where axial loads and flexural demands may be especially high.[21.19]

Columns supporting discontinued stiff members, such as walls or trusses, may develop considerable inelastic response. Therefore, it is required that these columns have special transverse reinforcement throughout their length. This covers all columns beneath the level at which the stiff member has been discontinued, unless the factored forces corresponding to earthquake effect are low (see 21.4.4.5).

Field observations have shown significant damage to columns in the unconfined region near the midheight. The requirements of 21.4.4.6 are to ensure a relatively uniform toughness of the column along its length.

R21.4.4.6 — The provisions of 21.4.4.6 were added to the 1989 code to provide reasonable protection and ductility to the midheight of columns between transverse reinforcement. Observations after earthquakes have shown significant damage to columns in the nonconfined region, and the minimum ties or spirals required should provide a more uniform toughness of the column along its length.

CODE

21.4.5 — Shear strength requirements

21.4.5.1 — Design forces

The design shear force, V_e, shall be determined from consideration of the maximum forces that can be generated at the faces of the joints at each end of the member. These joint forces shall be determined using the maximum probable moment strengths, M_{pr}, at each end of the member associated with the range of factored axial loads, P_u, acting on the member. The member shears need not exceed those determined from joint strengths based on M_{pr} of the transverse members framing into the joint. In no case shall V_e be less than the factored shear determined by analysis of the structure.

21.4.5.2 — Transverse reinforcement over the lengths ℓ_o, identified in 21.4.4.4, shall be proportioned to resist shear assuming $V_c = 0$ when both (a) and (b) occur:

(a) The earthquake-induced shear force, calculated in accordance with 21.4.5.1, represents one-half or more of the maximum required shear strength within ℓ_o;

(b) The factored axial compressive force, P_u, including earthquake effects is less than $A_g f_c'/20$.

21.5 — Joints of special moment frames

21.5.1 — General requirements

21.5.1.1 — Forces in longitudinal beam reinforcement at the joint face shall be determined by assuming that the stress in the flexural tensile reinforcement is $1.25f_y$.

21.5.1.2 — Strength of joint shall be governed by the appropriate ϕ factors in 9.3.4.

21.5.1.3 — Beam longitudinal reinforcement terminated in a column shall be extended to the far face of the confined column core and anchored in tension according to 21.5.4 and in compression according to Chapter 12.

21.5.1.4 — Where longitudinal beam reinforcement extends through a beam-column joint, the column dimension parallel to the beam reinforcement shall not be less than 20 times the diameter of the largest longitudinal beam bar for normalweight concrete. For lightweight concrete, the dimension shall be not less than 26 times the bar diameter.

COMMENTARY

R21.4.5 — Shear strength requirements

R21.4.5.1 — Design forces

The provisions of 21.3.4.1 also apply to members subjected to axial loads (for example, columns). Above the ground floor the moment at a joint may be limited by the flexural strength of the beams framing into the joint. Where beams frame into opposite sides of a joint, the combined strength may be the sum of the negative moment strength of the beam on one side of the joint and the positive moment strength of the beam on the other side of the joint. Moment strengths are to be determined using a strength reduction factor of 1.0 and reinforcing steel stress equal to at least $1.25f_y$. Distribution of the combined moment strength of the beams to the columns above and below the joint should be based on analysis. The value of M_{pr} in Fig. R21.3.4 may be computed from the flexural member strengths at the beam-column joints.

R21.5 — Joints of special moment frames

R21.5.1 — General requirements

Development of inelastic rotations at the faces of joints of reinforced concrete frames is associated with strains in the flexural reinforcement well in excess of the yield strain. Consequently, joint shear force generated by the flexural reinforcement is calculated for a stress of $1.25f_y$ in the reinforcement (see 21.5.1.1). A detailed explanation of the reasons for the possible development of stresses in excess of the yield strength in girder tensile reinforcement is provided in Reference 21.10.

R21.5.1.4 — Research[21.20-21.24] has shown that straight beam bars may slip within the beam-column joint during a series of large moment reversals. The bond stresses on these straight bars may be very large. To substantially reduce slip during the formation of adjacent beam hinging, it would be necessary to have a ratio of column dimension to bar diameter of approximately 1/32, which would result in very large joints. On reviewing the available tests, the limit of 1/20 of the column depth in the direction of loading for the maximum size of beam bars for normalweight concrete and

CODE | COMMENTARY

a limit of 1/26 for lightweight concrete were chosen. Due to the lack of specific data, the modification for lightweight concrete used a factor of 1.3 from Chapter 12. These limits provide reasonable control on the amount of potential slip of the beam bars in a beam-column joint, considering the number of anticipated inelastic excursions of the building frames during a major earthquake. A thorough treatment of this topic is given in Reference 21.25.

21.5.2 — Transverse reinforcement

21.5.2.1 — Transverse hoop reinforcement in 21.4.4 shall be provided within the joint, unless the joint is confined by structural members in 21.5.2.2.

21.5.2.2 — Within h of the shallowest framing member, transverse reinforcement equal to at least one-half the amount required by 21.4.4.1 shall be provided where members frame into all four sides of the joint and where each member width is at least three-fourths the column width. At these locations, the spacing required in 21.4.4.2 shall be permitted to be increased to 6 in.

21.5.2.3 — Transverse reinforcement as required by 21.4.4 shall be provided through the joint to provide confinement for longitudinal beam reinforcement outside the column core if such confinement is not provided by a beam framing into the joint.

R21.5.2 — Transverse reinforcement

However low the calculated shear force in a joint of a frame resisting earthquake-induced forces, confining reinforcement (see 21.4.4) should be provided through the joint around the column reinforcement (see 21.5.2.1). In 21.5.2.2, confining reinforcement may be reduced if horizontal members frame into all four sides of the joint. The 1989 code provided a maximum limit on spacing to these areas based on available data.[21.26-21.29]

Section 21.5.2.3 refers to a joint where the width of the girder exceeds the corresponding column dimension. In that case, girder reinforcement not confined by the column reinforcement should be provided lateral support either by a girder framing into the same joint or by transverse reinforcement.

21.5.3 — Shear strength

21.5.3.1 — V_n of the joint shall not be taken as greater than the values specified below for normal-weight concrete.

For joints confined on all four faces .... $20\sqrt{f_c'}A_j$

For joints confined on three faces or on two opposite faces .... $15\sqrt{f_c'}A_j$

For others .... $12\sqrt{f_c'}A_j$

A member that frames into a face is considered to provide confinement to the joint if at least three-quarters of the face of the joint is covered by the framing member. A joint is considered to be confined if such confining members frame into all faces of the joint.

A_j is the effective cross-sectional area within a joint computed from joint depth times effective joint width. Joint depth shall be the overall depth of the column. Effective joint width shall be the overall width of the column, except where a beam frames into a wider column, effective joint width shall not exceed the smaller of (a) and (b):

R21.5.3 — Shear strength

The requirements in Chapter 21 for proportioning joints are based on Reference 21.10 in that behavioral phenomena within the joint are interpreted in terms of a nominal shear strength of the joint. Because tests of joints[21.20] and deep beams[21.11] indicated that shear strength was not as sensitive to joint (shear) reinforcement as implied by the expression developed by ACI Committee 326[21.30] for beams and adopted to apply to joints by ACI Committee 352,[21.10] Committee 318 set the strength of the joint as a function of only the compressive strength of the concrete (see 21.5.3) and to require a minimum amount of transverse reinforcement in the joint (see 21.5.2). The effective area of joint A_j is illustrated in Fig. R21.5.3. In no case is A_j greater than the column cross-sectional area.

The three levels of shear strength required by 21.5.3.1 are based on the recommendation of ACI Committee 352.[21.10] Test data reviewed by the committee[21.28] indicate that the lower value given in 21.5.3.1 of the 1983 code was unconservative when applied to corner joints.

CODE

(a) beam width plus joint depth

(b) twice the smaller perpendicular distance from longitudinal axis of beam to column side.

21.5.3.2 — For lightweight aggregate concrete, the nominal shear strength of the joint shall not exceed three-quarters of the limits given in 21.5.3.1.

COMMENTARY

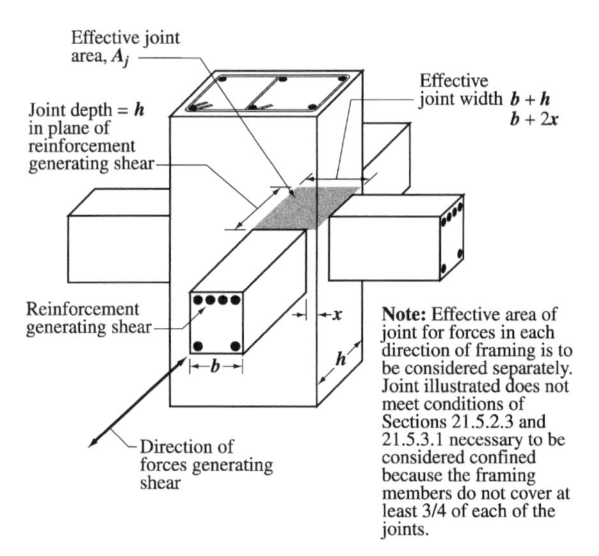

Fig. R21.5.3—Effective joint area

21.5.4 — Development length of bars in tension

21.5.4.1 — The development length, ℓ_{dh}, for a bar with a standard 90 degree hook in normalweight concrete shall not be less than the largest of $8d_b$, 6 in., and the length required by Eq. (21-6).

$$\ell_{dh} = f_y d_b / (65\sqrt{f_c'}) \quad (21\text{-}6)$$

for bar sizes No. 3 through No. 11.

For lightweight concrete, ℓ_{dh} for a bar with a standard 90 degree hook shall not be less than the largest of $10d_b$, 7-1/2 in., and 1.25 times the length required by Eq. (21-6).

The 90 degree hook shall be located within the confined core of a column or of a boundary element.

21.5.4.2 — For bar sizes No. 3 through No. 11, ℓ_d, the development length in tension for a straight bar, shall not be less than the larger of (a) and (b):

(a) 2.5 times the length required by 21.5.4.1 if the depth of the concrete cast in one lift beneath the bar does not exceed 12 in.;

(b) 3.25 times the length required by 21.5.4.1 if the depth of the concrete cast in one lift beneath the bar exceeds 12 in.

21.5.4.3 — Straight bars terminated at a joint shall pass through the confined core of a column or of a boundary element. Any portion of ℓ_d not within the confined core shall be increased by a factor of 1.6.

21.5.4.4 — If epoxy-coated reinforcement is used, the development lengths in 21.5.4.1 through 21.5.4.3 shall be multiplied by the applicable factor in 12.2.4 or 12.5.2.

R21.5.4 — Development length of bars in tension

Minimum development length in tension for deformed bars with standard hooks embedded in normalweight concrete is determined using Eq. (21-6). Eq. (21-6) is based on the requirements of 12.5. Because Chapter 21 stipulates that the hook is to be embedded in confined concrete, the coefficients 0.7 (for concrete cover) and 0.8 (for ties) have been incorporated in the constant used in Eq. (21-6). The development length that would be derived directly from 12.5 is increased to reflect the effect of load reversals.

The development length in tension of a deformed bar with a standard hook is defined as the distance, parallel to the bar, from the critical section (where the bar is to be developed) to a tangent drawn to the outside edge of the hook. The tangent is to be drawn perpendicular to the axis of the bar (see Fig. R12.5).

Factors such as the actual stress in the reinforcement being more than the yield force and the effective development length not necessarily starting at the face of the joint were implicitly considered in the development of the expression for basic development length that has been used as the basis for Eq. (21-6).

For lightweight concrete, the length required by Eq. (21-6) is to be increased by 25 percent to compensate for variability of bond characteristics of reinforcing bars in various types of lightweight concrete.

Section 21.5.4.2 specifies the minimum development length in tension for straight bars as a multiple of the length indicated by 21.5.4.1. Section 21.5.4.2(b) refers to top bars.

If the required straight embedment length of a reinforcing bar extends beyond the confined volume of concrete (as defined in 21.3.3, 21.4.4, or 21.5.2), the required development length is increased on the premise that the limiting bond stress outside the confined region is less than that inside.

CODE

COMMENTARY

$$\ell_{dm} = 1.6(\ell_d - \ell_{dcc}) + \ell_{dcc}$$

or

$$\ell_{dm} = 1.6\ell_d - 0.6\ell_{dcc}$$

where:

ℓ_{dm} = required development length if bar is not entirely embedded in confined concrete;

ℓ_d = required development length in tension for straight bar embedded in confined concrete (see 21.5.4.3);

ℓ_{dcc} = length of bar embedded in confined concrete.

Lack of reference to No. 14 and No. 18 bars in 21.5.4 is due to the paucity of information on anchorage of such bars subjected to load reversals simulating earthquake effects.

21.6 — Special moment frames constructed using precast concrete

21.6.1 — Special moment frames with ductile connections constructed using precast concrete shall satisfy (a) and (b) and all requirements for special moment frames constructed with cast-in-place concrete:

(a) V_n for connections computed according to 11.7.4 shall not be less than $2V_e$, where V_e is calculated according to 21.3.4.1 or 21.4.5.1;

(b) Mechanical splices of beam reinforcement shall be located not closer than $h/2$ from the joint face and shall meet the requirements of 21.2.6.

21.6.2 — Special moment frames with strong connections constructed using precast concrete shall satisfy all requirements for special moment frames constructed with cast-in-place concrete, as well as (a), (b), (c), and (d).

(a) Provisions of 21.3.1.2 shall apply to segments between locations where flexural yielding is intended to occur due to design displacements;

(b) Design strength of the strong connection, ϕS_n, shall be not less than S_e;

(c) Primary longitudinal reinforcement shall be made continuous across connections and shall be developed outside both the strong connection and the plastic hinge region; and

(d) For column-to-column connections, ϕS_n shall not be less than $1.4S_e$. At column-to-column connections, ϕM_n shall be not less than $0.4M_{pr}$ for the column within the story height, and ϕV_n of the connection shall be not less than V_e determined by 21.4.5.1.

R21.6 — Special moment frames constructed using precast concrete

The detailing provisions in 21.6.1 and 21.6.2 are intended to produce frames that respond to design displacements essentially like monolithic special moment frames.

Precast frame systems composed of concrete elements with ductile connections are expected to experience flexural yielding in connection regions. Reinforcement in ductile connections can be made continuous by using Type 2 mechanical splices or any other technique that provides development in tension or compression of at least 125 percent of the specified yield strength f_y of bars and the specified tensile strength of bars.[21.31-21.34] Requirements for mechanical splices are in addition to those in 21.2.6 and are intended to avoid strain concentrations over a short length of reinforcement adjacent to a splice device. Additional requirements for shear strength are provided in 21.6.1 to prevent sliding on connection faces. Precast frames composed of elements with ductile connections may be designed to promote yielding at locations not adjacent to the joints. Therefore, design shear, V_e, as computed according to 21.3.4.1 or 21.4.5.1, may be conservative.

Precast concrete frame systems composed of elements joined using strong connections are intended to experience flexural yielding outside the connections. Strong connections include the length of the coupler hardware as shown in Fig. R21.6.2. Capacity-design techniques are used in 21.6.2(b) to ensure the strong connection remains elastic following formation of plastic hinges. Additional column requirements are provided to avoid hinging and strength deterioration of column-to-column connections.

Strain concentrations have been observed to cause brittle fracture of reinforcing bars at the face of mechanical splices in laboratory tests of precast beam-column connections.[21.35] Designers should carefully select locations of strong connections or take other measures, such as debonding of reinforcing bars in highly stressed regions, to avoid strain concentrations that can result in premature fracture of reinforcement.

CODE

COMMENTARY

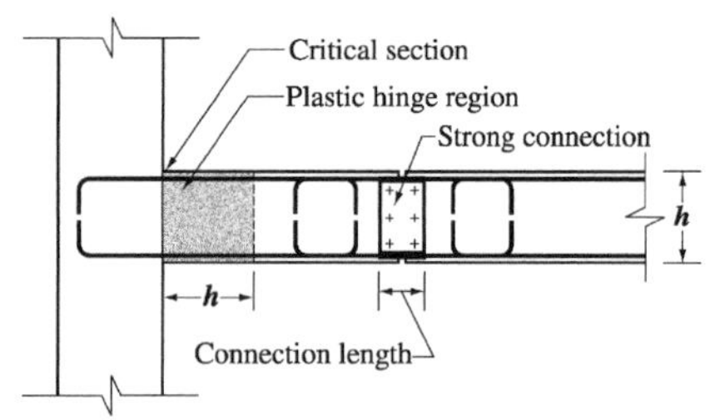

(a) Beam-to-beam connection

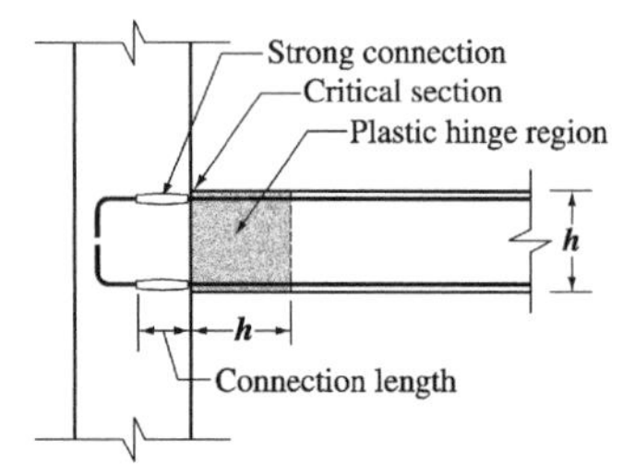

(b) Beam-to-column connection

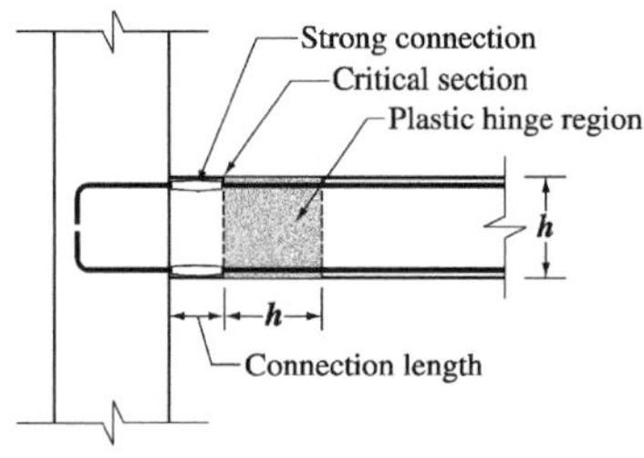

(c) Beam-to-column connection

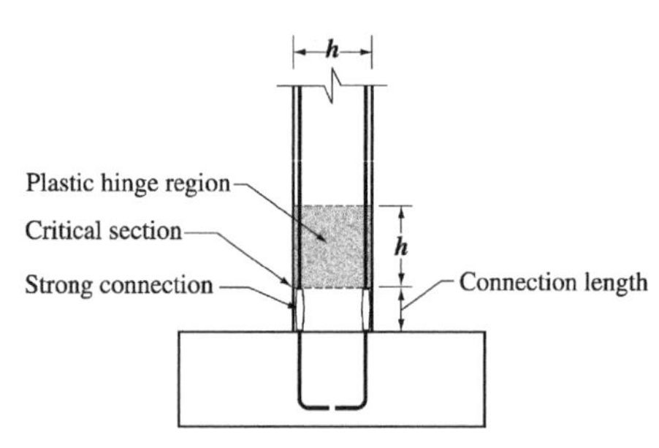

(d) Column-to-footing connection

Fig. R21.6.2 — Strong connection examples.

21.6.3 — Special moment frames constructed using precast concrete and not satisfying the requirements of 21.6.1 or 21.6.2 shall satisfy the requirements of ACI T1.1, "Acceptance Criteria for Moment Frames Based on Structural Testing," and the requirements of (a) and (b):

(a) Details and materials used in the test specimens shall be representative of those used in the structure; and

(b) The design procedure used to proportion the test specimens shall define the mechanism by which the frame resists gravity and earthquake effects, and shall establish acceptance values for sustaining that

R21.6.3 — Precast frame systems not satisfying the prescriptive requirements of Chapter 21 have been demonstrated in experimental studies to provide satisfactory seismic performance characteristics.[21.36,21.37] ACI T1.1 defines a protocol for establishing a design procedure, validated by analysis and laboratory tests, for such frames. The design procedure should identify the load path or mechanism by which the frame resists gravity and earthquake effects. The tests should be configured to test critical behaviors, and the measured quantities should establish upper-bound acceptance values for components of the load path, which may be in terms of limiting stresses, forces, strains, or other quantities. The design procedure used for the structure should not deviate from that used to design the test specimens, and acceptance values should not exceed values

CODE

mechanism. Portions of the mechanism that deviate from code requirements shall be contained in the test specimens and shall be tested to determine upper bounds for acceptance values.

21.7 — Special reinforced concrete structural walls and coupling beams

21.7.1 — Scope

The requirements of this section apply to special reinforced concrete structural walls and coupling beams serving as part of the earthquake force-resisting system.

21.7.2 — Reinforcement

21.7.2.1 — The distributed web reinforcement ratios, ρ_ℓ and ρ_t, for structural walls shall not be less than 0.0025, except that if V_u does not exceed $A_{cv}\sqrt{f_c'}$, ρ_ℓ and ρ_t shall be permitted to be reduced to the values required in 14.3. Reinforcement spacing each way in structural walls shall not exceed 18 in. Reinforcement contributing to V_n shall be continuous and shall be distributed across the shear plane.

21.7.2.2 — At least two curtains of reinforcement shall be used in a wall if V_u exceeds $2A_{cv}\sqrt{f_c'}$.

21.7.2.3 — Reinforcement in structural walls shall be developed or spliced for f_y in tension in accordance with Chapter 12, except:

(a) The effective depth of the member referenced in 12.10.3 shall be permitted to be $0.8\ell_w$ for walls.

(b) The requirements of 12.11, 12.12, and 12.13 need not be satisfied.

(c) At locations where yielding of longitudinal reinforcement is likely to occur as a result of lateral displacements, development lengths of longitudinal reinforcement shall be 1.25 times the values calculated for f_y in tension.

(d) Mechanical splices of reinforcement shall conform to 21.2.6 and welded splices of reinforcement shall conform to 21.2.7.

21.7.3 — Design forces

V_u shall be obtained from the lateral load analysis in accordance with the factored load combinations.

COMMENTARY

that were demonstrated by the tests to be acceptable. Materials and components used in the structure should be similar to those used in the tests. Deviations may be acceptable if the engineer can demonstrate that those deviations do not adversely affect the behavior of the framing system.

R21.7 — Special reinforced concrete structural walls and coupling beams

R21.7.1 — Scope

This section contains requirements for the dimensions and details of special reinforced concrete structural walls and coupling beams. In the 1995 code, 21.6 also contained provisions for diaphragms. Provisions for diaphragms are in 21.9.

R21.7.2 — Reinforcement

Minimum reinforcement requirements (see 21.7.2.1) follow from preceding codes. The uniform distribution requirement of the shear reinforcement is related to the intent to control the width of inclined cracks. The requirement for two layers of reinforcement in walls carrying substantial design shears (see 21.7.2.2) is based on the observation that, under ordinary construction conditions, the probability of maintaining a single layer of reinforcement near the middle of the wall section is quite low. Furthermore, presence of reinforcement close to the surface tends to inhibit fragmentation of the concrete in the event of severe cracking during an earthquake.

R21.7.2.3 — Requirements were modified in the 2005 code to remove the reference to beam-column joints in 21.5.4, which was unclear when applied to walls. Because actual forces in longitudinal reinforcement of structural walls may exceed calculated forces, reinforcement should be developed or spliced to reach the yield strength of the bar in tension. Requirements of 12.11, 12.12, and 12.13 address issues related to beams and do not apply to walls. At locations where yielding of longitudinal reinforcement is expected, a 1.25 multiplier is applied to account for the likelihood that the actual yield strength exceeds the specified yield strength of the bar, as well as the influence of strain hardening and cyclic load reversals. Where transverse reinforcement is used, development lengths for straight and hooked bars may be reduced as permitted in 12.2 and 12.5, respectively, because closely spaced transverse reinforcement improves the performance of splices and hooks subjected to repeated inelastic demands.[21.38]

R21.7.3 — Design forces

Design shears for structural walls are obtained from lateral load analysis with the appropriate load factors. However, the designer should consider the possibility of yielding in components of such structures, as in the portion of a wall between two window openings, in which case the actual shear may be in excess of the shear indicated by lateral load analysis based on factored design forces.

CODE

21.7.4 — Shear strength

21.7.4.1 — V_n of structural walls shall not exceed

$$V_n = A_{cv}(\alpha_c\sqrt{f_c'} + \rho_t f_y) \quad (21\text{-}7)$$

where the coefficient α_c is 3.0 for $h_w/\ell_w \leq 1.5$, is 2.0 for $h_w/\ell_w \geq 2.0$, and varies linearly between 3.0 and 2.0 for h_w/ℓ_w between 1.5 and 2.0.

21.7.4.2 — In 21.7.4.1, the value of ratio h_w/ℓ_w used for determining V_n for segments of a wall shall be the larger of the ratios for the entire wall and the segment of wall considered.

21.7.4.3 — Walls shall have distributed shear reinforcement providing resistance in two orthogonal directions in the plane of the wall. If h_w/ℓ_w does not exceed 2.0, reinforcement ratio ρ_ℓ shall not be less than reinforcement ratio ρ_t.

21.7.4.4 — For all wall piers sharing a common lateral force, V_n shall not be taken larger than $8A_{cv}\sqrt{f_c'}$, where A_{cv} is the gross area of concrete bounded by web thickness and length of section. For any one of the individual wall piers, V_n shall not be taken larger than $10A_{cw}\sqrt{f_c'}$, where A_{cw} is the area of concrete section of the individual pier considered.

21.7.4.5 — For horizontal wall segments and coupling beams, V_n shall not be taken larger than $10A_{cw}\sqrt{f_c'}$, where A_{cw} is the area of concrete section of a horizontal wall segment or coupling beam.

COMMENTARY

R21.7.4 — Shear strength

Eq. (21-7) recognizes the higher shear strength of walls with high shear-to-moment ratios.[21.10,21.30,21.39] The nominal shear strength is given in terms of the net area of the section resisting shear. For a rectangular section without openings, the term A_{cv} refers to the gross area of the cross section rather than to the product of the width and the effective depth. The definition of A_{cv} in Eq. (21-7) facilitates design calculations for walls with uniformly distributed reinforcement and walls with openings.

A wall segment refers to a part of a wall bounded by openings or by an opening and an edge. Traditionally, a vertical wall segment bounded by two window openings has been referred to as a pier. When designing an isolated wall or a vertical wall segment, ρ_t refers to horizontal reinforcement and ρ_ℓ refers to vertical reinforcement.

The ratio h_w/ℓ_w may refer to overall dimensions of a wall, or of a segment of the wall bounded by two openings, or an opening and an edge. The intent of 21.7.4.2 is to make certain that any segment of a wall is not assigned a unit strength larger than that for the whole wall. However, a wall segment with a ratio of h_w/ℓ_w higher than that of the entire wall should be proportioned for the unit strength associated with the ratio h_w/ℓ_w based on the dimensions for that segment.

To restrain the inclined cracks effectively, reinforcement included in ρ_t and ρ_ℓ should be appropriately distributed along the length and height of the wall (see 21.7.4.3). Chord reinforcement provided near wall edges in concentrated amounts for resisting bending moment is not to be included in determining ρ_t and ρ_ℓ. Within practical limits, shear reinforcement distribution should be uniform and at a small spacing.

If the factored shear force at a given level in a structure is resisted by several walls or several piers of a perforated wall, the average unit shear strength assumed for the total available cross-sectional area is limited to $8\sqrt{f_c'}$ with the additional requirement that the unit shear strength assigned to any single pier does not exceed $10\sqrt{f_c'}$. The upper limit of strength to be assigned to any one member is imposed to limit the degree of redistribution of shear force.

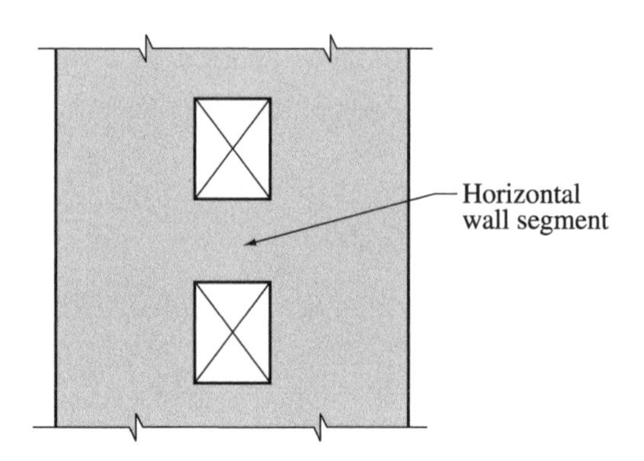

Fig. R21.7.4.5—Wall with openings

CODE / COMMENTARY

"Horizontal wall segments" in 21.7.4.5 refers to wall sections between two vertically aligned openings (see Fig. R21.7.4.5). It is, in effect, a pier rotated through 90 degrees. A horizontal wall segment is also referred to as a coupling beam when the openings are aligned vertically over the building height. When designing a horizontal wall segment or coupling beam, ρ_t refers to vertical reinforcement and ρ_ℓ refers to horizontal reinforcement.

21.7.5 — Design for flexure and axial loads

21.7.5.1 — Structural walls and portions of such walls subject to combined flexural and axial loads shall be designed in accordance with 10.2 and 10.3 except that 10.3.6 and the nonlinear strain requirements of 10.2.2 shall not apply. Concrete and developed longitudinal reinforcement within effective flange widths, boundary elements, and the wall web shall be considered effective. The effects of openings shall be considered.

21.7.5.2 — Unless a more detailed analysis is performed, effective flange widths of flanged sections shall extend from the face of the web a distance equal to the smaller of one-half the distance to an adjacent wall web and 25 percent of the total wall height.

R21.7.5 — Design for flexure and axial loads

R21.7.5.1 — Flexural strength of a wall or wall segment is determined according to procedures commonly used for columns. Strength should be determined considering the applied axial and lateral forces. Reinforcement concentrated in boundary elements and distributed in flanges and webs should be included in the strength computations based on a strain compatibility analysis. The foundation supporting the wall should be designed to develop the wall boundary and web forces. For walls with openings, the influence of the opening or openings on flexural and shear strengths is to be considered and a load path around the opening or openings should be verified. Capacity design concepts and strut-and-tie models may be useful for this purpose.[21.40]

R21.7.5.2 — Where wall sections intersect to form L-, T-, C-, or other cross-sectional shapes, the influence of the flange on the behavior of the wall should be considered by selecting appropriate flange widths. Tests[21.41] show that effective flange width increases with increasing drift level and the effectiveness of a flange in compression differs from that for a flange in tension. The value used for the effective compression flange width has little impact on the strength and deformation capacity of the wall; therefore, to simplify design, a single value of effective flange width based on an estimate of the effective tension flange width is used in both tension and compression.[21.41]

21.7.6 — Boundary elements of special reinforced concrete structural walls

21.7.6.1 — The need for special boundary elements at the edges of structural walls shall be evaluated in accordance with 21.7.6.2 or 21.7.6.3. The requirements of 21.7.6.4 and 21.7.6.5 also shall be satisfied.

21.7.6.2 — This section applies to walls or wall piers that are effectively continuous from the base of structure to top of wall and designed to have a single critical section for flexure and axial loads. Walls not

R21.7.6 — Boundary elements of special reinforced concrete structural walls

R21.7.6.1 — Two design approaches for evaluating detailing requirements at wall boundaries are included in 21.7.6.1. Section 21.7.6.2 allows the use of displacement-based design of walls, in which the structural details are determined directly on the basis of the expected lateral displacements of the wall. The provisions of 21.7.6.3 are similar to those of the 1995 code, and have been retained because they are conservative for assessing required transverse reinforcement at wall boundaries for many walls. Requirements of 21.7.6.4 and 21.7.6.5 apply to structural walls designed by either 21.7.6.2 or 21.7.6.3.

R21.7.6.2 — Section 21.7.6.2 is based on the assumption that inelastic response of the wall is dominated by flexural action at a critical, yielding section. The wall should be proportioned so that the critical section occurs where intended.

CODE

satisfying these requirements shall be designed by 21.7.6.3.

(a) Compression zones shall be reinforced with special boundary elements where:

$$c \geq \frac{\ell_w}{600(\delta_u/h_w)} \quad (21\text{-}8)$$

c in Eq. (21-8) corresponds to the largest neutral axis depth calculated for the factored axial force and nominal moment strength consistent with the design displacement δ_u. Ratio δ_u/h_w in Eq. (21-8) shall not be taken less than 0.007.

(b) Where special boundary elements are required by 21.7.6.2(a), the special boundary element reinforcement shall extend vertically from the critical section a distance not less than the larger of ℓ_w or $M_u/4V_u$.

21.7.6.3 — Structural walls not designed to the provisions of 21.7.6.2 shall have special boundary elements at boundaries and edges around openings of structural walls where the maximum extreme fiber compressive stress, corresponding to factored forces including E, load effects for earthquake effect, exceeds $0.2f_c'$. The special boundary element shall be permitted to be discontinued where the calculated compressive stress is less than $0.15f_c'$. Stresses shall be calculated for the factored forces using a linearly elastic model and gross section properties. For walls with flanges, an effective flange width as defined in 21.7.5.2 shall be used.

21.7.6.4 — Where special boundary elements are required by 21.7.6.2 or 21.7.6.3, (a) through (e) shall be satisfied:

(a) The boundary element shall extend horizontally from the extreme compression fiber a distance not less than the larger of $c - 0.1\ell_w$ and $c/2$, where c is the largest neutral axis depth calculated for the factored axial force and nominal moment strength consistent with δ_u;

(b) In flanged sections, the boundary element shall include the effective flange width in compression and shall extend at least 12 in. into the web;

(c) Special boundary element transverse reinforcement shall satisfy the requirements of 21.4.4.1 through 21.4.4.3, except Eq. (21-3) need not be satisfied;

COMMENTARY

Eq. (21-8) follows from a displacement-based approach.[21.42,21.43] The approach assumes that special boundary elements are required to confine the concrete where the strain at the extreme compression fiber of the wall exceeds a critical value when the wall is displaced to the design displacement. The horizontal dimension of the special boundary element is intended to extend at least over the length where the compression strain exceeds the critical value. The height of the special boundary element is based on upper bound estimates of plastic hinge length and extends beyond the zone over which concrete spalling is likely to occur. The lower limit of 0.007 on the quantity δ_u/h_w requires moderate wall deformation capacity for stiff buildings.

The neutral axis depth c in Eq. (21-8) is the depth calculated according to 10.2, except the nonlinear strain requirements of 10.2.2 need not apply, corresponding to development of nominal flexural strength of the wall when displaced in the same direction as δ_u. The axial load is the factored axial load that is consistent with the design load combination that produces the displacement δ_u.

R21.7.6.3 — By this procedure, the wall is considered to be acted on by gravity loads and the maximum shear and moment induced by earthquake in a given direction. Under this loading, the compressed boundary at the critical section resists the tributary gravity load plus the compressive resultant associated with the bending moment.

Recognizing that this loading condition may be repeated many times during the strong motion, the concrete is to be confined where the calculated compressive stresses exceed a nominal critical value equal to $0.2f_c'$. The stress is to be calculated for the factored forces on the section assuming linear response of the gross concrete section. The compressive stress of $0.2f_c'$ is used as an index value and does not necessarily describe the actual state of stress that may develop at the critical section under the influence of the actual inertia forces for the anticipated earthquake intensity.

R21.7.6.4 — The value of $c/2$ in 21.7.6.4(a) is to provide a minimum length of the special boundary element. Where flanges are heavily stressed in compression, the web-to-flange interface is likely to be heavily stressed and may sustain local crushing failure unless special boundary element reinforcement extends into the web. Eq. (21-3) does not apply to walls.

Because horizontal reinforcement is likely to act as web reinforcement in walls requiring boundary elements, it should be fully anchored in boundary elements that act as flanges (21.7.6.4). Achievement of this anchorage is difficult when large transverse cracks occur in the boundary elements. Therefore, standard 90 degree hooks or mechanical anchorage schemes are recommended instead of straight bar development.

CODE

(d) Special boundary element transverse reinforcement at the wall base shall extend into the support at least the development length of the largest longitudinal reinforcement in the special boundary element unless the special boundary element terminates on a footing or mat, where special boundary element transverse reinforcement shall extend at least 12 in. into the footing or mat;

(e) Horizontal reinforcement in the wall web shall be anchored to develop f_y within the confined core of the boundary element;

21.7.6.5 — Where special boundary elements are not required by 21.7.6.2 or 21.7.6.3, (a) and (b) shall be satisfied:

(a) If the longitudinal reinforcement ratio at the wall boundary is greater than $400/f_y$, boundary transverse reinforcement shall satisfy 21.4.4.1(c), 21.4.4.3, and 21.7.6.4(a). The maximum longitudinal spacing of transverse reinforcement in the boundary shall not exceed 8 in.;

(b) Except when V_u in the plane of the wall is less than $A_{cv}\sqrt{f_c'}$, horizontal reinforcement terminating at the edges of structural walls without boundary elements shall have a standard hook engaging the edge reinforcement or the edge reinforcement shall be enclosed in U-stirrups having the same size and spacing as, and spliced to, the horizontal reinforcement.

21.7.7 — Coupling beams

21.7.7.1 — Coupling beams with aspect ratio, $(\ell_n/h) \geq 4$, shall satisfy the requirements of 21.3. The provisions of 21.3.1.3 and 21.3.1.4 need not be satisfied if it can be shown by analysis that the beam has adequate lateral stability.

21.7.7.2 — Coupling beams with aspect ratio, $(\ell_n/h) < 4$, shall be permitted to be reinforced with two intersecting groups of diagonally placed bars symmetrical about the midspan.

21.7.7.3 — Coupling beams with aspect ratio, $(\ell_n/h) < 2$, and with V_u exceeding $4\sqrt{f_c'}A_{cw}$ shall be reinforced with two intersecting groups of diagonally placed bars symmetrical about the midspan, unless it can be shown that loss of stiffness and strength of the coupling beams will not impair the vertical load-carrying capacity of the

COMMENTARY

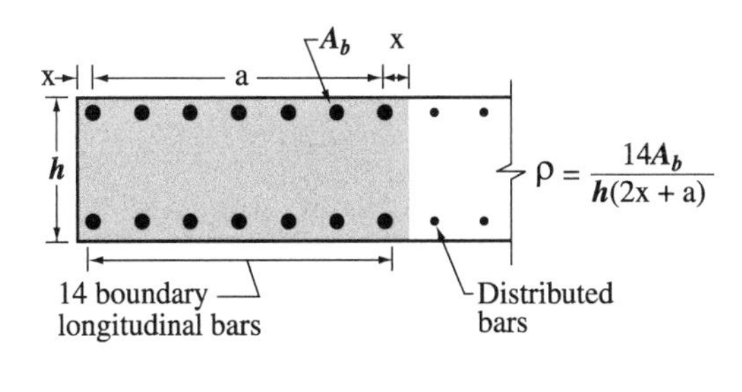

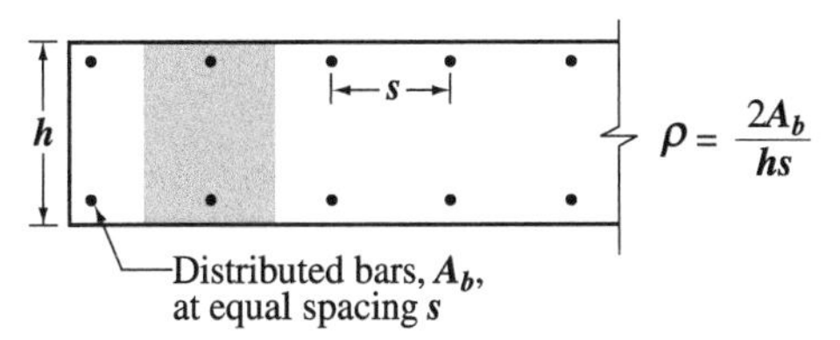

Fig. R21.7.6.5 — Longitudinal reinforcement ratios for typical wall boundary conditions

R21.7.6.5 — Cyclic load reversals may lead to buckling of boundary longitudinal reinforcement even in cases where the demands on the boundary of the wall do not require special boundary elements. For walls with moderate amounts of boundary longitudinal reinforcement, ties are required to inhibit buckling. The longitudinal reinforcement ratio is intended to include only the reinforcement at the wall boundary as indicated in Fig. R21.7.6.5. A larger spacing of ties relative to 21.7.6.4(c) is allowed due to the lower deformation demands on the walls.

The addition of hooks or U-stirrups at the ends of horizontal wall reinforcement provides anchorage so that the reinforcement will be effective in resisting shear forces. It will also tend to inhibit the buckling of the vertical edge reinforcement. In walls with low in-plane shear, the development of horizontal reinforcement is not necessary.

R21.7.7 — Coupling beams

Coupling beams connecting structural walls can provide stiffness and energy dissipation. In many cases, geometric limits result in coupling beams that are deep in relation to their clear span. Deep coupling beams may be controlled by shear and may be susceptible to strength and stiffness deterioration under earthquake loading. Test results[21.44,21.45] have shown that confined diagonal reinforcement provides adequate resistance in deep coupling beams.

Experiments show that diagonally oriented reinforcement is effective only if the bars are placed with a large inclination. Therefore, diagonally reinforced coupling beams are restricted to beams having aspect ratio $\ell_n/h < 4$.

Each diagonal element consists of a cage of longitudinal and transverse reinforcement as shown in Fig. R21.7.7. The cage contains at least four longitudinal bars and confines a

CODE

structure, or the egress from the structure, or the integrity of nonstructural components and their connections to the structure.

21.7.7.4 — Coupling beams reinforced with two intersecting groups of diagonally placed bars symmetrical about the midspan shall satisfy (a) through (f):

(a) Each group of diagonally placed bars shall consist of a minimum of four bars assembled in a core having sides measured to the outside of transverse reinforcement no smaller than $b_w/2$ perpendicular to the plane of the beam and $b_w/5$ in the plane of the beam and perpendicular to the diagonal bars;

(b) V_n shall be determined by

$$V_n = 2A_{vd}f_y\sin\alpha \leq 10\sqrt{f_c'}A_{cw} \qquad (21\text{-}9)$$

where α is the angle between the diagonally placed bars and the longitudinal axis of the coupling beam.

(c) Each group of diagonally placed bars shall be enclosed in transverse reinforcement satisfying 21.4.4.1 through 21.4.4.3. For the purpose of computing A_g for use in Eq. (10-5) and (21-3), the minimum concrete cover as required in 7.7 shall be assumed on all four sides of each group of diagonally placed reinforcing bars;

(d) The diagonally placed bars shall be developed for tension in the wall;

(e) The diagonally placed bars shall be considered to contribute to M_n of the coupling beam;

(f) Reinforcement parallel and transverse to the longitudinal axis shall be provided and, as a minimum, shall conform to 11.8.4 and 11.8.5.

21.7.8 — Construction joints

All construction joints in structural walls shall conform to 6.4 and contact surfaces shall be roughened as in 11.7.9.

21.7.9 — Discontinuous walls

Columns supporting discontinuous structural walls shall be reinforced in accordance with 21.4.4.5.

21.8 — Special structural walls constructed using precast concrete

21.8.1 — Special structural walls constructed using precast concrete shall satisfy all requirements of 21.7 for cast-in-place special structural walls in addition to 21.13.2 and 21.13.3.

COMMENTARY

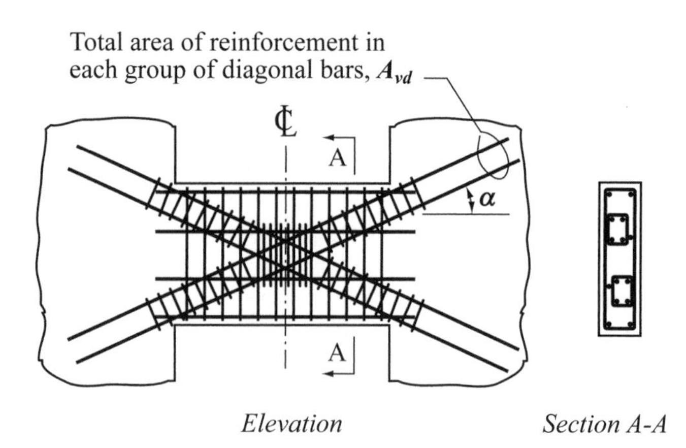

Fig. R21.7.7 — Coupling beam with diagonally oriented reinforcement

concrete core. The requirement on side dimensions of the cage and its core is to provide adequate toughness and stability to the cross section when the bars are loaded beyond yielding. The minimum dimensions and required reinforcement clearances may control the wall width.

When coupling beams are not used as part of the lateral force resisting system, the requirements for diagonal reinforcement may be waived. Nonprestressed coupling beams are permitted at locations where damage to these beams does not impair vertical load carrying capacity or egress of the structure, or integrity of the nonstructural components and their connections to the structure.

Tests in Reference 21.45 demonstrated that beams reinforced as described in Section 21.7.7 have adequate ductility at shear forces exceeding $10\sqrt{f_c'}b_wd$. Consequently, the use of a limit of $10\sqrt{f_c'}A_{cw}$ provides an acceptable upper limit.

When the diagonally oriented reinforcement is used, additional reinforcement in 21.7.7.4(f) is to contain the concrete outside the diagonal cores if the concrete is damaged by earthquake loading (Fig. R21.7.7).

CODE

21.9 — Structural diaphragms and trusses

21.9.1 — Scope

Floor and roof slabs acting as structural diaphragms to transmit design actions induced by earthquake ground motions shall be designed in accordance with this section. This section also applies to struts, ties, chords, and collector elements that transmit forces induced by earthquakes, as well as trusses serving as parts of the earthquake force-resisting systems.

21.9.2 — Cast-in-place composite-topping slab diaphragms

A composite-topping slab cast in place on a precast floor or roof shall be permitted to be used as a structural diaphragm provided the topping slab is reinforced and its connections are proportioned and detailed to provide for a complete transfer of forces to chords, collector elements, and the lateral-force-resisting system. The surface of the previously hardened concrete on which the topping slab is placed shall be clean, free of laitance, and intentionally roughened.

21.9.3 — Cast-in-place topping slab diaphragms

A cast-in-place noncomposite topping on a precast floor or roof shall be permitted to serve as a structural diaphragm, provided the cast-in-place topping acting alone is proportioned and detailed to resist the design forces.

21.9.4 — Minimum thickness of diaphragms

Concrete slabs and composite topping slabs serving as structural diaphragms used to transmit earthquake forces shall not be less than 2 in. thick. Topping slabs placed over precast floor or roof elements, acting as structural diaphragms and not relying on composite action with the precast elements to resist the design seismic forces, shall have thickness not less than 2-1/2 in.

21.9.5 — Reinforcement

21.9.5.1 — The minimum reinforcement ratio for structural diaphragms shall be in conformance with 7.12. Reinforcement spacing each way in nonpost-ten-

COMMENTARY

R21.9 — Structural diaphragms and trusses

R21.9.1 — Scope

Diaphragms as used in building construction are structural elements (such as a floor or roof) that provide some or all of the following functions:

(a) Support for building elements (such as walls, partitions, and cladding) resisting horizontal forces but not acting as part of the building vertical lateral-force-resisting system;

(b) Transfer of lateral forces from the point of application to the building vertical lateral-force-resisting system;

(c) Connection of various components of the building vertical lateral-force-resisting system with appropriate strength, stiffness, and toughness so the building responds as intended in the design.[21.46]

R21.9.2 — Cast-in-place composite-topping slab diaphragms

A bonded topping slab is required so that the floor or roof system can provide restraint against slab buckling. Reinforcement is required to ensure the continuity of the shear transfer across precast joints. The connection requirements are introduced to promote a complete system with necessary shear transfers.

R21.9.3 — Cast-in-place topping slab diaphragms

Composite action between the topping slab and the precast floor elements is not required, provided that the topping slab is designed to resist the design seismic forces.

R21.9.4 — Minimum thickness of diaphragms

The minimum thickness of concrete diaphragms reflects current practice in joist and waffle systems and composite topping slabs on precast floor and roof systems. Thicker slabs are required when the topping slab does not act compositely with the precast system to resist the design seismic forces.

R21.9.5 — Reinforcement

Minimum reinforcement ratios for diaphragms correspond to the required amount of temperature and shrinkage reinforcement (7.12). The maximum spacing for web reinforce-

CODE

sioned floor or roof systems shall not exceed 18 in. Where welded wire reinforcement is used as the distributed reinforcement to resist shear in topping slabs placed over precast floor and roof elements, the wires parallel to the span of the precast elements shall be spaced not less than 10 in. on center. Reinforcement provided for shear strength shall be continuous and shall be distributed uniformly across the shear plane.

21.9.5.2 — Bonded tendons used as primary reinforcement in diaphragm chords or collectors shall be proportioned such that the stress due to design seismic forces does not exceed 60,000 psi. Precompression from unbonded tendons shall be permitted to resist diaphragm design forces if a complete load path is provided.

21.9.5.3 — Structural truss elements, struts, ties, diaphragm chords, and collector elements with compressive stresses exceeding $0.2f_c'$ at any section shall have transverse reinforcement, as given in 21.4.4.1 through 21.4.4.3, over the length of the element. The special transverse reinforcement is permitted to be discontinued at a section where the calculated compressive stress is less than $0.15f_c'$. Stresses shall be calculated for the factored forces using a linearly elastic model and gross-section properties of the elements considered.

Where design forces have been amplified to account for the overstrength of the vertical elements of the seismic-force-resisting system, the limit of $0.2f_c'$ shall be increased to $0.5f_c'$, and the limit of $0.15f_c'$ shall be increased to $0.4f_c'$.

21.9.5.4 — All continuous reinforcement in diaphragms, trusses, struts, ties, chords, and collector elements shall be developed or spliced for f_y in tension.

21.9.5.5 — Type 2 splices are required where mechanical splices are used to transfer forces between the diaphragm and the vertical components of the lateral-force-resisting system.

21.9.6 — Design forces

The seismic design forces for structural diaphragms shall be obtained from the lateral load analysis in accordance with the design load combinations.

COMMENTARY

ment is intended to control the width of inclined cracks. Minimum average prestress requirements (7.12.3) are considered to be adequate to limit the crack widths in post-tensioned floor systems; therefore, the maximum spacing requirements do not apply to these systems.

The minimum spacing requirement for welded wire reinforcement in topping slabs on precast floor systems (see 21.9.5.1) is to avoid fracture of the distributed reinforcement during an earthquake. Cracks in the topping slab open immediately above the boundary between the flanges of adjacent precast members, and the wires crossing those cracks are restrained by the transverse wires.[21.47] Therefore, all the deformation associated with cracking should be accommodated in a distance not greater than the spacing of the transverse wires. A minimum spacing of 10 in. for the transverse wires is required in 21.9.5.1 to reduce the likelihood of fracture of the wires crossing the critical cracks during a design earthquake. The minimum spacing requirements do not apply to diaphragms reinforced with individual bars, because strains are distributed over a longer length.

In documents such as the 2000 NEHRP provisions (NEHRP),[21.1] SEI/ASCE 7-02,[21.48] the 2003 International Building Code (IBC),[21.49] and the 1997 Uniform Building Code (UBC),[21.2] collector elements of diaphragms are to be designed for forces amplified by a factor, Ω_0, to account for the overstrength in the vertical elements of the seismic-force-resisting systems. The amplication factor Ω_0 ranges between 2 and 3 for concrete structures, depending on the document selected and on the type of seismic system. In some documents, the factor can be calculated based on the maximum forces that can be developed by the elements of the vertical seismic-force-resisting system. The factor is not applied to diaphragm chords.

Compressive stress calculated for the factored forces on a linearly elastic model based on gross section of the structural diaphragm is used as an index value to determine whether confining reinforcement is required. A calculated compressive stress of $0.2f_c'$ in a member, or $0.5f_c'$ for forces amplified by Ω_0, is assumed to indicate that integrity of the entire structure depends on the ability of that member to resist substantial compressive force under severe cyclic loading. Therefore, transverse reinforcement in 21.4.4 is required in such members to provide confinement for the concrete and the reinforcement (21.9.5.3).

The dimensions of typical structural diaphragms often preclude the use of transverse reinforcement along the chords. Reducing the calculated compressive stress by reducing the span of the diaphragm is considered to be a solution.

Bar development and lap splices are designed according to requirements of Chapter 12 for reinforcement in tension. Reductions in development or splice length for calculated stresses less than f_y are not permitted, as indicated in 12.2.5.

CODE

21.9.7 — Shear strength

21.9.7.1 — V_n of structural diaphragms shall not exceed

$$V_n = A_{cv}(2\sqrt{f_c'} + \rho_t f_y) \quad (21\text{-}10)$$

21.9.7.2 — V_n of cast-in-place composite-topping slab diaphragms and cast-in-place noncomposite topping slab diaphragms on a precast floor or roof shall not exceed

$$V_n = A_{cv}\rho_t f_y \quad (21\text{-}11)$$

where A_{cv} is based on the thickness of the topping slab. The required web reinforcement shall be distributed uniformly in both directions.

21.9.7.3 — Nominal shear strength shall not exceed $8A_{cv}\sqrt{f_c'}$ where A_{cv} is the gross area of the diaphragm cross section.

21.9.8 — Boundary elements of structural diaphragms

21.9.8.1 — Boundary elements of structural diaphragms shall be proportioned to resist the sum of the factored axial forces acting in the plane of the diaphragm and the force obtained from dividing M_u at the section by the distance between the boundary elements of the diaphragm at that section.

21.9.8.2 — Splices of tension reinforcement in the chords and collector elements of diaphragms shall develop f_y. Mechanical and welded splices shall conform to 21.2.6 and 21.2.7, respectively.

21.9.8.3 — Reinforcement for chords and collectors at splices and anchorage zones shall satisfy either (a) or (b):

(a) A minimum center-to-center spacing of three longitudinal bar diameters, but not less than 1-1/2 in., and a minimum concrete clear cover of two and one-half longitudinal bar diameters, but not less than 2 in.; or

(b) Transverse reinforcement as required by 11.5.6.3, except as required in 21.9.5.3.

21.9.9 — Construction joints

All construction joints in diaphragms shall conform to 6.4 and contact surfaces shall be roughened as in 11.7.9.

COMMENTARY

R21.9.7 — Shear strength

The shear strength requirements for monolithic diaphragms, Eq. (21-10) in 21.9.7.1, are the same as those for slender structural walls. The term A_{cv} refers to the thickness times the width of the diaphragm. This corresponds to the gross area of the effective deep beam that forms the diaphragm. The shear reinforcement should be placed perpendicular to the span of the diaphragm.

The shear strength requirements for topping slab diaphragms are based on a shear friction model, and the contribution of the concrete to the nominal shear strength is not included in Eq. (21-9) for topping slabs placed over precast floor elements. Following typical construction practice, the topping slabs are roughened immediately above the boundary between the flanges of adjacent precast floor members to direct the paths of shrinkage cracks. As a result, critical sections of the diaphragm are cracked under service loads, and the contribution of the concrete to the shear capacity of the diaphragm may have already been reduced before the design earthquake occurs.

R21.9.8 — Boundary elements of structural diaphragms

For structural diaphragms, the design moments are assumed to be resisted entirely by chord forces acting at opposite edges of the diaphragm. Reinforcement located at the edges of collectors should be fully developed for its specified yield strength. Adequate confinement of lap splices is also required. If chord reinforcement is located within a wall, the joint between the diaphragm and the wall should be provided with adequate shear strength to transfer the shear forces.

Section 21.9.8.3 is intended to reduce the possibility of chord buckling in the vicinity of splices and anchorage zones.

CODE

21.10 — Foundations

21.10.1 — Scope

21.10.1.1 — Foundations resisting earthquake-induced forces or transferring earthquake-induced forces between structure and ground shall comply with 21.10 and other applicable code provisions.

21.10.1.2 — The provisions in this section for piles, drilled piers, caissons, and slabs on grade shall supplement other applicable code design and construction criteria. See 1.1.5 and 1.1.6.

21.10.2 — Footings, foundation mats, and pile caps

21.10.2.1 — Longitudinal reinforcement of columns and structural walls resisting forces induced by earthquake effects shall extend into the footing, mat, or pile cap, and shall be fully developed for tension at the interface.

21.10.2.2 — Columns designed assuming fixed-end conditions at the foundation shall comply with 21.10.2.1 and, if hooks are required, longitudinal reinforcement resisting flexure shall have 90 degree hooks near the bottom of the foundation with the free end of the bars oriented towards the center of the column.

21.10.2.3 — Columns or boundary elements of special reinforced concrete structural walls that have an edge within one-half the footing depth from an edge of the footing shall have transverse reinforcement in accordance with 21.4.4 provided below the top of the footing. This reinforcement shall extend into the footing a distance no less than the smaller of the depth of the footing, mat, or pile cap, or the development length in tension of the longitudinal reinforcement.

21.10.2.4 — Where earthquake effects create uplift forces in boundary elements of special reinforced concrete structural walls or columns, flexural reinforcement shall be provided in the top of the footing, mat or pile cap to resist the design load combinations, and shall not be less than required by 10.5.

21.10.2.5 — See 22.10 for use of plain concrete in footings and basement walls.

COMMENTARY

R21.10 — Foundations

R21.10.1 — Scope

Requirements for foundations supporting buildings assigned to high seismic performance or design categories were added to the 1999 code. They represent a consensus of a minimum level of good practice in designing and detailing concrete foundations including piles, drilled piers, and caissons. It is desirable that inelastic response in strong ground shaking occurs above the foundations, as repairs to foundations can be extremely difficult and expensive.

R21.10.2 — Footings, foundation mats, and pile caps

R21.10.2.2 — Tests[21.50] have demonstrated that flexural members terminating in a footing, slab or beam (a T-joint) should have their hooks turned inwards toward the axis of the member for the joint to be able to resist the flexure in the member forming the stem of the T.

R21.10.2.3 — Columns or boundary members supported close to the edge of the foundation, as often occurs near property lines, should be detailed to prevent an edge failure of the footing, pile cap, or mat.

R21.10.2.4 — The purpose of 21.10.2.4 is to alert the designer to provide top reinforcement as well as other required reinforcement.

R21.10.2.5 — In regions of high seismicity, it is desirable to reinforce foundations. Committee 318 recommends that foundation or basement walls be reinforced in regions of high seismicity.

CODE

21.10.3 — Grade beams and slabs on grade

21.10.3.1 — Grade beams designed to act as horizontal ties between pile caps or footings shall have continuous longitudinal reinforcement that shall be developed within or beyond the supported column or anchored within the pile cap or footing at all discontinuities.

21.10.3.2 — Grade beams designed to act as horizontal ties between pile caps or footings shall be proportioned such that the smallest cross-sectional dimension shall be equal to or greater than the clear spacing between connected columns divided by 20, but need not be greater than 18 in. Closed ties shall be provided at a spacing not to exceed the lesser of one-half the smallest orthogonal cross-sectional dimension or 12 in.

21.10.3.3 — Grade beams and beams that are part of a mat foundation subjected to flexure from columns that are part of the lateral-force-resisting system shall conform to 21.3.

21.10.3.4 — Slabs on grade that resist seismic forces from walls or columns that are part of the lateral-force-resisting system shall be designed as structural diaphragms in accordance with 21.9. The design drawings shall clearly state that the slab on grade is a structural diaphragm and part of the lateral-force-resisting system.

21.10.4 — Piles, piers, and caissons

21.10.4.1 — Provisions of 21.10.4 shall apply to concrete piles, piers, and caissons supporting structures designed for earthquake resistance.

21.10.4.2 — Piles, piers, or caissons resisting tension loads shall have continuous longitudinal reinforcement over the length resisting design tension forces. The longitudinal reinforcement shall be detailed to transfer tension forces within the pile cap to supported structural members.

21.10.4.3 — Where tension forces induced by earthquake effects are transferred between pile cap or mat foundation and precast pile by reinforcing bars grouted or post-installed in the top of the pile, the grouting system shall have been demonstrated by test to develop at least **$1.25f_y$** of the bar.

21.10.4.4 — Piles, piers, or caissons shall have transverse reinforcement in accordance with 21.4.4 at locations (a) and (b):

(a) At the top of the member for at least 5 times the member cross-sectional dimension, but not less than 6 ft below the bottom of the pile cap;

COMMENTARY

R21.10.3 — Grade beams and slabs on grade

For seismic conditions, slabs on grade (soil-supported slabs) are often part of the lateral-force-resisting system and should be designed in accordance with this code as well as other appropriate standards or guidelines. See 1.1.6.

R21.10.3.2 — Grade beams between pile caps or footings can be separate beams beneath the slab on grade or can be a thickened portion of the slab on grade. The cross-sectional limitation and minimum tie requirements provide reasonable proportions.

R21.10.3.3 — Grade beams resisting seismic flexural stresses from column moments should have reinforcement details similar to the beams of the frame above the foundation.

R21.10.3.4 — Slabs on grade often act as a diaphragm to hold the building together at the ground level and minimize the effects of out-of-phase ground motion that may occur over the footprint of the building. In these cases, the slab on grade should be adequately reinforced and detailed. The design drawings should clearly state that these slabs on grade are structural members so as to prohibit sawcutting of the slab.

R21.10.4 — Piles, piers, and caissons

Adequate performance of piles and caissons for seismic loadings requires that these provisions be met in addition to other applicable standards or guidelines. See R1.1.5.

R21.10.4.2 — A load path is necessary at pile caps to transfer tension forces from the reinforcing bars in the column or boundary member through the pile cap to the reinforcement of the pile or caisson.

R21.10.4.3 — Grouted dowels in a blockout in the top of a precast concrete pile need to be developed, and testing is a practical means of demonstrating capacity. Alternatively, reinforcing bars can be cast in the upper portion of the pile, exposed by chipping of concrete and mechanically spliced or welded to an extension.

R21.10.4.4 — During earthquakes, piles can be subjected to extremely high flexural demands at points of discontinuity, especially just below the pile cap and near the base of a soft or loose soil deposit. The 1999 code requirement for confinement reinforcement at the top of the pile is based on numerous failures observed at this location in recent earthquakes. Transverse reinforcement is required in

CODE

(b) For the portion of piles in soil that is not capable of providing lateral support, or in air and water, along the entire unsupported length plus the length required in 21.10.4.4(a).

21.10.4.5 — For precast concrete driven piles, the length of transverse reinforcement provided shall be sufficient to account for potential variations in the elevation in pile tips.

21.10.4.6 — Concrete piles, piers, or caissons in foundations supporting one- and two-story stud bearing wall construction are exempt from the transverse reinforcement requirements of 21.10.4.4 and 21.10.4.5.

21.10.4.7 — Pile caps incorporating batter piles shall be designed to resist the full compressive strength of the batter piles acting as short columns. The slenderness effects of batter piles shall be considered for the portion of the piles in soil that is not capable of providing lateral support, or in air or water.

21.11 — Members not designated as part of the lateral-force-resisting system

21.11.1 — Frame members assumed not to contribute to lateral resistance, except two-way slabs without beams, shall be detailed according to 21.11.2 or 21.11.3 depending on the magnitude of moments induced in those members when subjected to the design displacement δ_u. If effects of δ_u are not explicitly checked, it shall be permitted to apply the requirements of 21.11.3. For two-way slabs without beams, slab-column connections shall meet the requirements of 21.11.5.

21.11.2 — Where the induced moments and shears under design displacements, δ_u, of 21.11.1 combined with the factored gravity moments and shears do not exceed the design moment and shear strength of the frame member, the conditions of 21.11.2.1, 21.11.2.2, and 21.11.2.3 shall be satisfied. The gravity load combinations of $(1.2D + 1.0L + 0.2S)$ or $0.9D$, whichever is critical, shall be used. The load factor on the live load, L, shall be permitted to be reduced to 0.5 except for garages, areas occupied as places of public assembly, and all areas where L is greater than 100 lb/ft^2.

21.11.2.1 — Members with factored gravity axial forces not exceeding $A_g f_c'/10$ shall satisfy 21.3.2.1. Stirrups shall be spaced not more than $d/2$ throughout the length of the member.

COMMENTARY

this region to provide ductile performance. The designer should also consider possible inelastic action in the pile at abrupt changes in soil deposits, such as changes from soft to firm or loose to dense soil layers. Where precast piles are to be used, the potential for the pile tip to be driven to an elevation different than that specified in the drawings needs to be considered when detailing the pile. If the pile reaches refusal at a shallower depth, a longer length of pile will need to be cut off. If this possibility is not foreseen, the length of transverse reinforcement required by 21.10.4.4 may not be available after the excess pile length is cut off.

R21.10.4.7 — Extensive structural damage has often been observed at the junction of batter piles and the buildings. The pile cap and surrounding structure should be designed for the potentially large forces that can be developed in batter piles.

R21.11 — Members not designated as part of the lateral-force-resisting system

This section applies only to structures in regions of high seismic risk or for structures assigned to high seismic performance categories or design categories. Model building codes, such as the 2003 IBC and 1997 UBC, require all structural elements not designated as a part of the lateral-force-resisting system to be designed to support gravity loads while subjected to the design displacement. For concrete structures, the provisions of 21.11 satisfy this requirement for columns, beams, and slabs of the gravity system. The design displacement is defined in 21.1.

The principle behind the provisions of 21.11 is to allow flexural yielding of columns, beams, and slabs under the design displacement, and to provide sufficient confinement and shear strength in elements that yield. By the provisions of 21.11.1 through 21.11.3, columns and beams are assumed to yield if the combined effects of factored gravity loads and design displacements exceed the corresponding strengths, or if the effects of design displacements are not calculated. Requirements for transverse reinforcement and shear strength vary with the axial load on the member and whether or not the member yields under the design displacement.

Models used to determine design displacement of buildings should be chosen to produce results that conserva-

CODE

21.11.2.2 — Members with factored gravity axial forces exceeding $A_g f_c'/10$ shall satisfy 21.4.3, 21.4.4.1(c), 21.4.4.3, and 21.4.5. The maximum longitudinal spacing of ties shall be s_o for the full column height. Spacing s_o shall not exceed the smaller of six diameters of the smallest longitudinal bar enclosed and 6 in.

21.11.2.3 — Members with factored gravity axial forces exceeding $0.35P_o$ shall satisfy 21.11.2.2 and the amount of transverse reinforcement provided shall be one-half of that required by 21.4.4.1 but shall not be spaced greater than s_o for the full height of the column.

21.11.3 — If the induced moment or shear under design displacements, δ_u, of 21.11.1 exceeds ϕM_n or ϕV_n of the frame member, or if induced moments are not calculated, the conditions of 21.11.3.1, 21.11.3.2, and 21.11.3.3 shall be satisfied.

21.11.3.1 — Materials shall satisfy 21.2.4 and 21.2.5. Mechanical splices shall satisfy 21.2.6 and welded splices shall satisfy 21.2.7.1.

21.11.3.2 — Members with factored gravity axial forces not exceeding $A_g f_c'/10$ shall satisfy 21.3.2.1 and 21.3.4. Stirrups shall be spaced at not more than $d/2$ throughout the length of the member.

21.11.3.3 — Members with factored gravity axial forces exceeding $A_g f_c'/10$ shall satisfy 21.4.3.1, 21.4.4, 21.4.5, and 21.5.2.1.

21.11.4 — Precast concrete frame members assumed not to contribute to lateral resistance, including their connections, shall satisfy (a), (b), and (c), in addition to 21.11.1 through 21.11.3:

(a) Ties specified in 21.11.2.2 shall be provided over the entire column height, including the depth of the beams;

(b) Structural integrity reinforcement, as specified in 16.5, shall be provided; and

(c) Bearing length at support of a beam shall be at least 2 in. longer than determined from calculations using bearing strength values from 10.17.

21.11.5 — For slab-column connections of two-way slabs without beams, slab shear reinforcement satisfying the requirements of 11.12.3 and providing V_s not less than $3.5\sqrt{f_c'}\, b_o d$ shall extend at least four times the slab thickness from the face of the support, unless either (a) or (b) is satisfied:

COMMENTARY

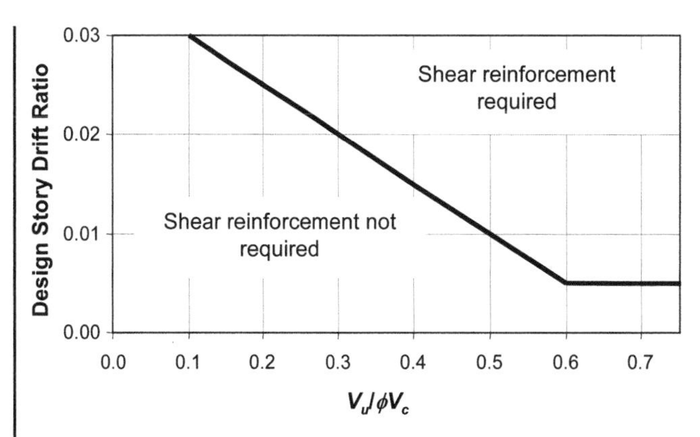

Fig. R21.11.5—Illustration of the criterion of 21.11.5(b)

tively bound the values expected during the design earthquake and should include, as appropriate, effects of concrete cracking, foundation flexibility, and deformation of floor and roof diaphragms.

R21.11.4 — Damage to some buildings with precast concrete gravity systems during the Northridge Earthquake was attributed to several factors addressed in 21.11.4. Columns should contain ties over their entire height, frame members not proportioned to resist earthquake forces should be tied together, and longer bearing lengths should be used to maintain integrity of the gravity system during shaking. The 2 in. increase in bearing length is based on an assumed 4 percent story drift ratio and 50 in. beam depth, and is considered to be conservative for the ground motions expected in high seismic zones. In addition to the provisions of 21.11.4, precast frame members assumed not to contribute to lateral resistance should also satisfy 21.11.1 through 21.11.3, as applicable.

R21.11.5 — Provisions for shear reinforcement at slab-column connections were added in the 2005 code to reduce the likelihood of slab punching shear failure. The shear reinforcement is required unless either 21.11.5(a) or (b) is satisfied.

Section 21.11.5(a) requires calculation of shear stress due to the factored shear force and induced moment according to 11.12.6.2. The induced moment is the moment that is calcu-

CODE

(a) The requirements of 11.12.6 using the design shear V_u and the induced moment transferred between the slab and column under the design displacement;

(b) The design story drift ratio does not exceed the larger of 0.005 and $[0.035 - 0.05(V_u/\phi V_c)]$.

Design story drift ratio shall be taken as the larger of the design story drift ratios of the adjacent stories above and below the slab-column connection. V_c is defined in 11.12.2. V_u is the factored shear force on the slab critical section for two-way action, calculated for the load combination $1.2D + 1.0L + 0.2S$. It shall be permitted to reduce the load factor on L to 0.5 in accordance with 9.2.1(a).

21.12 — Requirements for intermediate moment frames

21.12.1 — The requirements of this section apply to intermediate moment frames.

21.12.2 — Reinforcement details in a frame member shall satisfy 21.12.4 if the factored axial compressive load, P_u, for the member does not exceed $A_g f_c'/10$. If P_u is larger, frame reinforcement details shall satisfy 21.12.5 unless the member has spiral reinforcement according to Eq. (10-5). If a two-way slab system without beams is treated as part of a frame resisting E, load effects for earthquake effect, reinforcement details in any span resisting moments caused by lateral force shall satisfy 21.12.6.

21.12.3 — ϕV_n of beams, columns, and two-way slabs resisting earthquake effect, E, shall not be less than the smaller of (a) and (b):

(a) The sum of the shear associated with development of nominal moment strengths of the member at each restrained end of the clear span and the shear calculated for factored gravity loads;

(b) The maximum shear obtained from design load combinations that include E, with E assumed to be twice that prescribed by the governing code for earthquake-resistant design.

21.12.4 — Beams

21.12.4.1 — The positive moment strength at the face of the joint shall be not less than one-third the negative moment strength provided at that face of the joint. Neither the negative nor the positive moment strength at any section along the length of the member shall be less than one-fifth the maximum moment strength provided at the face of either joint.

COMMENTARY

lated to occur at the slab-column connection when subjected to the design displacement. Section 13.5.1.2 and the accompanying commentary provide guidance on selection of the stiffness of the slab-column connection for the purpose of this calculation.

Section 21.11.5(b) does not require the calculation of induced moments, and is based on research[21.51,21.52] that identifies the likelihood of punching shear failure considering the story drift ratio and shear due to gravity loads. Figure R21.11.5 illustrates the requirement. The requirement can be satisfied by adding slab shear reinforcement, increasing slab thickness, changing the design to reduce the design story drift ratio, or a combination of these.

If column capitals, drop panels, or other changes in slab thickness are used, the requirements of 21.11.5 are evaluated at all potential critical sections, as required by 11.12.1.2.

R21.12 — Requirements for intermediate moment frames

The objective of the requirements in 21.12.3 is to reduce the risk of failure in shear during an earthquake. The designer is given two options by which to determine the factored shear force.

According to option (a) of 21.12.3, the factored shear force is determined from the nominal moment strength of the member and the gravity load on it. Examples for a beam and a column are illustrated in Fig. R21.12.3.

To determine the maximum beam shear, it is assumed that its nominal moment strengths ($\phi = 1.0$) are developed simultaneously at both ends of its clear span. As indicated in Fig. R21.12.3, the shear associated with this condition $[(M_{nl} + M_{nr})/\ell_n]$ added algebraically to the effect of the factored gravity loads indicates the design shear for of the beam. For this example, both the dead load w_D and the live load w_L have been assumed to be uniformly distributed.

Determination of the design shear for a column is also illustrated for a particular example in Fig. R21.12.3. The factored axial force, P_u, should be chosen to develop the largest moment strength of the column.

In all applications of option (a) of 21.12.3, shears are required to be calculated for moment, acting clockwise and counterclockwise. Fig. R21.12.3 demonstrates only one of the two conditions that are to be considered for every member. Option (b) bases V_u on the load combination including the earthquake effect, E, which should be doubled. For example, the load combination defined by Eq. (9-5) would be:

$$U = 1.2D + 2.0E + 1.0L + 0.2S$$

where E is the value specified by the governing code.

CODE

COMMENTARY

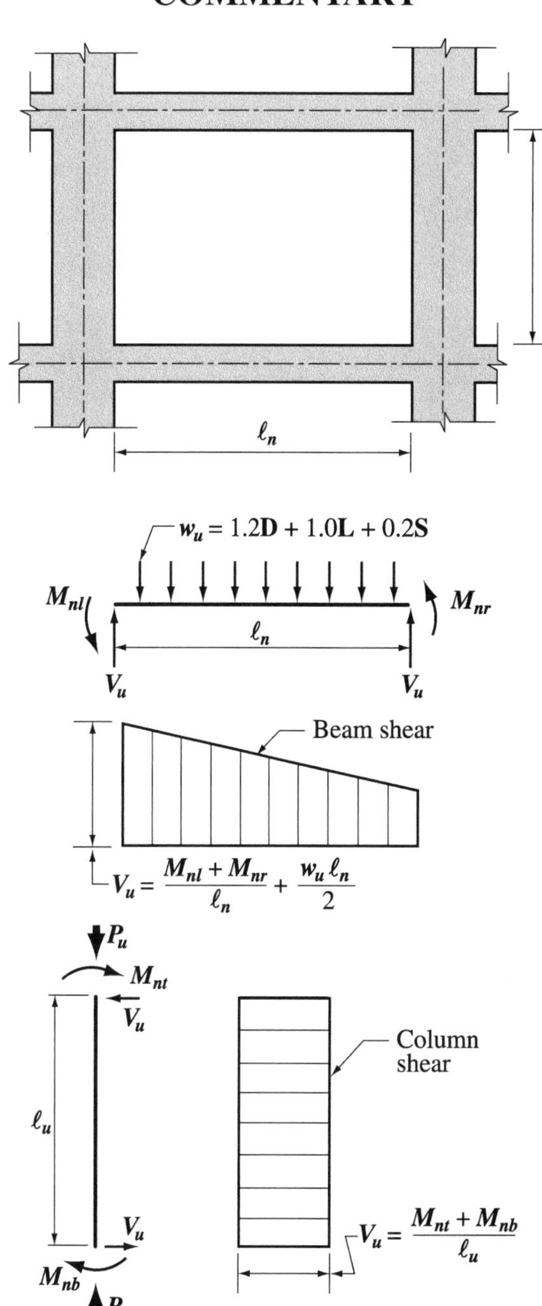

Fig. R21.12.3—Design shears for frames in regions of moderate seismic risk (see 21.12)

21.12.4.2 — At both ends of the member, hoops shall be provided over lengths equal $2h$ measured from the face of the supporting member toward midspan. The first hoop shall be located at not more than 2 in. from the face of the supporting member. Spacing of hoops shall not exceed the smallest of (a), (b), (c), and (d):

(a) $d/4$;

(b) Eight times the diameter of the smallest longitudinal bar enclosed;

Section 21.12.4 contains requirements for providing beams with a threshold level of toughness. Transverse reinforcement at the ends of the beam shall be hoops. In most cases, stirrups required by 21.12.3 for design shear force will be more than those required by 21.12.4. Requirements of 21.12.5 serve the same purpose for columns.

Section 21.12.6 applies to two-way slabs without beams, such as flat plates.

Using load combinations of Eq. (9-5) and (9-7) may result in moments requiring top and bottom reinforcement at the supports.

CODE

(c) 24 times the diameter of the hoop bar;

(d) 12 in.

21.12.4.3 — Stirrups shall be placed at not more than ***d*/2** throughout the length of the member.

21.12.5 — Columns

21.12.5.1 — Columns shall be spirally reinforced in accordance with 7.10.4 or shall conform with 21.12.5.2 through 21.12.5.4. Section 21.12.5.5 shall apply to all columns.

21.12.5.2 — At both ends of the member, hoops shall be provided at spacing s_o over a length ℓ_o measured from the joint face. Spacing s_o shall not exceed the smallest of (a), (b), (c), and (d):

(a) Eight times the diameter of the smallest longitudinal bar enclosed;

(b) 24 times the diameter of the hoop bar;

(c) One-half of the smallest cross-sectional dimension of the frame member;

(d) 12 in.

Length ℓ_o shall not be less than the largest of (e), (f), and (g):

(e) One-sixth of the clear span of the member;

(f) Maximum cross-sectional dimension of the member;

(g) 18 in.

21.12.5.3 — The first hoop shall be located at not more than $s_o/2$ from the joint face.

21.12.5.4 — Outside the length ℓ_o, spacing of transverse reinforcement shall conform to 7.10 and 11.5.5.1.

21.12.5.5 — Joint transverse reinforcement shall conform to 11.11.2.

21.12.6 — Two-way slabs without beams

21.12.6.1 — Factored slab moment at support related to earthquake effect, ***E***, shall be determined for load combinations given in Eq. (9-5) and (9-7). Reinforcement provided to resist M_{slab} shall be placed within the column strip defined in 13.2.1.

21.12.6.2 — Reinforcement placed within the effective width specified in 13.5.3.2 shall resist $\gamma_f M_{slab}$. Effective slab width for exterior and corner connections shall not extend beyond the column face a dis-

COMMENTARY

The moment M_{slab} refers, for a given design load combination with ***E*** acting in one horizontal direction, to that portion of the factored slab moment that is balanced by the supporting members at a joint. It is not necessarily equal to the total design moment at support for a load combination including earthquake effect. In accordance with 13.5.3.2, only a fraction of the moment M_{slab} is assigned to the slab effective width. For edge and corner connections, flexural reinforcement perpendicular to the edge is not considered fully effective unless it is placed within the effective slab width.[21.53,21.54] See Fig. 21.12.6.1.

Application of the provisions of 21.12.6 is illustrated in Fig. R21.12.6.2 and Fig. R21.12.6.3.

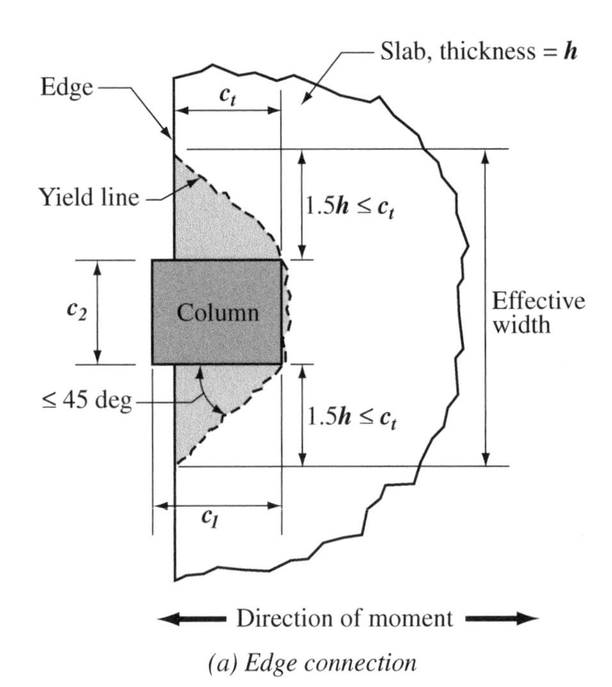

(a) Edge connection

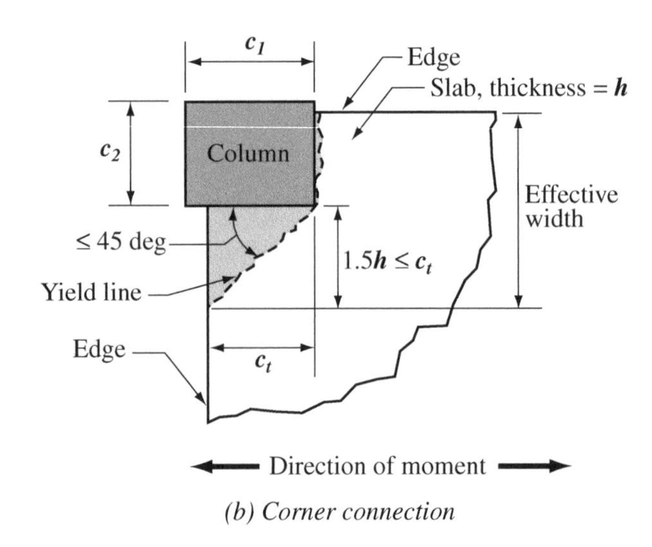

(b) Corner connection

Fig. R21.12.6.1 — Effective width for reinforcement placement in edge and corner connections

CODE

tance greater than c_t measured perpendicular to the slab span.

21.12.6.3 — Not less than one-half of the reinforcement in the column strip at support shall be placed within the effective slab width given in 13.5.3.2.

21.12.6.4 — Not less than one-quarter of the top reinforcement at the support in the column strip shall be continuous throughout the span.

21.12.6.5 — Continuous bottom reinforcement in the column strip shall be not less than one-third of the top reinforcement at the support in the column strip.

21.12.6.6 — Not less than one-half of all bottom middle strip reinforcement and all bottom column strip reinforcement at midspan shall be continuous and shall develop f_y at face of support as defined in 13.6.2.5.

21.12.6.7—At discontinuous edges of the slab all top and bottom reinforcement at support shall be developed at the face of support as defined in 13.6.2.5.

21.12.6.8 — At the critical sections for columns defined in 11.12.1.2, two-way shear caused by factored gravity loads shall not exceed $0.4\phi V_c$, where V_c shall be calculated as defined in 11.12.2.1 for nonprestressed slabs and in 11.12.2.2 for prestressed slabs. It shall be permitted to waive this requirement if the contribution of the earthquake-induced factored two-way shear stress transferred by eccentricity of shear in accordance with 11.12.6.1 and 11.12.6.2 at the point of maximum stress does not exceed one-half of the stress ϕv_n permitted by 11.12.6.2.

COMMENTARY

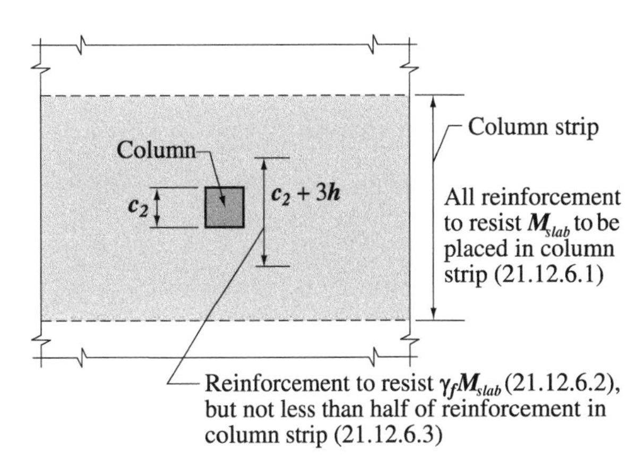

Fig. R21.12.6.2 — Location of reinforcement in slabs

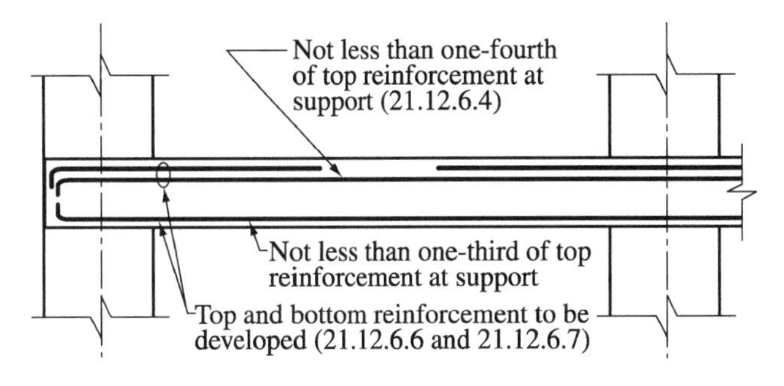

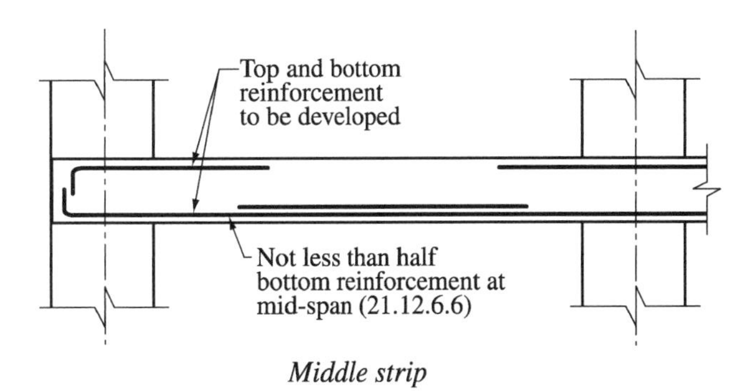

Fig. R21.12.6.3—Arrangement of reinforcement in slabs

R21.12.6.8 — The requirements apply to two-way slabs that are part of the primary lateral force-resisting system. Slab-column connections in laboratory tests[21.54] exhibited reduced lateral displacement ductility when the shear at the column connection exceeded the recommended limit.

CODE

21.13 — Intermediate precast structural walls

21.13.1 — The requirements of this section apply to intermediate precast structural walls used to resist forces induced by earthquake motions.

21.13.2 — In connections between wall panels, or between wall panels and the foundation, yielding shall be restricted to steel elements or reinforcement.

21.13.3 — Elements of the connection that are not designed to yield shall develop at least $1.5S_y$.

COMMENTARY

21.13 — Intermediate precast structural walls

Connections between precast wall panels or between wall panels and the foundation are required to resist forces induced by earthquake motions and to provide for yielding in the vicinity of connections. When Type 2 mechanical splices are used to directly connect primary reinforcement, the probable strength of the splice should be at least 1-1/2 times the specified yield strength of the reinforcement.

CHAPTER 22 — STRUCTURAL PLAIN CONCRETE

CODE

22.1 — Scope

22.1.1 — This chapter provides minimum requirements for design and construction of structural plain concrete members (cast-in-place or precast) except in 22.1.1.1 and 22.1.1.2.

22.1.1.1 — Structural plain concrete basement walls shall be exempt from the requirements for special exposure conditions of 4.2.2.

22.1.1.2 — Design and construction of soil-supported slabs, such as sidewalks and slabs on grade, shall not be governed by this code unless they transmit vertical loads or lateral forces from other parts of the structure to the soil.

22.1.2 — For special structures, such as arches, underground utility structures, gravity walls, and shielding walls, provisions of this chapter shall govern where applicable.

22.2 — Limitations

22.2.1 — Provisions of this chapter shall apply for design of structural plain concrete members. See 2.2.

22.2.2 — Use of structural plain concrete shall be limited to (a), (b), or (c):

(a) Members that are continuously supported by soil or supported by other structural members capable of providing continuous vertical support;

(b) Members for which arch action provides compression under all conditions of loading;

(c) Walls and pedestals. See 22.6 and 22.8.

The use of structural plain concrete columns shall not be permitted.

22.2.3 — This chapter shall not govern design and installation of cast-in-place concrete piles and piers embedded in ground.

COMMENTARY

R22.1 — Scope

Prior to the 1995 code, requirements for plain concrete were set forth in **"Building Code Requirements for Structural Plain Concrete (ACI 318.1-89) (Revised 1992)."** Requirements for plain concrete are now in the code.

R22.1.1.1 — Section 22.1.1.1 exempts structural plain concrete walls from the requirements for special exposure conditions because of the numerous successful uses of concrete with specified compressive strengths, f_c', of 2500 and 3000 psi in the basement walls of residences and other minor structures that did not meet the strength requirements of Table 4.2.2.

R22.1.1.2 — It is not within the scope of this code to provide design and construction requirements for nonstructural members of plain concrete such as soil-supported slabs (slabs on grade).

R22.2 — Limitations

R22.2.2 and R22.2.3 — Since the structural integrity of plain concrete members depends solely on the properties of the concrete, use of structural plain concrete members should be limited to members that are primarily in a state of compression, members that can tolerate random cracks without detriment to their structural integrity, and members where ductility is not an essential feature of design. The tensile strength of concrete can be recognized in design of members. Tensile stresses due to restraint from creep, shrinkage, or temperature effects are to be considered and sufficiently reduced by construction techniques to avoid uncontrolled cracks, or when uncontrolled cracks due to such restraint effects are anticipated to occur, they will not induce structural failure.

Plain concrete walls are permitted (see 22.6) without a height limitation. However, for multistory construction and other major structures, ACI Committee 318 encourages the use of walls designed in accordance with Chapter 14 (see R22.6).

CODE / COMMENTARY

Since plain concrete lacks the necessary ductility that columns should possess and because a random crack in an unreinforced column will most likely endanger its structural integrity, the code does not permit use of plain concrete for columns. It does allow its use for pedestals limited to a ratio of unsupported height to least lateral dimension of 3 or less (see 22.8.2).

Structural elements such as cast-in-place concrete piles and piers in ground or other material sufficiently stiff to provide adequate lateral support to prevent buckling are not covered by this code. Such elements are covered by the general building code.

22.2.4 — Minimum strength

Specified compressive strength of plain concrete to be used for structural purposes shall be not less than given in 1.1.1.

R22.2.4 — Minimum strength

A minimum specified compressive strength requirement for plain concrete construction is considered necessary because safety is based solely on strength and quality of concrete treated as a homogeneous material. Lean concrete mixtures may not produce adequately homogeneous material or well-formed surfaces.

22.3 — Joints

22.3.1 — Contraction or isolation joints shall be provided to divide structural plain concrete members into flexurally discontinuous elements. The size of each element shall limit control excessive buildup of internal stresses caused by restraint to movements from creep, shrinkage, and temperature effects.

22.3.2 — In determining the number and location of contraction or isolation joints, consideration shall be given to: influence of climatic conditions; selection and proportioning of materials; mixing, placing, and curing of concrete; degree of restraint to movement; stresses due to loads to which an element is subject; and construction techniques.

R22.3 — Joints

Joints in plain concrete construction are an important design consideration. In reinforced concrete, reinforcement is provided to resist the stresses due to restraint of creep, shrinkage, and temperature effects. In plain concrete, joints are the only means of controlling and thereby relieving the buildup of such tensile stresses. A plain concrete member, therefore, should be small enough, or divided into smaller elements by joints, to control the buildup of internal stresses. The joint may be a contraction joint or an isolation joint. A minimum 25 percent reduction of member thickness is considered sufficient for contraction joints to be effective. The jointing should be such that no axial tension or flexural tension can be developed across a joint after cracking, if applicable, a condition referred to as flexural discontinuity. Where random cracking due to creep, shrinkage, and temperature effects will not affect the structural integrity, and is otherwise acceptable, such as transverse cracks in a continuous wall footing, transverse contraction or isolation joints are not necessary.

22.4 — Design method

22.4.1 — Structural plain concrete members shall be designed for adequate strength in accordance with this code, using load factors and design strength.

22.4.2 — Factored loads and forces shall be in combinations as in 9.2.

22.4.3 — Where required strength exceeds design strength, reinforcement shall be provided and the member designed as a reinforced concrete member in accordance with appropriate design requirements of this code.

R22.4 — Design method

Plain concrete members are proportioned for adequate strength using factored loads and forces. When the design strength is exceeded, the section should be increased or the specified strength of concrete increased, or both, or the member designed as a reinforced concrete member in accordance with the code. The designer should note, however, that an increase in concrete section may have a detrimental effect; stress due to load will decrease but stresses due to creep, shrinkage, and temperature effects may increase.

CODE

22.4.4 — Strength design of structural plain concrete members for flexure and axial loads shall be based on a linear stress-strain relationship in both tension and compression.

22.4.5 — Tensile strength of concrete shall be permitted to be considered in design of plain concrete members when provisions of 22.3 have been followed.

22.4.6 — No strength shall be assigned to steel reinforcement that may be present.

22.4.7 — Tension shall not be transmitted through outside edges, construction joints, contraction joints, or isolation joints of an individual plain concrete element. No flexural continuity due to tension shall be assumed between adjacent structural plain concrete elements.

22.4.8 — When computing strength in flexure, combined flexure and axial load, and shear, the entire cross section of a member shall be considered in design, except for concrete cast against soil where overall thickness h shall be taken as 2 in. less than actual thickness.

22.5 — Strength design

22.5.1 — Design of cross sections subject to flexure shall be based on

$$\phi M_n \geq M_u \tag{22-1}$$

where

$$M_n = 5\sqrt{f_c'}S_m \tag{22-2}$$

if tension controls, and

$$M_n = 0.85 f_c' S_m \tag{22-3}$$

if compression controls, where S_m is the corresponding elastic section modulus.

22.5.2 — Design of cross sections subject to compression shall be based on

$$\phi P_n \geq P_u \tag{22-4}$$

where P_n is computed by

$$P_n = 0.60 f_c'\left[1 - \left(\frac{\ell_c}{32h}\right)^2\right]A_1 \tag{22-5}$$

COMMENTARY

R22.4.4 — Flexural tension may be considered in design of plain concrete members to sustain loads, provided the computed stress does not exceed the permissible stress, and construction, contraction, or isolation joints are provided to relieve the resulting tensile stresses due to restraint of creep, temperature, and shrinkage effects.

R22.4.8 — The reduced overall thickness h for concrete cast against earth is to allow for unevenness of excavation and for some contamination of the concrete adjacent to the soil.

R22.5 — Strength design

R22.5.2 — Eq. (22-5) is presented to reflect the general range of braced and restrained end conditions encountered in structural plain concrete elements. The effective length factor was omitted as a modifier of ℓ_c, the vertical distance between supports, since this is conservative for walls with assumed pin supports that are required to be braced against lateral translation as in 22.6.6.4.

CODE

and A_1 is the loaded area.

22.5.3 — Members subject to combined flexure and axial load in compression shall be proportioned such that on the compression face:

$$P_u/\phi P_n + M_u/\phi M_n \leq 1 \quad (22\text{-}6)$$

and on the tension face:

$$M_u/S_m - P_u/A_g \leq 5\phi\sqrt{f_c'} \quad (22\text{-}7)$$

22.5.4 — Design of rectangular cross sections subject to shear shall be based on

$$\phi V_n \geq V_u \quad (22\text{-}8)$$

where V_n is computed by

$$V_n = \frac{4}{3}\sqrt{f_c'}b_w h \quad (22\text{-}9)$$

for beam action and by

$$V_n = \left[\frac{4}{3} + \frac{8}{3\beta}\right]\sqrt{f_c'}b_o h \quad (22\text{-}10)$$

for two-way action, but not greater than $2.66\sqrt{f_c'}\,b_o h$. In Eq. (22-10), β corresponds to ratio of long side to short side of concentrated load or reaction area.

22.5.5 — Design of bearing areas subject to compression shall be based on

$$\phi B_n \geq B_u \quad (22\text{-}11)$$

where B_u is factored bearing load and B_n is nominal bearing strength of loaded area A_1 calculated by

$$B_n = 0.85 f_c' A_1 \quad (22\text{-}12)$$

COMMENTARY

R22.5.3 — Plain concrete members subject to combined flexure and axial compressive load are proportioned such that on the compression face:

$$\frac{P_u}{0.60\phi f_c'\left[1-\left(\frac{\ell_c}{32h}\right)^2\right]A_1} + \frac{M_u}{0.85\phi f_c' S_m} \leq 1$$

and that on the tension face:

$$\left(\begin{array}{c}\text{Calculated}\\ \text{bending stress}\end{array}\right) - \left(\begin{array}{c}\text{Calculated}\\ \text{axial stress}\end{array}\right) \leq 5\phi\sqrt{f_c'}$$

R22.5.4 — Proportions of plain concrete members usually are controlled by tensile strength rather than shear strength. Shear stress (as a substitute for principal tensile stress) rarely will control. However, since it is difficult to foresee all possible conditions where shear may have to be investigated (such as shear keys), Committee 318 maintains the investigation of this basic stress condition. An experienced designer will soon recognize where shear is not critical for plain concrete members and will adjust design procedures accordingly.

The shear requirements for plain concrete assume an uncracked section. Shear failure in plain concrete will be a diagonal tension failure, occurring when the principal tensile stress near the centroidal axis becomes equal to the tensile strength of the concrete. Since the major portion of the principal tensile stress comes from the shear, the code safeguards against tensile failure by limiting the permissible shear at the centroidal axis as calculated from the equation for a section of homogeneous material:

$$v = VQ/Ib$$

where v and V are the shear stress and shear force, respectively, at the section considered, Q is the statical moment of the area outside the section being considered about centroidal axis of the gross section, I is the moment of inertia of the gross section, and b is the width where shear stress is being computed.

CODE

except when the supporting surface is wider on all sides than the loaded area, B_n shall be multiplied by $\sqrt{A_2/A_1}$ but not more than 2.

22.5.6 — Lightweight concrete

22.5.6.1 — Provisions of 22.5 apply to normalweight concrete. When lightweight concrete is used, either (a) or (b) shall apply:

(a) When f_{ct} is specified and concrete is proportioned in accordance with 5.2, equations in 22.5 that include $\sqrt{f_c'}$ shall be modified by substituting $f_{ct}/6.7$ for $\sqrt{f_c'}$, but the value of $f_{ct}/6.7$ shall not exceed $\sqrt{f_c'}$;

(b) When f_{ct} is not specified, all values of $\sqrt{f_c'}$ in 22.5 shall be multiplied by 0.75 for all-lightweight concrete, and 0.85 for sand-lightweight concrete. Linear interpolation shall be permitted when partial sand replacement is used.

22.6 — Walls

22.6.1 — Structural plain concrete walls shall be continuously supported by soil, footings, foundation walls, grade beams, or other structural members capable of providing continuous vertical support.

22.6.2 — Structural plain concrete walls shall be designed for vertical, lateral, and other loads to which they are subjected.

22.6.3 — Structural plain concrete walls shall be designed for an eccentricity corresponding to the maximum moment that can accompany the axial load but not less than $0.10h$. If the resultant of all factored loads is located within the middle-third of the overall wall thickness, the design shall be in accordance with 22.5.3 or 22.6.5. Otherwise, walls shall be designed in accordance with 22.5.3.

22.6.4 — Design for shear shall be in accordance with 22.5.4.

22.6.5 — Empirical design method

22.6.5.1 — Structural plain concrete walls of solid rectangular cross section shall be permitted to be designed by Eq. (22-13) if the resultant of all factored loads is located within the middle-third of the overall thickness of wall.

COMMENTARY

R22.5.6 — Lightweight concrete

See R11.2

R22.6 — Walls

Plain concrete walls are commonly used for basement wall construction for residential and light commercial buildings in low or nonseismic areas. Although the code imposes no absolute maximum height limitation on the use of plain concrete walls, designers are cautioned against extrapolating the experience with relatively minor structures and using plain concrete walls in multistory construction and other major structures where differential settlement, wind, earthquake, or other unforeseen loading conditions require the walls to possess some ductility and ability to maintain their integrity when cracked. For such conditions, ACI Committee 318 strongly encourages the use of walls designed in accordance with Chapter 14.

The provisions for plain concrete walls are applicable only for walls laterally supported in such a manner as to prohibit relative lateral displacement at top and bottom of individual wall elements (see 22.6.6.4). The code does not cover walls without horizontal support to prohibit relative displacement at top and bottom of wall elements. Such laterally unsupported walls are to be designed as reinforced concrete members in accordance with the code.

R22.6.5 — Empirical design method

When the resultant load falls within the middle-third of the wall thickness (kern of wall section), plain concrete walls may be designed using the simplified Eq. (22-14). Eccentric loads and lateral forces are used to determine the total eccentricity of the factored axial force P_u. If the eccentricity does not exceed $h/6$, Eq. (22-14) may be applied, and design performed considering P_u as a concentric load. The factored

CODE

22.6.5.2 — Design of walls subject to axial loads in compression shall be based on

$$\phi P_n \geq P_u \quad (22\text{-}13)$$

where P_u is factored axial force and P_n is nominal axial strength calculated by

$$P_n = 0.45 f_c' A_g \left[1 - \left(\frac{\ell_c}{32h}\right)^2\right] \quad (22\text{-}14)$$

22.6.6 — Limitations

22.6.6.1 — Unless demonstrated by a detailed analysis, horizontal length of wall to be considered effective for each vertical concentrated load shall not exceed center-to-center distance between loads, nor width of bearing plus four times the wall thickness.

22.6.6.2 — Except as provided in 22.6.6.3, thickness of bearing walls shall be not less than 1/24 the unsupported height or length, whichever is shorter, nor less than 5-1/2 in.

22.6.6.3 — Thickness of exterior basement walls and foundation walls shall be not less than 7-1/2 in.

22.6.6.4 — Walls shall be braced against lateral translation. See 22.3 and 22.4.7.

22.6.6.5 — Not less than two No. 5 bars shall be provided around all window and door openings. Such bars shall extend at least 24 in. beyond the corners of openings.

22.7 — Footings

22.7.1 — Structural plain concrete footings shall be designed for factored loads and induced reactions in accordance with appropriate design requirements of this code and as provided in 22.7.2 through 22.7.8.

22.7.2 — Base area of footing shall be determined from unfactored forces and moments transmitted by footing to soil and permissible soil pressure selected through principles of soil mechanics.

22.7.3 — Plain concrete shall not be used for footings on piles.

22.7.4 — Thickness of structural plain concrete footings shall be not less than 8 in. See 22.4.8.

COMMENTARY

axial load P_u should not exceed the design axial load strength ϕP_n. Eq. (22-14) reflects the range of braced and restrained end conditions encountered in wall design. The limitations of 22.6.6 apply whether the wall is proportioned by 22.5.3 or by the empirical method of 22.6.5.

R22.7 — Footings

R22.7.4 — Thickness of plain concrete footings will be controlled by flexural strength (extreme fiber stress in tension not greater than $5\phi\sqrt{f_c'}$) rather than shear strength for the usual proportions of plain concrete footings. Shear rarely will control (see R22.5.4). For footings cast against soil, overall thickness h used for strength computations

CODE

COMMENTARY

should be taken as 2 in. less than actual thickness to allow for unevenness of excavation and contamination of the concrete adjacent to soil as required by 22.4.8. Thus, for a minimum footing thickness of 8 in., calculations for flexural and shear stresses must be based on an overall thickness h of 6 in.

22.7.5 — Maximum factored moment shall be computed at (a), (b), and (c):

(a) At the face of the column, pedestal, or wall, for footing supporting a concrete column, pedestal, or wall;

(b) Halfway between center and face of the wall, for footing supporting a masonry wall;

(c) Halfway between face of column and edge of steel base plate, for footing supporting a column with steel base plate.

22.7.6 — Shear in plain concrete footings

22.7.6.1 — V_u shall be computed in accordance with 22.7.6.2, with location of critical section measured from face of column, pedestal, or wall for footing supporting a column, pedestal, or wall. For footing supporting a column with steel base plates, the critical section shall be measured at location defined in 22.7.5(c).

22.7.6.2 — ϕV_n of structural plain concrete footings in the vicinity of concentrated loads or reactions shall be governed by the more severe of two conditions:

(a) Beam action for footing, with a critical section extending in a plane across the entire footing width and located at a distance h from face of concentrated load or reaction area. For this condition, the footing shall be designed in accordance with Eq. (22-9);

(b) Two-way action for footing, with a critical section perpendicular to plane of footing and located so that its perimeter b_o is a minimum, but need not approach closer than $h/2$ to perimeter of concentrated load or reaction area. For this condition, the footing shall be designed in accordance with Eq. (22-10).

22.7.7 — Circular or regular polygon shaped concrete columns or pedestals shall be permitted to be treated as square members with the same area for location of critical sections for moment and shear.

22.7.8 — Factored bearing load, B_u, on concrete at contact surface between supporting and supported member shall not exceed design bearing strength, ϕB_n, for either surface as given in 22.5.5.

CODE

22.8 — Pedestals

22.8.1 — Plain concrete pedestals shall be designed for vertical, lateral, and other loads to which they are subjected.

22.8.2 — Ratio of unsupported height to average least lateral dimension of plain concrete pedestals shall not exceed 3.

22.8.3 — Maximum factored axial load, P_u, applied to plain concrete pedestals shall not exceed design bearing strength, ϕB_n, given in 22.5.5.

22.9 — Precast members

22.9.1 — Design of precast plain concrete members shall consider all loading conditions from initial fabrication to completion of the structure, including form removal, storage, transportation, and erection.

22.9.2 — Limitations of 22.2 apply to precast members of plain concrete not only to the final condition but also during fabrication, transportation, and erection.

22.9.3 — Precast members shall be connected securely to transfer all lateral forces into a structural system capable of resisting such forces.

22.9.4 — Precast members shall be adequately braced and supported during erection to ensure proper alignment and structural integrity until permanent connections are completed.

22.10 — Plain concrete in earthquake-resisting structures

22.10.1 — Structures designed for earthquake-induced forces in regions of high seismic risk or assigned to high seismic performance or design categories shall not have foundation elements of structural plain concrete, except as follows:

(a) For detached one- and two-family dwellings three stories or less in height and constructed with stud bearing walls, plain concrete footings without longitudinal reinforcement supporting walls and isolated plain concrete footings supporting columns or pedestals are permitted;

(b) For all other structures, plain concrete footings supporting cast-in-place reinforced concrete or reinforced masonry walls are permitted provided the footings are reinforced longitudinally with not less than two continuous reinforcing bars. Bars shall not

COMMENTARY

R22.8 — Pedestals

The height-thickness limitation for plain concrete pedestals does not apply for portions of pedestals embedded in soil capable of providing lateral restraint.

R22.9 — Precast members

Precast structural plain concrete members are subject to all limitations and provisions for cast-in-place concrete contained in this chapter.

The approach to contraction or isolation joints is expected to be somewhat different than for cast-in-place concrete since the major portion of shrinkage stresses takes place prior to erection. To ensure stability, precast members should be connected to other members. The connection should transfer no tension.

CODE

be smaller than No. 4 and shall have a total area of not less than 0.002 times the gross cross-sectional area of the footing. Continuity of reinforcement shall be provided at corners and intersections;

(c) For detached one- and two-family dwellings three stories or less in height and constructed with stud bearing walls, plain concrete foundations or basement walls are permitted provided the wall is not less than 7-1/2 in. thick and retains no more than 4 ft of unbalanced fill.

COMMENTARY

Notes

APPENDIX A — STRUT-AND-TIE MODELS

CODE

A.1 — Definitions

B-region — A portion of a member in which the plane sections assumption of flexure theory from 10.2.2 can be applied.

Discontinuity — An abrupt change in geometry or loading.

D-region — The portion of a member within a distance, $\boldsymbol{h}$, from a force discontinuity or a geometric discontinuity.

Deep beam — See 10.7.1 and 11.8.1.

Node — The point in a joint in a strut-and-tie model where the axes of the struts, ties, and concentrated forces acting on the joint intersect.

COMMENTARY

RA.1 — Definitions

B-region — In general, any portion of a member outside of a D-region is a B-region.

Discontinuity — A discontinuity in the stress distribution occurs at a change in the geometry of a structural element or at a concentrated load or reaction. St. Venant's principle indicates that the stresses due to axial load and bending approach a linear distribution at a distance approximately equal to the overall height of the member, $\boldsymbol{h}$, away from the discontinuity. For this reason, discontinuities are assumed to extend a distance $\boldsymbol{h}$ from the section where the load or change in geometry occurs. Figure RA.1.1(a) shows typical geometric discontinuities, and Fig. RA.1.1(b) shows combined geometrical and loading discontinuities.

D-region — The shaded regions in Fig. RA.1.1(a) and (b) show typical D–regions.[A.1] The plane sections assumption of 10.2.2 is not applicable in such regions.

Each shear span of the beam in Fig. RA.1.2(a) is a D-region. If two D-regions overlap or meet as shown in Fig. RA.1.2 (b), they can be considered as a single D-region for design purposes. The maximum length-to-depth ratio of such a D-region would be approximately two. Thus, the smallest angle between the strut and the tie in a D-region is arctan 1/2 = 26.5 degrees, rounded to 25 degrees.

If there is a B-region between the D-regions in a shear span, as shown in Fig. RA.1.2(c), the strength of the shear span is governed by the strength of the B-region if the B- and D-regions have similar geometry and reinforcement.[A.2] This is because the shear strength of a B-region is less than the shear strength of a comparable D-region. Shear spans containing B-regions — the usual case in beam design — are designed for shear using the traditional shear design procedures from 11.1 through 11.5 ignoring D-regions.

Deep beam — See Fig. RA.1.2(a), RA.1.2(b), and RA.1.3, and Sections 10.7 and 11.8.

Node — For equilibrium, at least three forces should act on a node in a strut-and-tie model, as shown in Fig. RA.1.4. Nodes are classified according to the signs of these forces. A ***C-C-C*** node resists three compressive forces, a ***C-C-T*** node resists two compressive forces and one tensile force, and so on.

CODE

COMMENTARY

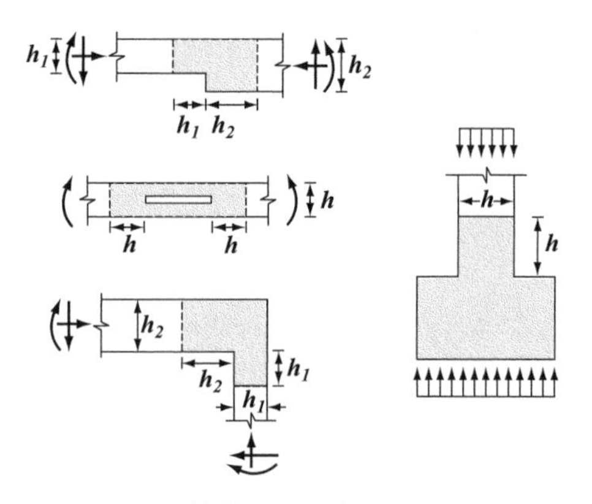

(a) Geometric discontinuities

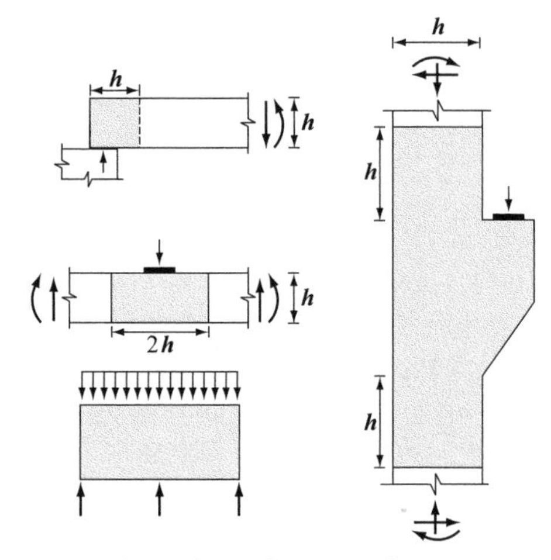

(b) Loading and geometric discontinuities

Fig. RA.1.1—D-regions and discontinuities

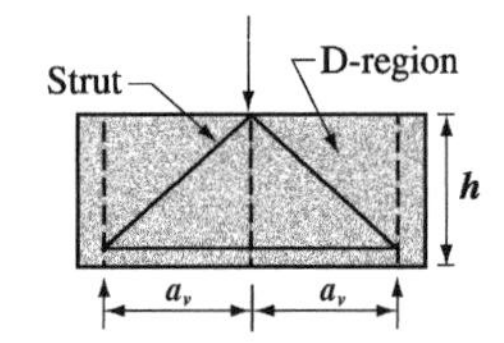

(a) Shear span, $a_v < 2h$, deep beam

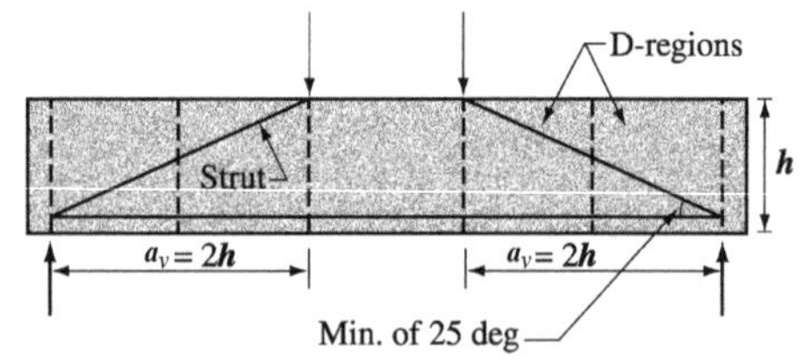

(b) Shear span, $a_v = 2h$, limit for a deep beam

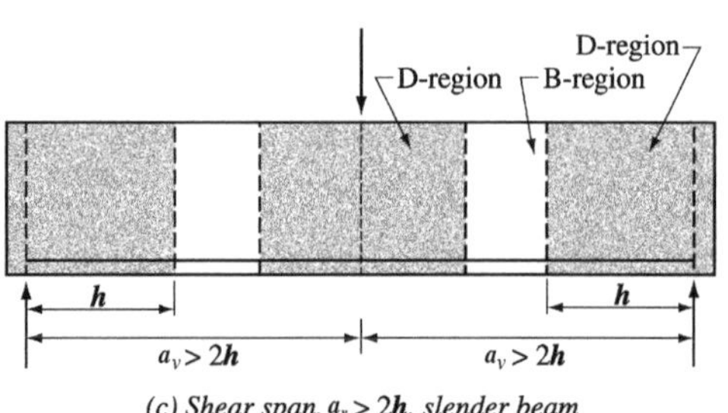

(c) Shear span, $a_v > 2h$, slender beam

Fig. RA.1.2—Description of deep and slender beams.

CODE

COMMENTARY

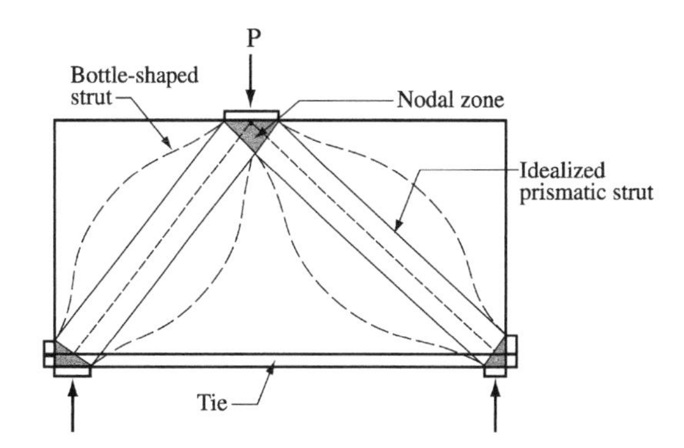

Fig. RA.1.3—Description of strut-and-tie model.

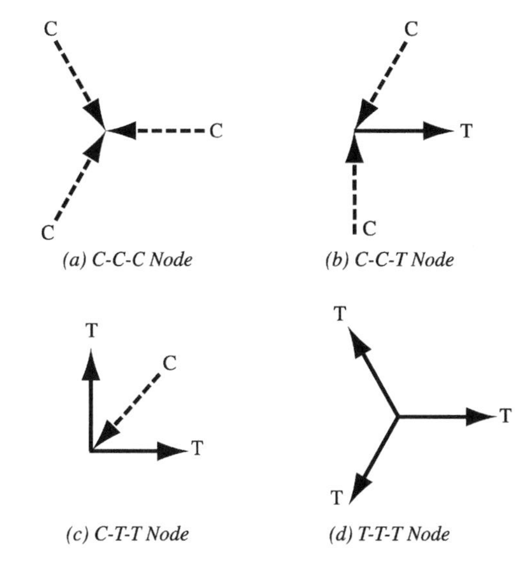

Fig. RA.1.4—Classification of nodes.

Nodal zone — The volume of concrete around a node that is assumed to transfer strut-and-tie forces through the node.

Nodal zone — Historically, hydrostatic nodal zones as shown in Fig. RA.1.5 were used. These were largely superseded by what are called extended nodal zones, shown in Fig. RA.1.6.

A ***hydrostatic nodal zone*** has loaded faces perpendicular to the axes of the struts and ties acting on the node and has equal stresses on the loaded faces. Figure RA.1.5(a) shows a ***C-C-C*** nodal zone. If the stresses on the face of the nodal zone are the same in all three struts, the ratios of the lengths of the sides of the nodal zone, w_{n1}: w_{n2}: w_{n3} are in the same proportions as the three forces C_1: C_2: C_3. The faces of a hydrostatic nodal zone are perpendicular to the axes of the struts and ties acting on the nodal zone.

These nodal zones are called hydrostatic nodal zones because the in-plane stresses are the same in all directions. Strictly speaking, this terminology is incorrect because the in-plane stresses are not equal to the out-of-plane stresses.

A ***C-C-T*** nodal zone can be represented as a hydrostatic nodal zone if the tie is assumed to extend through the node to be anchored by a plate on the far side of the node, as shown in Fig. RA.1.5(b), provided that the size of the plate results in

CODE

COMMENTARY

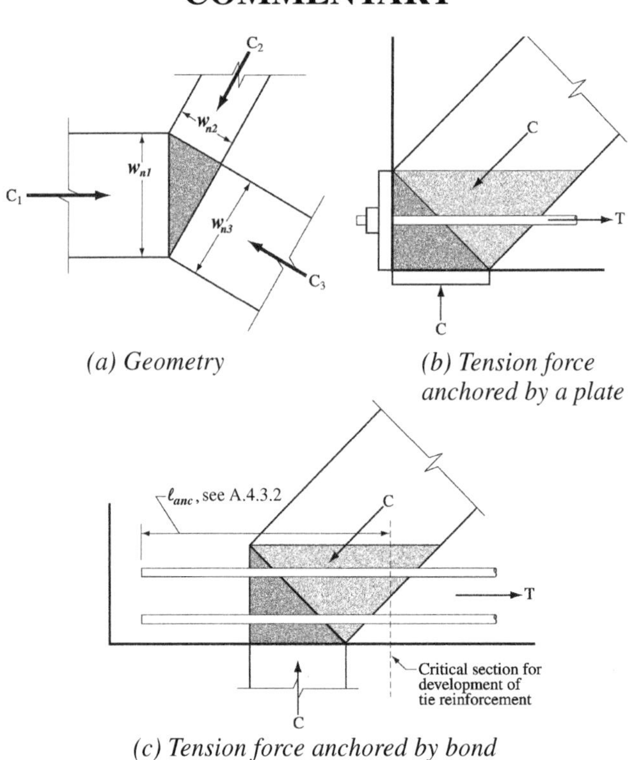

Fig. RA.1.5—Hydrostatic nodes.

bearing stresses that are equal to the stresses in the struts. The bearing plate on the left side of Fig. RA.1.5(b) is used to represent an actual tie anchorage. The tie force can be anchored by a plate, or through development of straight or hooked bars, as shown in Fig. RA.1.5(c).

The shaded areas in Fig. RA.1.6(a) and (b) are extended nodal zones. An ***extended nodal zone*** is that portion of a member bounded by the intersection of the effective strut width, w_s, and the effective tie width, w_t (see RA.4.2).

In the nodal zone shown in Fig. RA.1.7(a), the reaction R equilibrates the vertical components of the forces C_1 and C_2. Frequently, calculations are easier if the reaction R is divided into R_1, which equilibrates the vertical component of C_1 and R_2, which equilibrates the vertical component of the force C_2, as shown in Fig. RA1.7(b).

Strut — A compression member in a strut-and-tie model. A strut represents the resultant of a parallel or a fan-shaped compression field.

Strut — In design, struts are usually idealized as prismatic compression members, as shown by the straight line outlines of the struts in Fig. RA.1.2 and RA.1.3. If the effective compression strength f_{ce} differs at the two ends of a strut, due either to different nodal zone strengths at the two ends, or to different bearing lengths, the strut is idealized as a uniformly tapered compression member.

Bottle-shaped strut — A strut that is wider at midlength than at its ends.

Bottle-shaped struts — A bottle-shaped strut is a strut located in a part of a member where the width of the compressed concrete at midlength of the strut can spread laterally.[A.1,A.3]

CODE

COMMENTARY

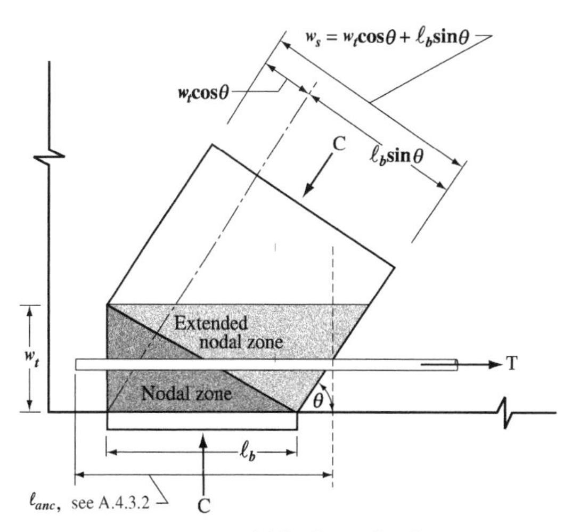

(a) One layer of steel

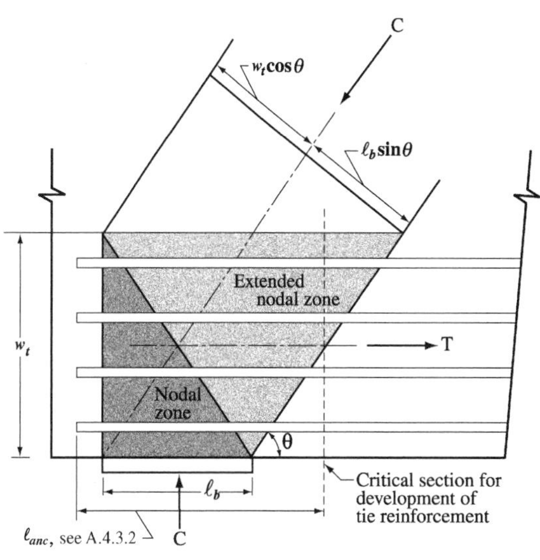

(b) Distributed steel

Fig. RA.1.6—Extended nodal zone showing the effect of the distribution of the force.

The curved dashed outlines of the struts in Fig. RA.1.3 and the curved solid outlines in Fig. RA.1.8 approximate the boundaries of bottle-shaped struts. A split cylinder test is an example of a bottle-shaped strut. The internal lateral spread of the applied compression force in such a test leads to a transverse tension that splits the specimen.

To simplify design, bottle-shaped struts are idealized either as prismatic or as uniformly tapered, and crack-control reinforcement from A.3.3 is provided to resist the transverse tension. The amount of confining transverse reinforcement can be computed using the strut-and-tie model shown in Fig. RA.1.8(b) with the struts that represent the spread of the compression force acting at a slope of 1:2 to the axis of the applied compressive force. Alternatively for f_c' not exceeding 6000 psi, Eq. (A-4) can be used. The cross-sectional area A_c of a bottle-shaped strut, is taken as the smaller of the cross-sectional areas at the two ends of the strut. See Fig. RA.1.8(a).

CODE

COMMENTARY

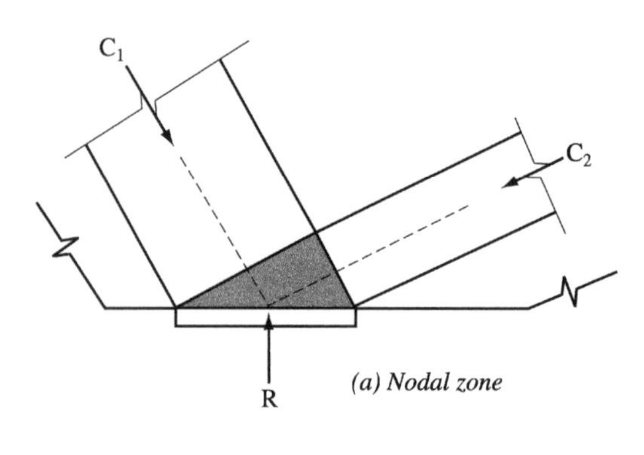

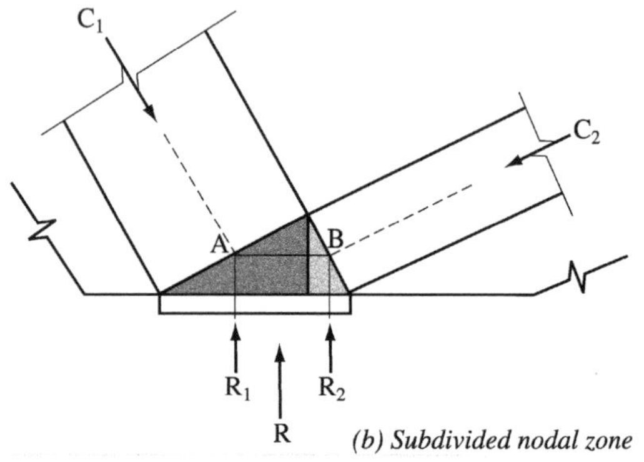

Fig. RA.1.7—Subdivision of nodal zone.

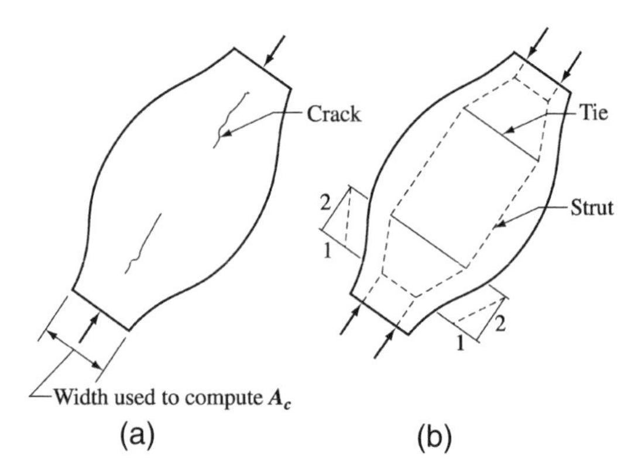

Fig. RA.1.8—Bottle-shaped strut: (a) cracking of a bottle-shaped strut; and (b) strut-and-tie model of a bottle-shaped strut.

Strut-and-tie model — A truss model of a structural member, or of a D-region in such a member, made up of struts and ties connected at nodes, capable of transferring the factored loads to the supports or to adjacent B-regions.

Strut-and-tie model — The components of a strut-and-tie model of a single-span deep beam loaded with a concentrated load are identified in Fig. RA.1.3. The cross-sectional dimensions of a strut or tie are designated as thickness and width, both perpendicular to the axis of the strut or tie. Thickness is perpendicular to the plane of the truss model, and width is in the plane of the truss model.

Tie — A tension member in a strut-and-tie model.

Tie — A tie consists of reinforcement or prestressing steel plus a portion of the surrounding concrete that is concentric with the axis of the tie. The surrounding concrete is included to define the zone in which the forces in the struts and ties are to be anchored. The concrete in a tie is not used to resist the axial force in the tie. Although not considered in design, the surrounding concrete will reduce the elongations of the tie, especially at service loads.

CODE

A.2 — Strut-and-tie model design procedure

A.2.1 — It shall be permitted to design structural concrete members, or D-regions in such members, by modeling the member or region as an idealized truss. The truss model shall contain struts, ties, and nodes as defined in A.1. The truss model shall be capable of transferring all factored loads to the supports or adjacent B-regions.

A.2.2 — The strut-and-tie model shall be in equilibrium with the applied loads and the reactions.

A.2.3 — In determining the geometry of the truss, the dimensions of the struts, ties, and nodal zones shall be taken into account.

COMMENTARY

RA.2 — Strut-and-tie model design procedure

RA.2.1 — The truss model described in A.2.1 is referred to as a strut-and-tie model. Details of the use of strut-and-tie models are given in References A.1 through A.7. The design of a D-region includes the following four steps:

1. Define and isolate each D-region;

2. Compute resultant forces on each D-region boundary;

3. Select a truss model to transfer the resultant forces across the D-region. The axes of the struts and ties, respectively, are chosen to approximately coincide with the axes of the compression and tension fields. The forces in the struts and ties are computed.

4. The effective widths of the struts and nodal zones are determined considering the forces from Step 3 and the effective concrete strengths defined in A.3.2 and A.5.2, and reinforcement is provided for the ties considering the steel strengths defined in A.4.1. The reinforcement should be anchored in the nodal zones.

Strut-and-tie models represent strength limit states and designers should also comply with the requirements for serviceability in the code. Deflections of deep beams or similar members can be estimated using an elastic analysis to analyze the strut-and-tie model. In addition, the crack widths in a tie can be checked using 10.6.4, assuming the tie is encased in a prism of concrete corresponding to the area of tie from RA.4.2.

RA.2.3 — The struts, ties, and nodal zones making up the strut-and-tie model all have finite widths that should be taken into account in selecting the dimensions of the truss. Figure RA.2.3(a) shows a node and the corresponding nodal zone. The vertical and horizontal forces equilibrate the force in the inclined strut. If the stresses are equal in all three struts, a hydrostatic nodal zone can be used and the widths of the struts will be in proportion to the forces in the struts.

If more than three forces act on a nodal zone in a two-dimensional structure, as shown in Fig. RA.2.3(b), it is generally necessary to resolve some of the forces to end up with three intersecting forces. The strut forces acting on Faces A–E and C–E in Fig. RA.2.3(b) can be replaced with one force acting on face A–C. This force passes through the node at D.

Alternatively, the strut-and-tie model could be analyzed assuming all the strut forces acted through the node at D, as shown in Fig. RA.2.3(c). In this case, the forces in the two struts on the right side of Node D can be resolved into a single force acting through Point D, as shown in Fig. RA.2.3(d).

CODE

COMMENTARY

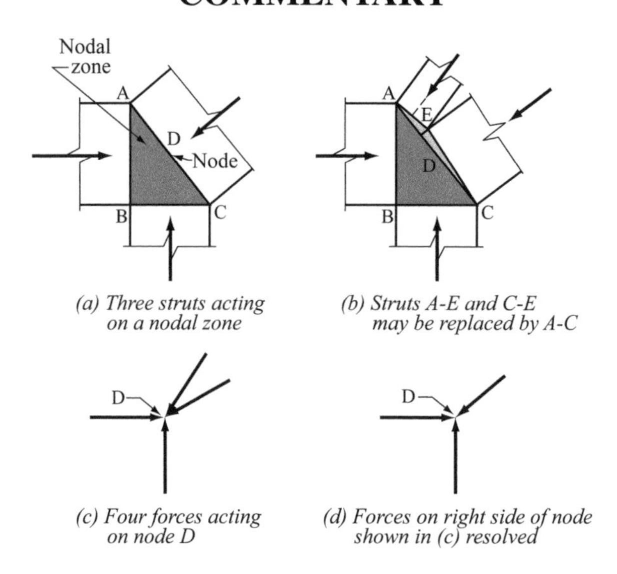

(a) Three struts acting on a nodal zone

(b) Struts A-E and C-E may be replaced by A-C

(c) Four forces acting on node D

(d) Forces on right side of node shown in (c) resolved

Fig. RA.2.3—Resolution of forces on a nodal zone.

If the width of the support in the direction perpendicular to the member is less than the width of the member, transverse reinforcement may be required to restrain vertical splitting in the plane of the node. This can be modeled using a transverse strut-and-tie model.

A.2.4 — Ties shall be permitted to cross struts. Struts shall cross or overlap only at nodes.

A.2.5 — The angle, θ, between the axes of any strut and any tie entering a single node shall not be taken as less than 25 degrees.

RA.2.5 — The angle between the axes of struts and ties acting on a node should be large enough to mitigate cracking and to avoid incompatibilities due to shortening of the struts and lengthening of the ties occurring in almost the same directions. This limitation on the angle prevents modeling the shear spans in slender beams using struts inclined at less than 25 degrees from the longitudinal steel. See Reference A.6.

A.2.6 — Design of struts, ties, and nodal zones shall be based on

$$\phi F_n \geq F_u \qquad \text{(A-1)}$$

where F_u is the factored force acting in a strut, in a tie, or on one face of a nodal zone; F_n is the nominal strength of the strut, tie, or nodal zone; and ϕ is specified in 9.3.2.6.

RA.2.6 — Factored loads are applied to the strut-and-tie model, and the forces in all the struts, ties, and nodal zones are computed. If several loading cases exist, each should be investigated. The strut-and-tie model, or models, are analyzed for the loading cases and, for a given strut, tie, or nodal zone, F_u is the largest force in that element for all loading cases.

A.3 — Strength of struts

A.3.1 — The nominal compressive strength of a strut without longitudinal reinforcement, F_{ns}, shall be taken as the smaller value of

$$F_{ns} = f_{ce} A_{cs} \qquad \text{(A-2)}$$

RA.3 — Strength of struts

RA.3.1 — The width of strut w_s used to compute A_{cs} is the smaller dimension perpendicular to the axis of the strut at the ends of the strut. This strut width is illustrated in Fig. RA.1.5(a) and Fig. RA.1.6(a) and (b). In two-dimensional structures, such as deep beams, the thickness of the struts may be taken as the width of the member.

CODE

at the two ends of the strut, where A_{cs} is the cross-sectional area at one end of the strut, and f_{ce} is the smaller of (a) and (b):

(a) the effective compressive strength of the concrete in the strut given in A.3.2;

(b) the effective compressive strength of the concrete in the nodal zone given in A.5.2.

A.3.2 — The effective compressive strength of the concrete, f_{ce}, in a strut shall be taken as

$$f_{ce} = 0.85\beta_s f_c' \qquad \text{(A-3)}$$

A.3.2.1 — For a strut of uniform cross-sectional area over its length .... $\beta_s = 1.0$

A.3.2.2 — For struts located such that the width of the midsection of the strut is larger than the width at the nodes (bottle-shaped struts):

(a) with reinforcement satisfying A.3.3... $\beta_s = 0.75$

(b) without reinforcement satisfying A.3.3.. $\beta_s = 0.60\lambda$

where λ is given in 11.7.4.3.

A.3.2.3 — For struts in tension members, or the tension flanges of members.......................... $\beta_s = 0.40$

A.3.2.4 — For all other cases.................. $\beta_s = 0.60$

COMMENTARY

RA.3.2 — The strength coefficient, $0.85f_c'$, in Eq. (A-3), represents the effective concrete strength under sustained compression, similar to that used in Eq. (10-1) and (10-2).

RA.3.2.1 — The value of β_s in A.3.2.1 applies to a strut equivalent to the rectangular stress block in a compression zone in a beam or column.

RA.3.2.2 — The value of β_s in A.3.2.2 applies to bottle-shaped struts as shown in Fig. RA.1.3. The internal lateral spread of the compression forces can lead to splitting parallel to the axis of the strut near the ends of the strut, as shown in Fig. RA.1.8. Reinforcement placed to resist the splitting force restrains crack width, allows the strut to resist more axial load, and permits some redistribution of force.

The value of β_s in A.3.2.2(b) includes the correction factor, λ, for lightweight concrete because the strength of a strut without transverse reinforcement is assumed to be limited to less than the load at which longitudinal cracking develops.

RA.3.2.3 — The value of β_s in A.3.2.3 applies, for example, to compression struts in a strut-and-tie model used to design the longitudinal and transverse reinforcement of the tension flanges of beams, box girders, and walls. The low value of β_s reflects that these struts need to transfer compression across cracks in a tension zone.

RA.3.2.4 — The value of β_s in A.3.2.4 applies to strut applications not included in A.3.2.1, A.3.2.2, and A.3.2.3. Examples are struts in a beam web compression field in the web of a beam where parallel diagonal cracks are likely to divide the web into inclined struts, and struts are likely to be crossed by cracks at an angle to the struts (see Fig. RA.3.2(a) and (b)). Section A.3.2.4 gives a reasonable lower limit on β_s except for struts described in A.3.2.2(b) and A.3.2.3.

CODE

COMMENTARY

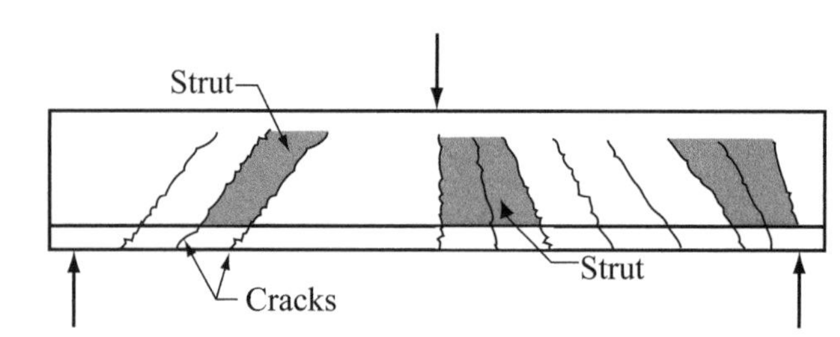

(a) Struts in a beam web with inclined cracks parallel to struts - Section A.3.2.4

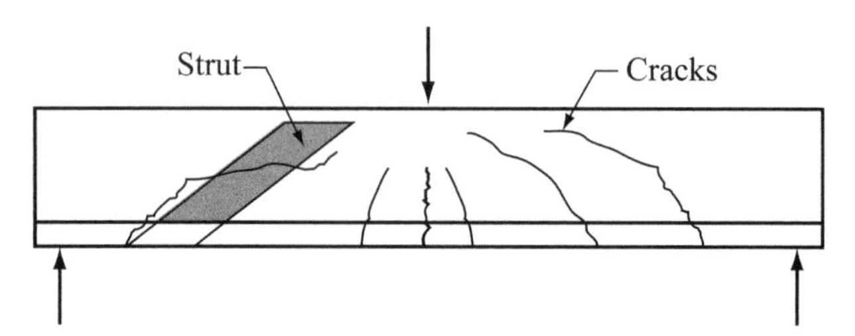

(b) Struts crossed by skew cracks - Section A.3.2.4

Fig. RA.3.2—Type of struts.

A.3.3 — If the value of β_s specified in A.3.2.2(a) is used, the axis of the strut shall be crossed by reinforcement proportioned to resist the transverse tensile force resulting from the compression force spreading in the strut. It shall be permitted to assume the compressive force in the strut spreads at a slope of 2 longitudinal to 1 transverse to the axis of the strut.

A.3.3.1 — For f_c' not greater than 6000 psi, the requirement of A.3.3 shall be permitted to be satisfied by the axis of the strut being crossed by layers of reinforcement that satisfy Eq. (A-4):

$$\Sigma \frac{A_{si}}{b_s s_i} \sin \alpha_i \geq 0.003 \quad \text{(A-4)}$$

where A_{si} is the total area of surface reinforcement at spacing s_i in the i-th layer of reinforcement crossing a strut at an angle α_i to the axis of the strut.

A.3.3.2 — The reinforcement required in A.3.3 shall be placed in either two orthogonal directions at angles α_1 and α_2 to the axis of the strut, or in one direction at an angle α to the axis of the strut. If the reinforcement is in only one direction, α shall not be less than 40 degrees.

RA.3.3 — The reinforcement required by A.3.3 is related to the tension force in the concrete due to the spreading of the strut, as shown in the strut-and-tie model in Fig. RA.1.8(b). Section RA.3.3 allows designers to use local strut-and-tie models to compute the amount of transverse reinforcement needed in a given strut. The compressive forces in the strut may be assumed to spread at a 2:1 slope, as shown in Fig. RA.1.8(b). For specified concrete compressive strengths not exceeding 6000 psi, the amount of reinforcement required by Eq. (A-4) is deemed to satisfy A.3.3.

Figure RA.3.3 shows two layers of reinforcement crossing a cracked strut. If the crack opens without shear slip along the crack, bars in layer i in the figure will cause a stress perpendicular to the strut of

$$\frac{A_{si} f_{si}}{b_s s_i} \sin \alpha_i$$

where the subscript i takes on the values of 1 and 2 for the vertical and horizontal bars, respectively, as shown in Fig. RA.3.3. Equation (A-4) is written in terms of a reinforcement ratio rather than a stress to simplify the calculation.

Often, the confinement reinforcement given in A.3.3 is difficult to place in three-dimensional structures such as pile caps. If this reinforcement is not provided, the value of f_{ce} given in A.3.2.2(b) is used.

RA.3.3.2 — In a corbel with a shear span-to-depth ratio less than 1.0, the confinement reinforcement required to satisfy A.3.3 is usually provided in the form of horizontal stirrups crossing the inclined compression strut, as shown in Fig. R.11.9.2.

CODE

COMMENTARY

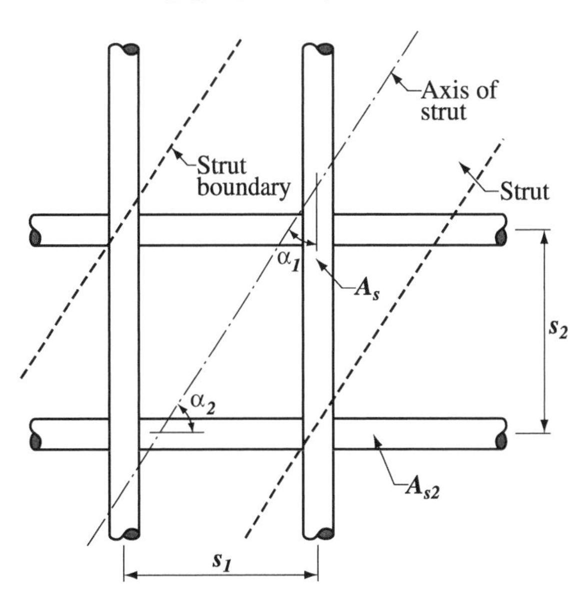

Fig. RA.3.3—Reinforcement crossing a strut.

A.3.4 — If documented by tests and analyses, it shall be permitted to use an increased effective compressive strength of a strut due to confining reinforcement.

RA.3.4 — The design of tendon anchorage zones for prestressed concrete sometimes uses confinement to enhance the compressive strength of the struts in the local zone. Confinement of struts is discussed in References A.4 and A.8.

A.3.5 — The use of compression reinforcement shall be permitted to increase the strength of a strut. Compression reinforcement shall be properly anchored, parallel to the axis of the strut, located within the strut, and enclosed in ties or spirals satisfying 7.10. In such cases, the nominal strength of a longitudinally reinforced strut is:

$$F_{ns} = f_{cu}A_c + A_s'f_s' \quad \text{(A-5)}$$

RA.3.5 — The strength added by the reinforcement is given by the last term in Eq. (A-5). The stress f_s' in the reinforcement in a strut at nominal strength can be obtained from the strains in the strut when the strut crushes. For Grade 40 or 60 reinforcement, f_s' can be taken as f_y.

A.4 — Strength of ties

RA.4 — Strength of ties

A.4.1 —The nominal strength of a tie, F_{nt}, shall be taken as

$$F_{nt} = A_{ts}f_y + A_{tp}(f_{se} + \Delta f_p) \quad \text{(A-6)}$$

where $(f_{se} + \Delta f_p)$ shall not exceed f_{py}, and A_{tp} is zero for nonprestressed members.

In Eq. (A–6), it shall be permitted to take Δf_p equal to 60,000 psi for bonded prestressed reinforcement, or 10,000 psi for unbonded prestressed reinforcement. Other values of Δf_p shall be permitted when justified by analysis.

A.4.2 — The axis of the reinforcement in a tie shall coincide with the axis of the tie in the strut-and-tie model.

RA.4.2 — The effective tie width assumed in design w_t can vary between the following limits, depending on the distribution of the tie reinforcement:

CODE | COMMENTARY

(a) If the bars in the tie are in one layer, the effective tie width can be taken as the diameter of the bars in the tie plus twice the cover to the surface of the bars, as shown in Fig. RA.1.6(a); and

(b) A practical upper limit of the tie width can be taken as the width corresponding to the width in a hydrostatic nodal zone, calculated as

$$w_{t,max} = F_{nt}/(f_{ce}b_s)$$

where f_{ce} is computed for the nodal zone in accordance with A.5.2. If the tie width exceeds the value from (a), the tie reinforcement should be distributed approximately uniformly over the width and thickness of the tie, as shown in Fig. RA.1.6(b).

A.4.3 — Tie reinforcement shall be anchored by mechanical devices, post-tensioning anchorage devices, standard hooks, or straight bar development as required by A.4.3.1 through A.4.3.4.

A.4.3.1 — Nodal zones shall develop the difference between the tie force on one side of the node and the tie force on the other side.

A.4.3.2 — At nodal zones anchoring one tie, the tie force shall be developed at the point where the centroid of the reinforcement in a tie leaves the extended nodal zone and enters the span.

A.4.3.3 — At nodal zones anchoring two or more ties, the tie force in each direction shall be developed at the point where the centroid of the reinforcement in the tie leaves the extended nodal zone.

A.4.3.4 — The transverse reinforcement required by A.3.3 shall be anchored in accordance with 12.13.

RA.4.3 — Anchorage of ties often requires special attention in nodal zones of corbels or in nodal zones adjacent to exterior supports of deep beams. The reinforcement in a tie should be anchored before it leaves the extended nodal zone at the point defined by the intersection of the centroid of the bars in the tie and the extensions of the outlines of either the strut or the bearing area. This length is ℓ_{anc}. In Fig. RA.1.6(a) and (b), this occurs where the outline of the extended nodal zone is crossed by the centroid of the reinforcement in the tie. Some of the anchorage may be achieved by extending the reinforcement through the nodal zone as shown in Fig. RA.1.5(c), and developing it beyond the nodal zone. If the tie is anchored using 90 deg hooks, the hooks should be confined within the reinforcement extending into the beam from the supporting member to avoid cracking along the outside of the hooks in the support region.

In deep beams, hairpin bars spliced with the tie reinforcement can be used to anchor the tension tie forces at exterior supports, provided the beam width is large enough to accommodate such bars.

Figure RA.4.3 shows two ties anchored at a nodal zone. Development is required where the centroid of the tie crosses the outline of the extended nodal zone.

The development length of the tie reinforcement can be reduced through hooks, mechanical devices, additional confinement, or by splicing it with several layers of smaller bars.

A.5 — Strength of nodal zones

A.5.1 — The nominal compression strength of a nodal zone, F_{nn}, shall be

$$F_{nn} = f_{ce}A_{nz} \quad \text{(A-7)}$$

where f_{ce} is the effective compressive strength of the concrete in the nodal zone as given in A.5.2 and A_{nz} is

RA.5 — Strength of nodal zones

RA.5.1 — If the stresses in all the struts meeting at a node are equal, a hydrostatic nodal zone can be used. The faces of such a nodal zone are perpendicular to the axes of the struts, and the widths of the faces of the nodal zone are proportional to the forces in the struts.

CODE

COMMENTARY

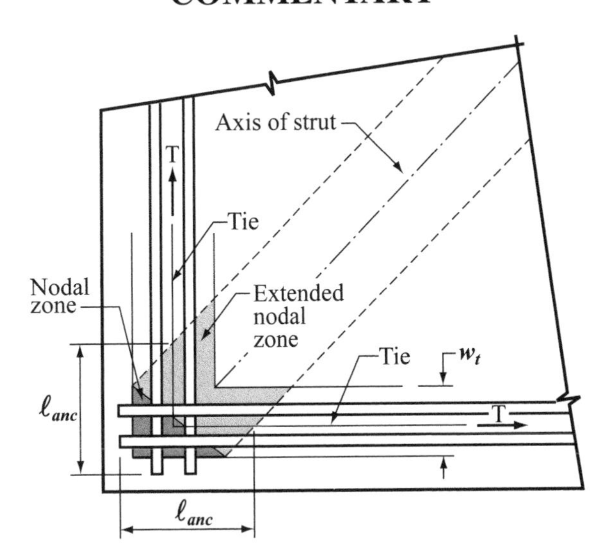

Fig. RA.4.3—Extended nodal zone anchoring two ties.

the smaller of (a) and (b):

(a) the area of the face of the nodal zone on which F_u acts, taken perpendicular to the line of action of F_u;

(b) the area of a section through the nodal zone, taken perpendicular to the line of action of the resultant force on the section.

Assuming the principal stresses in the struts and ties act parallel to the axes of the struts and ties, the stresses on faces perpendicular to these axes are principal stresses, and A5.1(a) is used. If, as shown in Fig. RA.1.6(b), the face of a nodal zone is not perpendicular to the axis of the strut, there will be both shear stresses and normal stresses on the face of the nodal zone. Typically, these stresses are replaced by the normal (principal compression) stress acting on the cross-sectional area A_c of the strut, taken perpendicular to the axis of the strut as given in A.5.1(a).

In some cases, A.5.1(b) requires that the stresses be checked on a section through a subdivided nodal zone. The stresses are checked on the least area section which is perpendicular to a resultant force in the nodal zone. In Fig. RA.1.7(b), the vertical face which divide the nodal zone into two parts is stressed by the resultant force acting along A-B. The design of the nodal zone is governed by the critical section from A.5.1(a) or A.5.1(b), whichever gives the highest stress.

A.5.2 — Unless confining reinforcement is provided within the nodal zone and its effect is supported by tests and analysis, the calculated effective compressive stress, f_{ce}, on a face of a nodal zone due to the strut-and-tie forces shall not exceed the value given by:

$$f_{ce} = 0.85\beta_n f_c' \quad \text{(A-8)}$$

where the value of β_n is given in A.5.2.1 through A.5.2.3.

A.5.2.1 — In nodal zones bounded by struts or bearing areas, or both.................... $\beta_n = 1.0$;

RA.5.2 — The nodes in two-dimensional members, such as deep beams, can be classified as ***C-C-C*** if all the members intersecting at the node are in compression; as ***C-C-T*** nodes if one of the members acting on the node is in tension; and so on, as shown in Fig. RA.1.4. The effective compressive strength of the nodal zone is given by Eq. (A-8), as modified by A.5.2.1 through A.5.2.3 apply to ***C-C-C*** nodes, ***C-C-T*** nodes, and ***C-T-T*** or ***T-T-T*** nodes, respectively.

The β_n values reflect the increasing degree of disruption of the nodal zones due to the incompatibility of tension strains in the ties and compression strains in the struts. The stress on any face of the nodal zone or on any section through the nodal zone should not exceed the value given by Eq. (A-8), as modified by A.5.2.1 through A.5.2.3.

CODE

A.5.2.2 — In nodal zones anchoring one tie $\beta_n = \mathbf{0.80}$;

or

A.5.2.3 — In nodal zones anchoring two or more ties .... $\beta_n = \mathbf{0.60}$.

A.5.3 — In a three-dimensional strut-and-tie model, the area of each face of a nodal zone shall not be less than that given in A.5.1, and the shape of each face of the nodal zones shall be similar to the shape of the projection of the end of the struts onto the corresponding faces of the nodal zones.

COMMENTARY

RA.5.3 — This description of the shape and orientation of the faces of the nodal zones is introduced to simplify the calculations of the geometry of a three-dimensional strut-and-tie model.

APPENDIX B — ALTERNATIVE PROVISIONS FOR REINFORCED AND PRESTRESSED CONCRETE FLEXURAL AND COMPRESSION MEMBERS

CODE

B.1 — Scope

Design for flexure and axial load by provisions of Appendix B shall be permitted. When Appendix B is used in design, B.8.4, B.8.4.1, B.8.4.2, and B.8.4.3 shall replace the corresponding numbered sections in Chapter 8; B.10.3.3 shall replace 10.3.3, 10.3.4, and 10.3.5, except 10.3.5.1 shall remain; B18.1.3, B.18.8.1, B.18.8.2, and B.18.8.3 shall replace the corresponding numbered sections in Chapter 18; B.18.10.4, B.18.10.4.1, B.18.10.4.2, and B.18.10.4.3 shall replace 18.10.4, 18.10.4.1, and 18.10.4.2. If any section in this appendix is used, all sections in this appendix shall be substituted in the body of the code, and all other sections in the body of the code shall be applicable.

B.8.4 — Redistribution of negative moments in continuous nonprestressed flexural members

For criteria on moment redistribution for prestressed concrete members, see B.18.10.4.

B.8.4.1 — Except where approximate values for moments are used, it shall be permitted to increase or decrease negative moments calculated by elastic theory at supports of continuous flexural members for any assumed loading arrangement by not more than

$$20\left(1-\frac{\rho-\rho'}{\rho_b}\right) \text{ percent}$$

B.8.4.2 — The modified negative moments shall be used for calculating moments at sections within the spans

B.8.4.3 — Redistribution of negative moments shall be made only when the section at which moment is reduced is so designed that ρ or $\rho - \rho'$ is not greater than $0.50\rho_b$, where

$$\rho_b = \frac{0.85\beta_1 f_c'}{f_y}\left(\frac{87{,}000}{87{,}000 + f_y}\right) \quad \text{(B-1)}$$

COMMENTARY

RB.1 — Scope

Reinforcement limits, strength reduction factors, ϕ, and moment redistribution in Appendix B differ from those in the main body of the code. Appendix B contains the reinforcement limits, strength reduction factors, ϕ, and moment redistribution used in the code for many years. Designs using the provisions of Appendix B satisfy the code, and are equally acceptable.

When this appendix is used, the corresponding commentary sections apply. The load factors and strength reduction factors of either Chapter 9 or Appendix C are applicable.

RB.8.4 — Redistribution of negative moments in continuous nonprestressed flexural members

Moment redistribution is dependent on adequate ductility in plastic hinge regions. These plastic hinge regions develop at points of maximum moment and cause a shift in the elastic moment diagram. The usual results are reduction in the values of negative moments in the plastic hinge region and an increase in the values of positive moments from those computed by elastic analysis. Since negative moments are determined for one loading arrangement and positive moments for another, each section has a reserve capacity that is not fully utilized for any one loading condition. The plastic hinges permit the utilization of the full capacity of more cross sections of a flexural member at ultimate loads. Using conservative values of ultimate concrete strains and lengths of plastic hinges derived from extensive tests, flexural members with small rotation capacity were analyzed for moment redistribution up to 20 percent, depending on the reinforcement ratio. The results were found to be conservative (see Fig. RB.8.4). Studies by Cohn[B.1] and Mattock[B.2] support this conclusion and indicate that cracking and deflection of beams designed for moment redistribution are not significantly greater at service loads than for beams designed by the elastic theory distribution of moments. Also, these studies indicated that adequate rotation capacity for the moment redistribution allowed by the code is available if the members satisfy the code requirements. This appendix maintains the same limit on redistribution as used in previous code editions.

Moment redistribution may not be used for slab systems designed by the Direct Design Method (see 13.6.1.7).

CODE

COMMENTARY

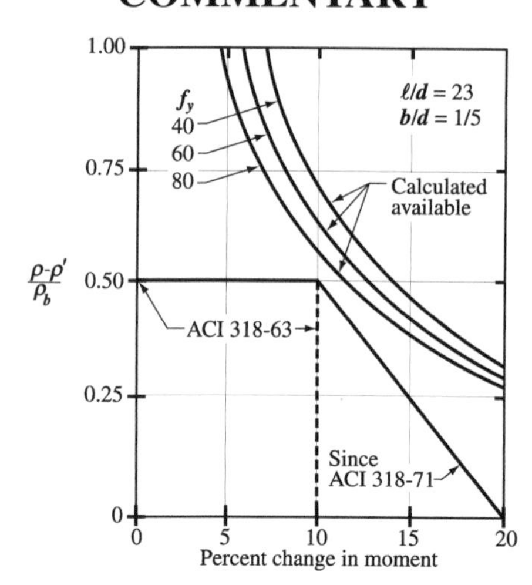

Fig. RB.8.4 — Permissible moment redistribution for minimum rotation capacity.

B.10.3 — General principles and requirements

B.10.3.3 — For flexural members and members subject to combined flexure and compressive axial load where ϕP_n is less than the smaller of $\mathbf{0.10 f_c' A_g}$ and ϕP_b, the ratio of reinforcement, ρ, provided shall not exceed 0.75 of the ratio ρ_b that would produce balanced strain conditions for the section under flexure without axial load. For members with compression reinforcement, the portion of ρ_b equalized by compression reinforcement need not be reduced by the 0.75 factor.

RB.10.3 — General principles and requirements

RB.10.3.3 — The maximum amount of tension reinforcement in flexural members is limited to ensure a level of ductile behavior.

The nominal flexural strength of a section is reached when the strain in the extreme compression fiber reaches the limiting strain in the concrete. At ultimate strain of the concrete, the strain in the tension reinforcement could just reach the strain at first yield, be less than the yield strain (elastic), or exceed the yield strain (inelastic). The steel strain that exists at limiting concrete strain depends on the relative proportion of steel to concrete and material strengths f_c' and f_y. If $\rho(f_y/f_c')$ is sufficiently low, the strain in the tension steel will greatly exceed the yield strain when the concrete strain reaches its limiting value, with large deflection and ample warning of impending failure (ductile failure condition). With a larger $\rho(f_y/f_c')$, the strain in the tension steel may not reach the yield strain when the concrete strain reaches its limiting value, with consequent small deflection and little warning of impending failure (brittle failure condition). For design it is considered more conservative to restrict the nominal strength condition so that a ductile failure mode can be expected.

Unless unusual amounts of ductility are required, the $\mathbf{0.75}\rho_b$ limitation will provide ductile behavior for most designs. One condition where greater ductile behavior is required is in design for redistribution of moments in continuous members and frames. Section B.8.4 permits negative moment redistribution. Since moment redistribution is dependent on adequate ductility in hinge regions, the amount of tension reinforcement in hinging regions is limited to $\mathbf{0.5}\rho_b$.

For ductile behavior of beams with compression reinforcement, only that portion of the total tension steel balanced by compression in the concrete need be limited; that portion of the total tension steel where force is balanced by compression reinforcement need not be limited by the 0.75 factor.

CODE

B.18.1 — Scope

B.18.1.3 — The following provisions of this code shall not apply to prestressed concrete, except as specifically noted: Sections 6.4.4, 7.6.5, B.8.4, 8.10.2, 8.10.3, 8.10.4, 8.11, B.10.3.3, 10.5, 10.6, 10.9.1, and 10.9.2; Chapter 13; and Sections 14.3, 14.5, and 14.6.

COMMENTARY

RB.18.1 — Scope

RB.18.1.3 — Some sections of the code are excluded from use in the design of prestressed concrete for specific reasons. The following discussion provides an explanation for such exclusions:

Section 6.4.4 — Tendons of continuous post-tensioned beams and slabs are usually stressed at a point along the span where the tendon profile is at or near the centroid of the concrete cross section. Therefore, interior construction joints are usually located within the end thirds of the span, rather than the middle third of the span as required by 6.4.4. Construction joints located as described in continuous post-tensioned beams and slabs have a long history of satisfactory performance. Thus, 6.4.4 is excluded from application to prestressed concrete.

Section 7.6.5 — Section 7.6.5 is excluded from application to prestressed concrete since the requirements for bonded reinforcement and unbonded tendons for cast-in-place members are provided in 18.9 and 18.12, respectively.

Section B.8.4 — Moment redistribution for prestressed concrete is provided in B.18.10.4.

Sections 8.10.2, 8.10.3, and 8.10.4 — The empirical provisions of 8.10.2, 8.10.3, and 8.10.4 for T-beams were developed for conventionally reinforced concrete and, if applied to prestressed concrete, would exclude many standard prestressed products in satisfactory use today. Hence, proof by experience permits variations.

By excluding 8.10.2, 8.10.3, and 8.10.4, no special requirements for prestressed concrete T-beams appear in the code. Instead, the determination of an effective width of flange is left to the experience and judgment of the engineer. Where possible, the flange widths in 8.10.2, 8.10.3, and 8.10.4 should be used unless experience has proven that variations are safe and satisfactory. It is not necessarily conservative in elastic analysis and design considerations to use the maximum flange width as permitted in 8.10.2.

Sections 8.10.1 and 8.10.5 provide general requirements for T-beams that are also applicable to prestressed concrete units. The spacing limitations for slab reinforcement are based on flange thickness, which for tapered flanges can be taken as the average thickness.

Section 8.11 — The empirical limits established for conventionally reinforced concrete joist floors are based on successful past performance of joist construction using "standard" joist forming systems. See R8.11. For prestressed joist construction, experience and judgment should be used. The provisions of 8.11 may be used as a guide.

CODE

COMMENTARY

Sections B.10.3.3, 10.5, 10.9.1, and 10.9.2 — For prestressed concrete, the limitations on reinforcement given in B.10.3.3, 10.5, 10.9.1, and 10.9.2 are replaced by those in B.18.8, 18.9, and 18.11.2.

Section 10.6 — When originally prepared, the provisions of 10.6 for distribution of flexural reinforcement were not intended for prestressed concrete members. The behavior of a prestressed member is considerably different from that a nonprestressed member. Experience and judgment should be used for proper distribution of reinforcement in a prestressed member.

Chapter 13 — The design of prestressed concrete slabs requires recognition of secondary moments induced by the undulating profile of the prestressing tendons. Also, volume changes due to the prestressing force can create additional loads on the structure that are not adequately covered in Chapter 13. Because of these unique properties associated with prestressing, many of the design procedures of Chapter 13 are not appropriate for prestressed concrete structures and are replaced by the provisions of 18.12.

Sections 14.5 and 14.6 — The requirements for wall design in 14.5 and 14.6 are largely empirical, utilizing considerations not intended to apply to prestressed concrete.

B.18.8 — Limits for reinforcement of flexural members

B.18.8.1 — Ratio of prestressed and nonprestressed reinforcement used for computation of moment strength of a member, except as provided in B.18.8.2, shall be such that ω_p, $[\omega_p + (d/d_p)(\omega - \omega')]$, or $[\omega_{pw} + (d/d_p)(\omega_w - \omega_w')]$ is not greater than $0.36\beta_1$, except as permitted in B.18.8.2.

Ratio ω_p is computed as $\rho_p f_{ps}/f_c'$. Ratios ω_w and ω_{pw} are computed as ω and ω_p, respectively, except that when computing ρ and ρ_p, b_w shall be used in place of b and the area of reinforcement or prestressing steel required to develop the compressive strength of the web only shall be used in place of A_s or A_{ps}. Ratio ω_w' is computed as ω', except that when computing ρ', b_w shall be used in place of b.

B.18.8.2 — When a reinforcement ratio exceeds the limit specified in B.18.8.1 is provided, design moment strength shall not exceed the moment strength based on the compression portion of the moment couple.

RB.18.8 — Limits for reinforcement of flexural members

RB.18.8.1 — The terms ω_p, $[\omega_p + (d/d_p)(\omega - \omega')]$ and $[\omega_{pw} + (d/d_p)(\omega_w - \omega_w')]$ are each equal to $0.85a/d_p$, where a is the depth of the equivalent rectangular stress block for the section under consideration, as defined in 10.2.7.1. Use of this relationship can simplify the calculations necessary to check compliance with RB.18.8.1.

RB.18.8.2 — Design moment strength of over-reinforced sections may be computed using strength equations similar to those for nonprestressed concrete members. The 1983 code provided strength equations for rectangular and flanged sections.

CODE

B.18.8.3 — Total amount of prestressed and nonprestressed reinforcement shall be adequate to develop a factored load at least 1.2 times the cracking load computed on the basis of the modulus of rupture f_r in 9.5.2.3. This provision shall be permitted to be waived for:

(a) two-way, unbonded post-tensioned slabs; and

(b) flexural members with shear and flexural strength at least twice that required by 9.2.

COMMENTARY

RB.18.8.3 — This provision is a precaution against abrupt flexural failure developing immediately after cracking. A flexural member designed according to code provisions requires considerable additional load beyond cracking to reach its flexural strength. This additional load should result in considerable deflection that would warn when the member nominal strength is being approached. If the flexural strength is reached shortly after cracking, the warning deflection would not occur.

Due to the very limited extent of initial cracking in the negative moment region near columns of two-way flat plates, deflection under load does not reflect any abrupt change in stiffness as the modulus of rupture of concrete is reached.

Only at load levels beyond the factored loads is the additional cracking extensive enough to cause an abrupt change in the deflection under load. Tests have shown that it is not possible to rupture (or even yield) unbonded post-tensioning tendons in two-way slabs before a punching shear failure.[B.3-B.8] The use of unbonded tendons in combination with the minimum bonded reinforcement requirements of 18.9.3 and 18.9.4 has been shown to ensure post-cracking ductility and that a brittle failure mode will not develop at first cracking.

B.18.10 — Statically indeterminate structures

RB.18.10 — Statically indeterminate structures

B.18.10.1 — Frames and continuous construction of prestressed concrete shall be designed for satisfactory performance at service load conditions and for adequate strength.

B.18.10.2 — Performance at service load conditions shall be determined by elastic analysis, considering reactions, moments, shears, and axial forces produced by prestressing, creep, shrinkage, temperature change, axial deformation, restraint of attached structural elements, and foundation settlement.

B.18.10.3 — Moments to be used to compute required strength shall be the sum of the moments due to reactions induced by prestressing (with a load factor of 1.0) and the moments due to factored loads. Adjustment of the sum of these moments shall be permitted as allowed in B.18.10.4.

RB.18.10.3 — For statically indeterminate structures, the moments due to reactions induced by prestressing forces, referred to as secondary moments, are significant in both the elastic and inelastic states. When hinges and full redistribution of moments occur to create a statically determinate structure, secondary moments disappear. However, the elastic deformations caused by a nonconcordant tendon change the amount of inelastic rotation required to obtain a given amount of moment redistribution. Conversely, for a beam with a given inelastic rotation capacity, the amount by which the moment at the support may be varied is changed by an amount equal to the secondary moment at the support due to prestressing. Thus, the code requires that secondary moments be included in determining design moments.

CODE

COMMENTARY

To determine the moments used in design, the order of calculation should be: (a) determine moments due to dead load and live load; (b) modify by algebraic addition of secondary moments; and (c) redistribute as permitted. A positive secondary moment at the support caused by a tendon transformed downward from a concordant profile will reduce the negative moments near the supports and increase the positive moments in the midspan regions. A tendon that is transformed upward will have the reverse effect.

B.18.10.4 — Redistribution of negative moments in continuous prestressed flexural members

B.18.10.4.1 — Where bonded reinforcement is provided at supports in accordance with 18.9, negative moments calculated by elastic theory for any assumed loading, arrangement shall be permitted to be increased or decreased by not more than

$$20\left[1-\frac{\omega_p+\frac{d}{d_p}(\omega-\omega')}{0.36\beta_1}\right]\text{ percent}$$

B.18.10.4.2 — The modified negative moments shall be used for calculating moments at sections within spans for the same loading arrangement.

B.18.10.4.3 — Redistribution of negative moments shall be made only when the section at which moment is reduced is so designed that ω_p, $[\omega_p + (d/d_p)(\omega - \omega')]$ or $[\omega_{pw} + (d/d_p)(\omega_w - \omega_w')]$, whichever is applicable, is not greater than $0.24\beta_1$.

RB.18.10.4 — Redistribution of negative moments in continuous prestressed flexural members

As member strength is approached, inelastic behavior at some sections can result in a redistribution of moments in prestressed concrete beams and slabs. Recognition of this behavior can be advantageous in design under certain circumstances. A rigorous design method for moment redistribution is complex. However, recognition of moment redistribution can be accomplished by permitting a reasonable adjustment of the sum of the elastically calculated factored gravity load moments and the unfactored secondary moments due to prestress. The amount of adjustment should be kept within predetermined safety limits.

The amount of redistribution allowed depends on the ability of the critical sections to deform inelastically by a sufficient amount. Serviceability is addressed in 18.4. The choice of $0.24\beta_1$ as the largest tension reinforcement index, ω_p, $[\omega_p + (d/d_p)(\omega - \omega')]$ or $[\omega_{pw} + (d/d_p)(\omega_w - \omega_w')]$, for which redistribution of moments is allowed, is in agreement with the requirements for nonprestressed concrete of $0.5\rho_b$ stated in B.8.4.

The terms ω_p, $[\omega_p + (d/d_p)(\omega - \omega')]$ and $[\omega_{pw} + (d/d_p)(\omega_w - \omega_w')]$ appear in B.18.10.4.1 and B.18.10.4.3 and are each equal to $0.85a/d_p$, where a is the depth of the equivalent rectangular stress distribution for the section under consideration, as defined in 10.2.7.1. Use of this relationship can simplify the calculations necessary to determine the amount of moment redistribution permitted by B.18.10.4.1 and to check compliance with the limitation on flexural reinforcement contained in B.18.10.4.3.

For the moment redistribution principles of B.18.10.4 to be applicable to beams and slabs with unbonded tendons, it is necessary that such beams and slabs contain sufficient bonded reinforcement to ensure that they act as flexural members after cracking and not as a series of tied arches. The minimum bonded reinforcement requirements of 18.9 serve this purpose.

APPENDIX C — ALTERNATIVE LOAD AND STRENGTH REDUCTION FACTORS

CODE

C.1 — General

C.1.1 — Structural concrete shall be permitted to be designed using the load combinations and strength reduction factors of Appendix C.

C.2 — Required strength

C.2.1 — Required strength U to resist dead load D and live load L shall not be less than

$$U = 1.4D + 1.7L \quad \text{(C-1)}$$

C.2.2 — For structures that also resist W, wind load, or E, the load effects of earthquake, U shall not be less than the larger of Eq. (C-1), (C-2), and (C-3):

$$U = 0.75(1.4D + 1.7L) + (1.6W \text{ or } 1.0E) \quad \text{(C-2)}$$

and

$$U = 0.9D + (1.6W \text{ or } 1.0E) \quad \text{(C-3)}$$

where W has not been reduced by a directionality factor, it shall be permitted to use $1.3W$ in place of $1.6W$ in Eq. (C-2) and (C-3). Where E is based on service-level seismic forces, $1.4E$ shall be used in place of $1.0E$ in Eq. (C-2) and (C-3).

C.2.3 — For structures that resist H, loads due to weight and pressure of soil, water in soil, or other related materials, U shall not be less than the larger of Eq. (C-1) and (C-4):

$$U = 1.4D + 1.7L + 1.7H \quad \text{(C-4)}$$

In Eq. (C-4), where D or L reduce the effect of H, $0.9D$ shall be substituted for $1.4D$, and zero value of L shall be used to determine the greatest required strength U.

C.2.4 — For structures that resist F, load due to weight and pressure of fluids with well-defined densities, the load factor for F shall be1.4, and F shall be added to all loading combinations that include L.

COMMENTARY

RC.1 — General

RC.1.1 — In the 2002 code, the load and strength reduction factors formerly in Chapter 9 were revised and moved to this appendix. They have evolved since the early 1960s and are considered to be reliable for concrete construction.

RC.2 — Required strength

The wind load equation in ASCE 7-98 and IBC 2000[C.1] includes a factor for wind directionality that is equal to 0.85 for buildings. The corresponding load factor for wind in the load combination equations was increased accordingly (1.3/0.85 = 1.53, rounded up to 1.6). The code allows use of the previous wind load factor of 1.3 when the design wind load is obtained from other sources that do not include the wind directionality factor.

Model building codes and design load references have converted earthquake forces to strength level, and reduced the earthquake load factor to 1.0 (ASCE 7-93[C.2]; BOCA/NBC 93[C.3]; SBC 94[C.4]; UBC 97[C.5]; and IBC 2000[C.1]). The code requires use of the previous load factor for earthquake loads, approximately 1.4, when service-level earthquake forces from earlier editions of these references are used.

RC.2.3 — If effects H caused by earth pressure, groundwater pressure, or pressure caused by granular materials are included in design, the required strength equations become:

$$U = 1.4D + 1.7L + 1.7H$$

and where D or L reduce the effect of H

$$U = 0.9D + 1.7H$$

but for any combination of D, L, or H

$$U = 1.4D + 1.7L$$

RC.2.4 — This section addresses the need to consider loading due to weight of liquid or liquid pressure. It specifies a load factor for such loadings with well-defined densities and controllable maximum heights equivalent to that used for dead load. Such reduced factors would not be appropriate where there is considerable uncertainty of pressures, as with groundwater pressures, or uncertainty as to the possible max-

CODE / COMMENTARY

imum liquid depth, as in ponding of water. See R8.2.

For well-defined fluid pressures, the required strength equations become:

$$U = 1.4D + 1.7L + 1.4F$$

and where D or L reduce the effect of F

$$U = 0.9D + 1.4F$$

but for any combination of D, L, or F

$$U = 1.4D + 1.7L$$

C.2.5 — If resistance to impact effects is taken into account in design, such effects shall be included with L.

RC.2.5 — If the live load is applied rapidly, as may be the case for parking structures, loading docks, warehouse floors, elevator shafts, etc., impact effects should be considered. In all equations, substitute (L + impact) for L when impact must be considered.

C.2.6 — Where structural effects of differential settlement, creep, shrinkage, expansion of shrinkage-compensating concrete, or temperature change, T, are significant, U shall not be less than the larger of Eq. (C-5) and (C-6):

$$U = 0.75(1.4D + 1.4T + 1.7L) \quad \text{(C-5)}$$

$$U = 1.4(D + T) \quad \text{(C-6)}$$

Estimations of differential settlement, creep, shrinkage, expansion of shrinkage-compensating concrete, or temperature change shall be based on realistic assessment of such effects occurring in service.

RC.2.6 — The designer should consider the effects of differential settlement, creep, shrinkage, temperature, and shrinkage-compensating concrete. The term "realistic assessment" is used to indicate that the most probable values, rather than the upper bound values, of the variables should be used.

Equation (C-6) is to prevent a design for load

$$U = 0.75\,(1.4D + 1.4T + 1.7L)$$

to approach

$$U = 1.05\,(D + T)$$

when live load is negligible.

C.2.7 — For post-tensioned anchorage zone design, a load factor of 1.2 shall be applied to the maximum prestressing steel jacking force.

RC.2.7 — The load factor of 1.2 applied to the maximum prestressing steel jacking force results in a design load of 113 percent of the specified yield strength of prestressing steel but not more than 96 percent of the nominal ultimate strength of the tendon. This compares well with a maximum attainable jacking force, which is limited by the anchor efficiency factor.

C.3 — Design strength

C.3.1 — Design strength provided by a member, its connections to other members, and its cross sections, in terms of flexure, axial load, shear, and torsion, shall be taken as the nominal strength calculated in accordance with requirements and assumptions of this code, multiplied by the ϕ factors in C.3.2, C.3.4, and C.3.5.

RC.3 — Design strength

RC.3.1 — The term "design strength" of a member, refers to the nominal strength calculated in accordance with the requirements stipulated in this code multiplied by a strength reduction factor ϕ that is always less than one.

The purposes of the strength reduction factor ϕ are: (1) to allow for the probability of understrength members due to variations in material strengths and dimensions; (2) to allow

CODE

COMMENTARY

for inaccuracies in the design equations; (3) to reflect the degree of ductility and required reliability of the member under the load effects being considered; and (4) to reflect the importance of the member in the structure. For example, a lower ϕ is used for columns than for beams because columns generally have less ductility, are more sensitive to variations in concrete strength, and generally support larger loaded areas than beams. Furthermore, spiral columns are assigned a higher ϕ than tied columns since they have greater ductility or toughness.

C.3.2 — Strength reduction factor ϕ shall be as follows:

C.3.2.1 — Tension-controlled sections, as defined in 10.3.4 (See also C.3.2.7) 0.90

RC.3.2.1 — In applying C.3.2.1 and C.3.2.2, the axial tensions and compressions to be considered are those caused by external forces. Effects of prestressing forces are not included.

C.3.2.2 — Compression-controlled sections, as defined in 10.3.3:

(a) Members with spiral reinforcement conforming to 10.9.3.. 0.75

(b) Other reinforced members 0.70

For sections in which the net tensile strain in the extreme tension steel at nominal strength, ε_t, is between the limits for compression-controlled and tension-controlled sections, ϕ shall be permitted to be linearly increased from that for compression-controlled sections to 0.90 as ε_t increases from the compression-controlled strain limit to 0.005.

Alternatively, when Appendix B is used, for members in which f_y does not exceed 60,000 psi, with symmetric reinforcement, and with $(d - d')/h$ not less than 0.70, ϕ shall be permitted to be increased linearly to 0.90 as ϕP_n decreases from $0.10f_c'A_g$ to zero. For other reinforced members, ϕ shall be permitted to be increased linearly to 0.90 as ϕP_n decreases from $0.10f_c'A_g$ or ϕP_b, whichever is smaller, to zero.

RC.3.2.2 — Before the 2002 edition, the code gave the magnitude of the ϕ-factor for cases of axial load or flexure, or both, in terms of the type of loading. For these cases, the ϕ-factor is now determined by the strain conditions at a cross section, at nominal strength.

A lower ϕ-factor is used for compression-controlled sections than is used for tension-controlled sections because compression-controlled sections have less ductility, are more sensitive to variations in concrete strength, and generally occur in members that support larger loaded areas than members with tension-controlled sections. Members with spiral reinforcement are assigned a higher ϕ than tied columns since they have greater ductility or toughness.

For sections subjected to axial load with flexure, design strengths are determined by multiplying both P_n and M_n by the appropriate single value of ϕ. Compression-controlled and tension-controlled sections are defined in 10.3.3 and 10.3.4 as those that have net tensile strain in the extreme tension steel at nominal strength less than or equal to the compression-controlled strain limit, and equal to or greater than 0.005, respectively. For sections with net tensile strain ε_t in the extreme tension steel at nominal strength between the above limits, the value of ϕ may be determined by linear interpolation, as shown in Fig. RC.3.2. The concept of net tensile strain ε_t is discussed in R10.3.3.

Since the compressive strain in the concrete at nominal strength is assumed in 10.2.3 to be 0.003, the net tensile strain limits for compression-controlled members may also be stated in terms of the ratio c/d_t, where c is the distance from the extreme compression fiber to the neutral axis at nominal strength, and d_t is the distance from the extreme compression fiber to the centroid of the extreme layer of longitudinal tension steel. The c/d_t limits for compression-controlled and tension-controlled sections are 0.6 and 0.375,

CODE

COMMENTARY

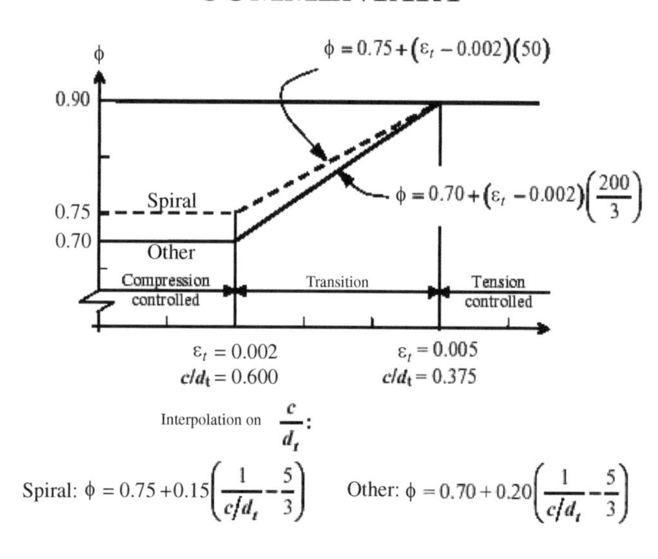

Fig. RC.3.2—Variation of ϕ with net tensile strain ε_t and c/d_t for Grade 60 reinforcement and for prestressing steel.

respectively. The 0.6 limit applies to sections reinforced with Grade 60 steel and to prestressed sections. Figure RC.3.2 also gives equations for ϕ as a function of c/d_t.

The net tensile strain limit for tension-controlled sections may also be stated in terms of the ρ/ρ_b as defined in the 1999 and earlier editions of the code. The net tensile strain limit of 0.005 corresponds to a ρ/ρ_b ratio of 0.63 for rectangular sections with Grade 60 reinforcement. For a comparison of these provisions with those of the body of the code, the 1999 ACI Code Section 9.3, see Reference C.6.

C.3.2.3 — Shear and torsion 0.85

C.3.2.4 — Bearing on concrete (except for post-tensioned anchorage zones and strut-and-tie models).. 0.70

C.3.2.5 — Post-tensioned anchorage zones 0.85

RC.3.2.5 — The ϕ-factor of 0.85 reflects the wide scatter of results of experimental anchorage zone studies. Since 18.13.4.2 limits the nominal compressive strength of unconfined concrete in the general zone to $0.7\lambda f'_{ci}$, the effective design strength for unconfined concrete is $0.85 \times 0.7\lambda f'_{ci} \approx 0.6\lambda f'_{ci}$.

C.3.2.6 — Strut-and-tie models (Appendix A), and struts, ties, nodal zones, and bearing areas in such models.. 0.85

C.3.2.7 — Flexure sections without axial load in pretensioned members where strand embedment is less than the development length as provided in 12.9.1.1 .... 0.85

RC.3.2.7 — If a critical section occurs in a region where strand is not fully developed, failure may be by bond slip. Such a failure resembles a brittle shear failure; hence the requirement for a reduced ϕ.

C.3.3 — Development lengths specified in Chapter 12 do not require a ϕ-factor.

CODE

C.3.4 — For structures that rely on special moment resisting frames or special reinforced concrete structural walls to resist ***E***, ϕ shall be modified as given in (a) through (c):

(a) For any structural member that is designed to resist ***E***, ϕ for shear shall be 0.60 if the nominal shear strength of the member is less than the shear corresponding to the development of the nominal flexural strength of the member. The nominal flexural strength shall be determined considering the most critical factored axial loads and including ***E***;

(b) For diaphragms, ϕ for shear shall not exceed the minimum ϕ for shear used for the vertical components of the primary lateral-force-resisting system;

(c) For joints and diagonally reinforced coupling beams, ϕ for shear shall be 0.85.

C.3.5 — In Chapter 22, ϕ shall be 0.65 for flexure, compression, shear, and bearing of structural plain concrete.

COMMENTARY

RC.3.4 — Strength-reduction factors in C.3.4 are intended to compensate for uncertainties in estimation of strength of structural members in buildings. They are based primarily on experience with constant or steadily increasing applied load. For construction in regions of high seismic risk, some of the strength reduction factors have been modified in C.3.4 to account for the effects of displacements on strength into the nonlinear range of response on strength.

Section C.3.4(a) refers to brittle members, such as low-rise walls or portions of walls between openings, or diaphragms that are impractical to reinforce to raise their nominal shear strength above nominal flexural strength for the pertinent loading conditions.

Short structural walls were the primary vertical elements of the lateral-force-resisting system in many of the parking structures that sustained damage during the 1994 Northridge earthquake. Section C.3.4(b) requires the shear strength reduction factor for diaphragms to be 0.60 if the shear strength reduction factor for the walls is 0.60.

RC.3.5 — The strength-reduction factor ϕ for structural plain concrete design is the same for all strength conditions. Since both flexural tension strength and shear strength for plain concrete depend on the tensile strength characteristics of the concrete, with no reserve strength or ductility possible due to the absence of reinforcement, equal strength reduction factors for both bending and shear are considered appropriate.

Notes

APPENDIX D — ANCHORING TO CONCRETE

CODE

D.1 — Definitions

Anchor — A steel element either cast into concrete or post-installed into a hardened concrete member and used to transmit applied loads, including headed bolts, hooked bolts (J- or L-bolt), headed studs, expansion anchors, or undercut anchors.

Anchor group — A number of anchors of approximately equal effective embedment depth with each anchor spaced at less than three times its embedment depth from one or more adjacent anchors.

Anchor pullout strength — The strength corresponding to the anchoring device or a major component of the device sliding out from the concrete without breaking out a substantial portion of the surrounding concrete.

Attachment — The structural assembly, external to the surface of the concrete, that transmits loads to or receives loads from the anchor.

Brittle steel element — An element with a tensile test elongation of less than 14 percent, or reduction in area of less than 30 percent, or both.

Cast-in anchor — A headed bolt, headed stud, or hooked bolt installed before placing concrete.

Concrete breakout strength — The strength corresponding to a volume of concrete surrounding the anchor or group of anchors separating from the member.

Concrete pryout strength — The strength corresponding to formation of a concrete spall behind short, stiff anchors displaced in the direction opposite to the applied shear force.

Distance sleeve — A sleeve that encases the center part of an undercut anchor, a torque-controlled expansion anchor, or a displacement-controlled expansion anchor, but does not expand.

Ductile steel element — An element with a tensile test elongation of at least 14 percent and reduction in area of at least 30 percent. A steel element meeting the requirements of ASTM A 307 shall be considered ductile.

COMMENTARY

RD.1 — Definitions

Brittle steel element and ductile steel element The 14 percent elongation should be measured over the gage length specified in the appropriate ASTM standard for the steel.

CODE

Edge distance — The distance from the edge of the concrete surface to the center of the nearest anchor.

Effective embedment depth — The overall depth through which the anchor transfers force to or from the surrounding concrete. The effective embedment depth will normally be the depth of the concrete failure surface in tension applications. For cast-in headed anchor bolts and headed studs, the effective embedment depth is measured from the bearing contact surface of the head.

Expansion anchor — A post-installed anchor, inserted into hardened concrete that transfers loads to or from the concrete by direct bearing or friction or both. Expansion anchors may be torque-controlled, where the expansion is achieved by a torque acting on the screw or bolt; or displacement-controlled, where the expansion is achieved by impact forces acting on a sleeve or plug and the expansion is controlled by the length of travel of the sleeve or plug.

Expansion sleeve — The outer part of an expansion anchor that is forced outward by the center part, either by applied torque or impact, to bear against the sides of the predrilled hole.

Five percent fractile — A statistical term meaning 90 percent confidence that there is 95 percent probability of the actual strength exceeding the nominal strength.

Hooked bolt — A cast-in anchor anchored mainly by mechanical interlock from the 90-deg bend (L-bolt) or 180-deg bend (J-bolt) at its lower end, having a minimum e_h of $3d_o$.

Headed stud — A steel anchor conforming to the requirements of AWS D1.1 and affixed to a plate or similar steel attachment by the stud arc welding process before casting.

Post-installed anchor — An anchor installed in hardened concrete. Expansion anchors and undercut anchors are examples of post-installed anchors.

Projected area — The area on the free surface of the concrete member that is used to represent the larger base of the assumed rectilinear failure surface.

COMMENTARY

Effective embedment depths for a variety of anchor types are shown in Fig. RD.1.

Five percent fractile — The determination of the coefficient K_{05} associated with the 5 percent fractile, $\bar{x} - K_{05}s_s$, depends on the number of tests, n, used to compute the sample mean, $\bar{x}$, and sample standard deviation, s_s. Values of K_{05} range, for example, from 1.645 for $n = \infty$, to 2.010 for $n = 40$, and 2.568 for $n = 10$. With this definition of the 5 percent fractile, the nominal strength in D.4.2 is the same as the characteristic strength in ACI 355.2.

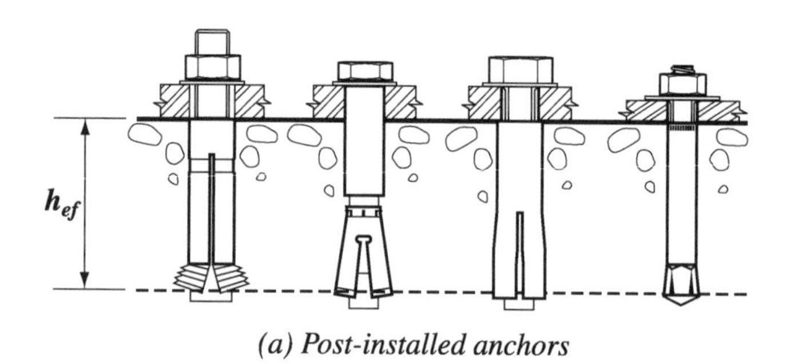

(a) Post-installed anchors

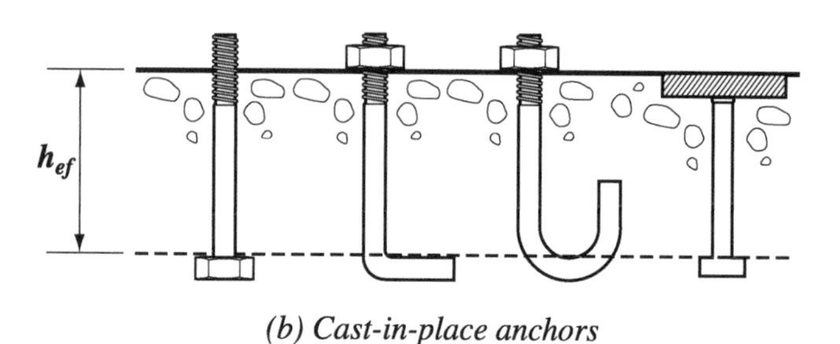

(b) Cast-in-place anchors

Fig. RD.1—Types of anchors.

CODE

Side-face blowout strength — The strength of anchors with deeper embedment but thinner side cover corresponding to concrete spalling on the side face around the embedded head while no major breakout occurs at the top concrete surface.

Specialty insert — Predesigned and prefabricated cast-in anchors specifically designed for attachment of bolted or slotted connections. Specialty inserts are often used for handling, transportation, and erection, but are also used for anchoring structural elements. Specialty inserts are not within the scope of this appendix.

Supplementary reinforcement — Reinforcement proportioned to tie a potential concrete failure prism to the structural member.

Undercut anchor — A post-installed anchor that develops its tensile strength from the mechanical interlock provided by undercutting of the concrete at the embedded end of the anchor. The undercutting is achieved with a special drill before installing the anchor or alternatively by the anchor itself during its installation.

D.2 — Scope

D.2.1 — This appendix provides design requirements for anchors in concrete used to transmit structural loads by means of tension, shear, or a combination of tension and shear between (a) connected structural elements; or (b) safety-related attachments and structural elements. Safety levels specified are intended for in-service conditions, rather than for short-term handling and construction conditions.

D.2.2 — This appendix applies to both cast-in anchors and post-installed anchors. Specialty inserts, through bolts, multiple anchors connected to a single steel plate at the embedded end of the anchors, adhesive or grouted anchors, and direct anchors such as powder or pneumatic actuated nails or bolts, are not included. Reinforcement used as part of the embedment shall be designed in accordance with other parts of this code.

D.2.3 — Headed studs and headed bolts having a geometry that has been demonstrated to result in a pullout strength in uncracked concrete equal or exceeding $\mathbf{1.4}\boldsymbol{N_p}$ (where $\boldsymbol{N_p}$ is given by Eq. (D-15)) are included. Hooked bolts that have a geometry that has been demonstrated to result in a pullout strength without the benefit of friction in uncracked concrete equal

COMMENTARY

RD.2 — Scope

RD.2.1 — Appendix D is restricted in scope to structural anchors that transmit structural loads related to strength, stability, or life safety. Two types of applications are envisioned. The first is connections between structural elements where the failure of an anchor or an anchor group could result in loss of equilibrium or stability of any portion of the structure. The second is where safety-related attachments that are not part of the structure (such as sprinkler systems, heavy suspended pipes, or barrier rails) are attached to structural elements. The levels of safety defined by the combinations of load factors and ϕ factors are appropriate for structural applications. Other standards may require more stringent safety levels during temporary handling.

RD.2.2 — The wide variety of shapes and configurations of specialty inserts makes it difficult to prescribe generalized tests and design equations for many insert types. Hence, they have been excluded from the scope of Appendix D. Adhesive anchors are widely used and can perform adequately. At this time, however, such anchors are outside the scope of this appendix.

RD.2.3 — Typical cast-in headed studs and headed bolts with geometries consistent with ANSI/ASME B1.1,[D.1] B18.2.1,[D.2] and B18.2.6[D.3] have been tested and proven to behave predictably, so calculated pullout values are acceptable. Post-installed anchors do not have predictable pullout capacities, and therefore are required to be tested. For a post-installed anchor to be used in conjunction with the

CODE

or exceeding $1.4N_p$ (where N_p is given by Eq. (D-16)) are included. Post-installed anchors that meet the assessment requirements of ACI 355.2 are included. The suitability of the post-installed anchor for use in concrete shall have been demonstrated by the ACI 355.2 prequalification tests.

D.2.4 — Load applications that are predominantly high cycle fatigue or impact loads are not covered by this appendix.

D.3 — General requirements

D.3.1 — Anchors and anchor groups shall be designed for critical effects of factored loads as determined by elastic analysis. Plastic analysis approaches are permitted where nominal strength is controlled by ductile steel elements, provided that deformational compatibility is taken into account.

D.3.2 — The design strength of anchors shall equal or exceed the largest required strength calculated from the applicable load combinations in 9.2.

D.3.3 — When anchor design includes seismic loads, the additional requirements of D.3.3.1 through D.3.3.5 shall apply.

COMMENTARY

requirements of this appendix, the results of the ACI 355.2 tests have to indicate that pullout failures exhibit an acceptable load-displacement characteristic or that pullout failures are precluded by another failure mode.

RD.2.4 — The exclusion from the scope of load applications producing high cycle fatigue or extremely short duration impact (such as blast or shock wave) are not meant to exclude seismic load effects. D.3.3 presents additional requirements for design when seismic loads are included.

RD.3 — General requirements

RD.3.1 — When the strength of an anchor group is governed by breakage of the concrete, the behavior is brittle and there is limited redistribution of the forces between the highly stressed and less stressed anchors. In this case, the theory of elasticity is required to be used assuming the attachment that distributes loads to the anchors is sufficiently stiff. The forces in the anchors are considered to be proportional to the external load and its distance from the neutral axis of the anchor group.

If anchor strength is governed by ductile yielding of the anchor steel, significant redistribution of anchor forces can occur. In this case, an analysis based on the theory of elasticity will be conservative. References D.4 to D.6 discuss nonlinear analysis, using theory of plasticity, for the determination of the capacities of ductile anchor groups.

RD.3.3 — Post-installed structural anchors are required to be qualified for moderate or high seismic risk zone usage by demonstrating the capacity to undergo large displacements through several cycles as specified in the seismic simulation tests of ACI 355.2. Because ACI 355.2 excludes plastic hinge zones, Appendix D is not applicable to the design of anchors in plastic hinge zones under seismic loads. In addition, the design of anchors in zones of moderate or high seismic risk is based on a more conservative approach by the introduction of 0.75 factor on the design strength ϕN_n and ϕV_n, and by requiring the system to have adequate ductility. Anchorage capacity should be governed by ductile yielding of a steel element. If the anchor cannot meet these ductility requirements, then the attachment is required to be designed so as to yield at a load well below the anchor capacity. In designing attachments for adequate ductility, the ratio of yield to ultimate load capacity should be considered. A connection element could yield only to result in a secondary failure as one or more elements strain harden and fail if the ultimate load capacity is excessive when compared to the yield capacity.

CODE

COMMENTARY

Under seismic conditions, the direction of shear loading may not be predictable. The full shear load should be assumed in any direction for a safe design.

D.3.3.1 —The provisions of Appendix D do not apply to the design of anchors in plastic hinge zones of concrete structures under seismic loads.

RD.3.3.1 —Section 3.1 of ACI 355.2 specifically states that the seismic test procedures do no simulate the behavior of anchors in plastic hinge zones. The possible higher level of cracking and spalling in plastic hinge zones are beyond the damage states for which Appendix D is applicable.

D.3.3.2 — In regions of moderate or high seismic risk, or for structures assigned to intermediate or high seismic performance or design categories, post-installed structural anchors for use under D.2.3 shall have passed the Simulated Seismic Tests of ACI 355.2.

D.3.3.3 — In regions of moderate or high seismic risk, or for structures assigned to intermediate or high seismic performance or design categories, the design strength of anchors shall be taken as $0.75\phi N_n$ and $0.75\phi V_n$, where ϕ is given in D.4.4 or D.4.5 and N_n and V_n are determined in accordance with D.4.1.

D.3.3.4 — In regions of moderate or high seismic risk, or for structures assigned to intermediate or high seismic performance or design categories, anchors shall be designed to be governed by tensile or shear strength of a ductile steel element, unless D.3.3.5 is satisfied.

D.3.3.5 — Instead of D.3.3.4, the attachment that the anchor is connecting to the structure shall be designed so that the attachment will undergo ductile yielding at a load level corresponding to anchor forces no greater than the design strength of anchors specified in D.3.3.3.

D.3.4 — All provisions for anchor axial tension and shear strength apply to normalweight concrete. If lightweight concrete is used, provisions for N_n and V_n shall be modified by multiplying all values of $\sqrt{f_c'}$ affecting N_n and V_n, by 0.75 for all-lightweight concrete and 0.85 for sand-lightweight concrete. Linear interpolation shall be permitted when partial sand replacement is used.

D.3.5 — The values of f_c' used for calculation purposes in this appendix shall not exceed 10,000 psi for cast-in anchors, and 8,000 psi for post-installed anchors. Testing is required for post-installed anchors when used in concrete with f_c' greater than 8,000 psi.

RD.3.5 — A limited number of tests of cast-in-place and post-installed anchors in high-strength concrete[D.7] indicate that the design procedures contained in this appendix become unconservative, particularly for cast-in anchors in concrete with compressive strengths in the range of 11,000 to 12,000 psi. Until further tests are available, an upper limit on f_c' of 10,000 psi has been imposed in the design of cast-in-place anchors. This is consistent with Chapters 11 and 12. The companion ACI 355.2 does not require testing of

CODE

COMMENTARY

post-installed anchors in concrete with f_c' greater than 8000 psi because some post-installed anchors may have difficulty expanding in very high-strength concretes. Because of this, f_c' is limited to 8000 psi in the design of post-installed anchors unless testing is performed.

D.4 — General requirements for strength of anchors

D.4.1 — Strength design of anchors shall be based either on computation using design models that satisfy the requirements of D.4.2, or on test evaluation using the 5 percent fractile of test results for the following:

(a) steel strength of anchor in tension (D.5.1);
(b) steel strength of anchor in shear (D.6.1);
(c) concrete breakout strength of anchor in tension (D.5.2);
(d) concrete breakout strength of anchor in shear (D.6.2);
(e) pullout strength of anchor in tension (D.5.3);
(f) concrete side-face blowout strength of anchor in tension (D.5.4); and
(g) concrete pryout strength of anchor in shear (D.6.3).

In addition, anchors shall satisfy the required edge distances, spacings, and thicknesses to preclude splitting failure, as required in D.8.

D.4.1.1 — For the design of anchors, except as required in D.3.3,

$$\phi N_n \geq N_{ua} \quad \text{(D-1)}$$

$$\phi V_n \geq V_{ua} \quad \text{(D-2)}$$

D.4.1.2 — In Eq. (D-1) and (D-2), ϕN_n and ϕV_n are the lowest design strengths determined from all appropriate failure modes. ϕN_n is the lowest design strength in tension of an anchor or group of anchors as determined from consideration of ϕN_{sa}, $\phi n N_{pn}$, either ϕN_{sb} or ϕN_{sbg}, and either ϕN_{cb} or ϕN_{cbg}. ϕV_n is the lowest design strength in shear of an anchor or a group of anchors as determined from consideration of: ϕV_{sa}, either ϕV_{cb} or ϕV_{cbg}, and either ϕV_{cp} or ϕV_{cpg}.

D.4.1.3 — When both N_{ua} and V_{ua} are present, interaction effects shall be considered in accordance with D.4.3.

D.4.2 — The nominal strength for any anchor or group of anchors shall be based on design models that result in predictions of strength in substantial agreement with results of comprehensive tests. The materials used in

RD.4 — General requirements for strength of anchors

RD.4.1 — This section provides requirements for establishing the strength of anchors to concrete. The various types of steel and concrete failure modes for anchors are shown in Fig. RD.4.1(a) and RD.4.1(b) Comprehensive discussions of anchor failure modes are included in References D.8 to D.10. Any model that complies with the requirements of D.4.2 and D.4.3 can be used to establish the concrete-related strengths. For anchors such as headed bolts, headed studs, and post-installed anchors, the concrete breakout design methods of D.5.2 and D.6.2 are acceptable. The anchor strength is also dependent on the pullout strength of D.5.3, the side-face blowout strength of D.5.4, and the minimum spacings and edge distances of D.8. The design of anchors for tension recognizes that the strength of anchors is sensitive to appropriate installation; installation requirements are included in D.9. Some post-installed anchors are less sensitive to installation errors and tolerances. This is reflected in varied ϕ factors based on the assessment criteria of ACI 355.2.

Test procedures can also be used to determine the single-anchor breakout strength in tension and in shear. The test results, however, are required to be evaluated on a basis statistically equivalent to that used to select the values for the concrete breakout method "considered to satisfy" provisions of D.4.2. The basic strength cannot be taken greater than the 5 percent fractile. The number of tests has to be sufficient for statistical validity and should be considered in the determination of the 5 percent fractile.

RD.4.2 and RD.4.3 — D.4.2 and D.4.3 establish the performance factors for which anchor design models are required to be verified. Many possible design approaches exist and the user is always permitted to "design by test" using D.4.2 as long as sufficient data are available to verify the model.

CODE

the tests shall be compatible with the materials used in the structure. The nominal strength shall be based on the 5 percent fractile of the basic individual anchor strength. For nominal strengths related to concrete strength, modifications for size effects, the number of anchors, the effects of close spacing of anchors, proximity to edges, depth of the concrete member, eccentric loadings of anchor groups, and presence or absence of cracking shall be taken into account. Limits on edge distances and anchor spacing in the design models shall be consistent with the tests that verified the model.

COMMENTARY

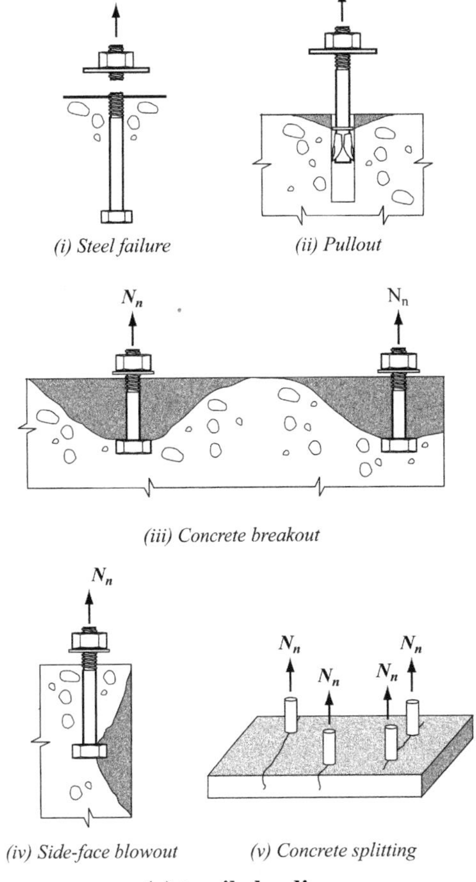

(a) tensile loading

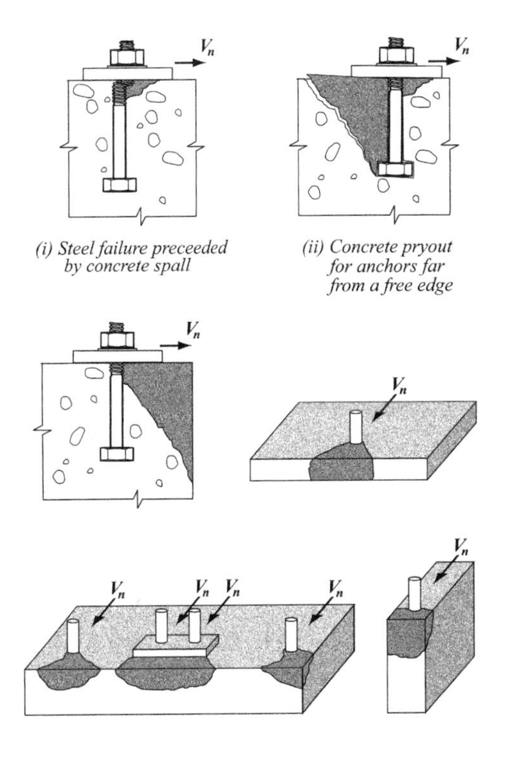

(b) shear loading

Fig. RD.4.1 — Failure modes for anchors.

CODE

D.4.2.1 — The effect of supplementary reinforcement provided to confine or restrain the concrete breakout, or both, shall be permitted to be included in the design models used to satisfy D.4.2.

D.4.2.2 — For anchors with diameters not exceeding 2 in., and tensile embedments not exceeding 25 in. in depth, the concrete breakout strength requirements shall be considered satisfied by the design procedure of D.5.2 and D.6.2.

D.4.3 — Resistance to combined tensile and shear loads shall be considered in design using an interaction expression that results in computation of strength in substantial agreement with results of comprehensive tests. This requirement shall be considered satisfied by D.7.

COMMENTARY

RD.4.2.1 — The addition of supplementary reinforcement in the direction of the load, confining reinforcement, or both, can greatly enhance the strength and ductility of the anchor connection. Such enhancement is practical with cast-in anchors such as those used in precast sections.

The shear strength of headed anchors located near the edge of a member can be significantly increased with appropriate supplementary reinforcement. References D.8, D.11, and D.12 provide substantial information on design of such reinforcement. The effect of such supplementary reinforcement is not included in the ACI 355.2 anchor acceptance tests or in the concrete breakout calculation method of D.5.2 and D.6.2. The designer has to rely on other test data and design theories in order to include the effects of supplementary reinforcement.

For anchors exceeding the limitations of D.4.2.2, or for situations where geometric restrictions limit breakout capacity, or both, reinforcement oriented in the direction of load and proportioned to resist the total load within the breakout prism, and fully anchored on both sides of the breakout planes, may be provided instead of calculating breakout capacity.

The breakout strength of an unreinforced connection can be taken as an indication of the load at which significant cracking will occur. Such cracking can represent a serviceability problem if not controlled. (See RD.6.2.1)

RD.4.2.2 — The method for concrete breakout design included as "considered to satisfy" D.4.2 was developed from the Concrete Capacity Design (CCD) Method,[D.9,D.10] which was an adaptation of the κ Method[D.13,D.14] and is considered to be accurate, relatively easy to apply, and capable of extension to irregular layouts. The CCD Method predicts the load capacity of an anchor or group of anchors by using a basic equation for tension, or for shear for a single anchor in cracked concrete, and multiplied by factors that account for the number of anchors, edge distance, spacing, eccentricity and absence of cracking. The limitations on anchor size and embedment length are based on the current range of test data.

The breakout strength calculations are based on a model suggested in the κ Method. It is consistent with a breakout prism angle of approximately 35 degrees [Fig. RD.4.2.2(a) and (b)].

CODE

COMMENTARY

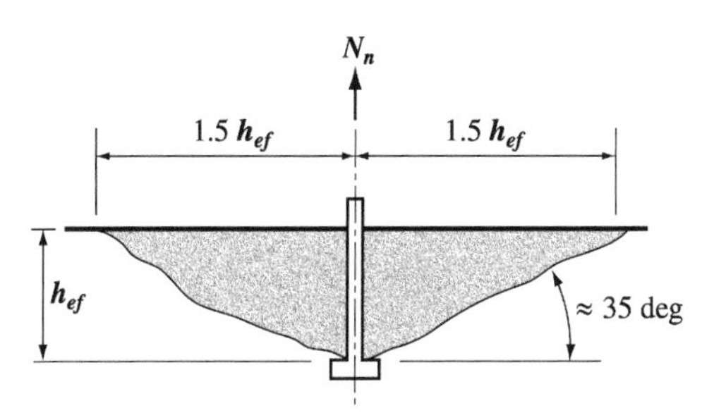

Fig. RD.4.2.2(a)—Breakout cone for tension

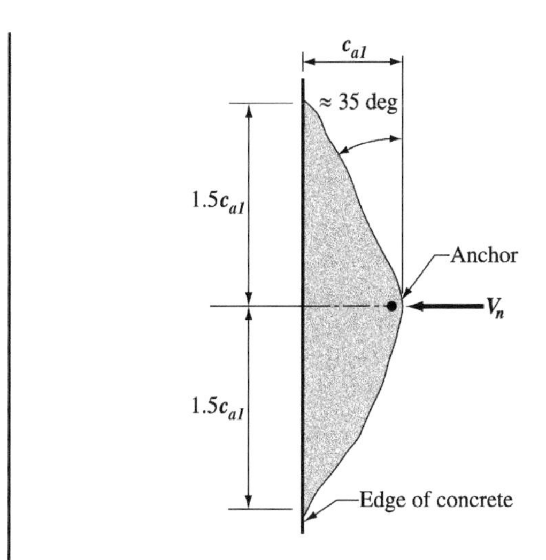

Fig. RD.4.2.2(b)—Breakout cone for shear

D.4.4 — Strength reduction factor ϕ for anchors in concrete shall be as follows when the load combinations of 9.2 are used:

a) Anchor governed by strength of a ductile steel element

i) Tension loads 0.75
ii) Shear loads 0.65

b) Anchor governed by strength of a brittle steel element

i) Tension loads 0.65
ii) Shear loads 0.60

c) Anchor governed by concrete breakout, side-face blowout, pullout, or pryout strength

	Condition A	Condition B
i) Shear loads	0.75	0.70
ii) Tension loads		
Cast-in headed studs, headed bolts, or hooked bolts	0.75	0.70

RD.4.4 — The ϕ factors for steel strength are based on using f_{uta} to determine the nominal strength of the anchor (see D.5.1 and D.6.1) rather than f_{ya} as used in the design of reinforced concrete members. Although the ϕ factors for use with f_{uta} appear low, they result in a level of safety consistent with the use of higher ϕ factors applied to f_{ya}. The smaller ϕ factors for shear than for tension do not reflect basic material differences but rather account for the possibility of a non-uniform distribution of shear in connections with multiple anchors. It is acceptable to have a ductile failure of a steel element in the attachment if the attachment is designed so that it will undergo ductile yielding at a load level no greater than 75 percent of the minimum design strength of an anchor (See D.3.3.4). For anchors governed by the more brittle concrete breakout or blowout failure, two conditions are recognized. If supplementary reinforcement is provided to tie the failure prism into the structural member (Condition A), more ductility is present than in the case where such supplementary reinforcement is not present (Condition B). Design of supplementary reinforcement is discussed in RD.4.2.1 and References D.8, D.11, D.12, and D.15. Further discussion of strength reduction factors is presented in RD.4.5.

The ACI 355.2 tests for sensitivity to installation procedures determine the category appropriate for a particular anchoring device. In the ACI 355.2 tests, the effects of variability in anchor torque during installation, tolerance on drilled hole size, energy level used in setting anchors, and for

CODE

Post-installed anchors with category as determined from ACI 355.2		
Category 1 (Low sensitivity to installation and high reliability)	0.75	0.65
Category 2 (Medium sensitivity to installation and medium reliability)	0.65	0.55
Category 3 (High sensitivity to installation and lower reliability)	0.55	0.45

Condition A applies where the potential concrete failure surfaces are crossed by supplementary reinforcement proportioned to tie the potential concrete failure prism into the structural member.

Condition B applies where such supplementary reinforcement is not provided, or where pullout or pryout strength governs.

D.4.5 — Strength reduction factor ϕ for anchors in concrete shall be as follows when the load combinations referenced in Appendix C are used:

a) Anchor governed by strength of a ductile steel element

i) Tension loads 0.80
ii) Shear loads 0.75

b) Anchor governed by strength of a brittle steel element

i) Tension loads 0.70
ii) Shear loads 0.65

c) Anchor governed by concrete breakout, side-face blowout, pullout, or pryout strength

	Condition A	Condition B
i) Shear loads	0.85	0.75
ii) Tension loads		
Cast-in headed studs, headed bolts, or hooked bolts	0.85	0.75

COMMENTARY

anchors approved for use in cracked concrete, increased crack widths are considered. The three categories of acceptable post-installed anchors are:

Category 1 — low sensitivity to installation and high reliability;

Category 2 — medium sensitivity to installation and medium reliability; and

Category 3 — high sensitivity to installation and lower reliability.

The capacities of anchors under shear loads are not as sensitive to installation errors and tolerances. Therefore, for shear calculations of all anchors, $\phi = 0.75$ for Condition A and $\phi = 0.70$ for Condition B.

RD.4.5 — As noted in R9.1, the 2002 code incorporated the load factors of ASCE 7-98 and the corresponding strength reduction factors provided in the 1999 Appendix C into 9.2 and 9.3, except that the factor for flexure has been increased. Developmental studies for the ϕ factors to be used for Appendix D were based on the 1999 9.2 and 9.3 load and strength reduction factors. The resulting ϕ factors are presented in D.4.5 for use with the load factors of the 2002 Appendix C. The ϕ factors for use with the load factors of the 1999 Appendix C were determined in a manner consistent with the other ϕ factors of the 1999 Appendix C. These ϕ factors are presented in D.4.4 for use with the load factors of 2002 9.2. Since developmental studies for ϕ factors to be used with Appendix D, for brittle concrete failure modes, were performed for the load and strength reduction factors now given in Appendix C, the discussion of the selection of these ϕ factors appears in this section.

Even though the ϕ factor for plain concrete in Appendix C uses a value of 0.65, the basic factor for brittle concrete failures ($\phi = 0.75$) was chosen based on results of probabilistic studies[D.16] that indicated the use of $\phi = 0.65$ with mean values of concrete-controlled failures produced adequate safety levels. Because the nominal resistance expressions used in this appendix and in the test requirements are based on the 5 percent fractiles, the $\phi = 0.65$ value would be overly conservative. Comparison with other design procedures and probabilistic studies[D.16] indicated that the choice of $\phi = 0.75$ was justified.

CODE

Post-installed anchors with category as determined from ACI 355.2

Category 1 (Low sensitivity to installation and high reliability)	0.85	0.75
Category 2 (Medium sensitivity to installation and medium reliability)	0.75	0.65
Category 3 (High sensitivity to installation and lower reliability)	0.65	0.55

Condition A applies where the potential concrete failure surfaces are crossed by supplementary reinforcement proportioned to tie the potential concrete failure prism into the structural member.

Condition B applies where such supplementary reinforcement is not provided, or where pullout or pryout strength governs.

D.5 — Design requirements for tensile loading

D.5.1 — Steel strength of anchor in tension

D.5.1.1 — The nominal strength of an anchor in tension as governed by the steel, N_{sa}, shall be evaluated by calculations based on the properties of the anchor material and the physical dimensions of the anchor.

D.5.1.2 — The nominal strength of a single anchor or group of anchors in tension, N_{sa}, shall not exceed:

$$N_{sa} = nA_{se} f_{uta} \qquad \text{(D-3)}$$

where n is the number of anchors in the group, and f_{uta} shall not be taken greater than the smaller of $1.9f_{ya}$ and 125,000 psi.

COMMENTARY

For applications with supplementary reinforcement and more ductile failures (Condition A), the ϕ factors are increased. The value of $\phi = 0.85$ is compatible with the level of safety for shear failures in concrete beams, and has been recommended in the PCI Design Handbook[D.17] and by ACI 349.[D.15]

RD.5 — Design requirements for tensile loading

RD.5.1 — Steel strength of anchor in tension

RD.5.1.2 — The nominal tension strength of anchors is best represented by $A_{se}f_{uta}$ rather than $A_{se}f_{ya}$ because the large majority of anchor materials do not exhibit a well-defined yield point. The American Institute of Steel Construction (AISC) has based tension strength of anchors on $A_{se}f_{uta}$ since the 1986 edition of their specifications. The use of Eq. (D-3) with 9.2 load factors and the ϕ factors of D.4.4 give design strengths consistent with the AISC Load and Resistance Factor Design Specifications.[D.18]

The limitation of $1.9f_{ya}$ on f_{uta} is to ensure that under service load conditions the anchor does not exceed f_{ya}. The limit on f_{uta} of $1.9f_{ya}$ was determined by converting the LRFD provisions to corresponding service level conditions. For Section 9.2, the average load factor of 1.4 (from $1.2D + 1.6L$) divided by the highest ϕ factor (0.75 for tension) results in a limit of f_{uta}/f_{ya} of 1.4/0.75 = 1.87. For Appendix C, the average load factor of 1.55 (from $1.4D + 1.7L$),

CODE

COMMENTARY

divided by the highest ϕ factor (0.80 for tension), results in a limit of f_{uta}/f_{ya} of 1.55/0.8 = 1.94. For consistent results, the serviceability limitation of f_{uta} was taken as $1.9f_{ya}$. If the ratio of f_{uta} to f_{ya} exceeds this value, the anchoring may be subjected to service loads above f_{ya} under service loads. Although not a concern for standard structural steel anchors (maximum value of f_{uta}/f_{ya} is 1.6 for ASTM A 307), the limitation is applicable to some stainless steels.

The effective cross-sectional area of an anchor should be provided by the manufacturer of expansion anchors with reduced cross-sectional area for the expansion mechanism. For threaded bolts, ANSI/ASME B1.1[D.1] defines A_{se} as:

$$A_{se} = \frac{\pi}{4}\left(d_o - \frac{0.9743}{n_t}\right)^2$$

where n_t is the number of threads per in.

D.5.2 — Concrete breakout strength of anchor in tension

D.5.2.1 — The nominal concrete breakout strength, N_{cb} or N_{cbg}, of a single anchor or group of anchors in tension shall not exceed

(a) for a single anchor:

$$N_{cb} = \frac{A_{Nc}}{A_{Nco}}\psi_{ed,N}\psi_{c,N}\psi_{cp,N}N_b \quad \text{(D-4)}$$

(b) for a group of anchors:

$$N_{cbg} = \frac{A_{Nc}}{A_{Nco}}\psi_{ec,N}\psi_{ed,N}\psi_{c,N}\psi_{cp,N}N_b \quad \text{(D-5)}$$

Factors $\psi_{ec,N}$, $\psi_{ed,N}$, $\psi_{c,N}$, and $\psi_{cp,N}$ are defined in D.5.2.4, D.5.2.5, D.5.2.6, and D.5.2.7, respectively. A_{Nc} is the projected concrete failure area of a single anchor or group of anchors that shall be approximated as the base of the rectilinear geometrical figure that results from projecting the failure surface outward $1.5h_{ef}$ from the centerlines of the anchor, or in the case of a group of anchors, from a line through a row of adjacent anchors. A_{Nc} shall not exceed nA_{Nco}, where n is the number of tensioned anchors in the group. A_{Nco} is the projected concrete failure area of a single anchor with an edge distance equal to or greater than $1.5h_{ef}$:

$$A_{Nco} = 9h_{ef}^2 \quad \text{(D-6)}$$

RD.5.2 — Concrete breakout strength of anchor in tension

RD.5.2.1 — The effects of multiple anchors, spacing of anchors, and edge distance on the nominal concrete breakout strength in tension are included by applying the modification factors A_{Nc}/A_{Nco} and $\psi_{ed,N}$ in Eq. (D-4) and (D-5).

Figure RD.5.2.1(a) shows A_{Nco} and the development of Eq. (D-6). A_{Nco} is the maximum projected area for a single anchor. Figure RD.5.2.1(b) shows examples of the projected areas for various single-anchor and multiple-anchor arrangements. Because A_{Nc} is the total projected area for a group of anchors, and A_{Nco} is the area for a single anchor, there is no need to include n, the number of anchors, in Eq. (D-4) or (D-5). If anchor groups are positioned in such a way that their projected areas overlap, the value of A_{Nc} is required to be reduced accordingly.

CODE

COMMENTARY

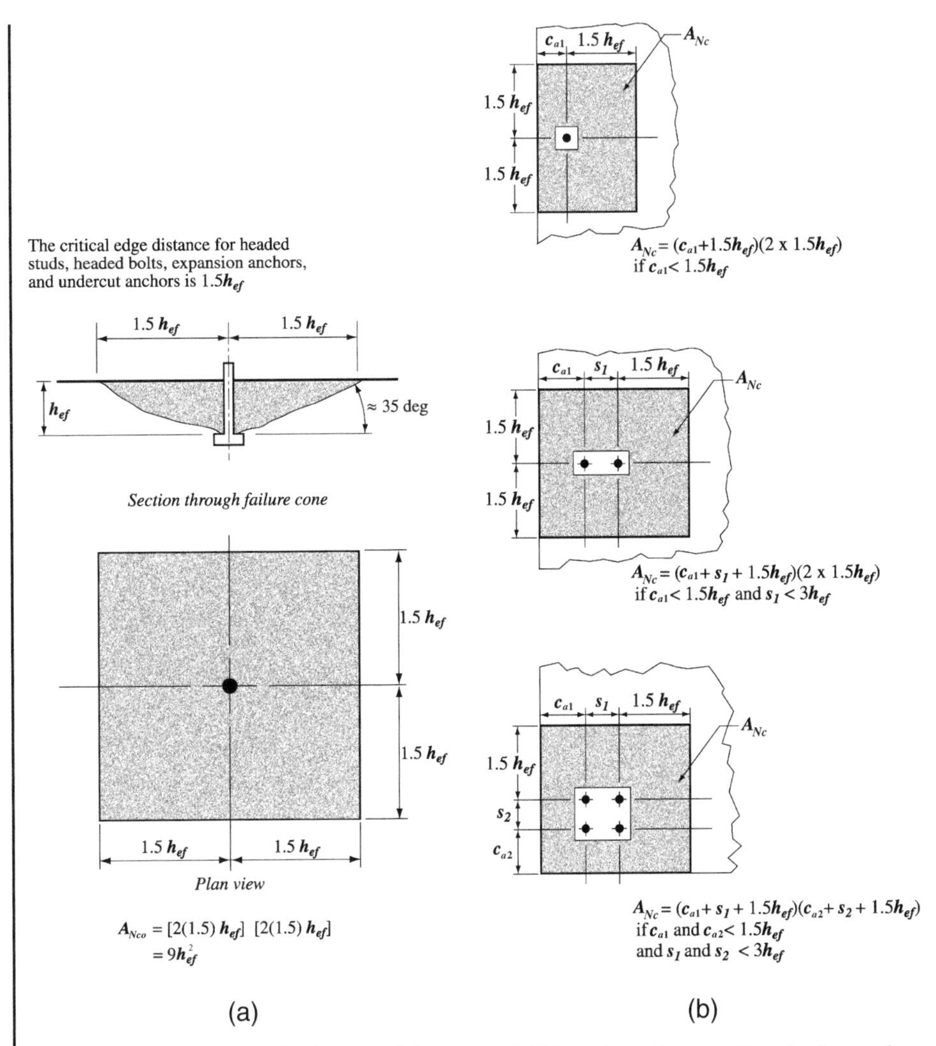

Fig. RD.5.2.1(a) — Calculation of $\mathbf{A_{Nco}}$*; and (b) projected areas for single anchors and groups of anchors and calculation of* $\mathbf{A_{Nc}}$.

D.5.2.2 — The basic concrete breakout strength of a single anchor in tension in cracked concrete, N_b, shall not exceed

$$N_b = k_c\sqrt{f_c'}\,h_{ef}^{1.5} \qquad \text{(D-7)}$$

where

k_c = 24 for cast-in anchors; and

k_c = 17 for post-installed anchors.

The value of k_c for post-installed anchors shall be permitted to be increased above 17 based on ACI 355.2 product-specific tests, but shall in no case exceed 24.

Alternatively, for cast-in headed studs and headed bolts with 11 in. ≤ h_{ef} ≤ 25 in., N_b shall not exceed

$$N_b = 16\sqrt{f_c'}\,h_{ef}^{5/3} \qquad \text{(D-8)}$$

RD.5.2.2 — The basic equation for anchor capacity was derived[D.9-D.11,D.14] assuming a concrete failure prism with an angle of about 35 degrees, considering fracture mechanics concepts.

The values of k_c in Eq. (D-7) were determined from a large database of test results in uncracked concrete[D.9] at the 5 percent fractile. The values were adjusted to corresponding k_c values for cracked concrete.[D.10,D.19] Higher k_c values for post-installed anchors may be permitted, provided they have been determined from product approval testing in accordance with ACI 355.2. For anchors with a deep embedment (h_{ef} > 11 in.), test evidence indicates the use of $h_{ef}^{1.5}$ can be overly conservative for some cases. Often such tests have been with selected aggregates for special applications. An alternative expression (Eq. (D-8)) is provided using $h_{ef}^{5/3}$ for evaluation of cast-in anchors with 11 in. ≤ h_{ef} ≤ 25 in. The limit of 25 in. corresponds to the upper range of test data. This expression can also be appropriate for some

CODE

D.5.2.3 — Where anchors are located less than $1.5h_{ef}$ from three or more edges, the value of h_{ef} used in Eq. (D-4) through (D-11) shall be the greater of $c_{a,max}/1.5$ and one-third of the maximum spacing between anchors within the group.

COMMENTARY

undercut post-installed anchors. However, for such anchors, the use of Eq. (D-8) should be justified by test results in accordance with D.4.2.

RD.5.2.3 — For anchors located less than $1.5h_{ef}$ from three or more edges, the tensile breakout strength computed by the CCD Method, which is the basis for Eq. (D-4) to (D-11), gives overly conservative results.[D.20] This occurs because the ordinary definitions of A_{Nc}/A_{Nco} do not correctly reflect the edge effects. This problem is corrected by limiting the the value of h_{ef} used in Eq. (D-4) through (D-11) to $c_{a,max}/1.5$, where $c_{a,max}$ is the largest of the influencing edge distances that are less than or equal to the actual $1.5h_{ef}$. In no case should $c_{a,max}$ be taken less than one-third of the maximum spacing between anchors within the group. The limit on h_{ef} of at least one-third of the maximum spacing between anchors within the group prevents the designer from using a calculated strength based on individual breakout prisms for a group anchor configuration.

This approach is illustrated in Fig. RD.5.2.3. In this example, the proposed limit on the value of h_{ef} to be used in the computations where $h_{ef} = c_{a,max}/1.5$, results in $h_{ef} = h'_{ef} = 4$ in. For this example, this would be the proper value to be used for h_{ef} in computing the resistance even if the actual embedment depth is larger.

The requirement of D.5.2.3 may be visualized by moving the actual concrete breakout surface, which originates at the actual h_{ef}, toward the surface of the concrete parallel to the applied tension load. The value of h_{ef} used in Eq. (D-4) to (D-11) is determined when either: (a) the outer boundaries of the failure surface first intersect a free edge; or (b) the intersection of the breakout surface between anchors within the group first intersects the surface of the concrete. For the example shown in Fig. RD.5.2.3, point "A" defines the intersection of the assumed failure surface for limiting h_{ef} with the concrete surface.

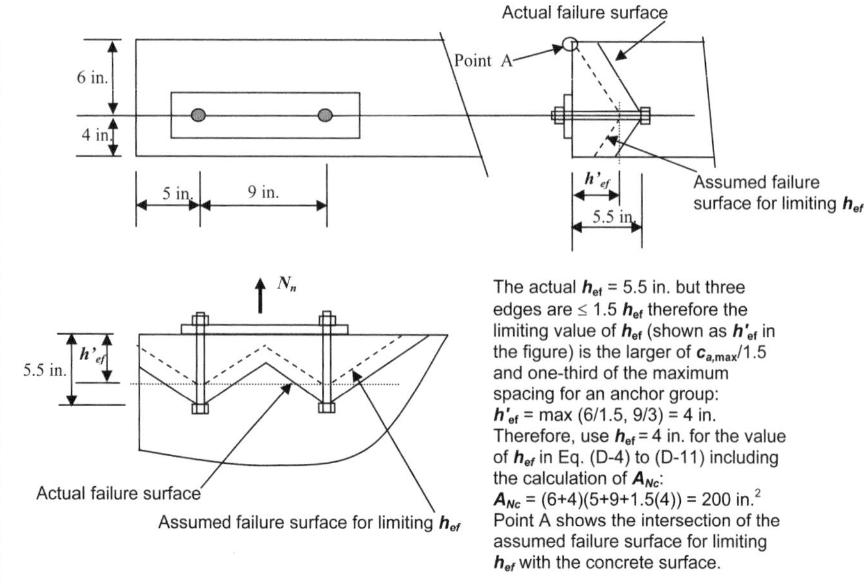

Fig. RD.5.2.3 — Tension in narrow members.

CODE

COMMENTARY

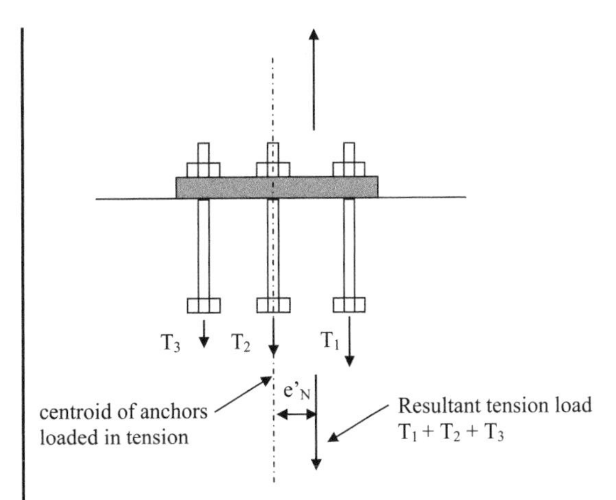

(a) When all anchors in a group are in tension

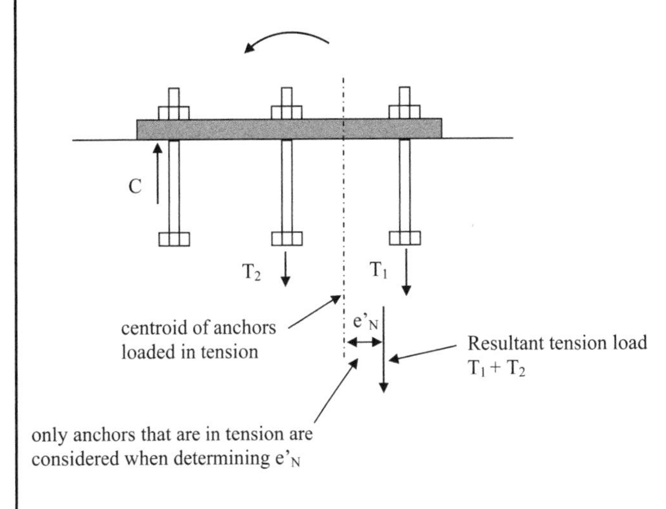

(b) When only some anchors in a group are in tension

Fig. RD.5.2.4 — Definition of $\mathbf{e}_N'$ *for group anchors.*

D.5.2.4 — The modification factor for anchor groups loaded eccentrically in tension is:

$$\psi_{ec,N} = \frac{1}{\left(1 + \frac{2e_N'}{3h_{ef}}\right)} \leq 1.0 \qquad \text{(D-9)}$$

If the loading on an anchor group is such that only some anchors are in tension, only those anchors that are in tension shall be considered when determining the eccentricity e_N' for use in Eq. (D-9) and for the calculation of N_{cbg} in Eq. (D-5).

In the case where eccentric loading exists about two axes, the modification factor, $\psi_{ec,N}$, shall be computed for each axis individually and the product of these factors used as $\psi_{ec,N}$ in Eq. (D-5).

RD.5.2.4 — Figure RD.5.2.4(a) shows a group of anchors that are all in tension but the resultant force is eccentric with respect to the centroid of the anchor group. Groups of anchors can also be loaded in such a way that only some of the anchors are in tension [Fig. RD.5.2.4(b)]. In this case, only the anchors in tension are to be considered in the determination of e_N'. The anchor loading has to be determined as the resultant anchor tension at an eccentricity with respect to the center of gravity of the anchors in tension.

CODE

D.5.2.5 — The modification factor for edge effects for single anchors or anchor groups loaded in tension is:

$$\psi_{ed,N} = 1 \text{ if } c_{a,min} \geq 1.5h_{ef} \quad \text{(D-10)}$$

$$\psi_{ed,N} = 0.7 + 0.3\frac{c_{a,min}}{1.5h_{ef}} \text{ if } c_{a,min} < 1.5h_{ef} \quad \text{(D-11)}$$

D.5.2.6 — For anchors located in a region of a concrete member where analysis indicates no cracking at service load levels, the following modification factor shall be permitted:

$\psi_{c,N}$ = 1.25 for cast-in anchors; and

$\psi_{c,N}$ = 1.4 for post-installed anchors, where the value of k_c used in Eq. (D-7) is 17

Where the value of k_c used in Eq. (D-7) is taken from the ACI 355.2 product evaluation report for post-installed anchors qualified for use in both cracked and uncracked concrete, the values of k_c and $\psi_{c,N}$ shall be based on the ACI 355.2 product evaluation report.

Where the value of k_c used in Eq. (D-7) is taken from the ACI 355.2 product evaluation report for post-installed anchors qualified for use only in uncracked concrete, $\psi_{c,N}$ shall be taken as 1.0.

When analysis indicates cracking at service load levels, $\psi_{c,N}$ shall be taken as 1.0 for both cast-in anchors and post-installed anchors. Post-installed anchors shall be qualified for use in cracked concrete in accordance with ACI 355.2. The cracking in the concrete shall be controlled by flexural reinforcement distributed in accordance with 10.6.4, or equivalent crack control shall be provided by confining reinforcement.

D5.2.7 — The modification factor for post-installed anchors designed for uncracked concrete in accordance with D.5.2.6 without supplementary reinforcement to control splitting is:

$$\psi_{cp,N} = 1.0 \text{ if } c_{a,min} \geq c_{ac} \quad \text{(D-12)}$$

$$\psi_{cp,N} = \frac{c_{a,min}}{c_{ac}} \geq \frac{1.5h_{ef}}{c_{ac}} \text{ if } c_{a,min} < c_{ac} \quad \text{(D-13)}$$

where the critical distance c_{ac} is defined in D.8.6.

For all other cases, including cast-in anchors, $\psi_{cp,N}$ shall be taken as 1.0

COMMENTARY

RD.5.2.5 — If anchors are located close to an edge so that there is not enough space for a complete breakout prism to develop, the load-bearing capacity of the anchor is further reduced beyond that reflected in A_{Nc}/A_{Nco}. If the smallest side cover distance is greater than or equal to $1.5h_{ef}$, a complete prism can form and there is no reduction ($\psi_{ed,N} = 1$). If the side cover is less than $1.5h_{ef}$, the factor $\psi_{ed,N}$ is required to adjust for the edge effect.[D.9]

RD.5.2.6 — Post-installed and cast-in anchors that have not met the requirements for use in cracked concrete according to ACI 355.2 should be used in uncracked regions only. The analysis for the determination of crack formation should include the effects of restrained shrinkage (see 7.12.1.2). The anchor qualification tests of ACI 355.2 require that anchors in cracked concrete zones perform well in a crack that is 0.012 in. wide. If wider cracks are expected, confining reinforcement to control the crack width to about 0.012 in. should be provided.

The concrete breakout strengths given by Eq. (D-7) and (D-8) assume cracked concrete (that is, $\psi_{c,N}$ = 1.0) with $\psi_{c,N}k_c$ = 24 for cast-in-place, and 17 for post-installed (cast-in 40 percent higher). When the uncracked concrete $\psi_{c,N}$ factors are applied (1.25 for cast-in, and 1.4 for post-installed), the results are $\psi_{c,N}k_c$ factors of 30 for cast-in and 24 for post-installed (25 percent higher for cast-in). This agrees with field observations and tests that show cast-in anchor strength exceeds that of post-installed for both cracked and uncracked concrete.

RD.5.2.7 — The design provisions in D.5 are based on the assumption that the basic concrete breakout strength can be achieved if the minimum edge distance, $c_{a,min}$, equals $1.5h_{ef}$. However, test results[D.21] indicate that many torque-controlled and displacement-controlled expansion anchors and some undercut anchors require minimum edge distances exceeding $1.5h_{ef}$ to achieve the basic concrete breakout strength when tested in uncracked concrete without supplementary reinforcement to control splitting. When a tension load is applied, the resulting tensile stresses at the embedded end of the anchor are added to the tensile stresses induced due to anchor installation, and splitting failure may occur before reaching the concrete breakout strength defined in D.5.2.1. To account for this potential splitting

CODE | COMMENTARY

mode of failure, the basic concrete breakout strength is reduced by a factor $\psi_{cp,N}$ if $c_{a,min}$ is less than the critical edge distance c_{ac}. If supplementary reinforcement to control splitting is present or if the anchors are located in a region where analysis indicates cracking of the concrete at service loads, then the reduction factor $\psi_{cp,N}$ is taken as 1.0. The presence of supplementary reinforcement to control splitting does not affect the selection of Condition A or B in D.4.4 or D.4.5.

D.5.2.8 — Where an additional plate or washer is added at the head of the anchor, it shall be permitted to calculate the projected area of the failure surface by projecting the failure surface outward $1.5h_{ef}$ from the effective perimeter of the plate or washer. The effective perimeter shall not exceed the value at a section projected outward more than the thickness of the washer or plate from the outer edge of the head of the anchor.

D.5.3 — Pullout strength of anchor in tension

RD.5.3 — Pullout strength of anchor in tension

D.5.3.1 — The nominal pullout strength of a single anchor in tension, N_{pn}, shall not exceed:

$$N_{pn} = \psi_{c,P} N_p \quad \text{(D-14)}$$

where $\psi_{c,P}$ is defined in D.5.3.6.

D.5.3.2 — For post-installed expansion and undercut anchors, the values of N_p shall be based on the 5 percent fractile of results of tests performed and evaluated according to ACI 355.2. It is not permissible to calculate the pullout strength in tension for such anchors.

RD.5.3.2 — The pullout strength equations given in D.5.3.4 and D.5.3.5 are only applicable to cast-in headed and hooked anchors;[D.8,D.22] they are not applicable to expansion and undercut anchors that use various mechanisms for end anchorage unless the validity of the pullout strength equations are verified by tests.

D.5.3.3 — For single cast-in headed studs and headed bolts, it shall be permitted to evaluate the pullout strength in tension using D.5.3.4. For single J- or L-bolts, it shall be permitted to evaluate the pullout strength in tension using D.5.3.5. Alternatively, it shall be permitted to use values of N_p based on the 5 percent fractile of tests performed and evaluated in the same manner as the ACI 355.2 procedures but without the benefit of friction.

RD.5.3.3 — The pullout strength in tension of headed studs or headed bolts can be increased by providing confining reinforcement, such as closely spaced spirals, throughout the head region. This increase can be demonstrated by tests.

CODE

D.5.3.4 — The pullout strength in tension of a single headed stud or headed bolt, N_p, for use in Eq. (D-14), shall not exceed:

$$N_p = 8A_{brg}f_c' \quad \text{(D-15)}$$

D.5.3.5 — The pullout strength in tension of a single hooked bolt, N_p, for use in Eq. (D-14) shall not exceed:

$$N_p = 0.9f_c'e_h d_o \quad \text{(D-16)}$$

where $3d_o \leq e_h \leq 4.5d_o$.

D.5.3.6 — For an anchor located in a region of a concrete member where analysis indicates no cracking at service load levels, the following modification factor shall be permitted

$$\psi_{c,P} = 1.4$$

where analysis indicates cracking at service load levels, $\psi_{c,P}$ shall be taken as 1.0.

D.5.4 — Concrete side-face blowout strength of a headed anchor in tension

D.5.4.1 — For a single headed anchor with deep embedment close to an edge ($c_{a1} < 0.4h_{ef}$), the nominal side-face blowout strength, N_{sb},shall not exceed:

$$N_{sb} = 160c_{a1}\sqrt{A_{brg}}\sqrt{f_c'} \quad \text{(D-17)}$$

If c_{a2} for the single headed anchor is less than $3c_{a1}$, the value of N_{sb} shall be multiplied by the factor $(1 + c_{a2}/c_{a1})/4$ where $1.0 \leq c_{a2}/c_{a1} \leq 3.0$.

D.5.4.2 — For multiple headed anchors with deep embedment close to an edge ($c_{a1} < 0.4h_{ef}$) and anchor spacing less than $6c_{a1}$, the nominal strength of the group of anchors for a side-face blowout failure N_{sbg} shall not exceed:

$$N_{sbg} = \left(1 + \frac{s}{6c_{a1}}\right)N_{sb} \quad \text{(D-18)}$$

where s is spacing of the outer anchors along the edge in the group; and N_{sb} is obtained from Eq. (D-17) without modification for a perpendicular edge distance.

COMMENTARY

RD.5.3.4 — Equation (D-15) corresponds to the load at which the concrete under the anchor head begins to crush.[D.8,D.15] It is not the load required to pull the anchor completely out of the concrete, so the equation contains no term relating to embedment depth. The designer should be aware that local crushing under the head will greatly reduce the stiffness of the connection, and generally will be the beginning of a pullout failure.

RD.5.3.5 — Equation (D-16) for hooked bolts was developed by Lutz based on the results of Reference D.22. Reliance is placed on the bearing component only, neglecting any frictional component because crushing inside the hook will greatly reduce the stiffness of the connection, and generally will be the beginning of pullout failure. The limits on e_h are based on the range of variables used in the three tests programs reported in Reference D.22.

RD.5.4 — Concrete side-face blowout strength of a headed anchor in tension

The design requirements for side-face blowout are based on the recommendations of Reference D.23. These requirements are applicable to headed anchors that usually are cast-in anchors. Splitting during installation rather than side-face blowout generally governs post-installed anchors, and is evaluated by the ACI 355.2 requirements.

CODE

COMMENTARY

D.6 — Design requirements for shear loading

D.6.1 — Steel strength of anchor in shear

D.6.1.1 — The nominal strength of an anchor in shear as governed by steel, V_{sa}, shall be evaluated by calculations based on the properties of the anchor material and the physical dimensions of the anchor.

D.6.1.2 — The nominal strength of a single anchor or group of anchors in shear, V_{sa}, shall not exceed (a) through (c):

(a) for cast-in headed stud anchors

$$V_{sa} = nA_{se}f_{uta} \quad \text{(D-19)}$$

where n is the number of anchors in the group and f_{uta} shall not be taken greater than the smaller of $1.9f_{ya}$ and 125,000 psi.

(b) for cast-in headed bolt and hooked bolt anchors and for post-installed anchors where sleeves do not extend through the shear plane

$$V_{sa} = n0.6A_{se}f_{uta} \quad \text{(D-20)}$$

where n is the number of anchors in the group and f_{uta} shall not be taken greater than the smaller of $1.9f_{ya}$ and 125,000 psi.

(c) for post-installed anchors where sleeves extend through the shear plane, V_{sa} shall be based on the results of tests performed and evaluated according to ACI 355.2. Alternatively, Eq. (D-20) shall be permitted to be used.

D.6.1.3 — Where anchors are used with built-up grout pads, the nominal strengths of D.6.1.2 shall be multiplied by a 0.80 factor.

D.6.2 — Concrete breakout strength of anchor in shear

D.6.2.1 — The nominal concrete breakout strength, V_{cb} or V_{cbg}, in shear of a single anchor or group of anchors shall not exceed:

(a) for shear force perpendicular to the edge on a single anchor:

$$V_{cb} = \frac{A_{Vc}}{A_{Vco}}\psi_{ed,V}\psi_{c,V}V_b \quad \text{(D-21)}$$

RD.6 — Design requirements for shear loading

RD.6.1 — Steel strength of anchor in shear

RD.6.1.2 — The nominal shear strength of anchors is best represented by $A_{se}f_{uta}$ for headed stud anchors and $0.6A_{se}f_{uta}$ for other anchors rather than a function of $A_{se}f_{ya}$ because typical anchor materials do not exhibit a well-defined yield point. The use of Eq. (D-19) and (D-20) with 9.2 load factors and the ϕ factors of D.4.4 give design strengths consistent with the AISC Load and Resistance Factor Design Specifications.[D.18]

The limitation of $1.9f_{ya}$ on f_{uta} is to ensure that under service load conditions the anchor stress does not exceed f_{ya}. The limit on f_{uta} a of $1.9f_{ya}$ was determined by converting the LRFD provisions to corresponding service level conditions as discussed in RD.5.1.2.

The effective cross-sectional area of an anchor should be provided by the manufacturer of expansion anchors with reduced cross-sectional area for the expansion mechanism. For threaded bolts, ANSI/ASME B1.1[D.1] defines A_{se} as:

$$A_{se} = \frac{\pi}{4}\left(d_o - \frac{0.9743}{n_t}\right)^2$$

where n_t is the number of threads per in.

RD.6.2 — Concrete breakout strength of anchor in shear

RD.6.2.1 — The shear strength equations were developed from the CCD method. They assume a breakout cone angle of approximately 35 degrees (See Fig. RD.4.2.2(b)), and consider fracture mechanics theory. The effects of multiple anchors, spacing of anchors, edge distance, and thickness of the concrete member on nominal concrete breakout strength in shear are included by applying the reduction factor A_{Vc}/A_{Vco} in Eq. (D-21) and (D-22), and $\psi_{ec,V}$ in Eq. (D-22). For anchors far from the edge, D.6.2 usually will not govern. For these cases, D.6.1 and D.6.3 often govern.

CODE

(b) for shear force perpendicular to the edge on a group of anchors:

$$V_{cbg} = \frac{A_{Vc}}{A_{Vco}} \psi_{ec,V}\psi_{ed,V}\psi_{c,V}V_b \qquad \text{(D-22)}$$

(c) for shear force parallel to an edge, V_{cb} or V_{cbg} shall be permitted to be twice the value of the shear force determined from Eq. (D-21) or (D-22), respectively, with the shear force assumed to act perpendicular to the edge and with $\psi_{ed,V}$ taken equal to 1.0.

(d) for anchors located at a corner, the limiting nominal concrete breakout strength shall be determined for each edge, and the minimum value shall be used.

Factors $\psi_{ec,V}$, $\psi_{ed,V}$, and $\psi_{c,V}$ are defined in D.6.2.5, D.6.2.6, and D.6.2.7, respectively. V_b is the basic concrete breakout strength value for a single anchor. A_{Vc} is the projected area of the failure surface on the side of the concrete member at its edge for a single anchor or a group of anchors. It shall be permitted to evaluate A_{Vc} as the base of a truncated half pyramid projected on the side face of the member where the top of the half pyramid is given by the axis of the anchor row selected as critical. The value of c_{a1} shall be taken as the distance from the edge to this axis. A_{Vc} shall not exceed nA_{Vco}, where n is the number of anchors in the group.

A_{Vco} is the projected area for a single anchor in a deep member with a distance from edges equal or greater than $1.5c_{a1}$ the direction perpendicular to the shear force. It shall be permitted to evaluate A_{Vco} as the base of a half pyramid with a side length parallel to the edge of $3c_{a1}$ and a depth of $1.5c_{a1}$:

$$A_{Vco} = 4.5(c_{a1})^2 \qquad \text{(D-23)}$$

Where anchors are located at varying distances from the edge and the anchors are welded to the attachment so as to distribute the force to all anchors, it shall be permitted to evaluate the strength based on the distance to the farthest row of anchors from the edge. In this case, it shall be permitted to base the value of c_{a1} on the distance from the edge to the axis of the farthest anchor row that is selected as critical, and all of the shear shall be assumed to be carried by this critical anchor row alone.

COMMENTARY

Fig. RD.6.2.1(a) shows A_{Vco} and the development of Eq. (D-23). A_{Vco} is the maximum projected area for a single anchor that approximates the surface area of the full breakout prism or cone for an anchor unaffected by edge distance, spacing or depth of member. Fig. RD.6.2.1(b) shows examples of the projected areas for various single anchor and multiple anchor arrangements. A_{Vc} approximates the full surface area of the breakout cone for the particular arrangement of anchors. Because A_{Vc} is the total projected area for a group of anchors, and A_{Vco} is the area for a single anchor, there is no need to include the number of anchors in the equation.

When using Eq. (D-22) for anchor groups loaded in shear, both assumptions for load distribution illustrated in examples on the right side of Fig. RD.6.2.1(b) should be considered because the anchors nearest the edge could fail first or the whole group could fail as a unit with the failure surface originating from the anchors farthest from the edge. If the anchors are welded to a common plate, when the anchor nearest the front edge begins to form a failure cone, shear load would be transferred to the stiffer and stronger rear anchor. For this reason, anchors welded to a common plate do not need to consider the failure mode shown in the upper right figure of Fig. RD.6.2.1(b). The *PCI Design Handbook* approach[D.17] suggests in Section 6.5.2.2 that the capacity of the anchors away from the edge be considered. Because this is a reasonable approach, assuming that the anchors are spaced far enough apart so that the shear failure surfaces do not intersect,[D.11] D.6.2 allows such a procedure. If the failure surfaces do not intersect, as would generally occur if the anchor spacing s is equal to or greater than $1.5c_{a1}$, then after formation of the near-edge failure surface, the higher capacity of the farther anchor would resist most of the load. As shown in the bottom right example in Fig. RD.6.2.1(b), it would be appropriate to consider the full shear capacity to be provided by this anchor with its much larger resisting failure surface. No contribution of the anchor near the edge is then considered. Checking the near-edge anchor condition is advisable to preclude undesirable cracking at service load conditions. Further discussion of design for multiple anchors is given in Reference D.8.

For the case of anchors near a corner subjected to a shear force with components normal to each edge, a satisfactory solution is to check independently the connection for each component of the shear force. Other specialized cases, such as the shear resistance of anchor groups where all anchors do not have the same edge distance, are treated in Reference D.11.

The detailed provisions of D.6.2.1(a) apply to the case of shear force directed towards an edge. When the shear force is directed away from the edge, the strength will usually be governed by D.6.1 or D.6.3.

CODE

D.6.2.2 — The basic concrete breakout strength in shear of a single anchor in cracked concrete, V_b, shall not exceed

$$V_b = 7\left(\frac{\ell_e}{d_o}\right)^{0.2}\sqrt{d_o}\sqrt{f_c'}(c_{a1})^{1.5} \tag{D-24}$$

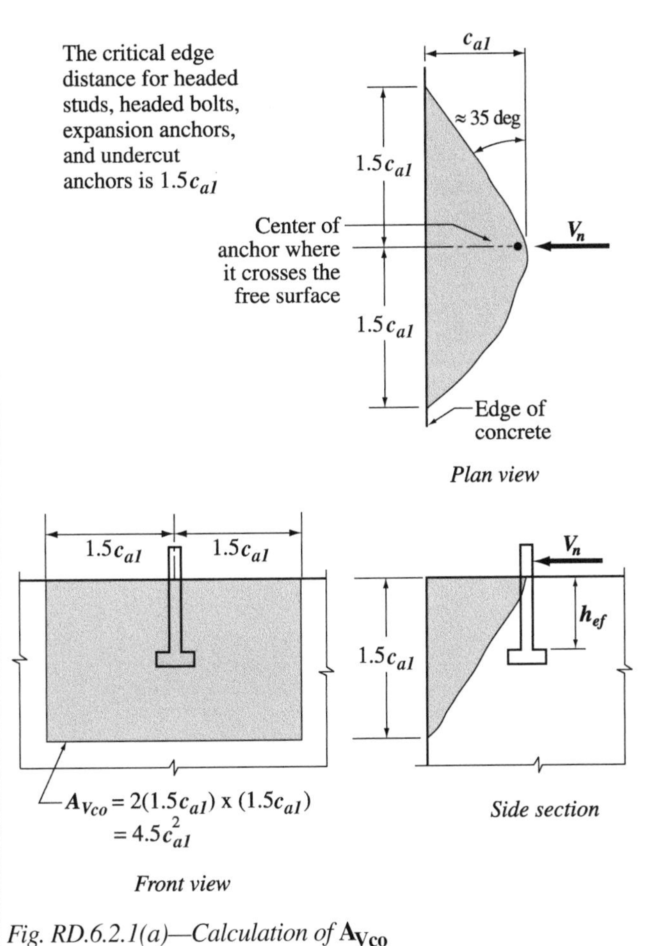

Fig. RD.6.2.1(a)—Calculation of A_{Vco}

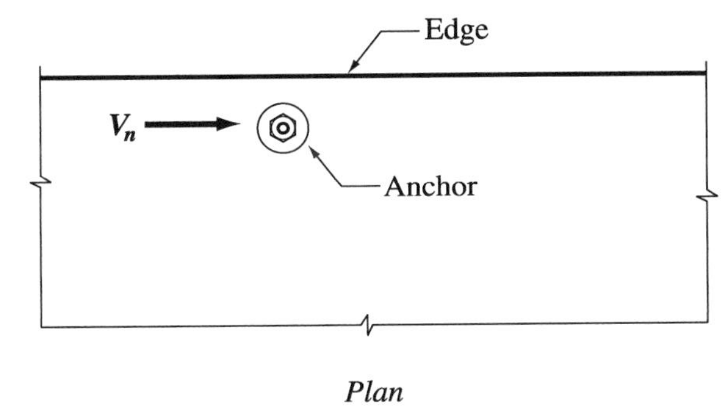

Fig. RD.6.2.1(c) — Shear force parallel to an edge

COMMENTARY

The case of shear force parallel to an edge is shown in Fig. RD.6.2.1(c). A special case can arise with shear force parallel to the edge near a corner. In the example of a single anchor near a corner (See Fig. RD.6.2.1(d)), the provisions for shear force applied perpendicular to the edge should be checked in addition to the provisions for shear force applied parallel to the edge.

RD.6.2.2 — Like the concrete breakout tensile capacity, the concrete breakout shear capacity does not increase with the failure surface, which is proportional to $(c_{a1})^2$. Instead the capacity increases proportionally to $(c_{a1})^{1.5}$ due to size effect. The capacity is also influenced by the anchor stiffness and the anchor diameter.[D.9-D.11,D.14]

if $h_a < 1.5c_{a1}$: $A_{vc} = 2(1.5c_{a1})h_a$

if $h_a < 1.5c_{a1}$: $A_{vc} = 2(1.5c_{a1})h_a$

Note: One assumption of the distribution of forces indicates that half the shear would be critical on front anchor and its projected area.

if $c_{a2} < 1.5c_{a1}$: $A_{vc} = 1.5c_{a1}(1.5c_{a1} + c_{a2})$

if $h_a < 1.5c_{a1}$: $A_{vc} = 2(1.5c_{a1})h_a$

if $h_a < 1.5c_{a1}$ and $s_1 < 3c_{a1}$: $A_{vc} = [2(1.5c_{a1}) + s_1]h_a$

Note: Another assumption of the distribution of forces indicates that the total shear would be critical on the rear anchor and its projected area. Only this assumption needs to be considered when anchors are rigidly connected to the attachment.

Fig. RD.6.2.1(b) — Projected area for single anchors and groups of anchors and calculation of A_{Vc}

Plan

Fig. RD.6.2.1(d) — Shear force near a corner.

CODE

where ℓ_e is the load bearing length of the anchor for shear:

$\ell_e = h_{ef}$ for anchors with a constant stiffness over the full length of embedded section, such as headed studs and post-installed anchors with one tubular shell over full length of the embedment depth,

$\ell_e = 2d_o$ for torque-controlled expansion anchors with a distance sleeve separated from expansion sleeve, and

in no case shall ℓ_e exceed $8d_o$.

D.6.2.3 — For cast-in headed studs, headed bolts, or hooked bolts that are continuously welded to steel attachments having a minimum thickness equal to the greater of 3/8 in. and half of the anchor diameter, the basic concrete breakout strength in shear of a single anchor in cracked concrete, V_b, shall not exceed

$$V_b = 8\left(\frac{\ell_e}{d_o}\right)^{0.2}\sqrt{d_o}\sqrt{f_c'}(c_{a1})^{1.5} \quad \text{(D-25)}$$

where ℓ_e is defined in D.6.2.2.

provided that:

(a) for groups of anchors, the strength is determined based on the strength of the row of anchors farthest from the edge;

(b) anchor spacing, s, is not less than 2.5 in.; and

(c) supplementary reinforcement is provided at the corners if $c_{a2} \le 1.5h_{ef}$.

D.6.2.4 — Where anchors are influenced by three or more edges, the value of c_{a1} used in Eq. (D-23) through (D-28) shall not exceed the greatest of: $c_{a2}/1.5$ in either direction, $h_a/1.5$; and one-third of the maximum spacing between anchors within the group.

COMMENTARY

The constant, 7, in the shear strength equation was determined from test data reported in Reference D.9 at the 5 percent fractile adjusted for cracking.

RD.6.2.3 — For the special case of cast-in headed bolts continuously welded to an attachment, test data[D.24] show that somewhat higher shear capacity exists, possibly due to the stiff welding connection clamping the bolt more effectively than an attachment with an anchor gap. Because of this, the basic shear value for such anchors is increased. Limits are imposed to ensure sufficient rigidity. The design of supplementary reinforcement is discussed in References D.8, D.11, and D.12.

RD.6.2.4 — For anchors influenced by three or more edges where any edge distance is less than $1.5c_{a1}$, the shear breakout strength computed by the basic CCD Method, which is the basis for Eq. (D-21) though (D-28), gives safe but overly conservative results. These special cases were studied for the κ Method[D.14] and the problem was pointed out by Lutz.[D.20] Similarly, the approach used for tensile breakouts in D.5.2.3, a correct evaluation of the capacity is determined if the value of c_{a1} used in Eq. (D-21) to (D-28) is limited to the maximum of $c_{a2}/1.5$ in each direction, $h_a/1.5$, and one-third of the maximum spacing between anchors within the group. The limit on c_{a1} of at least one-third of the maximum spacing between anchors within the group prevents the designer from using a calculated strength based on individual breakout prisms for a group anchor configuration.

CODE | **COMMENTARY**

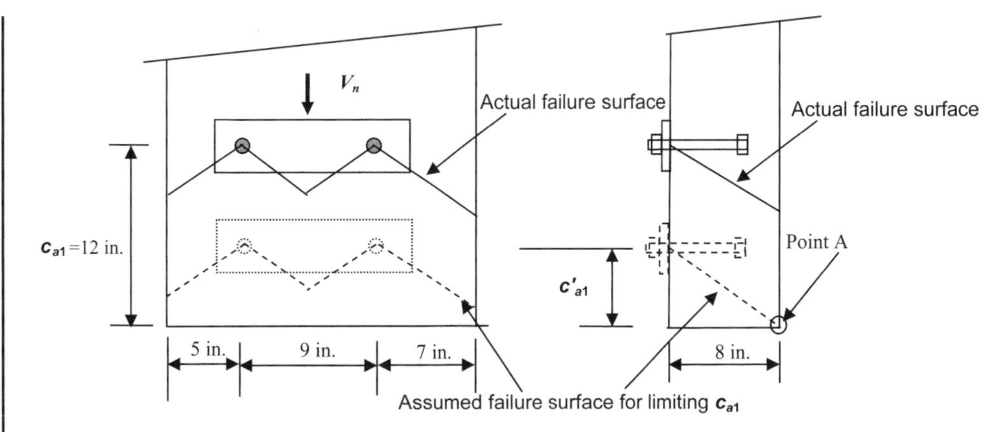

The actual c_{a1} = 12 in. but two orthogonal edges c_{a2} and h_a are ≤ 1.5 c_{a1} therefore the limiting value of c_{a1} (shown as c'_{a1} in the figure) is the larger of $c_{a2,max}$**/1.5**, h_a**/1.5** and one-third of the maximum spacing for an anchor group:
c'_{a1} = max (7/1.5, 8/1.5, 9/3) = 5.33 in.
Therefore, use c'_{a1} = 5.33 in. in Eq. (D-21) to (D-28) including the calculation of A_{vc}:
A_{vc} = (5 + 9 + 7)(1.5(5.33)) = 168 in.^{2}
Point A shows the intersection of the assumed failure surface for limiting c_{a1} with the concrete surface.

Fig. RD.6.2.4—Shear when anchors are influenced by three or more degrees.

This approach is illustrated in Fig. RD.6.2.4. In this example, the limit on the value of c_{a1} is the largest of c_{a2}**/1.5** in either direction, h_a**/1.5**, and one-third the maximum spacing between anchors for anchor groups results in c'_{a1} = 5.33 in. For this example, this would be the proper value to be used for c_{a1} in computing V_{cb} or V_{cbg}, even if the actual edge distance that the shear is directed toward is larger. The requirement of D.6.2.4 may be visualized by moving the actual concrete breakout surface originating at the actual c_{a1} toward the surface of the concrete in the direction of the applied shear load. The value of c_{a1} used in Eq. (D-21) to (D-28) is determined when either: (a) the outer boundaries of the failure surface first intersect a free edge; or (b) the intersection of the breakout surface between anchors within the group first intersects the surface of the concrete. For the example shown in Fig. RD.6.2.4, point "A" shows the intersection of the assumed failure surface for limiting c_{a1} with the concrete surface.

D.6.2.5 — The modification factor for anchor groups loaded eccentrically in shear is

$$\psi_{ec,V} = \frac{1}{\left(1 + \frac{2e_V'}{3c_{a1}}\right)} \leq 1 \qquad \text{(D-26)}$$

If the loading on an anchor group is such that only some anchors are loaded in shear in the same direction, only those anchors that are loaded in shear in the same direction shall be considered when determining the eccentricity of e_V' for use in Eq. (D-26) and for the calculation of V_{cbg} in Eq. (D-22).

RD.6.2.5 — This section provides a modification factor for an eccentric shear force towards an edge on a group of anchors. If the shear force originates above the plane of the concrete surface, the shear should first be resolved as a shear in the plane of the concrete surface, with a moment that may or may not also cause tension in the anchors, depending on the normal force. Figure RD.6.2.5 defines the term e_v' for calculating the $\psi_{ec,V}$ modification factor that accounts for the fact that more shear is applied to one anchor than others, tending to split the concrete near an edge.

CODE

D.6.2.6 — The modification factor for edge effect for a single anchor or group of anchors loaded in shear is:

$$\psi_{ed,V} = 1.0 \quad \text{(D-27)}$$

if $c_{a2} \geq 1.5c_{a1}$

$$\psi_{ed,V} = 0.7 + 0.3\frac{c_{a2}}{1.5c_{a1}} \quad \text{(D-28)}$$

if $c_{a2} < 1.5c_{a1}$

D.6.2.7 — For anchors located in a region of a concrete member where analysis indicates no cracking at service loads, the following modification factor shall be permitted

$$\psi_{c,V} = 1.4$$

For anchors located in a region of a concrete member where analysis indicates cracking at service load levels, the following modification factors shall be permitted:

$\psi_{c,V}$ = **1.0** for anchors in cracked concrete with no supplementary reinforcement or edge reinforcement smaller than a No. 4 bar;

$\psi_{c,V}$ = **1.2** for anchors in cracked concrete with supplementary reinforcement of a No. 4 bar or greater between the anchor and the edge; and

$\psi_{c,V}$ = **1.4** for anchors in cracked concrete with supplementary reinforcement of a No. 4 bar or greater between the anchor and the edge, and with the supplementary reinforcement enclosed within stirrups spaced at not more than 4 in.

D.6.3 — Concrete pryout strength of anchor in shear

D.6.3.1 — The nominal pryout strength, V_{cp} or V_{cpg} shall not exceed:

(a) for a single anchor:

$$V_{cp} = k_{cp}N_{cb} \quad \text{(D-29)}$$

(b) for a group of anchors:

$$V_{cpg} = k_{cp}N_{cbg} \quad \text{(D-30)}$$

where:

COMMENTARY

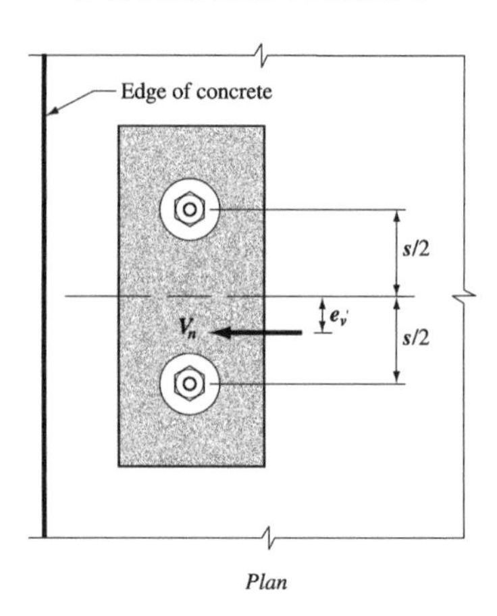

Fig. RD.6.2.5 — Definition of dimensions e_V'.

RD.6.2.7 — Torque-controlled and displacement-controlled expansion anchors are permitted in cracked concrete under pure shear loadings.

RD.6.3 — Concrete pryout strength of anchor in shear

Reference D.9 indicates that the pryout shear resistance can be approximated as one to two times the anchor tensile resistance with the lower value appropriate for h_{ef} less than 2.5 in.

CODE

k_{cp} = **1.0** for h_{ef} < 2.5 in.; and
k_{cp} = **2.0** for $h_{ef} \geq$ 2.5 in.

N_{cb} and N_{cbg} shall be determined from Eq. (D-4) and (D-5), respectively.

COMMENTARY

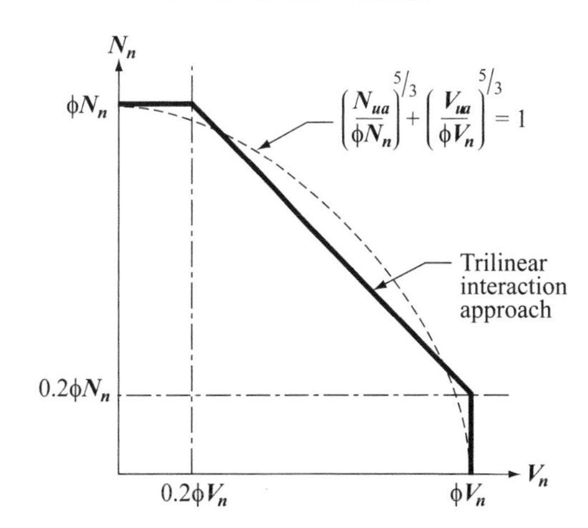

Fig. RD.7 — Shear and tensile load interaction equation.

D.7 — Interaction of tensile and shear forces

Unless determined in accordance with D.4.3, anchors or groups of anchors that are subjected to both shear and axial loads shall be designed to satisfy the requirements of D.7.1 through D.7.3. The value of ϕN_n shall be as required in D.4.1.2. The value of ϕV_n shall be as defined in D.4.1.2.

D.7.1 — If $V_{ua} \leq 0.2\phi V_n$, then full strength in tension shall be permitted: $\phi N_n \geq N_{ua}$.

D.7.2 — If $N_{ua} \leq 0.2\phi N_n$, then full strength in shear shall be permitted: $\phi V_n \geq V_{ua}$.

D.7.3 — If $V_{ua} > 0.2\phi V_n$ and $N_{ua} > 0.2\phi N_n$, then:

$$\frac{N_{ua}}{\phi N_n} + \frac{V_{ua}}{\phi V_n} \leq 1.2 \quad \text{(D-31)}$$

RD.7 — Interaction of tensile and shear forces

The shear-tension interaction expression has traditionally been expressed as:

$$\left(\frac{N_{ua}}{N_n}\right)^{\zeta} + \left(\frac{V_{ua}}{V_n}\right)^{\zeta} \leq 1.0$$

where ζ varies from 1 to 2. The current trilinear recommendation is a simplification of the expression where ζ **= 5/3** (Fig. RD.7). The limits were chosen to eliminate the requirement for computation of interaction effects where very small values of the second force are present. Any other interaction expression that is verified by test data, however, can be used to satisfy D.4.3.

D.8 — Required edge distances, spacings, and thicknesses to preclude splitting failure

Minimum spacings and edge distances for anchors and minimum thicknesses of members shall conform to D.8.1 through D.8.6, unless supplementary reinforcement is provided to control splitting. Lesser values from product-specific tests performed in accordance with ACI 355.2 shall be permitted.

D.8.1 — Unless determined in accordance with D.8.4, minimum center-to-center spacing of anchors shall be $4d_o$ for untorqued cast-in anchors, and $6d_o$ for torqued cast-in anchors and post-installed anchors.

RD.8 — Required edge distances, spacings, and thicknesses to preclude splitting failure

The minimum spacings, edge distances, and thicknesses are very dependent on the anchor characteristics. Installation forces and torques in post-installed anchors can cause splitting of the surrounding concrete. Such splitting also can be produced in subsequent torquing during connection of attachments to anchors including cast-in anchors. The primary source of values for minimum spacings, edge distances, and thicknesses of post-installed anchors should be the product-specific tests of ACI 355.2. In some cases, however, specific products are not known in the design stage. Approximate values are provided for use in design.

CODE

D.8.2 — Unless determined in accordance with D.8.4, minimum edge distances for cast-in headed anchors that will not be torqued shall be based on minimum cover requirements for reinforcement in 7.7. For cast-in headed anchors that will be torqued, the minimum edge distances shall be $6d_o$.

D.8.3 — Unless determined in accordance with D.8.4, minimum edge distances for post-installed anchors shall be based on the greater of the minimum cover requirements for reinforcement in 7.7, or the minimum edge distance requirements for the products as determined by tests in accordance with ACI 355.2, and shall not be less than 2.0 times the maximum aggregate size. In the absence of product-specific ACI 355.2 test information, the minimum edge distance shall be taken as not less than:

Undercut anchors .... $6d_o$
Torque-controlled anchors .... $8d_o$
Displacement-controlled anchors .... $10d_o$

D.8.4 — For anchors where installation does not produce a splitting force and that will remain untorqued, if the edge distance or spacing is less than those specified in D.8.1 to D.8.3, calculations shall be performed by substituting for d_o a smaller value d_o' that meets the requirements of D.8.1 to D.8.3. Calculated forces applied to the anchor shall be limited to the values corresponding to an anchor having a diameter of d_o'.

D.8.5 — The value of h_{ef} for an expansion or undercut post-installed anchor shall not exceed the greater of 2/3 of the member thickness and the member thickness less 4 in.

D.8.6 — Unless determined from tension tests in accordance with ACI 355.2, the critical edge distance, c_{ac}, shall not be taken less than:

Undercut anchors	$2.5h_{ef}$
Torque-controlled anchors	$4h_{ef}$
Displacement-controlled anchors	$4h_{ef}$

D.8.7 — Project drawings and project specifications shall specify use of anchors with a minimum edge distance as assumed in design.

COMMENTARY

RD.8.2 — Because the edge cover over a deep embedment close to the edge can have a significant effect on the side-face blowout strength of D.5.4, in addition to the normal concrete cover requirements, the designer may wish to use larger cover to increase the side-face blowout strength.

RD.8.3 — Drilling holes for post-installed anchors can cause microcracking. The requirement for a minimum edge distance twice the maximum aggregate size is to minimize the effects of such microcracking.

RD.8.4 — In some cases, it may be desirable to use a larger-diameter anchor than the requirements on D.8.1 to D.8.3 permit. In these cases, it is permissible to use a larger-diameter anchor provided the design strength of the anchor is based on a smaller assumed anchor diameter, d_o'.

RD.8.5 — This minimum thickness requirement is not applicable to through-bolts because they are outside the scope of Appendix D. In addition, splitting failures are caused by the load transfer between the bolt and the concrete. Because through-bolts transfer their load differently than cast-in or expansion and undercut anchors, they would not be subject to the same member thickness requirements. Post-installed anchors should not be embedded deeper than 2/3 of the member thickness.

RD.8.6 — The critical edge distance c_{ac} is determined by the corner test in ACI 355.2. Research has indicated that the corner-test requirements are not met with $c_{a,min} = 1.5h_{ef}$ for many expansion anchors and some undercut anchors because installation of these types of anchors introduces splitting tensile stresses in the concrete that are increased during load application, potentially resulting in a premature splitting failure. To permit the design of these types of anchors when product-specific information is not available, conservative default values for c_{ac} are provided.

CODE

D.9 — Installation of anchors

D.9.1 — Anchors shall be installed in accordance with the project drawings and project specifications.

COMMENTARY

RD.9 — Installation of anchors

Many anchor performance characteristics depend on proper installation of the anchor. Anchor capacity and deformations can be assessed by acceptance testing under ACI 355.2. These tests are carried out assuming that the manufacturer's installation directions will be followed. Certain types of anchors can be sensitive to variations in hole diameter, cleaning conditions, orientation of the axis, magnitude of the installation torque, crack width, and other variables. Some of this sensitivity is indirectly reflected in the assigned ϕ values for the different anchor categories, which depend in part on the results of the installation safety tests. Gross deviations from the ACI 355.2 acceptance testing results could occur if anchor components are incorrectly exchanged, or if anchor installation criteria and procedures vary from those recommended. Project specifications should require that anchors be installed according to the manufacturer's recommendations.

Notes

APPENDIX E — STEEL REINFORCEMENT INFORMATION

As an aid to users of the ACI Building Code, information on sizes, areas, and weights of various steel reinforcement is presented.

ASTM STANDARD REINFORCING BARS

Bar size, no.	Nominal diameter, in.	Nominal area, in.^{2}	Nominal weight, lb/ft
3	0.375	0.11	0.376
4	0.500	0.20	0.668
5	0.625	0.31	1.043
6	0.750	0.44	1.502
7	0.875	0.60	2.044
8	1.000	0.79	2.670
9	1.128	1.00	3.400
10	1.270	1.27	4.303
11	1.410	1.56	5.313
14	1.693	2.25	7.650
18	2.257	4.00	13.600

ASTM STANDARD PRESTRESSING TENDONS

Type*	Nominal diameter, in.	Nominal area, in.^{2}	Nominal weight, lb/ft
Seven-wire strand (Grade 250)	1/4 (0.250)	0.036	0.122
	5/16 (0.313)	0.058	0.197
	3/8 (0.375)	0.080	0.272
	7/16 (0.438)	0.108	0.367
	1/2 (0.500)	0.144	0.490
	(0.600)	0.216	0.737
Seven-wire strand (Grade 270)	3/8 (0.375)	0.085	0.290
	7/16 (0.438)	0.115	0.390
	1/2 (0.500)	0.153	0.520
	(0.600)	0.217	0.740
Prestressing wire	0.192	0.029	0.098
	0.196	0.030	0.100
	0.250	0.049	0.170
	0.276	0.060	0.200
Prestressing bars (plain)	3/4	0.44	1.50
	7/8	0.60	2.04
	1	0.78	2.67
	1-1/8	0.99	3.38
	1-1/4	1.23	4.17
	1-3/8	1.48	5.05
Prestressing bars (deformed)	5/8	0.28	0.98
	3/4	0.42	1.49
	1	0.85	3.01
	1-1/4	1.25	4.39
	1-3/8	1.58	5.56

* Availability of some tendon sizes should be investigated in advance.

WRI STANDARD WIRE REINFORCEMENT*

W & D size		Nominal	Nominal	Nominal	Area, in.^{2}/ft of width for various spacings						
					Center-to-center spacing, in.						
Plain	Deformed	diameter, in.	area, in.^{2}	weight, lb/ft	2	3	4	6	8	10	12
W31	D31	0.628	0.310	1.054	1.86	1.24	0.93	0.62	0.46	0.37	0.31
W30	D30	0.618	0.300	1.020	1.80	1.20	0.90	0.60	0.45	0.36	0.30
W28	D28	0.597	0.280	0.952	1.68	1.12	0.84	0.56	0.42	0.33	0.28
W26	D26	0.575	0.260	0.934	1.56	1.04	0.78	0.52	0.39	0.31	0.26
W24	D24	0.553	0.240	0.816	1.44	0.96	0.72	0.48	0.36	0.28	0.24
W22	D22	0.529	0.220	0.748	1.32	0.88	0.66	0.44	0.33	0.26	0.22
W20	D20	0.504	0.200	0.680	1.20	0.80	0.60	0.40	0.30	0.24	0.20
W18	D18	0.478	0.180	0.612	1.08	0.72	0.54	0.36	0.27	0.21	0.18
W16	D16	0.451	0.160	0.544	0.96	0.64	0.48	0.32	0.24	0.19	0.16
W14	D14	0.422	0.140	0.476	0.84	0.56	0.42	0.28	0.21	0.16	0.14
W12	D12	0.390	0.120	0.408	0.72	0.48	0.36	0.24	0.18	0.14	0.12
W11	D11	0.374	0.110	0.374	0.66	0.44	0.33	0.22	0.16	0.13	0.11
W10.5		0.366	0.105	0.357	0.63	0.42	0.315	0.21	0.15	0.12	0.105
W10	D10	0.356	0.100	0.340	0.60	0.40	0.30	0.20	0.15	0.12	0.10
W9.5		0.348	0.095	0.323	0.57	0.38	0.285	0.19	0.14	0.11	0.095
W9	D9	0.338	0.090	0.306	0.54	0.36	0.27	0.18	0.13	0.10	0.09
W8.5		0.329	0.085	0.289	0.51	0.34	0.255	0.17	0.12	0.10	0.085
W8	D8	0.319	0.080	0.272	0.48	0.32	0.24	0.16	0.12	0.09	0.08
W7.5		0.309	0.075	0.255	0.45	0.30	0.225	0.15	0.11	0.09	0.075
W7	D7	0.298	0.070	0.238	0.42	0.28	0.21	0.14	0.10	0.08	0.07
W6.5		0.288	0.065	0.221	0.39	0.26	0.195	0.13	0.09	0.07	0.065
W6	D6	0.276	0.060	0.204	0.36	0.24	0.18	0.12	0.09	0.07	0.06
W5.5		0.264	0.055	0.187	0.33	0.22	0.165	0.11	0.08	0.06	0.055
W5	D5	0.252	0.050	0.170	0.30	0.20	0.15	0.10	0.07	0.06	0.05
W4.5		0.240	0.045	0.153	0.27	0.18	0.135	0.09	0.06	0.05	0.045
W4	D4	0.225	0.040	0.136	0.24	0.16	0.12	0.08	0.06	0.04	0.04
W3.5		0.211	0.035	0.119	0.21	0.14	0.105	0.07	0.05	0.04	0.035
W3		0.195	0.030	0.102	0.18	0.12	0.09	0.06	0.04	0.03	0.03
W2.9		0.192	0.029	0.098	0.174	0.116	0.087	0.058	0.04	0.03	0.029
W2.5		0.178	0.025	0.085	0.15	0.10	0.075	0.05	0.03	0.03	0.025
W2		0.159	0.020	0.068	0.12	0.08	0.06	0.04	0.03	0.02	0.02
W1.4		0.135	0.014	0.049	0.084	0.056	0.042	0.028	0.02	0.01	0.014

*Reference "Structural Welded Wire Reinforcement Manual of Standard Practice," Wire Reinforcement Institute, Hartford, CT, 6th Edition, Apr. 2001, 38 pp.

COMMENTARY REFERENCES

References, Chapter 1

1.1. ACI Committee 307, "Standard Practice for the Design and Construction of Cast-in-Place Reinforced Concrete Chimneys (ACI 307-98)," American Concrete Institute, Farmington Hills, MI, 1998, 32 pp. Also *ACI Manual of Concrete Practice.*

1.2. ACI Committee 313, "Standard Practice for Design and Construction of Concrete Silos and Stacking Tubes for Storing Granular Materials (ACI 313-97)," American Concrete Institute, Farmington Hills, MI, 1997, 22 pp. Also *ACI Manual of Concrete Practice.*

1.3. ACI Committee 350, "Environmental Engineering Concrete Structures (ACI 350R-89)," American Concrete Institute, Farmington Hills, MI, 1989, 20 pp. Also *ACI Manual of Concrete Practice.*

1.4. ACI Committee 349, "Code Requirements for Nuclear Safety Related Concrete Structures (ACI 349-97)," American Concrete Institute, Farmington Hills, MI, 1997, 129 pp., plus 1997 Supplement. Also *ACI Manual of Concrete Practice.*

1.5. ACI-ASME Committee 359, "Code for Concrete Reactor Vessels and Containments (ACI 359-92)," American Concrete Institute, Farmington Hills, MI, 1992.

1.6. ACI Committee 543, "Recommendations for Design, Manufacture, and Installation of Concrete Piles, (ACI 543R-74) (Reapproved 1980)," ACI JOURNAL, *Proceedings* V. 71, No. 10, Oct. 1974, pp. 477-492.

1.7. ACI Committee 336, "Design and Construction of Drilled Piers (ACI 336.3R-93)," American Concrete Institute, Farmington Hills, MI, 1993, 30 pp. Also *ACI Manual of Concrete Practice.*

1.8. "Recommended Practice for Design, Manufacture and Installation of Prestressed Concrete Piling," *PCI Journal*, V. 38, No. 2, Mar.-Apr. 1993, pp. 14-41.

1.9. ACI Committee 360, "Design of Slabs on Grade (ACI 360R-92 [Reapproved 1997])," American Concrete Institute, Farmington Hills, MI, 1997, 57 pp. Also *ACI Manual of Concrete Practice.*

1.10. PTI, "Design of Post-Tensioned Slabs-on-Ground," 3rd Edition, Post-Tensioning Institute, Phoenix, AZ, 2004, 106 pp.

1.11. ANSI/ASCE 3-91, "Standard for the Structural Design of Composite Slabs," ASCE, Reston, VA, 1994.

1.12. ANSI/ASCE 9-91, "Standard Practice for the Construction and Inspection of Composite Slabs," American Society of Civil Engineers, Reston, VA, 1994.

1.13. "The BOCA National Building Code, 13th Edition," Building Officials and Code Administration International, Inc., Country Club Hills, IL, 1996, 357 pp.

1.14. "Standard Building Code," Southern Building Code Congress International, Inc., Birmingham, AL, 1996, 656 pp.

1.15. "NEHRP Recommended Provisions for Seismic Regulations for New Buildings and Other Structures," Part 1: Provisions (FEMA 222, 199 pp.) and Part 2: Commentary (FEMA 223, 237pp.), Building Seismic Safety Council, Washington D.C., 1997.

1.16. "International Building Code," International Code Council, Falls Church, VA, 2000.

1.17. "International Building Code," International Code Council, Falls Church, VA, 2003.

1.18. "Building Construction and Safety Code—NFPA 5000," National Fire Protection Association, Quincy, MA, 2003.

1.19. "Minimum Design Loads for Buildings and Other Structures (SEI/ASCE 7-02)," ASCE, Reston, VA, 2002, 376 pp.

1.20. *Uniform Building Code, V. 2, Structural Engineering Design Provisions*, 1997 Edition, International Conference of Building Officials, Whittier, CA, 1997, 492 pp.

1.21. ACI Committee 311, "Guide for Concrete Inspection (ACI 311.4R-95)," American Concrete Institute, Farmington Hills, MI, 1995, 11 pp. Also *ACI Manual of Concrete Practice.*

1.22. ACI Committee 311, *ACI Manual of Concrete Inspection*, SP-2, 8th Edition, American Concrete Institute, Farmington Hills, MI, 1992, 200 pp.

References, Chapter 2

2.1. ACI Committee 116, "Cement and Concrete Terminology (ACI 116R-90)," American Concrete Institute, Farmington Hills, MI, 1990, 58 pp. Also *ACI Manual of Concrete Practice.*

References, Chapter 3

3.1. ACI Committee 214, "Recommended Practice for Evaluation of Strength Test Results of Concrete (ACI 214-77) (Reapproved 1989)," (ANSI/ACI 214-77), American Concrete Institute, Farmington Hills, MI, 1977, 14 pp. Also *ACI Manual of Concrete Practice.*

3.2 ACI Committee 440, "Guide for the Design and Construction of Concrete Reinforced with FRP Bars (ACI 440.1R-03)," American Concrete Institute, Farmington Hills, MI, 42 pp. Also *ACI Manual of Concrete Practice.*

3.3 ACI Committee 440, "Guide for the Design and Construction of Externally Bonded FRP Systems for Strengthening of Concrete Structures (ACI 440.2R-02)," American Concrete Institute, Farmington Hills, MI, 45 pp. Also *ACI Manual of Concrete Practice.*

3.4. Gustafson, D. P., and Felder, A. L., "Questions and Answers on ASTM A 706 Reinforcing Bars," *Concrete International*, V. 13, No. 7, July 1991, pp. 54-57.

3.5. ACI Committee 223, "Standard Practice for the Use of Shrinkage-Compensating Concrete (ACI 223-98)," American Concrete Institute, Farmington Hills, MI, 29 pp. Also *ACI Manual of Concrete Practice.*

References, Chapter 4

4.1. Dikeou, J. T., "Fly Ash Increases Resistance of Concrete to Sulfate Attack," *Research Report* No. C-1224, Concrete and Structures Branch, Division of Research, U.S. Bureau of Reclamation, Jan. 1967, 25 pp.

4.2. ASTM C 1012-89, "Test Method for Length Change of Hydraulic-Cement Mortars Exposed to a Sulfate Solution," *ASTM Book of Standards*, Part 04.01, ASTM, West Conshohocken, PA, 5 pp.

4.3. ACI Committee 211, "Standard Practice for Selecting Proportions for Normal, Heavyweight, and Mass Concrete (ACI 211.1-91)," American Concrete Institute, Farmington Hills, MI, 1991, 38 pp. Also *ACI Manual of Concrete Practice*.

4.4. Drahushak-Crow, R., "Freeze-Thaw Durability of Fly Ash Concrete," *EPRI Proceedings*, Eighth International Ash Utilization Symposium, V. 2, Oct. 1987, p. 37-1.

4.5. Sivasundaram, V.; Carette, G. G.; and Malhotra, V. M., "Properties of Concrete Incorporating Low Quantity of Cement and High Volumes of Low-Calcium Fly Ash," *Fly Ash, Silica Fume, Slag, and Natural Pozzolans in Concrete*, SP-114, American Concrete Institute, Farmington Hills, MI, 1989, pp. 45-71.

4.6. Whiting, D., "Deicer Scaling and Resistance of Lean Concretes Containing Fly Ash," *Fly Ash, Silica Fume, Slag, and Natural Pozzolans in Concrete*, SP-114, American Concrete Institute, Farmington Hills, MI, 1989, pp. 349-372.

4.7. Rosenberg, A., and Hanson, C. M., "Mechanisms of Corrosion of Steel in Concrete," *Materials Science in Concrete I*, American Ceramic Society, Westerville, OH, 1989, p. 285.

4.8. Berry, E. E., and Malhotra, V. M., *Fly Ash in Concrete*, CANMET, Ottawa, Ontario, Canada, 1985.

4.9. Li, S., and Roy, D. M., "Investigation of Relations between Porosity, Pore Structure and CL Diffusion of Fly Ash and Blended Cement Pastes," *Cement and Concrete Research*, V. 16, No. 5, Sept. 1986, pp. 749-759.

4.10. ACI Committee 201, "Guide to Durable Concrete (ACI 201.2R-92)," American Concrete Institute, Farmington Hills, MI, 1992, 39 pp. Also *ACI Manual of Concrete Practice*.

4.11. ACI Committee 222, "Corrosion of Metals in Concrete (ACI 222R-96)," American Concrete Institute, Farmington Hills, MI, 1996, 30 pp. Also *ACI Manual of Concrete Practice*.

4.12. Ozyildirim, C., and Halstead, W., "Resistance to Chloride Ion Penetration of Concretes Containing Fly Ash, Silica Fume, or Slag," *Permeability of Concrete,* SP-108, American Concrete Institute, Farmington Hills, MI, 1988, pp. 35-61.

4.13. ASTM C 1202-97, "Standard Test Method for Electrical Indication of Concrete's Ability to Resist Chloride Ion Penetration," *ASTM Book of Standards,* Part 04.02, ASTM, West Conshohocken, PA, 6 pp.

References, Chapter 5

5.1. ACI Committee 211, "Standard Practice for Selecting Proportions for Normal, Heavyweight, and Mass Concrete (ACI 211.1-98)," American Concrete Institute, Farmington Hills, MI, 1998, 38 pp. Also *ACI Manual of Concrete Practice*.

5.2 ACI Committee 211, "Standard Practice for Selecting Proportions for Structural Lightweight Concrete (ACI 211.2-91)," American Concrete Institute, Farmington Hills, MI, 1991, 18 pp. Also, *ACI Manual of Concrete Practice*.

5.3. ASTM C 1077-92, "Standard Practice for Laboratories Testing Concrete and Concrete Aggregates for Use in Construction and Criteria for Laboratory Evaluation," ASTM, West Conshohocken, PA, 5 pp.

5.4. ASTM D 3665-99, "Standard Practice for Random Sampling of Construction Materials," ASTM, West Conshohocken, PA, 5 pp.

5.5. Bloem, D. L., "Concrete Strength Measurement—Cores vs. Cylinders," *Proceedings*, ASTM, V. 65, 1965, pp. 668-696.

5.6. Bloem, Delmar L., "Concrete Strength in Structures," ACI JOURNAL, *Proceedings* V. 65, No. 3, Mar. 1968, pp. 176-187.

5.7. Malhotra, V. M., *Testing Hardened Concrete: Nondestructive Methods,* ACI Monograph No. 9, American Concrete Institute/Iowa State University Press, Farmington Hills, MI, 1976, 188 pp.

5.8. Malhotra, V. M., "Contract Strength Requirements—Cores Versus In Situ Evaluation," ACI JOURNAL, *Proceedings* V. 74, No. 4, Apr. 1977, pp. 163-172.

5.9 Bartlett, M.F., and MacGregor, J.G., "Effect of Moisture Condition on Concrete Core Strengths," *ACI Materials Journal,* V. 91, No. 3, May-June, 1994, pp. 227-236.

5.10. ACI Committee 304, "Guide for Measuring, Mixing, Transporting, and Placing Concrete (ACI 304R-89)," American Concrete Institute, Farmington Hills, MI, 1989, 49 pp. Also *ACI Manual of Concrete Practice*.

5.11. Newlon, H., Jr., and Ozol, A., "Delayed Expansion of Concrete Delivered by Pumping through Aluminum Pipe Line," *Concrete Case Study* No. 20; Virginia Highway Research Council, Oct. 1969, 39 pp.

5.12. ACI Committee 309, "Guide for Consolidation of Concrete (ACI 309R-96)," American Concrete Institute, Farmington Hills, MI, 1996, 40 pp. Also *ACI Manual of Concrete Practice*.

5.13. ACI Committee 308, "Guide to Curing Concrete (ACI 308R-01)," American Concrete Institute, Farmington Hills, MI, 2001, 31 pp. Also *ACI Manual of Concrete Practice*.

5.14. ACI Committee 306, "Cold Weather Concreting (ACI 306R-88)," American Concrete Institute, Farmington Hills, MI, 1988, 23 pp. Also *ACI Manual of Concrete Practice*.

5.15. ACI Committee 305, "Hot Weather Concreting (ACI 305R-91)," American Concrete Institute, Farmington Hills, MI, 1991, 17 pp. Also *ACI Manual of Concrete Practice*.

References, Chapter 6

6.1. ACI Committee 347, "Guide to Formwork for Concrete (ACI 347R-94)," American Concrete Institute, Farmington Hills, MI, 1994, 33 pp. Also *ACI Manual of Concrete Practice*.

6.2. Hurd, M. K., and ACI Committee 347, *Formwork for Concrete*, SP-4, 5th Edition, American Concrete Institute, Farmington Hills, MI, 1989, 475 pp.

6.3. Liu, X. L.; Lee, H. M.; and Chen, W. F., "Shoring and Reshoring of High-Rise Buildings," *Concrete International*, V. 10, No. 1, Jan. 1989, pp. 64-68.

6.4. ASTM C 873-99, "Standard Test Method for Compressive Strength of Concrete Cylinders Cast-in-Place in Cylindrical

Molds," ASTM, West Conshohocken, PA, 4 pp.

6.5. ASTM C 803/C 803M-97$^{\varepsilon 1}$, "Test Method for Penetration Resistance of Hardened Concrete," ASTM, West Conshohocken, PA, 4 pp.

6.6. ASTM C 900, "Standard Test Method for Pullout Strength of Hardened Concrete," ASTM, West Conshohocken, PA, 5 pp.

6.7. ASTM C 1074-87, "Estimating Concrete Strength by the Maturity Method," ASTM, West Conshohocken, PA.

6.8. "Power Piping (ANSI/ASME B 31.1-1992)," American Society of Mechanical Engineers, New York, 1992.

6.9. "Chemical Plant and Petroleum Refinery Piping (ANSI/ASME B 31.3-1990)," American Society of Mechanical Engineers, New York, 1990.

References, Chapter 7

7.1. ACI Committee 315, *ACI Detailing Manual—1994*, SP-66, American Concrete Institute, Farmington Hills, MI, 1994, 244 pp. Also "Details and Detailing of Concrete Reinforcement (ACI 315-92)," and "Manual of Engineering and Placing Drawings for Reinforced Structures (ACI 315R-94)." Also *ACI Manual of Concrete Practice.*

7.2. Black, William C., "Field Corrections to Partially Embedded Reinforcing Bars," ACI JOURNAL, *Proceedings* V. 70, No. 10, Oct. 1973, pp. 690-691.

7.3. Stecich, J.; Hanson, J. M.; and Rice, P. F.; "Bending and Straightening of Grade 60 Reinforcing Bars," *Concrete International: Design & Construction*, V. 6, No. 8, Aug. 1984, pp. 14-23.

7.4. Kemp, E. L.; Brezny, F. S.; and Unterspan, J. A., "Effect of Rust and Scale on the Bond Characteristics of Deformed Reinforcing Bars," ACI JOURNAL, *Proceedings* V. 65, No. 9, Sept. 1968, pp. 743-756.

7.5. Sason, A. S. "Evaluation of Degree of Rusting on Prestressed Concrete Strand," *PCI Journal*, V. 37, No. 3, May-June 1992, pp. 25-30.

7.6. ACI Committee 117, "Standard Tolerances for Concrete Construction and Materials (ACI 117-90)," American Concrete Institute, Farmington Hills, MI, 22 pp. Also *ACI Manual of Concrete Practice.*

7.7. *PCI Design Handbook: Precast and Prestressed Concrete*, 4th Edition, Precast/Prestressed Concrete Institute, Chicago, 1992, 580 pp.

7.8. ACI Committee 408, "Bond Stress—The State of the Art," ACI JOURNAL, *Proceedings* V. 63, No. 11, Nov. 1966, pp. 1161-1188.

7.9. "Standard Specifications for Highway Bridges," 15th Edition, American Association of State Highway and Transportation Officials, Washington, D.C., 1992, 686 pp.

7.10. Deatherage, J. H., Burdette, E. G. and Chew, C. K., "Development Length and Lateral Spacing Requirements of Prestressing Strand for Prestressed Concrete Bridge Girders," *PCI Journal*, V. 39, No. 1, Jan.-Feb. 1994, pp. 70-83.

7.11. Russell, B. W., and Burns, N. H. "Measured Transfer Lengths of 0.5 and 0.6 in. Strands in Pretensioned Concrete," *PCI Journal*, V. 41, No. 5, Sept.-Oct. 1996, pp. 44-65.

7.12. ACI Committee 362, "Design of Parking Structures (ACI 362.1R-97)," American Concrete Institute, Farmington Hills, MI, 1997, 40 pp.

7.13. Hanson, N. W., and Conner, H. W., "Seismic Resistance of Reinforced Concrete Beam-Column Joints," *Proceedings*, ASCE, V. 93, ST5, Oct. 1967, pp. 533-560.

7.14. ACI-ASCE Committee 352, "Recommendations for Design of Beam-Column Joints in Monolithic Reinforced Concrete Structures (ACI 352R-91)," American Concrete Institute, Farmington Hills, MI, 1991, 18 pp. Also *ACI Manual of Concrete Practice.*

7.15. Pfister, J. F., "Influence of Ties on the Behavior of Reinforced Concrete Columns," ACI JOURNAL, *Proceedings* V. 61, No. 5, May 1964, pp. 521-537.

7.16. Gilbert, R. I., "Shrinkage Cracking in Fully Restrained Concrete Members," *ACI Structural Journa*l, V. 89, No. 2, Mar.-Apr. 1992, pp. 141-149.

7.17. "Design and Typical Details of Connections for Precast and Prestressed Concrete," MNL-123-88, Precast/Prestressed Concrete Institute, Chicago, 1988, 270 pp.

7.18. PCI Building Code Committee, "Proposed Design Requirements for Precast Concrete," *PCI Journal*, V. 31, No. 6, Nov.-Dec. 1986, pp. 32-47.

References, Chapter 8

8.1. Fintel, M.; Ghosh, S. K.; and Iyengar, H., *Column Shortening in Tall Buildings—Prediction and Compensation,* EB108D, Portland Cement Association, Skokie, Ill., 1986, 34 pp.

8.2. Cohn, M. Z., "Rotational Compatibility in the Limit Design of Reinforced Concrete Continuous Beams," *Flexural Mechanics of Reinforced Concrete*, SP-12, American Concrete Institute/American Society of Civil Engineers, Farmington Hills, MI, 1965, pp. 359-382.

8.3. Mattock, A. H., "Redistribution of Design Bending Moments in Reinforced Concrete Continuous Beams," *Proceedings*, Institution of Civil Engineers (London), V. 13, 1959, pp. 35-46.

8.4. Mast, R.F., "Unified Design Provision for Reinforced and Prestressed Concrete Flexural and Compression Members," *ACI Structural Journal,* V. 89, No. 2, Mar.-Apr., 1992, pp. 185-199.

8.5. Pauw, Adrian, "Static Modulus of Elasticity of Concrete as Affected by Density," ACI JOURNAL, *Proceedings* V. 57, No. 6, Dec. 1960, pp. 679-687.

8.6. ASTM C 469-03, "Test Method for Static Modulus of Elasticity and Poisson's Ratio of Concrete in Compression," ASTM, West Conshohocken, PA.

8.7. "Handbook of Frame Constants," Portland Cement Association, Skokie, IL, 1972, 34 pp.

8.8. "Continuity in Concrete Building Frames," Portland Cement Association, Skokie, IL, 1959, 56 pp.

References, Chapter 9

9.1. "Minimum Design Loads for Buildings and Other Structures," SEI/ASCE 7-02, American Society of Civil Engineers, Reston, VA, 2002, 376 pp.

9.2. "International Building Code," International Code Council, Falls Church, VA, 2003.

9.3. "Minimum Design Loads for Buildings and Other Structures (ASCE 7-93)," ASCE, New York, 1993, 134 pp.

9.4. "BOCA National Building Code, 13th Edition," Building Officials and Code Administration International, Inc., Country Club Hills, IL, 1993, 357 pp.

9.5. "Standard Building Code," Southern Building Code Congress International, Inc., Birmingham, AL, 1994, 656 pp.

9.6. "Uniform Building Code, V. 2, Structural Engineering Design Provisions," International Conference of Building Officials, Whittier, CA, 1997, 492 pp.

9.7. MacGregor, J. G., "Safety and Limit States Design for Reinforced Concrete," *Canadian Journal of Civil Engineering*, V. 3, No. 4, Dec. 1976, pp. 484-513.

9.8. Winter, G., "Safety and Serviceability Provisions in the ACI Building Code," *Concrete Design: U.S. and European Practices*, SP-59, American Concrete Institute, Farmington Hills, MI, 1979, pp. 35-49.

9.9. Nowak, A. S., and Szerszen, M. M., "Reliability-Based Calibration for Structural Concrete," *Report UMCEE 01-04,* Department of Civil and Environmental Engineering, University of Michigan, Ann Arbor, MI, Nov. 2001.

9.10. Mast, R.F., "Unified Design Provision for Reinforced and Prestressed Concrete Flexural and Compression Members," *ACI Structural Journal,* V. 89, No. 2, Mar.-Apr., 1992, pp. 185-199.

9.11. *Deflections of Concrete Structures*, SP-43, American Concrete Institute, Farmington Hills, MI, 1974, 637 pp.

9.12. ACI Committee 213, "Guide for Structural Lightweight Aggregate Concrete (ACI 213R-87)," American Concrete Institute, Farmington Hills, MI, 1987, 27 pp. Also *ACI Manual of Concrete Practice.*

9.13. Branson, D. E., "Instantaneous and Time-Dependent Deflections on Simple and Continuous Reinforced Concrete Beams," *HPR Report* No. 7, Part 1, Alabama Highway Department, Bureau of Public Roads, Aug. 1965, pp. 1-78.

9.14. ACI Committee 435, "Deflections of Reinforced Concrete Flexural Members (ACI 435.2R-66) (Reapproved 1989)," ACI JOURNAL, *Proceedings* V. 63, No. 6, June 1966, pp. 637-674. Also *ACI Manual of Concrete Practice.*

9.15. Subcommittee 1, ACI Committee 435, "Allowable Deflections (ACI 435.3R-68) (Reapproved 1989)," ACI JOURNAL, *Proceedings* V. 65, No. 6, June 1968, pp. 433-444. Also *ACI Manual of Concrete Practice.*

9.16. Subcommittee 2, ACI Committee 209, "Prediction of Creep, Shrinkage, and Temperature Effects in Concrete Structures (ACI 209R-92)," *Designing for the Effects of Creep, Shrinkage, and Temperature in Concrete Structures*, SP-27, American Concrete Institute, Farmington Hills, MI, 1971, pp. 51-93.

9.17. ACI Committee 435, "Deflections of Continuous Concrete Beams (ACI 435.5R-73)(Reapproved 1989)," American Concrete Institute, Farmington Hills, MI, 1973, 7 pp. Also *ACI Manual of Concrete Practice.*

9.18. ACI Committee 435, "Proposed Revisions by Committee 435 to ACI Building Code and Commentary Provisions on Deflections," ACI JOURNAL, *Proceedings* V. 75, No. 6, June 1978, pp. 229-238.

9.19. Branson, D. E., "Compression Steel Effect on Long-Time Deflections," ACI JOURNAL, *Proceedings* V. 68, No. 8, Aug. 1971, pp. 555-559.

9.20. Branson, D. E., *Deformation of Concrete Structures,* McGraw-Hill Book Co., New York, 1977, 546 pp.

9.21. *PCI Design Handbook — Precast and Prestressed Concrete,* 5th Edition, Precast/Prestressed Concrete Institute, Chicago, IL, 1998, pp. 4-68 to 4-72.

9.22. Mast, R.F., "Analysis of Cracked Prestressed Concrete Sections: A Practical Approach," *PCI Journal,* V. 43, No. 4, July-Aug., 1998, pp. 80-91.

9.23. Shaikh, A. F., and Branson, D. E., "Non-Tensioned Steel in Prestressed Concrete Beams," *Journal of the Prestressed Concrete Institute*, V. 15, No. 1, Feb. 1970, pp. 14-36.

9.24. Branson, D. E., discussion of "Proposed Revision of ACI 318-63: Building Code Requirements for Reinforced Concrete," by ACI Committee 318, ACI JOURNAL, *Proceedings* V. 67, No. 9, Sept. 1970, pp. 692-695.

9.25. Subcommittee 5, ACI Committee 435, "Deflections of Prestressed Concrete Members (ACI 435.1R-63)(Reapproved 1989)," ACI JOURNAL, *Proceedings* V. 60, No. 12, Dec. 1963, pp. 1697-1728.

9.26. Branson, D. E.; Meyers, B. L.; and Kripanarayanan, K. M., "Time-Dependent Deformation of Noncomposite and Composite Prestressed Concrete Structures," *Symposium on Concrete Deformation*, Highway Research Record 324, Highway Research Board, 1970, pp. 15-43.

9.27. Ghali, A., and Favre, R., *Concrete Structures: Stresses and Deformations*, Chapman and Hall, New York, 1986, 348 pp.

References, Chapter 10

10.1. Leslie, K. E.; Rajagopalan, K. S.; and Everard, N. J., "Flexural Behavior of High-Strength Concrete Beams," ACI JOURNAL, *Proceedings* V. 73, No. 9, Sept. 1976, pp. 517-521.

10.2. Karr, P. H.; Hanson, N. W; and Capell, H. T.; "Stress-Strain Characteristics of High Strength Concrete," *Douglas McHenry International Symposium on Concrete and Concrete Structures*, SP-55, American Concrete Institute, Farmington Hills, MI, 1978, pp. 161-185.

10.3. Mattock, A. H.; Kriz, L. B.; and Hognestad, E., "Rectangular Concrete Stress Distribution in Ultimate Strength Design," ACI JOURNAL, *Proceedings* V. 57, No. 8, Feb. 1961, pp. 875-928.

10.4. *ACI Design Handbook—Columns*, SP-17(97), American Concrete Institute, Farmington Hills, MI, 1997, 482 pp.

10.5. *CRSI Handbook*, 9th Edition, Concrete Reinforcing Steel Institute, Schaumberg, IL, 2002, 648 pp.

10.6. Bresler, B., "Design Criteria for Reinforced Concrete Columns under Axial Load and Biaxial Bending," ACI JOURNAL, *Proceedings* V. 57, No. 5, Nov. 1960, pp. 481-490.

10.7. Parme, A. L.; Nieves, J. M.; and Gouwens, A., "Capacity of Reinforced Rectangular Columns Subjected to Biaxial Bending," ACI JOURNAL, *Proceedings* V. 63, No. 9, Sept. 1966, pp. 911-923.

10.8. Heimdahl, P. D., and Bianchini, A. C., "Ultimate Strength of Biaxially Eccentrically Loaded Concrete Columns Reinforced with High Strength Steel," *Reinforced Concrete Columns,* SP-50, American Concrete Institute, Farmington Hills, MI, 1975, pp. 100-101.

10.9. Furlong, R. W., "Concrete Columns Under Biaxially Eccentric Thrust," ACI JOURNAL, *Proceedings* V. 76, No. 10, Oct. 1979, pp. 1093-1118.

10.10. Hansell, W., and Winter, G., "Lateral Stability of Reinforced Concrete Beams," ACI JOURNAL, *Proceedings* V. 56, No. 3, Sept. 1959, pp. 193-214.

10.11. Sant, J. K., and Bletzacker, R. W., "Experimental Study of Lateral Stability of Reinforced Concrete Beams," ACI JOURNAL, *Proceedings* V. 58, No. 6, Dec. 1961, pp. 713-736.

10.12. Gergely, P., and Lutz, L. A., "Maximum Crack Width in Reinforced Concrete Flexural Members," *Causes, Mechanism, and Control of Cracking in Concrete*, SP-20, American Concrete Institute, Farmington Hills, MI, 1968, pp. 87-117.

10.13. Kaar, P. H., "High Strength Bars as Concrete Reinforcement, Part 8: Similitude in Flexural Cracking of T-Beam Flanges," *Journal*, PCA Research and Development Laboratories, V. 8, No. 2, May 1966, pp. 2-12.

10.14. Base, G. D.; Reed, J. B.; Beeby, A. W.; and Taylor, H. P. J., "An Investigation of the Crack Control Characteristics of Various Types of Bar in Reinforced Concrete Beams," *Research Report* No. 18, Cement and Concrete Association, London, Dec. 1966, 44 pp.

10.15. Beeby, A. W., "The Prediction of Crack Widths in Hardened Concrete," *The Structural Engineer*, V. 57A, No. 1, Jan. 1979, pp. 9-17.

10.16. Frosch, R. J., "Another Look at Cracking and Crack Control in Reinforced Concrete," *ACI Structural Journal*, V. 96, No. 3, May-June 1999, pp. 437-442.

10.17. ACI Committee 318, "Closure to Public Comments on ACI 318-99," *Concrete International*, May 1999, pp. 318-1 to 318-50.

10.18. Darwin, D., et al., "Debate: Crack Width, Cover, and Corrosion," *Concrete International*, V. 7, No. 5, May 1985, American Concrete Institute, Farmington Hills, MI, pp. 20-35.

10.19. Oesterle, R. G., "The Role of Concrete Cover in Crack Control Criteria and Corrosion Protection," RD Serial No. 2054, Portland Cement Association, Skokie, IL, 1997.

10.20. Frantz, G. C., and Breen, J. E., "Cracking on the Side Faces of Large Reinforced Concrete Beams," ACI JOURNAL, *Proceedings* V. 77, No. 5, Sept.-Oct. 1980, pp. 307-313.

10.21. Frosch, R. J., "Modeling and Control of Side Face Beam Cracking," *ACI Structural Journal*, V. 99, No. 3, May-June 2002, pp. 376-385.

10.22. Chow, L.; Conway, H.; and Winter, G., "Stresses in Deep Beams," *Transactions,* ASCE, V. 118, 1953, pp. 686-708.

10.23. "Design of Deep Girders," IS079D, Portland Cement Association, Skokie, IL, 1946, 10 pp.

10.24. Park, R., and Paulay, T., *Reinforced Concrete Structures,* Wiley-Inter-Science, New York, 1975, 769 pp.

10.25. Furlong, R. W., "Column Slenderness and Charts for Design," ACI JOURNAL, *Proceedings* V. 68, No. 1, Jan. 1971, pp. 9-18.

10.26. "Reinforced Concrete Column Investigation—Tentative Final Report of Committee 105," ACI JOURNAL, *Proceedings* V. 29, No. 5, Feb. 1933, pp. 275-282.

10.27. Saatcioglu, M., and Razvi, S. R., "Displacement-Based Design of Reinforced Concrete Columns for Confinement," *ACI Structural Journal*, V. 99, No. 1, Jan.-Feb. 2002, pp. 3-11.

10.28. Pessiki, S.; Graybeal, B.; and Mudlock, M., "Proposed Design of High-Strength Spiral Reinforcement in Compression Members," *ACI Structural Journal*, V. 98, No. 6, Nov.-Dec. 2001, pp. 799-810.

10.29. Richart, F. E.; Brandzaeg, A.; and Brown, R. L., "The Failure of Plain and Spirally Reinforced Concrete in Compression," *Bulletin No. 190*, University of Illinois Engineering Experiment Station, Apr. 1929, 74 pp.

10.30. MacGregor, J. G., "Design of Slender Concrete Columns—Revisited," *ACI Structural Journal*, V. 90, No. 3, May-June 1993, pp. 302-309.

10.31. MacGregor, J. G.; Breen, J. E.; and Pfrang, E. O., "Design of Slender Concrete Columns," ACI JOURNAL, *Proceedings* V. 67, No. 1, Jan. 1970, pp. 6-28.

10.32. Ford, J. S.; Chang, D. C.; and Breen, J. E., "Design Indications from Tests of Unbraced Multipanel Concrete Frames," *Concrete International: Design and Construction*, V. 3, No. 3, Mar. 1981, pp. 37-47.

10.33. MacGregor, J. G., and Hage, S. E., "Stability Analysis and Design Concrete," *Proceedings*, ASCE, V. 103, No. ST 10, Oct. 1977.

10.34. Grossman, J. S., "Slender Concrete Structures—The New Edge," *ACI Structural Journal*, V. 87, No. 1, Jan.-Feb. 1990, pp. 39-52.

10.35. Grossman, J. S., "Reinforced Concrete Design," *Building Structural Design Handbook*, R. N. White and C. G. Salmon, eds., John Wiley and Sons, New York, 1987.

10.36. "Guide to Design Criteria for Metal Compression Members," 2nd Edition, Column Research Council, Fritz Engineering Laboratory, Lehigh University, Bethlehem, PA, 1966.

10.37. ACI Committee 340, *Design of Structural Reinforced Concrete Elements in Accordance with Strength Design Method of ACI 318-95*, SP-17(97), American Concrete Institute, Farmington Hills, MI, 1997, 432 pp.

10.38. "Code of Practice for the Structural Use of Concrete, Part 1. Design Materials and Workmanship," CP110: Part 1, British Standards Institution, London, Nov. 1972, 154 pp.

10.39. Cranston, W. B., "Analysis and Design of Reinforced Concrete Columns," *Research Report* No. 20, Paper 41.020, Cement and Concrete Association, London, 1972, 54 pp.

10.40. Mirza, S. A.; Lee, P. M.; and Morgan, D. L, "ACI Stability Resistance Factor for RC Columns," *ASCE Structural Engineering*, American Society of Civil Engineers, V. 113, No. 9, Sept. 1987, pp. 1963-1976.

10.41. Mirza, S. A., "Flexural Stiffness of Rectangular Reinforced Concrete Columns," *ACI Structural Journal*, V. 87, No. 4, July-Aug. 1990, pp. 425-435.

10.42. Lai, S. M. A., and MacGregor, J. G., "Geometric Nonlinearities in Unbraced Multistory Frames," *ASCE Structural Engineering,* American Society of Civil Engineers, V. 109, No. 11, Nov. 1983, pp. 2528-2545.

10.43. Bianchini, A. C.; Woods, Robert E.; and Kesler, C. E., "Effect of Floor Concrete Strength on Column Strength," ACI JOURNAL, *Proceedings* V. 56, No. 11, May 1960, pp. 1149-1169.

10.44. Ospina, C. E., and Alexander, S. D. B., "Transmission of Interior Concrete Column Loads through Floors," *ASCE Journal of Structural Engineering*, V. 124, No. 6., 1998.

10.45. Everard, N. J., and Cohen, E., "Ultimate Strength Design of Reinforced Concrete Columns," SP-7, American Concrete Institute, Farmington Hills, MI, 1964, 182 pp.

10.46. Hawkins, N. M., "Bearing Strength of Concrete Loaded through Rigid Plates," *Magazine of Concrete Research* (London), V. 20, No. 62, Mar. 1968, pp. 31-40.

References, Chapter 11

11.1. ACI-ASCE Committee 426, "Shear Strength of Reinforced Concrete Members (ACI 426R-74) (Reapproved 1980)," *Proceedings*, ASCE, V. 99, No. ST6, June 1973, pp. 1148-1157.

11.2. MacGregor, J. G., and Hanson, J. M., "Proposed Changes in Shear Provisions for Reinforced and Prestressed Concrete Beams," ACI JOURNAL, *Proceedings* V. 66, No. 4, Apr. 1969, pp. 276-288.

11.3. ACI-ASCE Committee 326 (now 426), "Shear and Diagonal Tension," ACI JOURNAL, *Proceedings* V. 59, No. 1, Jan. 1962, pp. 1-30; No. 2, Feb. 1962, pp. 277-334; and No. 3, Mar. 1962, pp. 352-396.

11.4. Barney, G. B.; Corley, W. G.; Hanson, J. M.; and Parmelee, R. A., "Behavior and Design of Prestressed Concrete Beams with Large Web Openings," *Journal of the Prestressed Concrete Institute,* V. 22, No. 6, Nov.-Dec. 1977, pp. 32-61.

11.5. Schlaich, J.; Schafer, K.; and Jennewein, M., "Toward a Consistent Design of Structural Concrete," *Journal of the Prestressed Concrete Institute*, V. 32, No. 3, May-June 1987, pp. 74-150.

11.6. Joint Committee, "Recommended Practice and Standard Specification for Concrete and Reinforced Concrete," *Proceedings*, ASCE, V. 66, No. 6, Part 2, June 1940, 81 pp.

11.7. Mphonde, A. G., and Frantz, G. C., "Shear Tests of High- and Low-Strength Concrete Beams without Stirrups," ACI JOURNAL, *Proceedings* V. 81, No. 4, July-Aug. 1984, pp. 350-357.

11.8. Elzanaty, A. H.; Nilson, A. H.; and Slate, F. O., "Shear Capacity of Reinforced Concrete Beams Using High Strength Concrete," ACI JOURNAL, *Proceedings* V. 83, No. 2, Mar.-Apr. 1986, pp. 290-296.

11.9. Roller, J. J., and Russell, H. G., "Shear Strength of High-Strength Concrete Beams with Web Reinforcement," *ACI Structural Journal*, V. 87, No. 2, Mar.-Apr. 1990, pp. 191-198.

11.10. Johnson, M.K., and Ramirez, J.A., "Minimum Amount of Shear Reinforcement in High Strength Concrete Members," *ACI Structural Journal,* V. 86, No. 4, July-Aug. 1989, pp. 376-382.

11.11. Ozcebe, G.; Ersoy, U.; and Tankut, T., "Evaluation of Minimum Shear Reinforcement for Higher Strength Concrete," *ACI Structural Journal*, V. 96, No., 3, May-June 1999, pp. 361-368.

11.12. Ivey, D. L., and Buth, E., "Shear Capacity of Lightweight Concrete Beams," ACI JOURNAL, *Proceedings* V. 64, No. 10, Oct. 1967, pp. 634-643.

11.13. Hanson, J. A., "Tensile Strength and Diagonal Tension Resistance of Structural Lightweight Concrete," ACI JOURNAL, *Proceedings* V. 58, No. 1, July 1961, pp. 1-40.

11.14. Kani, G. N. J., "Basic Facts Concerning Shear Failure," ACI JOURNAL, *Proceedings* V. 63, No. 6, June 1966, pp. 675-692.

11.15. Kani, G. N. J., "How Safe Are Our Large Reinforced Concrete Beams," ACI JOURNAL, *Proceedings* V. 64, No. 3, Mar. 1967, pp. 128-141.

11.16. Faradji, M. J., and Diaz de Cossio, R., "Diagonal Tension in Concrete Members of Circular Section" (in Spanish) Institut de Ingenieria, Mexico (translation by Portland Cement Association, Foreign Literature Study No. 466).

11.17. Khalifa, J. U., and Collins, M. P., "Circular Reinforced Concrete Members Subjected to Shear," Publications No. 81-08, Department of Civil Engineering, University of Toronto, Dec. 1981.

11.18. *PCI Design Handbook—Precast and Prestressed Concrete*, 4th Edition, Precast/Prestressed Concrete Institute, Chicago, 1992, 580 pp.

11.19. ACI Committee 318, "Commentary on Building Code Requirements for Reinforced Concrete (ACI 318-63)," SP-10, American Concrete Institute, Farmington Hills, MI, 1965, pp. 78-84.

11.20. Guimares, G. N.; Kreger, M. E.; and Jirsa, J. O., "Evaluation of Joint-Shear Provisions for Interior Beam-Column-Slab Connections Using High Strength Materials," *ACI Structural Journal*, V. 89, No. 1, Jan.-Feb. 1992, pp. 89-98.

11.21. Griezic, A.; Cook, W. D.; and Mitchell, D., "Tests to Determine Performance of Deformed Welded-Wire Fabric Stirrups," *ACI Structural Journal*, V. 91, No. 2, Mar.-Apr. 1994, pp. 211-220.

11.22. Furlong, R. W.; Fenves, G. L.; and Kasl, E. P., "Welded Structural Wire Reinforcement for Columns," *ACI Structural Journal,* V. 88, No. 5, Sept.-Oct. 1991, pp. 585-591.

11.23. Angelakos, D.; Bentz, E. C.; and Collins, M. D., "Effect of Concrete Strength and Minimum Stirrups on Shear Strength of Large Members," *ACI Structural Journal*, V. 98, No. 3, May-June 2001, pp. 290-300.

11.24. Becker, R. J., and Buettner, D. R., "Shear Tests of Extruded Hollow Core Slabs," *PCI Journal*, V. 30, No. 2, Mar.-Apr. 1985.

11.25. Anderson, A. R., "Shear Strength of Hollow Core Members," *Technical Bulletin* 78-81, Concrete Technology Associates, Tacoma, WA, Apr. 1978, 33 pp.

11.26. Olesen, S. E.; Sozen, M. A.; and Siess, C. P., "Investigation of Prestressed Reinforced Concrete for Highway Bridges, Part IV: Strength in Shear of Beams with Web Reinforcement," *Bulletin* No. 493, University of Illinois, Engineering Experiment Station, Urbana, 1967.

11.27. Anderson, N. S., and Ramirez, J. A., "Detailing of Stirrup

Reinforcement," *ACI Structural Journal*, V. 86, No. 5, Sept.-Oct. 1989, pp. 507-515.

11.28. Leonhardt, F., and Walther, R., "The Stuttgart Shear Tests," *C&CA Translation*, No. 111, Cement and Concrete Association, 1964, London, 134 pp.

11.29. MacGregor, J. G., and Ghoneim, M. G., "Design for Torsion," *ACI Structural Journal,* V. 92, No. 2, Mar.-Apr. 1995, pp. 211-218.

11.30. Hsu, T. T. C., "ACI Shear and Torsion Provisions for Prestressed Hollow Girders," *ACI Structural Journal,* V. 94, No. 6, Nov.-Dec. 1997, pp. 787-799.

11.31. Hsu, T. T. C., "Torsion of Structural Concrete—Behavior of Reinforced Concrete Rectangular Members," *Torsion of Structural Concrete*, SP-18, American Concrete Institute, Farmington Hills, MI, 1968, pp. 291-306.

11.32. Collins, M. P., and Lampert, P., "Redistribution of Moments at Cracking—The Key to Simpler Torsion Design?" *Analysis of Structural Systems for Torsion*, SP-35, American Concrete Institute, Farmington Hills, MI, 1973, pp. 343-383.

11.33. Hsu, T. T. C., and Burton, K. T., "Design of Reinforced Concrete Spandrel Beams," *Proceedings*, ASCE, V. 100, No. ST1, Jan. 1974, pp. 209-229.

11.34. Hsu, T. C., "Shear Flow Zone in Torsion of Reinforced Concrete," *ASCE Structural Engineering*, American Society of Civil Engineers, V. 116, No. 11, Nov. 1990, pp. 3206-3226.

11.35. Mitchell, D., and Collins, M. P., "Detailing for Torsion," ACI JOURNAL, *Proceedings* V. 73, No. 9, Sept. 1976, pp. 506-511.

11.36. Behera, U., and Rajagopalan, K. S., "Two-Piece U-Stirrups in Reinforced Concrete Beams," ACI JOURNAL, *Proceedings* V. 66, No. 7, July 1969, pp. 522-524.

11.37. Zia, P., and McGee, W. D., "Torsion Design of Prestressed Concrete," *PCI Journal*, V. 19, No. 2, Mar.-Apr. 1974.

11.38. Zia, P., and Hsu, T. T. C., "Design for Torsion and Shear in Prestressed Concrete Flexural Members," *PCI Journal*, V. 49, No. 3, May-June 2004.

11.39. Collins, M. P., and Mitchell, D., "Shear and Torsion Design of Prestressed and Non-Prestressed Concrete Beams," *PCI Journal*, V. 25, No. 4, Sept.-Oct. 1980.

11.40. PCI, *PCI Design Handbook—Precast and Prestressed Concrete*, 4th Edition, Precast/Prestressed Concrete Institute, Chicago, Ill., 1992.

11.41. Klein, G. J., "Design of Spandrel Beams," *PCI Specially Funded Research Project* No. 5, Precast/Prestressed Concrete Institute, Chicago, Ill., 1986.

11.42. Birkeland, P. W., and Birkeland, H. W., "Connections in Precast Concrete Construction," ACI JOURNAL, *Proceedings* V. 63, No. 3, Mar. 1966, pp. 345-368.

11.43. Mattock, A. H., and Hawkins, N. M., "Shear Transfer in Reinforced Concrete—Recent Research," *Journal of the Prestressed Concrete Institute*, V. 17, No. 2, Mar.-Apr. 1972, pp. 55-75.

11.44. Mattock, A. H.; Li, W. K.; and Want, T. C., "Shear Transfer in Lightweight Reinforced Concrete," *Journal of the Prestressed Concrete Institute*, V. 21, No. 1, Jan.-Feb. 1976, pp. 20-39.

11.45. Mattock, A. H., "Shear Transfer in Concrete Having Reinforcement at an Angle to the Shear Plane," *Shear in Reinforced Concrete*, SP-42, American Concrete Institute, Farmington Hills, MI, 1974, pp. 17-42.

11.46. Mattock, A. H., discussion of "Considerations for the Design of Precast Concrete Bearing Wall Buildings to Withstand Abnormal Loads," by PCI Committee on Precast Concrete Bearing Wall Buildings, *Journal of the Prestressed Concrete Institute*, V. 22, No. 3, May-June 1977, pp. 105-106.

11.47. "Chapter 1—Composite Members," *Load and Resistance Factor Design Specification for Structural Steel for Buildings,* American Institute of Steel Construction, Chicago, Sept. 1986, pp. 51-58.

11.48. Mattock, A. H.; Johal, L.; and Chow, H. C., "Shear Transfer in Reinforced Concrete with Moment or Tension Acting Across the Shear Plane," *Journal of the Prestressed Concrete Institute*, V. 20, No. 4, July-Aug. 1975, pp. 76-93.

11.49. Rogowsky, D. M., and MacGregor, J. G., "Design of Reinforced Concrete Deep Beams," *Concrete International: Design and Construction,* V. 8, No. 8, Aug. 1986, pp. 46-58.

11.50. Marti, P., "Basic Tools of Reinforced Concrete Beam Design," ACI JOURNAL, *Proceedings* V. 82, No. 1, Jan.-Feb. 1985, pp. 46-56.

11.51. Crist, R. A., "Shear Behavior of Deep Reinforced Concrete Beams," *Proceedings*, Symposium on the Effects of Repeated Loading of Materials and Structural Elements (Mexico City, 1966), V. 4, RILEM, Paris, 31 pp.

11.52. Kriz, L. B., and Raths, C. H., "Connections in Precast Concrete Structures—Strength of Corbels," *Journal of the Prestressed Concrete Institute*, V. 10, No. 1, Feb. 1965, pp. 16-47.

11.53. Mattock, A. H.; Chen, K. C.; and Soongswang, K., "The Behavior of Reinforced Concrete Corbels," *Journal of the Prestressed Concrete Institute*, V. 21, No. 2, Mar.-Apr. 1976, pp. 52-77.

11.54. Cardenas, A. E.; Hanson, J. M.; Corley, W. G.; and Hognestad, E., "Design Provisions for Shear Walls," ACI JOURNAL, *Proceedings* V. 70, No. 3, Mar. 1973, pp. 221-230.

11.55. Barda, F.; Hanson, J. M.; and Corley, W. G., "Shear Strength of Low-Rise Walls with Boundary Elements," *Reinforced Concrete Structures in Seismic Zones,* SP-53, American Concrete Institute, Farmington Hills, MI, 1977, pp. 149-202.

11.56. Hanson, N. W., and Conner, H. W., "Seismic Resistance of Reinforced Concrete Beam-Column Joints," *Proceedings*, ASCE, V. 93, ST5, Oct. 1967, pp. 533-560.

11.57. ACI-ASCE Committee 352, "Recommendations for Design of Beam-Column Joints in Monolithic Reinforced Concrete Structures (ACI 352R-91)," American Concrete Institute, Farmington Hills, MI, 1991, 18 pp. Also *ACI Manual of Concrete Practice*.

11.58. ACI-ASCE Committee 426, "The Shear Strength of Reinforced Concrete Members," *Proceedings*, ASCE, V. 100, No. ST8, Aug. 1974, pp. 1543-1591.

11.59. Vanderbilt, M. D., "Shear Strength of Continuous Plates," *Journal of the Structural Division*, ASCE, V. 98, No. ST5, May 1972, pp. 961-973.

11.60. ACI-ASCE Committee 423, "Recommendations for Concrete Members Prestressed with Unbonded Tendons (ACI 423.3R-89)," American Concrete Institute, Farmington Hills, MI, 18 pp. Also *ACI Manual of Concrete Practice*.

11.61. Burns, N. H., and Hemakom, R., "Test of Scale Model of Post-Tensioned Flat Plate," *Proceedings*, ASCE, V. 103, ST6, June 1977, pp. 1237-1255.

11.62. Hawkins, N. M., "Shear Strength of Slabs with Shear Reinforcement," *Shear in Reinforced Concrete*, SP-42, V. 2, American Concrete Institute, Farmington Hills, MI, 1974, pp. 785-815.

11.63. Broms, C.E., "Shear Reinforcement for Deflection Ductility of Flat Plates," *ACI Structural Journal,* V. 87, No. 6, Nov.-Dec. 1990, pp. 696-705.

11.64. Yamada, T.; Nanni, A.; and Endo, K., "Punching Shear Resistance of Flat Slabs: Influence of Reinforcement Type and Ratio," *ACI Structural Journal,* V. 88, No. 4, July-Aug. 1991, pp. 555-563.

11.65. Hawkins, N. M.; Mitchell, D.; and Hannah, S. N., "The Effects of Shear Reinforcement on Reversed Cyclic Loading Behavior of Flat Plate Structures," *Canadian Journal of Civil Engineering* (Ottawa), V. 2, 1975, pp. 572-582.

11.66. ACI-ASCE Committee 421, "Shear Reinforcement for Slabs (ACI 421.1R-99)," American Concrete Institute, Farmington Hills, MI, 1999, 15 pp.

11.67. Corley, W. G., and Hawkins. N. M., "Shearhead Reinforcement for Slabs," ACI JOURNAL, *Proceedings* V. 65, No. 10, Oct. 1968, pp. 811-824.

11.68. Hanson, N. W., and Hanson, J. M., "Shear and Moment Transfer between Concrete Slabs and Columns," *Journal*, PCA Research and Development Laboratories, V. 10, No. 1, Jan. 1968, pp. 2-16.

11.69. Hawkins, N. M., "Lateral Load Resistance of Unbonded Post-Tensioned Flat Plate Construction," *Journal of the Prestressed Concrete Institute*, V. 26, No. 1, Jan.-Feb. 1981, pp. 94-115.

11.70. Hawkins, N. M., and Corley, W. G., "Moment Transfer to Columns in Slabs with Shearhead Reinforcement," *Shear in Reinforced Concrete*, SP-42, American Concrete Institute, Farmington Hills, MI, 1974, pp. 847-879.

References, Chapter 12

12.1. ACI Committee 408, "Bond Stress—The State of the Art," ACI JOURNAL, *Proceedings* V. 63, No. 11, Nov. 1966, pp. 1161-1188.

12.2. ACI Committee 408, "Suggested Development, Splice, and Standard Hook Provisions for Deformed Bars in Tension," (ACI 408.1R-90), American Concrete Institute, Farmington Hills, MI, 1990, 3 pp. Also *ACI Manual of Concrete Practice.*

12.3. Jirsa, J. O.; Lutz, L. A.; and Gergely, P., "Rationale for Suggested Development, Splice, and Standard Hook Provisions for Deformed Bars in Tension," *Concrete International: Design & Construction*, V. 1, No. 7, July 1979, pp. 47-61.

12.4. Jirsa, J. O., and Breen, J. E., "Influence of Casting Position and Shear on Development and Splice Length—Design Recommendations," *Research Report* 242-3F, Center for Transportation Research, Bureau of Engineering Research, University of Texas at Austin, Nov. 1981.

12.5. Jeanty, P. R.; Mitchell, D.; and Mirza, M. S., "Investigation of 'Top Bar' Effects in Beams," *ACI Structural Journal* V. 85, No. 3, May-June 1988, pp. 251-257.

12.6. Treece, R. A., and Jirsa, J. O., "Bond Strength of Epoxy-Coated Reinforcing Bars," *ACI Materials Journal*, V. 86, No. 2, Mar.-Apr. 1989, pp. 167-174.

12.7. Johnston, D. W., and Zia, P., "Bond Characteristics of Epoxy-Coated Reinforcing Bars," Department of Civil Engineering, North Carolina State University, *Report* No. FHWA/NC/82-002, Aug. 1982.

12.8. Mathey, R. G., and Clifton, J. R., "Bond of Coated Reinforcing Bars in Concrete," *Journal of the Structural Division*, ASCE, V. 102, No. ST1, Jan. 1976, pp. 215-228.

12.9. Orangun, C. O.; Jirsa, J. O.; and Breen, J. E., "A Reevaluation of Test Data on Development Length and Splices," ACI JOURNAL, *Proceedings* V. 74, No. 3, Mar. 1977, pp. 114-122.

12.10. Azizinamini, A.; Pavel, R.; Hatfield, E.; and Ghosh, S. K., "Behavior of Spliced Reinforcing Bars Embedded in High-Strength Concrete," *ACI Structural Journal,* V. 96, No. 5, Sept.-Oct. 1999, pp. 826-835.

12.11. Azizinamini, A.; Darwin, D.; Eligehausen, R.; Pavel, R.; and Ghosh, S. K., "Proposed Modifications to ACI 318-95 Development and Splice Provisions for High-Strength Concrete," *ACI Structural Journal,* V. 96, No. 6, Nov.-Dec. 1999, pp. 922-926.

12.12. Jirsa, J. O., and Marques, J. L. G., "A Study of Hooked Bar Anchorages in Beam-Column Joints," ACI JOURNAL, *Proceedings* V. 72, No. 5, May 1975, pp. 198-200.

12.13. Hamad, B. S.; Jirsa, J. O.; and D'Abreu, N. I., "Anchorage Strength of Epoxy-Coated Hooked Bars," *ACI Structural Journal,* V. 90, No. 2, Mar.-Apr. 1993, pp. 210-217.

12.14. Bartoletti, S. J., and Jirsa, J. O., "Effects of Epoxy-Coating on Anchorage and Development of Welded Wire Fabric," *ACI Structural Journal*, V. 92, No. 6, Nov.-Dec. 1995, pp. 757-764.

12.15. Rose, D. R., and Russell, B. W., 1997, "Investigation of Standardized Tests to Measure the Bond Performance of Prestressing Strand," *PCI Journal,* V. 42, No. 4, Jul.-Aug., 1997, pp. 56-80.

12.16. Logan, D. R., "Acceptance Criteria for Bond Quality of Strand for Pretensioned Prestressed Concrete Applications," *PCI Journal,* V. 42, No. 2, Mar.-Apr., 1997, pp. 52-90.

12.17. Martin, L., and Korkosz, W., "Strength of Prestressed Members at Sections Where Strands Are Not Fully Developed," *PCI Journal,* V. 40, No. 5, Sept.-Oct. 1995, pp. 58-66.

12.18. *PCI Design Handbook — Precast and Prestressed Concrete,* 5th Edition, Precast/Prestressed Concrete Institute, Chicago, IL, 1998, pp. 4-27 to 4-29.

12.19. Kaar, P., and Magura, D., "Effect of Strand Blanketing on Performance of Pretensioned Girders," *Journal of the Prestressed Concrete Institute*, V. 10, No. 6, Dec. 1965, pp. 20-34.

12.20. Hanson, N. W., and Kaar, P. H., "Flexural Bond Tests Pretensioned Beams," ACI JOURNAL, *Proceedings* V. 55, No. 7, Jan. 1959, pp. 783-802.

12.21. Kaar, P. H.; La Fraugh, R. W.; and Mass, M. A., "Influence of Concrete Strength on Strand Transfer Length," *Journal of the Prestressed Concrete Institute*, V. 8, No. 5, Oct. 1963, pp. 47-67.

12.22. Rabbat, B. G.; Kaar, P. H.; Russell, H. G.; and Bruce, R. N., Jr., "Fatigue Tests of Pretensioned Girders with Blanketed and Draped Strands," *Journal of the Prestressed Concrete Institute,* V. 24. No. 4, July-Aug. 1979, pp. 88-114.

12.23. Rogowsky, D. M., and MacGregor, J. G., "Design of Reinforced Concrete Deep Beams," *Concrete International: Design & Construction*, V. 8, No. 8, Aug. 1986, pp. 46-58.

12.24. Joint PCI/WRI Ad Hoc Committee on Welded Wire Fabric for Shear Reinforcement, "Welded Wire Fabric for Shear Reinforcement," *Journal of the Prestressed Concrete Institute*, V. 25, No. 4, July-Aug. 1980, pp. 32-36.

12.25. Pfister, J. F., and Mattock, A. H., "High Strength Bars as Concrete Reinforcement, Part 5: Lapped Splices in Concentrically Loaded Columns," *Journal*, PCA Research and Development Laboratories, V. 5, No. 2, May 1963, pp. 27-40.

12.26. Lloyd, J. P., and Kesler, C. E., "Behavior of One-Way Slabs Reinforced with Deformed Wire and Deformed Wire Fabric," *T&AM Report* No. 323, University of Illinois, 1969, 129 pp.

12.27. Lloyd, J. P., "Splice Requirements for One-Way Slabs Reinforced with Smooth Welded Wire Fabric," *Publication* No. R(S)4, Civil Engineering, Oklahoma State University, June 1971, 37 pp.

References, Chapter 13

13.1. Hatcher, D. S.; Sozen, M. A.; and Siess, C. P., "Test of a Reinforced Concrete Flat Plate," *Proceedings*, ASCE, V. 91, ST5, Oct. 1965, pp. 205-231.

13.2. Guralnick, S. A., and LaFraugh, R. W., "Laboratory Study of a Forty-Five-Foot Square Flat Plate Structure," ACI JOURNAL, *Proceedings* V. 60, No. 9, Sept. 1963, pp. 1107-1185.

13.3. Hatcher, D. S.; Sozen, M. A.; and Siess, C. P., "Test of a Reinforced Concrete Flat Slab," *Proceedings*, ASCE, V. 95, No. ST6, June 1969, pp. 1051-1072.

13.4. Jirsa, J. O.; Sozen, M. A.; and Siess, C. P., "Test of a Flat Slab Reinforced with Welded Wire Fabric," *Proceedings*, ASCE, V. 92, No. ST3, June 1966, pp. 199-224.

13.5. Gamble, W. L.; Sozen, M. A.; and Siess, C. P., "Tests of a Two-Way Reinforced Concrete Floor Slab," *Proceedings*, ASCE, V. 95, No. ST6, June 1969, pp. 1073-1096.

13.6. Vanderbilt, M. D.; Sozen, M. A.; and Siess, C. P., "Test of a Modified Reinforced Concrete Two-Way Slab," *Proceedings*, ASCE, V. 95, No. ST6, June 1969, pp. 1097-1116.

13.7. Xanthakis, M., and Sozen, M. A., "An Experimental Study of Limit Design in Reinforced Concrete Flat Slabs," Civil Engineering Studies, *Structural Research Series* No. 277, University of Illinois, Dec. 1963, 159 pp.

13.8. *ACI Design Handbook, V. 3—Two-Way Slabs*, SP-17(91)(S), American Concrete Institute, Farmington Hills, MI, 1991, 104 pp.

13.9. Mitchell, D., and Cook, W. D., "Preventing Progressive Collapse of Slab Structures," *Journal of Structural Engineering*, V. 110, No. 7, July 1984, pp. 1513-1532.

13.10. Carpenter, J. E.; Kaar, P. H.; and Corley, W. G., "Design of Ductile Flat-Plate Structures to Resist Earthquakes," *Proceedings*, Fifth World Conference on Earthquake Engineering Rome, June 1973, International Association for Earthquake Engineering, V. 2, pp. 2016-2019.

13.11. Morrison, D. G., and Sozen, M. A., "Response to Reinforced Concrete Plate-Column Connections to Dynamic and Static Horizontal Loads," Civil Engineering Studies, *Structural Research Series* No. 490, University of Illinois, Apr. 1981, 249 pp.

13.12. Vanderbilt, M. D., and Corley, W. G., "Frame Analysis of Concrete Buildings," *Concrete International: Design and Construction,* V. 5, No. 12, Dec. 1983, pp. 33-43.

13.13. Grossman, J. S., "Code Procedures, History, and Shortcomings: Column-Slab Connections," *Concrete International,* V. 11, No. 9, Sept. 1989, pp. 73-77.

13.14. Moehle, J. P., "Strength of Slab-Column Edge Connections," *ACI Structural Journal*, V. 85, No. 1, Jan.-Feb. 1988, pp. 89-98.

13.15. ACI-ASCE Committee 352,"Recommendations for Design of Slab-Column Connections in Monolithic Reinforced Concrete Structures (ACI 352.1R-89)," *ACI Structural Journal*, V. 85, No. 6, Nov.-Dec. 1988, pp. 675-696.

13.16. Jirsa, J. O.; Sozen, M. A.; and Siess, C. P., "Pattern Loadings on Reinforced Concrete Floor Slabs," *Proceedings*, ASCE, V. 95, No. ST6, June 1969, pp. 1117-1137.

13.17. Nichols, J. R., "Statical Limitations upon the Steel Requirement in Reinforced Concrete Flat Slab Floors," *Transactions*, ASCE, V. 77, 1914, pp. 1670-1736.

13.18. Corley, W. G.; Sozen, M. A.; and Siess, C. P., "Equivalent-Frame Analysis for Reinforced Concrete Slabs," Civil Engineering Studies, *Structural Research Series* No. 218, University of Illinois, June 1961, 166 pp.

13.19. Jirsa, J. O.; Sozen, M. A.; and Siess, C. P., "Effects of Pattern Loadings on Reinforced Concrete Floor Slabs," Civil Engineering Studies, *Structural Research Series* No. 269, University of Illinois, July 1963.

13.20. Corley, W. G., and Jirsa, J. O., "Equivalent Frame Analysis for Slab Design," ACI JOURNAL, *Proceedings* V. 67, No. 11, Nov. 1970, pp. 875-884.

13.21. Gamble, W. L., "Moments in Beam Supported Slabs," ACI JOURNAL, *Proceedings* V. 69, No. 3, Mar. 1972, pp. 149-157.

References, Chapter 14

14.1. Oberlander, G. D., and Everard, N. J., "Investigation of Reinforced Concrete Walls," ACI JOURNAL, *Proceedings* V. 74, No. 6, June 1977, pp. 256-263.

14.2. Kripanarayanan, K. M., "Interesting Aspects of the Empirical Wall Design Equation," ACI JOURNAL, *Proceedings* V. 74, No. 5, May 1977, pp. 204-207.

14.3. *Uniform Building Code,* V. 2, "Structural Engineering Design Provisions," International Conference of Building Officials, Whittier, CA, 1997, 492 pp.

14.4. Athey, J. W., ed., "Test Report on Slender Walls," Southern California Chapter of the American Concrete Institute and Structural Engineers Association of Southern California, Los Angeles, CA, 1982, 129 pp.

14.5. ACI Committee 551, "Tilt-Up Concrete Structures (ACI 551R-92)," American Concrete Institute, Farmington Hills, MI, 1992, 46 pp. Also *ACI Manual of Concrete Practice*.

14.6. Carter III, J. W., Hawkins, N. M., and Wood, S. L. "Seismic Response of Tilt-Up Construction," Civil Engineering Series, SRS No. 581, University of Illinois, Urbana, IL, Dec. 1993, 224 pp.

References, Chapter 15

15.1. ACI Committee 336, "Suggested Analysis and Design Procedures for Combined Footings and Mats (ACI 336.2R-88)," American Concrete Institute, Farmington Hills, MI, 1988, 21 pp. Also *ACI Manual of Concrete Practice*.

15.2. Kramrisch, F., and Rogers, P., "Simplified Design of Combined Footings," *Proceedings*, ASCE, V. 87, No. SM5, Oct. 1961, p. 19.

15.3 Adebar, P.; Kuchma, D.; and Collins, M. P., "Strut-and-Tie Models for the Design of Pile Caps: An Experimental Study," *ACI Structural Journal,* V. 87, No. 1, Jan.-Feb. 1990, pp. 81-92.

15.4. *CRSI Handbook*, 7th Edition, Concrete Reinforcing Steel Institute, Schaumburg, IL, 1992, 840 pp.

References, Chapter 16

16.1. *Industrialization in Concrete Building Construction*, SP-48, American Concrete Institute, Farmington Hills, MI, 1975, 240 pp.

16.2. Waddell, J. J., "Precast Concrete: Handling and Erection," *Monograph* No. 8, American Concrete Institute, Farmington Hills, MI, 1974, 146 pp.

16.3. "Design and Typical Details of Connections for Precast and Prestressed Concrete," MNL-123-88, 2nd Edition, Precast/Prestressed Concrete Institute, Chicago, 1988, 270 pp.

16.4. *PCI Design Handbook—Precast and Prestressed Concrete*, MNL-120-92, 4th Edition, Precast/Prestressed Concrete Institute, Chicago, 1992, 580 pp.

16.5. "Design of Prefabricated Concrete Buildings for Earthquake Loads," *Proceedings of Workshop*, Apr. 27-29, 1981, ATC-8, Applied Technology Council, Redwood City, CA, 717 pp.

16.6. PCI Committee on Building Code and PCI Technical Activities Committee, "Proposed Design Requirements for Precast Concrete," *PCI Journal*, V. 31, No. 6, Nov.-Dec. 1986, pp. 32-47.

16.7. ACI-ASCE Committee 550, "Design Recommendations for Precast Concrete Structures (ACI 550R-93)," *ACI Structural Journal,* V. 90, No. 1, Jan.-Feb. 1993, pp. 115-121. Also *ACI Manual of Concrete Practice*.

16.8. ACI Committee 551, "Tilt-Up Concrete Structures (ACI 551R-92)," American Concrete Institute, Farmington Hills, MI, 1992, 46 pp. Also *ACI Manual of Concrete Practice*.

16.9. *Manual for Quality Control for Plants and Production of Precast and Prestressed Concrete Products*, MNL-116-85, 3rd Edition, Precast/Prestressed Concrete Institute, Chicago, 1985, 123 pp.

16.10. "Manual for Quality Control for Plants and Production of Architectural Precast Concrete," MNL-117-77, Precast/Prestressed Concrete Institute, Chicago, 1977, 226 pp.

16.11. PCI Committee on Tolerances, "Tolerances for Precast and Prestressed Concrete," *PCI Journal,* V. 30, No. 1, Jan.-Feb. 1985, pp. 26-112.

16.12. ACI Committee 117, "Standard Specifications for Tolerances for Concrete Construction and Materials (ACI 117-90) and Commentary (117R-90)," American Concrete Institute, Farmington Hills, MI, 1990. Also *ACI Manual of Concrete Practice*.

16.13. LaGue, D. J., "Load Distribution Tests on Precast Prestressed Hollow-Core Slab Construction," *PCI Journal*, V. 16, No. 6, Nov.-Dec. 1971, pp. 10-18.

16.14. Johnson, T., and Ghadiali, Z., "Load Distribution Test on Precast Hollow Core Slabs with Openings," *PCI Journal*, V. 17, No. 5, Sept.-Oct. 1972, pp. 9-19.

16.15. Pfeifer, D. W., and Nelson, T. A., "Tests to Determine the Lateral Distribution of Vertical Loads in a Long-Span Hollow-Core Floor Assembly," *PCI Journal*, V. 28, No. 6, Nov.-Dec. 1983, pp. 42-57.

16.16. Stanton, J., "Proposed Design Rules for Load Distribution in Precast Concrete Decks," *ACI Structural Journal*, V. 84, No. 5, Sept.-Oct. 1987, pp. 371-382.

16.17. *PCI Manual for the Design of Hollow Core Slabs*, MNL-126-85, Precast/Prestressed Concrete Institute, Chicago, 1985, 120 pp.

16.18. Stanton, J. F., "Response of Hollow-Core Floors to Concentrated Loads," *PCI Journal,* V. 37, No. 4, July-Aug. 1992, pp. 98-113.

16.19. Aswad, A., and Jacques, F. J., "Behavior of Hollow-Core Slabs Subject to Edge Loads," *PCI Journal*, V. 37, No. 2, Mar.-Apr. 1992, pp. 72-84.

16.20. "Design of Concrete Structures for Buildings," CAN3-A23.3-M84, and "Precast Concrete Materials and Construction," CAN3-A23.4-M84, Canadian Standards Association, Rexdale, Ontario, Canada.

16.21. "Design and Construction of Large-Panel Concrete Structures," six reports, 762 pp., 1976-1980, EB 100D; three studies, 300 pp., 1980, EB 102D, Portland Cement Association, Skokie, IL

16.22. PCI Committee on Precast Concrete Bearing Wall Buildings, "Considerations for the Design of Precast Concrete Bearing Wall Buildings to Withstand Abnormal Loads," *PCI Journal*, V. 21, No. 2, Mar.-Apr. 1976, pp. 18-51.

16.23. Salmons, J. R., and McCrate, T. E., "Bond Characteristics of Untensioned Prestressing Strand," *PCI Journal*, V. 22, No. 1, Jan.-Feb. 1977, pp. 52-65.

16.24. PCI Committee on Quality Control and Performance Criteria, "Fabrication and Shipment Cracks in Prestressed Hollow-Core Slabs and Double Tees," *PCI Journal*, V. 28, No. 1, Jan.-Feb. 1983, pp. 18-39.

16.25. PCI Committee on Quality Control and Performance Criteria, "Fabrication and Shipment Cracks in Precast or Prestressed Beams and Columns," *PCI Journal*, V. 30, No. 3, May-June 1985, pp. 24-49.

References, Chapter 17

17.1. "Specification for Structural Steel Buildings—Allowable Stress Design and Plastic Design, with Commentary" June 1989, and "Load and Resistance Factor Design Specification for Structural Steel Buildings," Sept. 1986, American Institute of Steel Construction, Chicago.

17.2. Kaar, P. H.; Kriz, L. B.; and Hognestad, E., "Precast-Prestressed Concrete Bridges: (1) Pilot Tests of Continuous Girders," *Journal*, PCA Research and Development Laboratories, V. 2, No. 2, May 1960, pp. 21-37.

17.3. Saemann, J. C., and Washa, G. W., "Horizontal Shear Connections between Precast Beams and Cast-in-Place Slabs," ACI JOURNAL, *Proceedings* V. 61, No. 11, Nov. 1964, pp. 1383-1409. Also see discussion, ACI JOURNAL, June 1965.

17.4. Hanson, N. W., "Precast-Prestressed Concrete Bridges: Horizontal Shear Connections," *Journal*, PCA Research and Development Laboratories, V. 2, No. 2, May 1960, pp. 38-58.

17.5. Grossfield, B., and Birnstiel, C., "Tests of T-Beams with Precast Webs and Cast-in-Place Flanges," ACI JOURNAL, *Proceedings* V. 59, No. 6, June 1962, pp. 843-851.

17.6. Mast, R. F., "Auxiliary Reinforcement in Concrete Connections," *Proceedings*, ASCE, V. 94, No. ST6, June 1968, pp. 1485-1504.

References, Chapter 18

18.1. Mast, R. F., "Analysis of Cracked Prestressed Concrete Sections: A Practical Approach," *PCI Journal,* V. 43, No. 4, Jul.-Aug., 1998.

18.2. *PCI Design Handbook—Precast and Prestressed Concrete*, 4th Edition, Precast/Prestressed Concrete Institute, Chicago, 1992, pp. 4-42 through 4-44.

18.3. ACI-ASCE Committee 423, "Tentative Recommendations for Prestressed Concrete," ACI JOURNAL, *Proceedings* V. 54, No. 7, Jan. 1958, pp. 545-578.

18.4. ACI Committee 435, "Deflections of Prestressed Concrete Members (ACI 435.1R-63)(Reapproved 1989)," ACI JOURNAL, *Proceedings* V. 60, No. 12, Dec. 1963, pp. 1697-1728. Also *ACI Manual of Concrete Practice*.

18.5. PCI Committee on Prestress Losses, "Recommendations for Estimating Prestress Losses," *Journal of the Prestressed Concrete Institute*, V. 20, No. 4, July-Aug. 1975, pp. 43-75.

18.6. Zia, P.; Preston, H. K.; Scott, N. L.; and Workman, E. B., "Estimating Prestress Losses," *Concrete International: Design & Construction,* V. 1, No. 6, June 1979, pp. 32-38.

18.7. Mojtahedi, S., and Gamble, W. L., "Ultimate Steel Stresses in Unbonded Prestressed Concrete," *Proceedings*, ASCE, V. 104, ST7, July 1978, pp. 1159-1165.

18.8. Mattock, A. H.; Yamazaki, J.; and Kattula, B. T., "Comparative Study of Prestressed Concrete Beams, with and without Bond," ACI JOURNAL, *Proceedings* V. 68, No. 2, Feb. 1971, pp. 116-125.

18.9. ACI-ASCE Committee 423, "Recommendations for Concrete Members Prestressed with Unbonded Tendons (ACI 423.3R-89)," *ACI Structural Journal*, V. 86, No. 3, May-June 1989, pp. 301-318. Also *ACI Manual of Concrete Practice.*

18.10. Odello, R. J., and Mehta, B. M., "Behavior of a Continuous Prestressed Concrete Slab with Drop Panels," *Report*, Division of Structural Engineering and Structural Mechanics, University of California, Berkeley, 1967.

18.11. Smith, S. W., and Burns, N. H., "Post-Tensioned Flat Plate to Column Connection Behavior," *Journal of the Prestressed Concrete Institute*, V. 19, No. 3, May-June 1974, pp. 74-91.

18.12. Burns, N. H., and Hemakom, R., "Test of Scale Model Post-Tensioned Flat Plate," *Proceedings*, ASCE, V. 103, ST6, June 1977, pp. 1237-1255.

18.13. Hawkins, N. M., "Lateral Load Resistance of Unbonded Post-Tensioned Flat Plate Construction," *Journal of the Prestressed Concrete Institute*, V. 26, No. 1, Jan.-Feb. 1981, pp. 94-116.

18.14. "Guide Specifications for Post-Tensioning Materials," *Post-Tensioning Manual*, 5th Edition, Post-Tensioning Institute, Phoenix, Ariz., 1990, pp. 208-216.

18.15. Foutch, D. A.; Gamble, W. L.; and Sunidja, H., "Tests of Post-Tensioned Concrete Slab-Edge Column Connections," *ACI Structural Journal*, V. 87, No. 2, Mar.-Apr. 1990, pp. 167-179.

18.16. Bondy, K. B., "Moment Redistribution: Principles and Practice Using ACI 318-02," *PTI Journal*, V. 1, No. 1, Post-Tensioned Institute, Phoenix, AZ, Jan. 2003, pp. 3-21.

18.17. Lin, T. Y., and Thornton, K., "Secondary Moment and Moment Redistribution in Continuous Prestressed Beams," *PCI Journal*, V. 17, No. 1, Jan.-Feb. 1972, pp. 8-20 and comments by A. H. Mattock and author's closure, *PCI Journal*, V. 17, No. 4, July-Aug. 1972, pp. 86-88.

18.18. Collins, M. P., and Mitchell, D., *Prestressed Concrete Structures*, Response Publications, Canada, 1997, pp. 517-518.

18.19. Mast, R.F., "Unified Design Provision for Reinforced and Prestressed Concrete Flexural and Compression Members," *ACI Structural Journal,* V. 89, No. 2, Mar.-Apr., 1992, pp. 185-199.

18.20. "Design of Post-Tensioned Slabs," Post-Tensioning Institute, Phoenix, Ariz., 1984, 54 pp.

18.21. Gerber, L. L., and Burns, N. H., "Ultimate Strength Tests of Post-Tensioned Flat Plates," *Journal of the Prestressed Concrete Institute*, V. 16, No. 6, Nov.-Dec. 1971, pp. 40-58.

18.22. Scordelis, A. C.; Lin, T. Y.; and Itaya, R., "Behavior of a Continuous Slab Prestressed in Two Directions," ACI JOURNAL, *Proceedings* V. 56, No. 6, Dec. 1959, pp. 441-459.

18.23. American Association of State Highway and Transportation Officials, "Standard Specifications for Highway Bridges," 17th Edition, 2002.

18.24. Breen, J. E.; Burdet, O.; Roberts, C.; Sanders, D.; Wollmann, G.; and Falconer, B., "Anchorage Zone Requirements for Post-Tensioned Concrete Girders," NCHRP *Report* 356, Transportation Research Board, National Academy Press, Washington, D.C., 1994.

18.25. ACI-ASCE Committee 423, "Recommendations for Concrete Members Prestressed with Unbonded Tendons," *ACI Structural Journal*, V. 86, No. 3, May-June 1989, p. 312.

18.26. "Specification for Unbonded Single Strand Tendons," revised 1993, Post-Tensioning Institute, Phoenix, AZ, 1993, 20 pp.

18.27. "Guide Specifications for Design and Construction of Segmental Concrete Bridges, AASHTO, Washington, DC, 1989, 50 pp.

18.28. Gerwick, B. C. Jr., "Protection of Tendon Ducts," *Construction of Prestressed Concrete Structures*, John Wiley and Sons, Inc., New York, 1971, 411 pp.

18.29. "Recommended Practice for Grouting of Post-Tensioned Prestressed Concrete," *Post-Tensioning Manual*, 5th Edition, Post-Tensioning Institute, Phoenix, AZ, 1990, pp. 230-236.

18.30. *Manual for Quality Control for Plants and Production of Precast and Prestressed Concrete Products*, 3rd Edition, MNL-116-85, Precast/Prestressed Concrete Institute, Chicago, 1985, 123 pp.

18.31. ACI Committee 301, "Standard Specifications for Structural Concrete for Buildings (ACI 301-96)," American Concrete Institute, Farmington Hills, MI, 1996, 34 pp. Also *ACI Manual of Concrete Practice*.

18.32. Salmons, J. R., and McCrate, T. E., "Bond Characteristics of Untensioned Prestressing Strand," *Journal of the Prestressed Concrete Institute*, V. 22, No. 1, Jan.-Feb. 1977, pp. 52-65.

18.33. ACI Committee 215, "Considerations for Design of Concrete Structures Subjected to Fatigue Loading (ACI 215R-74)(Revised 1992)," American Concrete Institute, Farmington Hills, MI, 1992, 24 pp. Also *ACI Manual of Concrete Practice*.

18.34. Barth, F., "Unbonded Post-Tensioning in Building Construction," *Concrete Construction Engineering Handbook*, CRC Press, 1997, pp. 12.32-12.47.

References, Chapter 19

19.1. ACI Committee 334, "Concrete Shell Structures—Practice and Commentary (ACI 334.1R-92)," American Concrete Institute, Farmington Hills, MI, 14 pp. Also *ACI Manual of Concrete Practice*.

19.2. IASS Working Group No. 5, "Recommendations for Reinforced Concrete Shells and Folded Plates," International Association for Shell and Spatial Structures, Madrid, Spain, 1979, 66 pp.

19.3. Tedesko, A., "How Have Concrete Shell Structures Performed?" *Bulletin*, International Association for Shell and Spatial Structures, Madrid, Spain, No. 73, Aug. 1980, pp. 3-13.

19.4. ACI Committee 334, "Reinforced Concrete Cooling Tower Shells—Practice and Commentary (ACI 334.2R-91)," American Concrete Institute, Farmington Hills, MI, 1991, 9 pp. Also *ACI Manual of Concrete Practice*.

19.5. ACI Committee 373R, "Design and Construction of Circular Prestressed Concrete Structures with Circumferential Tendons (ACI 373R-97)," American Concrete Institute, Farmington Hills, MI, 1997, 26 pp. Also *ACI Manual of Concrete Practice*.

19.6. Billington, D. P., *Thin Shell Concrete Structures*, 2nd Edition, McGraw-Hill Book Co., New York, 1982, 373 pp.

19.7. "Phase I Report on Folded Plate Construction," ASCE Task Committee, ASCE, *Journal of Structural Division*, V. 89, No. ST6 1963, pp. 365-406.

19.8. *Concrete Thin Shells*, SP-28, American Concrete Institute, Farmington Hills, MI, 1971, 424 pp.

19.9. Esquillan N., "The Shell Vault of the Exposition Palace, Paris," ASCE, *Journal of Structural Division*, V. 86, No. ST1, Jan. 1960, pp. 41-70.

19.10. *Hyperbolic Paraboloid Shells*, SP-110, American Concrete Institute, Farmington Hills, MI, 1988, 184 pp.

19.11. Billington, D. P., "Thin Shell Structures," *Structural Engineering Handbook*, Gaylord and Gaylord, eds., McGraw-Hill, New York, 1990, pp. 24.1-24.57.

19.12. Scordelis, A. C., "Non-Linear Material, Geometric, and Time Dependent Analysis of Reinforced and Prestressed Concrete Shells," *Bulletin*, International Association for Shells and Spatial Structures, Madrid, Spain, No. 102, Apr. 1990, pp. 57-90.

19.13. Schnobrich, W. C., "Reflections on the Behavior of Reinforced Concrete Shells," *Engineering Structures*, Butterworth, Heinemann, Ltd., Oxford, V. 13, No. 2, Apr. 1991, pp. 199-210.

19.14. Sabnis, G. M.; Harris, H. G.; and Mirza, M. S., *Structural Modeling and Experimental Techniques*, Prentice-Hall, Inc., Englewood Cliffs, NJ, 1983.

19.15. *Concrete Shell Buckling*, SP-67, American Concrete Institute, Farmington Hills, MI, 1981, 234 pp.

19.16. Gupta, A. K., "Membrane Reinforcement in Concrete Shells: A Review," *Nuclear Engineering and Design*, Nofi-Holland Publishing, Amsterdam, V. 82, Oct. 1984, pp. 63-75.

19.17. Vecchio, F. J., and Collins, M. P., "Modified Compression-Field Theory for Reinforced Concrete Beams Subjected to Shear," ACI JOURNAL, *Proceedings* V. 83, No. 2, Mar.-Apr. 1986, pp. 219-223.

19.18. Fialkow, M. N., "Compatible Stress and Cracking in Reinforced Concrete Membranes with Multidirectional Reinforcement," *ACI Structural Journal*, V. 88, No. 4, July-Aug. 1991, pp. 445-457.

19.19. Medwadowski, S., "Multidirectional Membrane Reinforcement," *ACI Structural Journal*, V. 86, No. 5, Sept.-Oct. 1989, pp. 563-569.

19.20. ACI Committee 224, "Control of Cracking in Concrete Structures (ACI 224R-90)," American Concrete Institute, Farmington Hills, MI, 1990, 43 pp. Also *ACI Manual of Concrete Practice*.

19.21. Gupta, A. K., "Combined Membrane and Flexural Reinforcement in Plates and Shells," *Structural Engineering,* ASCE, V. 112, No. 3, Mar, 1986, pp. 550-557.

19.22. Tedesko, A., "Construction Aspects of Thin Shell Structures," ACI JOURNAL, *Proceedings* V. 49, No. 6, Feb. 1953, pp. 505-520.

19.23. Huber, R. W., "Air Supported Forming—Will it Work?" *Concrete International*, V. 8, No. 1, Jan. 1986, pp. 13-17.

References, Chapter 21

21.1. "NEHRP Recommended Provisions for Seismic Regulations for New Buildings and Other Structures," Part 1: Provisions (FEMA 302, 353 pp.) and Part 2: Commentary (FEMA 303, 335 pp.), Building Seismic Safety Council, Washington, D. C., 1997.

21.2. *Uniform Building Code*, V. 2, "Structural Engineering Design Provisions," 1997 Edition, International Conference of Building Officials, Whittier, CA, 1997, 492 pp.

21.3. "BOCA National Building Code," 13th Edition, Building Officials and Code Administration International, Inc., Country Club Hills, IL, 1996, 357 pp.

21.4. "Standard Building Code," Southern Building Code Congress International, Inc., Birmingham, Ala., 1996, 656 pp.

21.5. "International Building Code," International Code Council, Falls Church, VA, International Council, 2000.

21.6. Blume, J. A.; Newmark, N. M.; and Corning, L. H., *Design of Multistory Reinforced Concrete Buildings for Earthquake Motions*, Portland Cement Association, Skokie, IL, 1961, 318 pp.

21.7. Clough, R. W., "Dynamic Effects of Earthquakes," *Proceedings*, ASCE, V. 86, ST4, Apr. 1960, pp. 49-65.

21.8. Gulkan, P., and Sozen, M. A., "Inelastic Response of Reinforced Concrete Structures to Earthquake Motions," ACI JOURNAL, *Proceedings* V. 71, No. 12, Dec. 1974., pp. 604-610.

21.9. "NEHRP Recommended Provisions for Seismic Regulations for New Buildings and Other Structures: Part 1: Provisions (FEMA 368, 374 pp.); and Part 2: Commentary (FEMA 369, 444 pp.), Building Seismic Safety Council, Washington, DC, 2000.

21.10. ACI-ASCE Committee 352, "Recommendations for Design of Beam-Column Joints in Monolithic Reinforced Concrete Structures (ACI 352R-91)," American Concrete Institute, Farmington Hills, MI, 1991, 18 pp. Also *ACI Manual of Concrete Practice*.

21.11. Hirosawa, M., "Strength and Ductility of Reinforced Concrete Members," *Report* No. 76, Building Research Institute, Ministry of Construction, Tokyo, Mar. 1977 (in Japanese). Also, data in Civil Engineering Studies, *Structural Research Series* No. 452, University of Illinois, 1978.

21.12. "Recommended Lateral Force Requirements and Commentary," Seismology Committee of the Structural Engineers Association of California, Sacramento, CA, 6th Edition, 504 pp.

21.13. Popov, E. P.; Bertero, V. V.; and Krawinkler, H., "Cyclic Behavior of Three R/C Flexural Members with High Shear," EERC *Report* No. 72-5, Earthquake Engineering Research Center, University of California, Berkeley, Oct. 1972.

21.14. Wight, J. K., and Sozen, M. A., "Shear Strength Decay of RC Columns under Shear Reversals," *Proceedings*, ASCE, V. 101, ST5, May 1975, pp. 1053-1065.

21.15. French, C. W., and Moehle, J. P., "Effect of Floor Slab on Behavior of Slab-Beam-Column Connections," ACI SP-123, *Design of Beam-Column Joints for Seismic Resistance*, American Concrete Institute, Farmington Hills, MI, 1991, pp. 225-258.

21.16. Sivakumar, B.; Gergely, P.; White, R. N., "Suggestions for the Design of R/C Lapped Splices for Seismic Loading," *Concrete International*, V. 5, No. 2, Feb. 1983, pp. 46-50.

21.17. Sakai, K., and Sheikh, S. A., "What Do We Know about Confinement in Reinforced Concrete Columns? (A Critical Review of Previous Work and Code Provisions)," *ACI Structural Journal*, V. 86, No. 2, Mar.-Apr. 1989, pp. 192-207.

21.18. Park, R. "Ductile Design Approach for Reinforced Concrete Frames," *Earthquake Spectra*, V. 2, No. 3, May 1986, pp. 565-619.

21.19. Watson, S.; Zahn, F. A.; and Park, R., "Confining Reinforcement for Concrete Columns," *Journal of Structural Engineering*, V. 120, No. 6, June 1994, pp. 1798-1824.

21.20. Meinheit, D. F., and Jirsa, J. O., "Shear Strength of Reinforced Concrete Beam-Column Joints," *Report* No. 77-1, Department of Civil Engineering, Structures Research Laboratory, University of Texas at Austin, Jan. 1977.

21.21. Briss, G. R.; Paulay, T; and Park, R., "Elastic Behavior of Earthquake Resistant R. C. Interior Beam-Column Joints," *Report* 78-13, University of Canterbury, Department of Civil Engineering, Christchurch, New Zealand, Feb. 1978.

21.22. Ehsani, M. R., "Behavior of Exterior Reinforced Concrete Beam to Column Connections Subjected to Earthquake Type Loading," *Report* No. UMEE 82R5, Department of Civil Engineering, University of Michigan, July 1982, 275 pp.

21.23. Durrani, A. J., and Wight, J. K., "Experimental and Analytical Study of Internal Beam to Column Connections Subjected to Reversed Cyclic Loading," *Report* No. UMEE 82R3, Department of Civil Engineering, University of Michigan, July 1982, 275 pp.

21.24. Leon, R. T., "Interior Joints with Variable Anchorage Lengths," *Journal of Structural Engineering*, ASCE, V. 115, No. 9, Sept. 1989, pp. 2261- 2275.

21.25. Zhu, S., and Jirsa, J. O., "Study of Bond Deterioration in Reinforced Concrete Beam-Column Joints," PMFSEL *Report* No. 83-1, Department of Civil Engineering, University of Texas at Austin, July 1983.

21.26. Meinheit, D. F., and Jirsa, J. O., "Shear Strength of R/C Beam-Column Connections," *Journal of the Structural Division*, ASCE, V. 107, No. ST11, Nov. 1982, pp. 2227-2244.

21.27. Ehsani, M. R., and Wight, J. K., "Effect of Transverse Beams and Slab on Behavior of Reinforced Concrete Beam to Column Connections," ACI JOURNAL, *Proceedings* V. 82, No. 2, Mar.-Apr. 1985, pp. 188-195.

21.28. Ehsani, M. R., "Behavior of Exterior Reinforced Concrete Beam to Column Connections Subjected to Earthquake Type Loading," ACI JOURNAL, *Proceedings* V. 82, No. 4, July-Aug. 1985, pp. 492-499.

21.29. Durrani, A. J., and Wight, J. K., "Behavior of Interior Beam to Column Connections under Earthquake Type Loading," ACI JOURNAL, *Proceedings* V. 82, No. 3, May-June 1985, pp. 343-349.

21.30. ACI-ASCE Committee 326, "Shear and Diagonal Tension," ACI JOURNAL, *Proceedings* V. 59, No. 1, Jan. 1962, pp. 1-30; No. 2, Feb. 1962, pp. 277-334; and No. 3, Mar. 1962, pp. 352-396.

21.31. Yoshioka, K., and Sekine, M., "Experimental Study of Prefabricated Beam-Column Subassemblages," *Design of Beam-Column Joints for Seismic Resistance*, SP-123, American Concrete

Institute, Farmington Hills, MI, 1991, pp. 465-492.

21.32. Kurose, Y,; Nagami, K.; and Saito, Y., "Beam-Column Joints in Precast Concrete Construction in Japan," *Design of Beam-Column Joints for Seismic Resistance,* SP-123, American Concrete Institute, 1991, pp. 493-514.

21.33. Restrepo, J.; Park, R.; and Buchanan, A., "Tests on Connections of Earthquake Resisting Precast Reinforced Concrete Perimeter Frames," *Precast/Prestressed Concrete Institute Journal,* V. 40, No. 5, pp. 44-61.

21.34. Restrepo, J.; Park, R.; and Buchanan, A., "Design of Connections of Earthquake Resisting Precast Reinforced Concrete Perimeter Frames," *Precast/Prestressed Concrete Institute Journal,* V. 40, No. 5, 1995, pp. 68-80.

21.35. Palmieri, L.; Saqan, E.; French, C.; and Kreger, M., "Ductile Connections for Precast Concrete Frame Systems," *Mete A. Sozen Symposium, ACI SP-162,* American Concrete Institute, Farmington Hills, MI, 1996, pp. 315-335.

21.36. Stone, W.; Cheok, G.; and Stanton, J., "Performance of Hybrid Moment-Resisting Precast Beam-Column Concrete Connections Subjected to Cyclic Loading," *ACI Structural Journal*, V. 92, No. 2, Mar.-Apr. 1995, pp. 229-249.

21.37. Nakaki, S. D.; Stanton, J.F.; and Sritharan, S., "An Overview of the PRESSS Five-Story Precast Test Building," *Precast/Prestressed Concrete Institute Journal,* V. 44, No. 2, pp. 26-39.

21.38. ACI Committee 408, "State-of-the-Art Report on Bond under Cyclic Loads (ACI 408.2R-92 [Reapproved 1999]," American Concrete Institute, Farmington Hills, MI, 1999, 5 pp.

21.39. Barda, F.; Hanson, J. M.; and Corley, W. G., "Shear Strength of Low-Rise Walls with Boundary Elements," *Reinforced Concrete Structures in Seismic Zones,* SP-53, American Concrete Institute, Farmington Hills, MI, 1977, pp. 149-202.

21.40. Taylor, C. P.; Cote, P. A.; and Wallace, J. W., "Design of Slender RC Walls with Openings," *ACI Structural Journal*, V. 95, No. 4, July-Aug. 1998, pp. 420-433.

21.41. Wallace, J. W., "Evaluation of UBC-94 Provisions for Seismic Design of RC Structural Walls," *Earthquake Spectra*, V. 12, No. 2, May 1996, pp. 327-348.

21.42. Moehle, J. P., "Displacement-Based Design of RC Structures Subjected to Earthquakes," *Earthquake Spectra*, V. 8, No. 3, Aug. 1992, pp. 403-428.

21.43. Wallace, J. W., and Orakcal, K., "ACI 318-99 Provisions for Seismic Design of Structural Walls," *ACI Structural Journal*, V. 99, No. 4, July-Aug. 2002, pp. 499-508.

21.44. Paulay, T., and Binney, J. R., "Diagonally Reinforced Coupling Beams of Shear Walls," *Shear in Reinforced Concrete*, SP-42, American Concrete Institute, Farmington Hills, MI, 1974, pp. 579-598.

21.45. Barney, G. G. et al., *Behavior of Coupling Beams under Load Reversals* (RD068.01B), Portland Cement Association, Skokie, IL, 1980.

21.46. Wyllie, L. A., Jr., "Structural Walls and Diaphragms — How They Function," *Building Structural Design Handbook*, R. N. White, and C. G. Salmon, eds., John Wiley & Sons, 1987, pp. 188-215.

21.47. Wood, S. L., Stanton, J. F., and Hawkins, N. M., "Development of New Seismic Design Provisions for Diaphragms Based on the Observed Behavior of Precast Concrete Parking Garages during the 1994 Northridge Earthquake," *Journal*, Precast/Prestressed Concrete Institute, V. 45, No. 1, Jan.-Feb. 2000, pp. 50-65.

21.48. "Minimum Design Loads for Buildings and Other Structures," SEI/ASCE 7-02, American Society of Civil Engineers, Reston, VA, 2002, 337 pp.

21.49. "International Building Code," International Code Council, Falls Church, VA, 2003.

21.50. Nilsson, I. H. E., and Losberg, A., "Reinforced Concrete Corners and Joints Subjected to Bending Moment," *Journal of the Structural Division*, ASCE, V. 102, No. ST6, June 1976, pp. 1229-1254.

21.51. Megally, S., and Ghali, A., "Punching Shear Design of Earthquake-Resistant Slab-Column Connections," *ACI Structural Journal*, V. 97, No. 5, Sept.-Oct. 2002, pp. 720-730.

21.52. Moehle, J. P., "Seismic Design Considerations for Flat Plate Construction," *Mete A. Sozen Symposium: A Tribute from his Students*, SP-162, J. K. Wight and M. E. Kreger, eds., American Concrete Institute, Farmington Hills, MI, pp. 1-35.

21.53. ACI-ASCE Committee 352, "Recommendations for Design of Slab-Column Connections in Monolithic Reinforced Concrete Structures (ACI 352.1R-89)," American Concrete Institute, Farmington Hills, MI, 1989.

21.54. Pan, A., and Moehle, J. P., "Lateral Displacement Ductility of Reinforced Concrete Flat Plates," *ACI Structural Journal,* V. 86, No. 3, May-June, 1989, pp. 250-258.

References, Appendix A

A.1. Schlaich, J.; Schäfer, K.; and Jennewein, M., "Toward a Consistent Design of Structural Concrete," *PCI Journal*, V. 32, No. 3, May-June, 1987, pp 74-150.

A.2. Collins, M. P., and Mitchell, D., *Prestressed Concrete Structures*, Prentice Hall Inc., Englewood Cliffs, NJ, 1991, 766 pp.

A.3. MacGregor, J. G., *Reinforced Concrete: Mechanics and Design, 3rd Edition.*, Prentice Hall, Englewood Cliffs, NJ, 1997, 939 pp.

A.4. *FIP Recommendations, Practical Design of Structural Concrete*, FIP-Commission 3, "Practical Design," Sept. 1996, Pub.: SETO, London, Sept. 1999.

A.5. Menn, C, *Prestressed Concrete Bridges,* Birkhäuser, Basle, 535 pp.

A.6. Muttoni, A; Schwartz, J.; and Thürlimann, B., *Design of Concrete Structures with Stress Fields*, Birkhauser, Boston, Mass., 1997, 143 pp.

A.7. Joint ACI-ASCE Committee 445, "Recent Approaches to Shear Design of Structural Concrete," *ASCE Journal of Structural Engineering*, Dec., 1998, pp 1375-1417.

A.8. Bergmeister, K.; Breen, J. E.; and Jirsa, J. O., "Dimensioning of the Nodes and Development of Reinforcement," *IABSE Colloquium Stuttgart 1991,* International Association for Bridge and Structural Engineering, Zurich, 1991, pp. 551-556.

References, Appendix B

B.1. Cohn, M. A., "Rotational Compatibility in the Limit Design of Reinforced Concrete Continuous Beams," *Flexural Mechanics of Reinforced Concrete, ACI SP-12,* American Concrete Institute/American Society of Civil Engineers, Farmington Hills, MI, 1965, pp. 35-46.

B.2. Mattock, A. H., "Redistribution of Design Bending Moments in Reinforced Concrete Continuous Beams," *Proceedings,* Institution of Civil Engineers, London, V. 13, 1959, pp. 35-46.

B.3. "Design of Post-Tensioned Slabs," Post-Tensioning Institute, Phoenix, Ariz., 1984, 54 pp.

B.4. Gerber, L. L., and Burns, N. H., "Ultimate Strength Tests of Post-Tensioned Flat Plates," *Journal of the Prestressed Concrete Institute,* V. 16, No. 6, Nov.-Dec. 1971, pp. 40-58.

B.5. Smith, S. W., and Burns, N. H., "Post-Tensioned Flat Plate to Column Connection Behavior," *Journal of the Prestressed Concrete Institute,* V. 19, No. 3, May-June, 1974, pp. 74-91.

B.6. Burns, N. H., and Hemakom, R., "Test of Scale Model Post-Tensioned Flat Plate," *Proceedings,* ASCE, V. 103, ST6, June 1977, pp. 1237-1255.

B.7. Burns, N. H., and Hemakom, R., "Test of Flat Plate with Bonded Tendons," *Proceedings,* ASCE, V. 111, No. 9, Sept. 1985, pp. 1899-1915.

B.8. Kosut, G. M.; Burns, N. H.; and Winter, C. V., "Test of Four-Panel Post-Tensioned Flat Plate," *Proceedings,* ASCE, V. 111, No. 9, Sept. 1985, pp. 1916-1929.

References, Appendix C

C.1. "International Building Code," International Code Council, Falls Church, VA, 2003.

C.2. "Minimum Design Loads for Buildings and Other Structures (SEI/ASCE 7-02)," ASCE, New York, 2002, 376 pp.

C.3. "BOCA National Building Code," 12th Edition, Building Officials and Code Administration International, Inc., Country Club Hills, IL, 1993, 357 pp.

C.4. "Standard Building Code, 1994 Edition," Southern Building Code Congress International, Inc., Birmingham, AL, 1994, 656 pp.

C.5. "Uniform Building Code, V. 2, Structural Engineering Design Provisions," International Conference of Building Officials, Whittier, CA, 1997, 492 pp.

C.6. Mast, R. F., "Unified Design Provisions for Reinforced and Prestressed Concrete Flexural and Compression Members," *ACI Structural Journal,* V. 89, No. 2, Mar.-Apr. 1992, pp. 185-199.

References, Appendix D

D.1. ANSI/ASME B1.1, "Unified Inch Screw Threads (UN and UNR Thread Form), ASME, Fairfield, N.J., 1989.

D.2. ANSI/ASME B18.2.1, "Square and Hex Bolts and Screws, Inch Series," ASME, Fairfield, N.J., 1996.

D.3. ANSI/ASME B18.2.6, "Fasteners for Use in Structural Applications," ASME, Fairfield, N.J., 1996.

D.4. Cook, R. A., and Klingner, R. E., "Behavior of Ductile Multiple-Anchor Steel-to-Concrete Connections with Surface-Mounted Baseplates," *Anchors in Concrete: Design and Behavior*, SP-130, 1992, American Concrete Institute, Farmington Hills, MI, pp. 61-122.

D.5. Cook, R. A., and Klingner, R. E., "Ductile Multiple-Anchor Steel-to-Concrete Connections," *Journal of Structural Engineering*, ASCE, V. 118, No. 6, June 1992, pp. 1645–1665.

D.6. Lotze, D., and Klingner, R.E., "Behavior of Multiple-Anchor Attachments to Concrete from the Perspective of Plastic Theory," *Report PMFSEL 96-4*, Ferguson Structural Engineering Laboratory, The University of Texas at Austin, Mar., 1997.

D.7. Primavera, E. J.; Pinelli, J.-P.; and Kalajian, E. H., "Tensile Behavior of Cast-in-Place and Undercut Anchors in High-Strength Concrete," *ACI Structural Journal*, V. 94, No. 5, Sept.-Oct. 1997, pp. 583-594.

D.8. *Design of Fastenings in Concrete*, Comite Euro-International du Beton (CEB), Thomas Telford Services Ltd., London, Jan. 1997.

D.9. Fuchs, W.; Eligehausen, R.; and Breen, J., "Concrete Capacity Design (CCD) Approach for Fastening to Concrete," *ACI Structural Journal*, V. 92, No. 1, Jan.-Feb., 1995, pp. 73–93. Also discussion, *ACI Structural Journal*, V. 92, No. 6, Nov.-Dec., 1995, pp. 787-802.

D.10. Eligehausen, R., and Balogh, T., "Behavior of Fasteners Loaded in Tension in Cracked Reinforced Concrete," *ACI Structural Journal*, V. 92, No. 3, May-June 1995, pp. 365-379.

D.11. "Fastenings to Concrete and Masonry Structures, State of the Art Report," Comite Euro-International du Beton, (CEB), *Bulletin* No. 216, Thomas Telford Services Ltd., London, 1994.

D.12. Klingner, R.; Mendonca, J.; and Malik, J., "Effect of Reinforcing Details on the Shear Resistance of Anchor Bolts under Reversed Cyclic Loading," ACI JOURNAL, *Proceedings* V. 79, No. 1, Jan.-Feb. 1982, pp. 3-12.

D.13. Eligehausen, R.; Fuchs, W.; and Mayer, B., "Load Bearing Behavior of Anchor Fastenings in Tension," Betonwerk + Fertigteiltechnik, 12/1987, pp. 826–832, and 1/1988, pp. 29-35.

D.14. Eligehausen, R., and Fuchs, W., "Load Bearing Behavior of Anchor Fastenings under Shear, Combined Tension and Shear or Flexural Loadings," Betonwerk + Fertigteiltechnik, 2/1988, pp. 48-56.

D.15. ACI Committee 349, "Code Requirements for Nuclear Safety Related Concrete Structures (ACI 349-85)," See also *ACI Manual of Concrete Practice*, Part 4, 1987.

D.16. Farrow, C.B., and Klingner, R.E., "Tensile Capacity of Anchors with Partial or Overlapping Failure Surfaces: Evaluation of Existing Formulas on an LRFD Basis," *ACI Structural Journal*, V. 92, No. 6, Nov.-Dec. 1995, pp. 698-710.

D.17. *PCI Design Handbook, 5th Edition*, Precast/Prestressed Concrete Institute, Chicago, 1999.

D.18. "AISC Load and Resistance Factor Design Specifications for Structural Steel Buildings," Dec. 1999, 327 pp.

D.19. Zhang, Y., "Dynamic Behavior of Multiple Anchor Connections in Cracked Concrete," PhD dissertation, The University of

Texas at Austin, Aug. 1997.

D.20. Lutz, L., "Discussion to Concrete Capacity Design (CCD) Approach for Fastening to Concrete," *ACI Structural Journal*, Nov.-Dec. 1995, pp. 791-792. Also authors' closure, pp. 798-799.

D.21. Asmus, J., "Verhalten von Befestigungen bei der Versagensart Spalten des Betons (Behavior of Fastenings with the Failure Mode Splitting of Concrete)," dissertation, Universität Stuttgart, Germany, 1999.

D.22. Kuhn, D., and Shaikh, F., "Slip-Pullout Strength of Hooked Anchors," *Research Report*, University of Wisconsin-Milwaukee, submitted to the National Codes and Standards Council, 1996.

D.23. Furche, J., and Eligehausen, R., "Lateral Blow-out Failure of Headed Studs Near a Free Edge," *Anchors in Concrete-Design and Behavior*, SP-130, American Concrete Institute, Farmington Hills, MI, 1991, pp. 235–252.

D.24. Shaikh, A. F., and Yi, W., "In-Place Strength of Welded Studs," *PCI Journal*, V.30, No. 2, Mar.-Apr. 1985.

INDEX